Human Anatomy
Laboratory Manual and Lecture Workbook

Third Edition

William W. Farrar
Suzanne Byrd
Eastern Kentucky University

The McGraw-Hill Companies, Inc.
Primis Custom Publishing

New York St. Louis San Francisco Auckland Bogotá
Caracas Lisbon London Madrid Mexico Milan Montreal
New Delhi Paris San Juan Singapore Sydney Tokyo Toronto

McGraw-Hill
A Division of The McGraw-Hill Companies

Human Anatomy

Laboratory Manual and Lecture Workbook

McGraw-Hill's Primis Custom Series consists of products that are produced from camera-ready copy. Peer review, class testing, and accuracy are primarily the responsibility of the author(s).

7 8 9 0 GDP GDP 0 9 8 7 6 5 4 3 2 1

ISBN 0-07-289253-6

Editor: Jan Scipio
Cover Design: Catherine L. Bradish
Printer/Binder: Greyden Press

TABLE OF CONTENTS

LECTURE

GENERAL BODY PLAN

CYTOLOGY AND HISTOLOGY

THE REPRODUCTIVE SYSTEMS

SYSTEMS INVOLVED IN SUPPORT AND MOVEMENT

OPEN SYSTEMS THAT ALLOW FOR THE EXCHANGE OF SUBSTANCES BETWEEN THE ENVIRONMENT AND THE BLOOD

THE INTEGUMENTARY SYSTEM

THE SPECIAL SENSATIONS

SYSTEMS THAT CONNECT AND / OR REGULATE THE SYSTEMS OF THE BODY

LABORATORY

HOW TO USE YOUR MANUAL

Forseeing the problems that you might encounter in this course dictates at least a few words of introduction, especially for a manual as practical as this one. On the following pages are numerous diagrams, and approximately 2000 questions pertaining to the information that you will encounter in this course, which, we are hopeful, you will use to help those to whom you have dedicated your life's work. Regardless of your professional aims and goals, however, it is simply useful, practical and good to know and understand how your own body is constructed and how it functions.

Much of the burden of learning in this course will belong to you. As a way of easing your burden, we would like to offer the following suggestions and guidelines: 1) Read carefully the general introductions to the lecture sections before you attend lecture as these introductions will allow you to at least become familiar with the upcoming information; 2) Your attention to the PRELABORATORY PREPARATION sections that accompany the laboratories is absolutely essential. Be absolutely certain that you fill in the blanks in these sections for not only will it compel you to look ahead at the diagrams, but they will represent the structures / areas on the laboratory models for which you will be responsible on laboratory examinations.

We would like to take this opportunity to thank Ms. Mary C. Baker for her diligent reading, numerous corrections, and useful suggestions and help on much of this manual. Nevertheless, the errors herein are ours. We ask only that when you encounter any error that you please let us know. And of course your thoughtful suggestions for improving this manual for future classes are always welcome.

INTRODUCTION TO THE GENERAL PLAN OF THE HUMAN BODY

I. General Introduction

Anatomically, the human body is **bilaterally symmetrical.** That is, the left and right sides of the body are mirror images of one another and look essentially the same. We therefore, in most instances, have two of most organs: Two upper limbs, two lower limbs, two kidneys, two lungs, two adrenal glands, two sets of identical teeth on either side of the maxilla and mandible, two testes (males), two ovaries, uterine tubes, and mammary glands (females), two eyes, two ears, two thumbs, two sets of fingers (digits). Most, though not all, bones occur in pairs.

There are of course several organs and structures that are unpaired. To name a few, there is the brain, spinal cord, several bones in the face and skull, some parts of the larynx (voice box), the heart, spleen, liver, gall bladder, stomach, intestinal tract, penis (males), vagina and uterus (females), and urinary bladder. But even then, we can still in many of these organs detect symmetrical left and right sides. As examples, there are right and left sides to the brain that look very similar, a left and right side of the spinal cord, a left and right side to the heart, a right and left side of the tongue, a right and left side of the nose, and so on. We do encounter some asymmetry in the digestive system in which it is difficult to see that the right and left sides are identical.

The bilateral symmetry that we see is not perfect, however. Even a cursory examination of facial photographs will reveal that the right and left sides of the face look different. Measurements of various dimensions of the limbs reveal slight differences between them, and many people need a larger shoe size for one foot versus the other. Moreover, most of us have experienced that a ring that fits perfectly on the ring finger of one hand will be too large or too small for the ring finger of the other hand.

These variations in our anatomical bilateral symmetry are subtle and we all tend to accept them. We can all hope that a few inches here and there matters little. But you might want to consider again the differences in the right and left sides of the face. There must be some significance in this difference for indeed there is a special area in our brains devoted to remembering and recognizing faces and facial differences.

Functionally, our bodies are in many important ways **bilaterally asymmetrical,** for in contrast to the slight variations in the anatomy of our bilateral symmetries, the functional variations between right and left can be startling. We see profound differences in functional preference of right over left, and vice-versa, even though right and left may be identical anatomically. This preference is most noticable in which hand we prefer to use and this preference is referred to as "handedness". But when we stop to think about it we can decern many functional differences between our anatomical identical right and left sides of several organs. The right cerebral hemisphere of the brain, for example, is apparently emotional and artistic, while the left cerebral hemisphere is apparently adept at mathematical manipulations, is where we have our verbal abilities, and the hemisphere in which we perform our analytical thinking. Our ability to speak is also a function of either the right or left cerebral hemisphere, not both. In addition, the right side of the adult heart pumps blood to the lungs, but the left side pumps blood out over to the entire body (except the lungs). We use both eyes to see; yet, when we need to line up objects we will use either the right or left eye to do so, even though both are open. Which ear do you prefer to use during a telephone conversation ? Most of us are more adept at using the muscles on one side of the body. This is true not only for our manipulations using the "preferred" hand, but for all the muscles on the side of the body that we prefer. In general, if we are "left-handed", then we tend to also use our left lower limb moreso than the right one.

So, in general, we are **anatomically symmetrical** , but **functionally asymmetrical,** in a bilateral way. Perhaps, however, bilateral symmetry does not always lead to functional differences. Although it may not be true, we do seem to use both lungs lungs equally, both kidneys equally, both ovaries equally, both testes equally.

In this section we are going to take a general look at the organization of the human body. We will go over **planes** and **directions** because many structures and areas are named using nomenclature and definitions of planes and directions. **Cavities** are spaces in the body in which are located the various organs. After this introduction we will venture on into how the body is organized with increasing complexity beginning with cells and progressing through systems. The systems are important to generally master because we take a "systems approach" rather than a regional approach. Finally, in this section we will discover how the body increases the area of a structure without increasing its volume, a clever trick.

1. What does the term bilateral mean ?

2. What does being bilaterally symmetrical mean ?

3. Name some structures and organs that are paired.

4. Name some structures and organs that are unpaired.

5. what does being functionally bilaterally asymmetrical mean ?

6. List some organs that are unpaired but nevertheless have symmetrical right and left sides.

7. Name some unpaired, but anatomically symmetrical, organs that are functionally asymmetrical.

8. In what way are the following organs functionally asymmetrical: The cerebral hemispheres of the brain; the adult heart; the eyes; the muscles of the body; the face.

9. Which eye do **you** use to line up two objects with both eyes open ? You can easily test which one by holding up your pencil vertically at arms length and then lining it up with some other vertical structure that is further away. Do this with both eyes open. With the pencil still in front of you close your right eye. Then close your left eye and open your right eye. Which eye did you use to line your pencil with the more distant object ? Are you left handed or right handed ?

II. Planes, Directions, and Cavities

A. Planes

1. Definition :

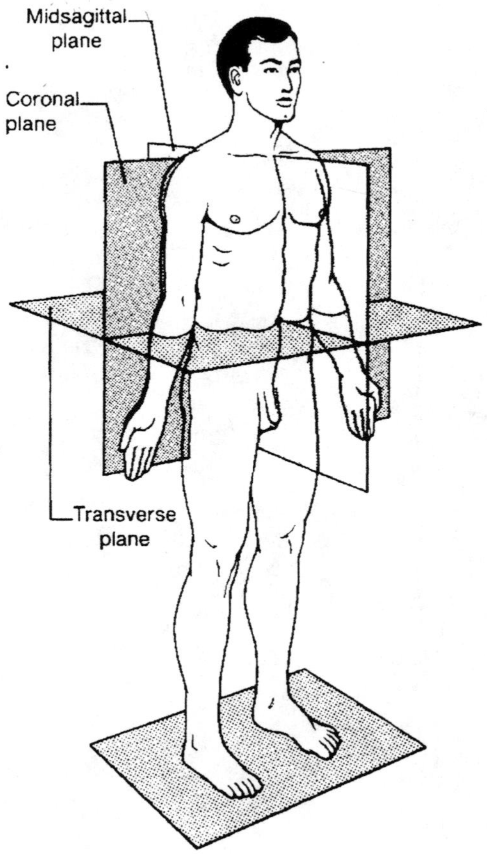

2. Directions

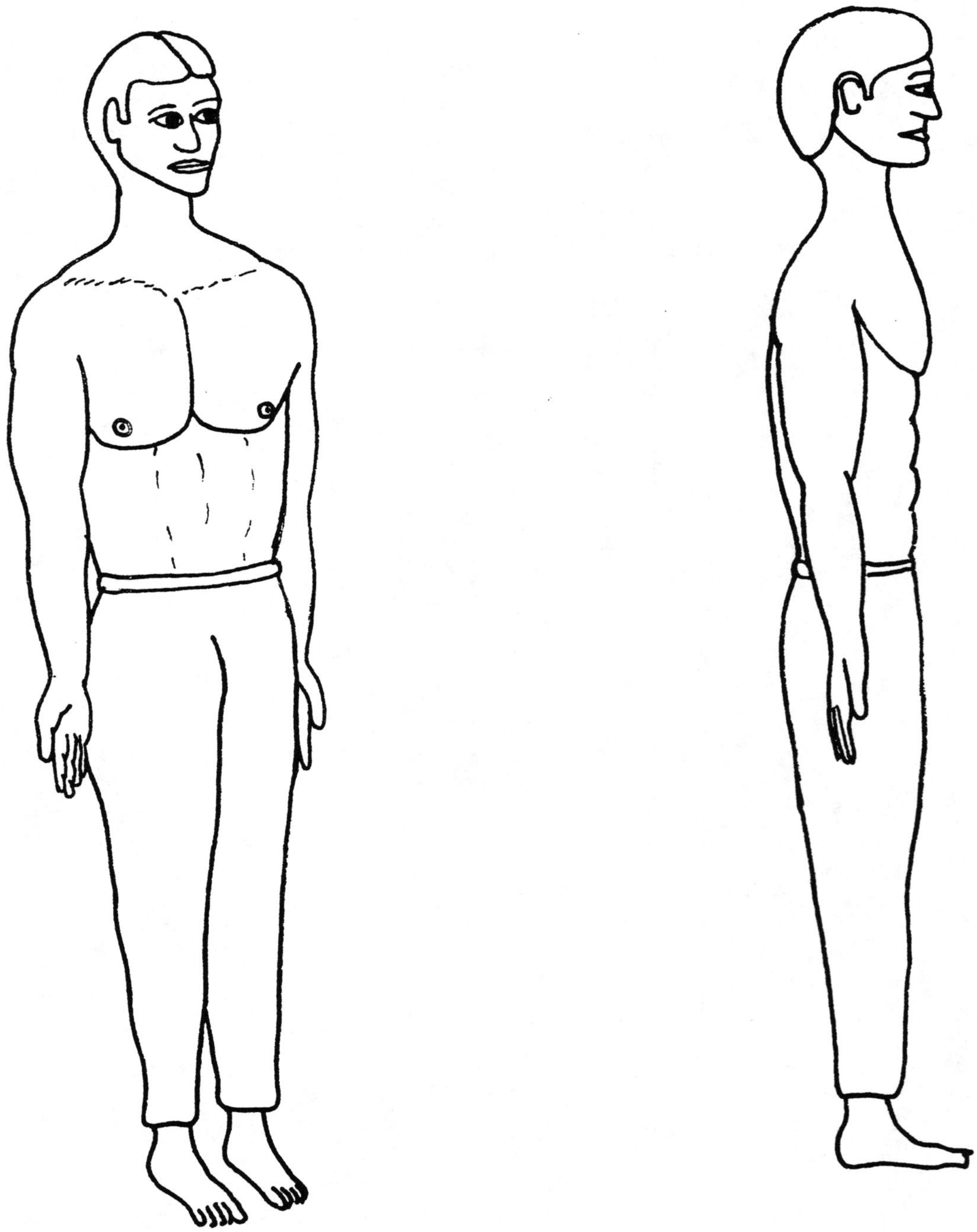

3. Cavities

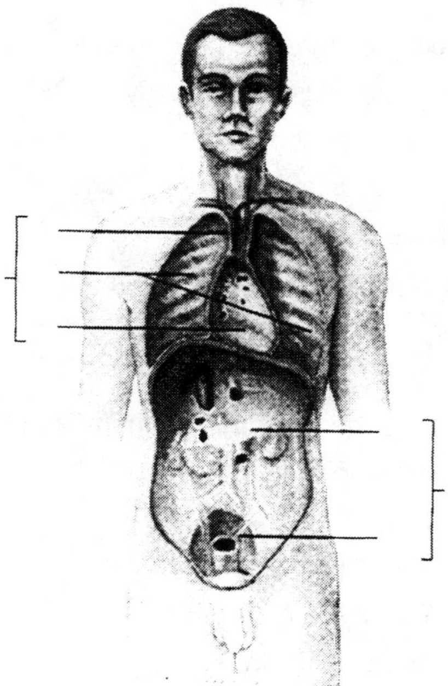

(b)

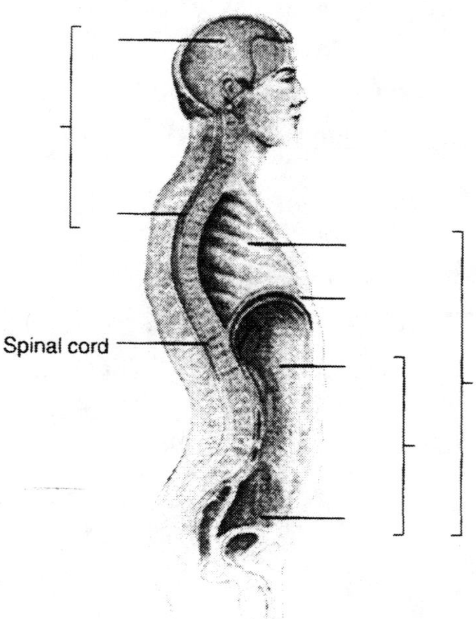

(a)

10. In these diagrams can you see that the body is bilaterally symmetrical ?

11. What does the term sagittal refer to ?

12. What is the difference between midsagittal and sagittal ?

13. A sagittal plane through the body divides the body into what type of sides ?

14. A transverse plane through the body divides the body into what two general areas ?

15. A frontal plane through the body will divide the body into what two general areas ?

16. Define the following terms: Superior, inferior, caudal, cranial, dorsal, ventral, anterior, posterior, lateral, medial, proximal, distal.

17. Rearrange the following structures from superior to inferior: Knee, umbilicus, big toe, larynx, nipples, brain, thigh, teeth.

18. Describe the anatomical position.

19. In the anatomical position which is more lateral, the thumb or the little finger ?

20. In the anatomical position which is more dorsal, the palm of the hand or the back of the hand ?

21. What cavities are in the dorsal cavity ?

22. What organs / structures are located in the dorsal cavity ?

23. What cavities make up the ventral cavity ?

24. What two cavities make up the thoracic cavity ?

25. What organs are located in the pleural cavities ? In the pericardial cavity ?

26. What structure separates the thoracic cavity from the abdominal cavity ?

27. Name some structures located in the abdominal cavity ?

28. Name some structures located in the pelvic cavity in females. In males.

II. Cells, Tissues, Organs, Systems

A. Cells

1. Are microscopic in size
2. Are the basic units of the body
3. Structure of a "typical" cell

B. Tissues

1. Definition of a tissue:

2. Types, locations, and examples

    a. Epithelial

    b. Connective

    c. Muscular

    d. Nervous

C. Organs

1. Definition of an organ:

2. Some organs and their functions

    a. Stomach

    b. Brain

    c. Lungs

    d. Muscles

e. Liver

f. Heart

g. Skin

h. Uterus

i. Testes

D. Systems

1. Definition of a system:

2. Systems of fertilization and development (Reproduction)

a. **FEMALE REPRODUCTIVE SYSTEM**

1) Components

2) Functions of components

b. **MALE REPRODUCTIVE SYSTEM**

1) Components

2) Functions of components

2. Systems that support and cause movement

a. **SKELETAL SYSTEM**

b. **MUSCULAR SYSTEM**

3. Open systems that allow exchange of substances between the environment and blood

a. **RESPIRATORY SYSTEM**

1) Organ components

2) Functions of component organs

b. **DIGESTIVE SYSTEM**

1) Organ components

2) Functions of component organs

c. **URINARY SYSTEM**

1) Components

2) Functions of components

4. Systems that connect and / or regulate systems of the body

a. **NERVOUS SYSTEM**

1) Components

2) Functions of components

b. **ENDOCRINE SYSTEM**

1) Components

2) Functions of components

c. **CARDIOVASCULAR SYSTEM**

1) Components

2) Functions of components

d. **LYMPHATIC SYSTEM**

1) Components

2) Functions of components

5. **INTEGUMENTARY SYSTEM**

C. Increasing the surface area without increasing the size (volume)

1. Why is more surface area needed ?

2. Where are folds found ?
    a. Gyri and sulci in the cerebral hemispheres

    b. Rugae and the expansion of the
        1) Stomach

        2) Uterus

        3) Vagina

        4) Gall bladder

        5) Urinary bladder

    c. Plicae circulares in the small intestine

    d. Folding of the skin of the scrotum

29. Describe the general structure of a typical cell.
30. Define a tissue.
31. Name the four types of tissues and give their locations.
32. Define an organ.
33. Define a system.
34. For each of the following organs give a function and state which system the organ is in: Brain; Stomach; Heart; Lungs; Larynx; Uterus; Testes; Ovaries; Liver; Skin; Muscles; Urinary bladder; Gall bladder.
35. Give the location and functions of rugae, plicae circulares, and sulci / gyri.
36. Explain why it is necessary to have the skin of the scrotum wrinkled.

CYTOLOGY AND HISTOLOGY

I. General Introduction

The anatomical and functional unit of a tissue, and therefore of an organ, a system, and of the entire body is the **CELL**. The study of cell structure and function is called **CYTOLOGY**. As it turns out, most cells are just too small to be seen without the aid of a microscope. The human egg cell can perhaps be seen without the aid of lenses, but it seems to be the only one.

As we will discover with the reproductive systems, each of us started our lives as a single cell, the fertilized egg (called a **zygote**), most likely in the ampulla region of one of the fallopian (uterine, or ovarian) tubes in the individual who we will recognize and refer to for all of our lives by a variety of names including mother, mommy, mom, ma, etc. The individual who contributed the cell (spermatazoan, or sperm cell) that fertilized the egg which created the zygote and started us on our way will become the individual who we will recognize and refer to for all of our lives by a variety of names including father, dad, daddy, pa, etc.

The zygote represents an humble beginning, but nevertheless we all started that way. Mileva Einstein, You, Attilla the Hun, Toni Morrison, Dr. Martin Luther King, Thomas Merton, Malcolm X, Albert Einstein, Maude Menten, Rosa Lee Parks, Alexander the Great, all the Presidents, Adolph Hitler, Helen Keller, Clara Barton, Mary York, George Washington Carver, Booker T. Washington, King George IV, and all others have at least one thing in common and that is that we all began our lives as a single cell.

Once the zygote is produced by fertilization of the egg by a sperm, there is created a strange potential and that is the ability of the cell to divide, by a process called **MITOSIS**, into two genetically identical cells. Then each of these cells will undergo mitosis to produce four genetically identical cells, then further mitosis produces eight genetically identical cells, and so on.

The results of all of these mitotic divisions during embryonic and fetal life are two: 1) the production of trillions of genetically identical cells and 2) the formation of hundreds of different types of cells. It is true that we probably can not see a zygote without a microscope, or perhaps not even the two-cell, or four-cell individual. But we can certainly see, without the aid of a microscope, a conglomeration of trillions of cells. In fact, we give names to each conglomeration of human cells; there are conglomerations named Jennifer, Mary, Lars, Chung He, Sarah, Einar, George, Gary, Anatole, Natasha, Moon Beam, Sunshine, Pierre, Abdul, Herschel, Bill, Mao, Ivan, and so on. Obviously, reproductive activity among humans has created more individuals than we could possibly give different names to. There are lots of Jennifers and Marys and Ivans, a few Moon Beams perhaps, but, oddly, no two are identical, even if the names are the same.

One of the most amazing and least understood processes during development is how the different types of cells are formed, a process referred to as **DIFFERENTIATION** . But think about it. Just how is it that a single cell, the zygote, undergoes mitosis and along the way some of the resulting cells differentiate into bone cells, some into muscle cells, into neurons (nerve cells), kidney cells, skin cells, rods and cones of the retina, hair cells of the inner ear, erythrocytes (red blood cells), leucocytes (white blood cells), fibroblasts, parietal cells of the stomach, ova (eggs), spermatazoa, and so on ? The result of intrauterine development is an individual who, thanks to mitosis, has trillions of cells, and, thanks to differentiation, has hundreds of different types of cells. Moreover, because different cells perform different functions it is the process of differentiation that has allowed for the creation of different functions of tissues, organs, and systems. Bone cells form bone, muscle cells allow the individual to move, rods and cones allow for vision, hair cells will let the individual to hear, erythrocytes carry oxygen to all parts of the body, fibroblasts produce protein fibers that support other cells and produce ligaments and tendons, parietal cells will produce hydrochloric acid for digestion in the stomach, adipose cells store fat, and ova and sperm cells will allow the individual to reproduce.

The process of differentiation ensures that cells perform different and specialized types of functions. For example, nerve cells conduct an electrochemical impulse over their surfaces, liver cells make plasma proteins, muscle cells can shorten their lengths (contract), erythrocytes transport oxygen, some leucocytes produce antibodies, rods and cones are two types of cells that respond to light waves, and parietal cells in the stomach make and secrete hydrochloric acid. Differentiation ensures that the human body has different types of cells with unique functions, and mitosis ensures that in the body there are multiple copies of each cell type.

Copies of similar cells with similar functions get grouped together during development to form **TISSUES**. There are four types of tissues, although some texts will recognize five types. General types of cells get grouped to form **EPITHELIAL TISSUES**. These tissues are located on surfaces (skin is an epithelial tissue, for instance), line body cavities, and line tubules in the body. Epithelial tissues carry out absorption and secretion functions in skin, the digestive system and the urinary system to name a few. **MUSCULAR TISSUES** allow for movement, **CONNECTIVE TISSUES** are basically support tissues that form networks to support cells and keep them in place, to support the body (bone) and connect the body regions with fluids (blood and lymph). **NERVOUS TISSUE** is widespread and forms a network of nerves and nerve fibers throughout the body that controls and regulates the functions of the various other tissues, and therefore the functions of organs and systems. The study of the structures and functions of tissues is called **HISTOLOGY.**

1. What is the study of cells called ?
2. What is the study of tissues called ?
3. What is a fertilized egg called ?
4. What is mitosis ?
5. What is differentiation ?
6. What are the two results of mitosis and differentiation ?
7. Name some different types of cells found in the human body and give their functions.
8. Describe what a tissue is.
9. List the four tissues and give their general locations and functions.

II. CYTOLOGY

A. Fertilization

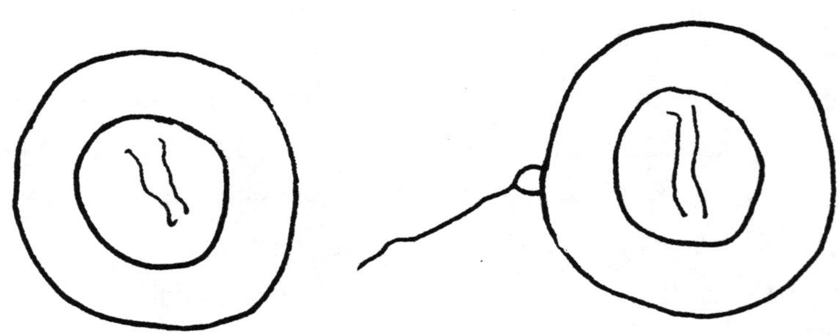

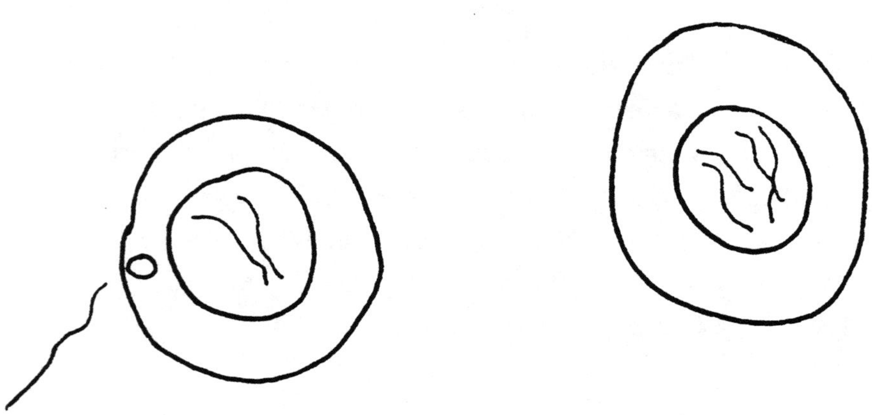

B. General structure of a "typical" cell

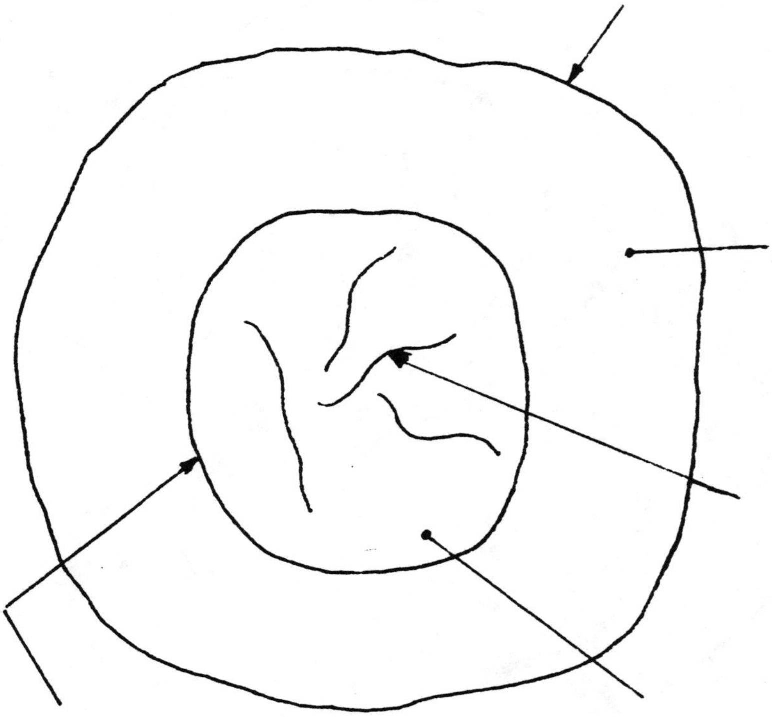

C. Mitosis and differentiation

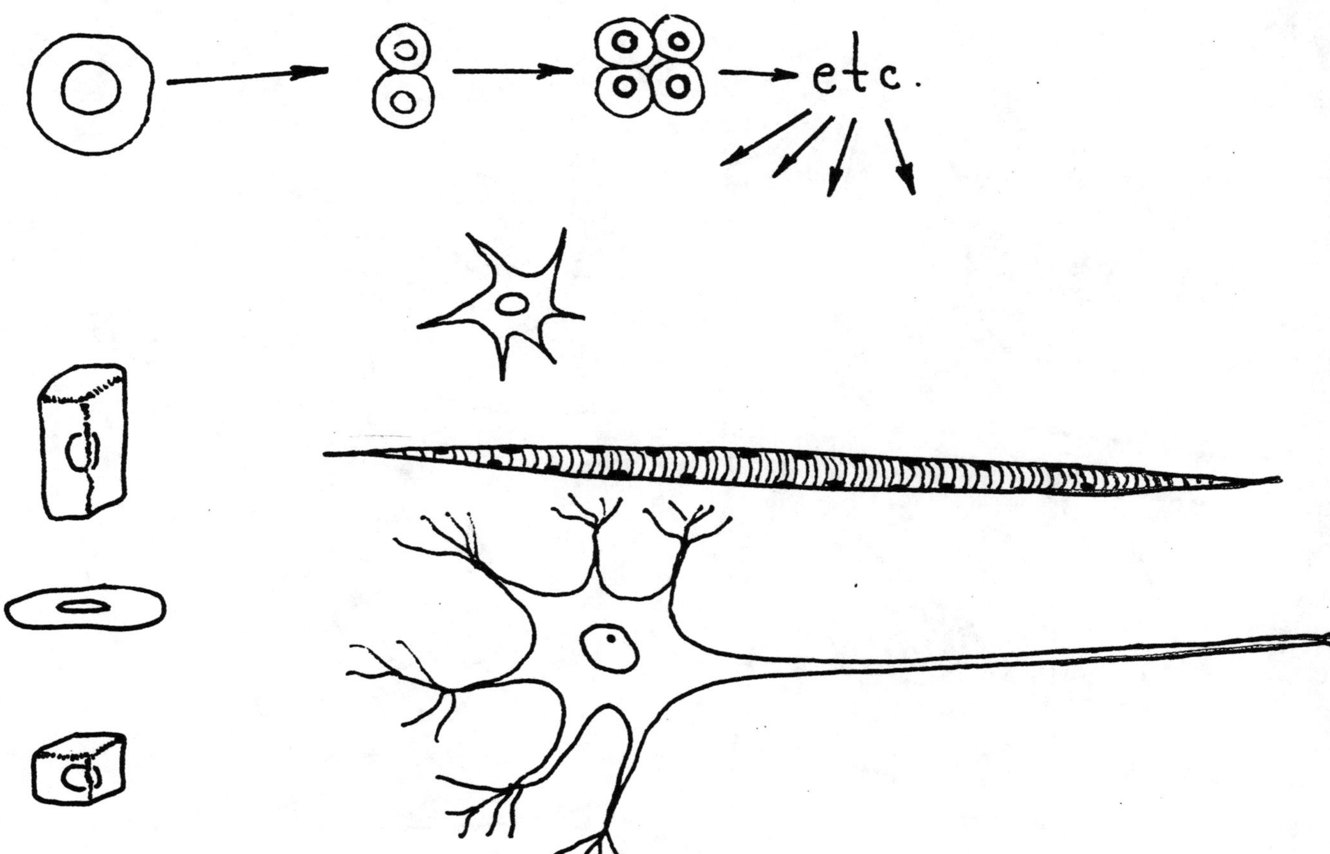

D. The cells' environments and the creation of the three "fluid spaces" in the body

1. Blood

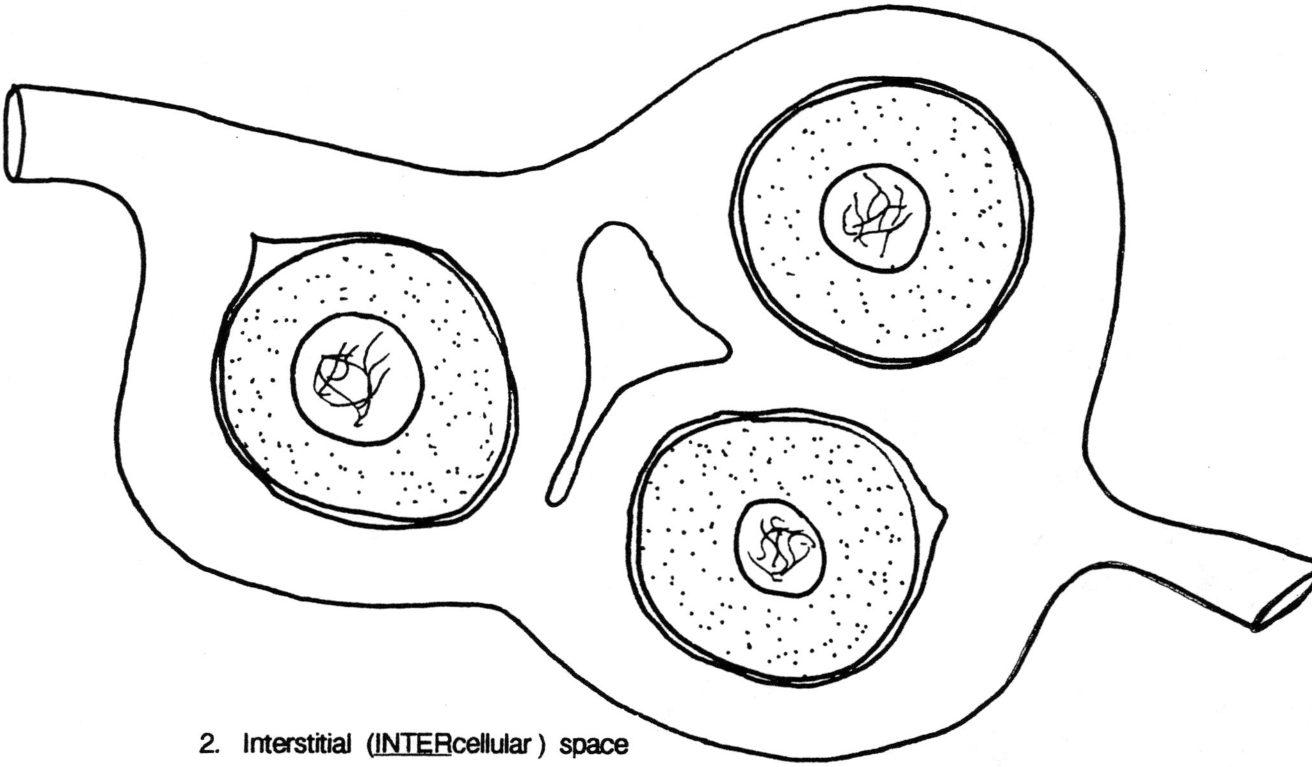

2. Interstitial (INTERcellular) space

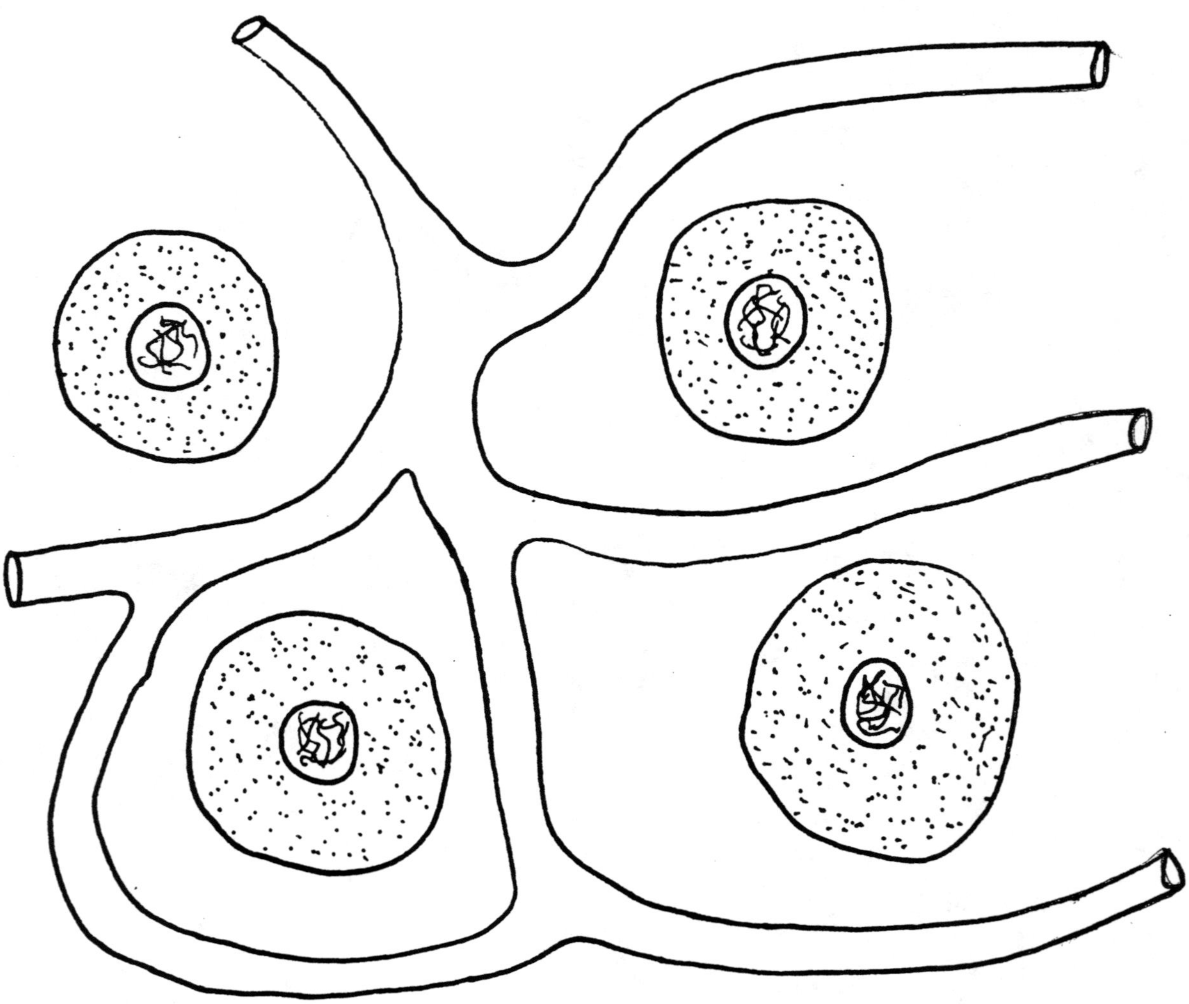

3. Where is the fluid in your body located ?

   a. In the blood in your blood vessels and heart = Cardiovascular space

   b. Between your body cells = Interstitial (INTERcellular) space

   c. Inside the trillions of your cells = INTRAcellular space

4. Edema =

5. Blood pressure

10. What is fertilization and what is the name of the resultant cell ?
11. What is the plasma membrane and what is its function ?
12. What is the cytoplasm ?
13. What is located in the nucleus ?
14. What is mitosis ?
15. What is differentiation ?
16. What are the two results of mitosis and differentiation ?
17. What are the three fluid spaces in the body ?
18. What is edema, and what are some of its causes ?
19. What is blood pressure ?

E. What all cells do regardless of their types

1. **Selectively transport** nutrients across their plasma membranes from either the blood or the interstitial space.

2. Breakdown sugars (glucose) to produce **Carbon Dioxide (CO_2)** and to make a form of chemical energy called **Adenosine Triphosphate (ATP).**

3. To absorb and use **Oxygen (O_2)** for the formation of **ATP**

4. To use the energy in the ATP formed from the breakdown of glucose and oxygen to make **protein molecules,** to make **fats**, and to make its **DNA** molecules during mitosis.

F. The cell performs its functions of selective transport, utilization of oxygen for ATP formation, of production (synthesis) of proteins, fats, and DNA , as well as other functions, by having small structures called **ORGANELLES** within their cytoplasm and nucleus that perform these functions.

1. Definition of an **ORGANELLE** =

2. Organelles of the cell

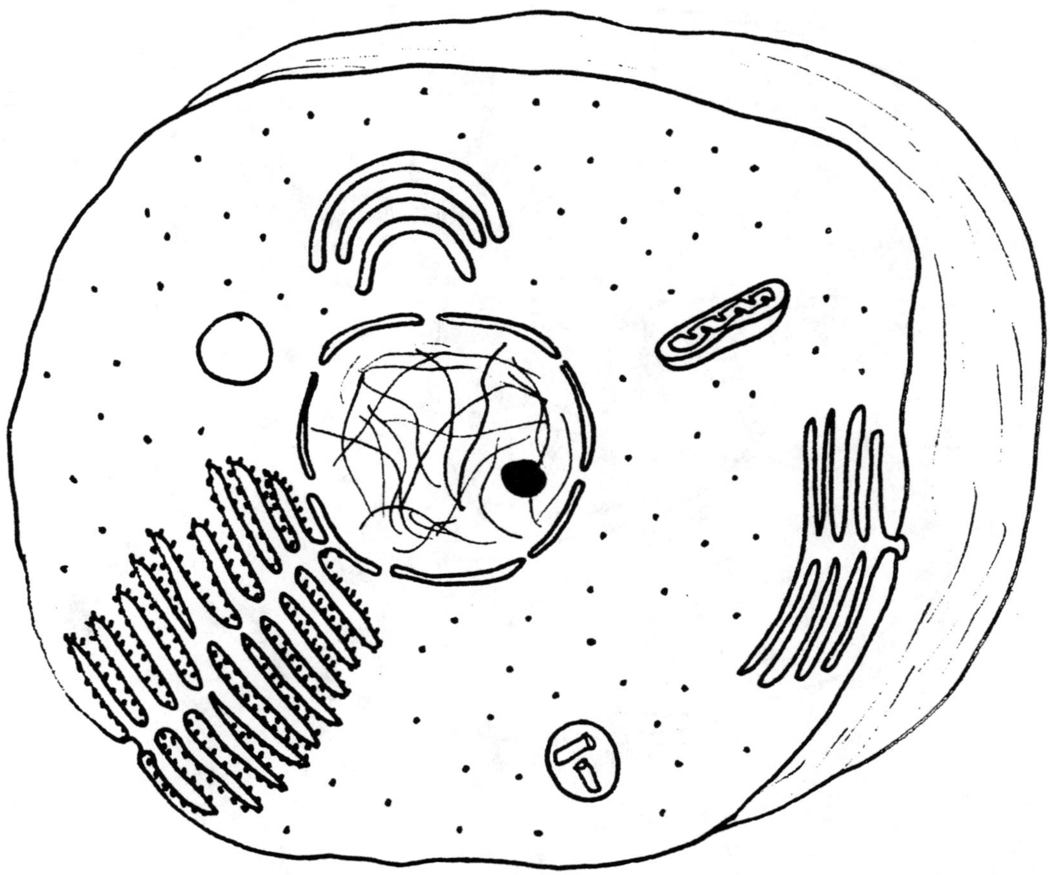

3. Selective (semipermeable) transport of substances from either the blood (cardiovascular space) or the interstitial space across the plasma membrane into the cell (Intracellular space).

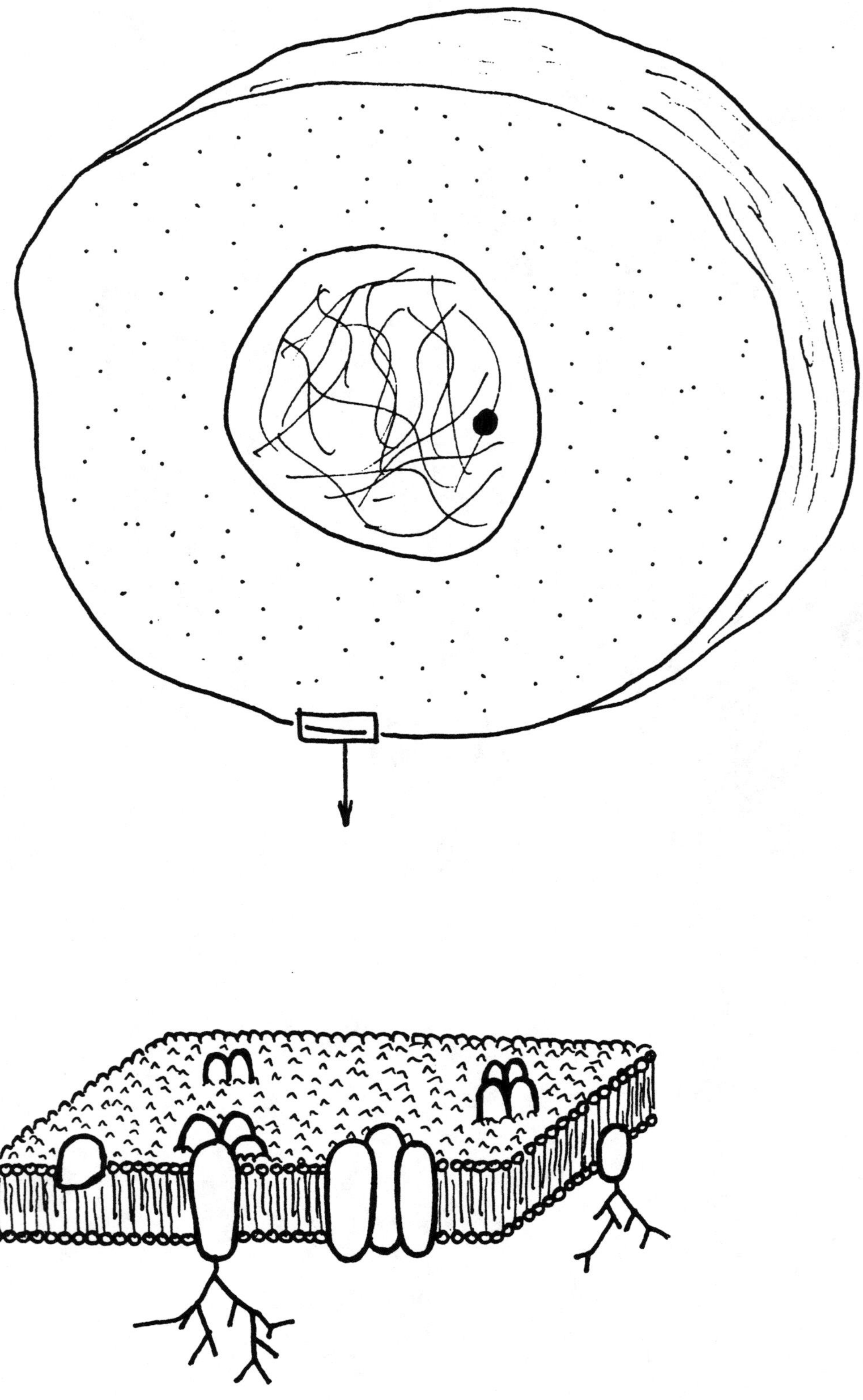

4. **Cytoplasmic organelles**

a. Organelles in general are of two types - those that have a fluid mosaic membrane as part of their structure (membranous organelles), and those that do not have a membrane as part of their structure (nonmembranous organelles).

b. **MEMBRANOUS ORGANELLES**

1) The **MITOCHONDRION**

a). Mitochondria = more than one mitochondrion

b). **FUNCTION** = to use O_2 and couple that utilization to make ATP. The Krebs citric acid cylcle occurs in the matrix of the mitochondrion, and the electron transport system is located in the inner mitochondrial membrane.

c). Their numbers vary from one cell type to another according to how much energy the cell type requires. A bone cell is fairly inactive and has only a few hundred mitochondria, but a muscle cell is active and therefore has several thousand mitochondria. A liver cell makes lots of protein molecules which requires ATP so a liver cell also contains several thousand mitochondria.

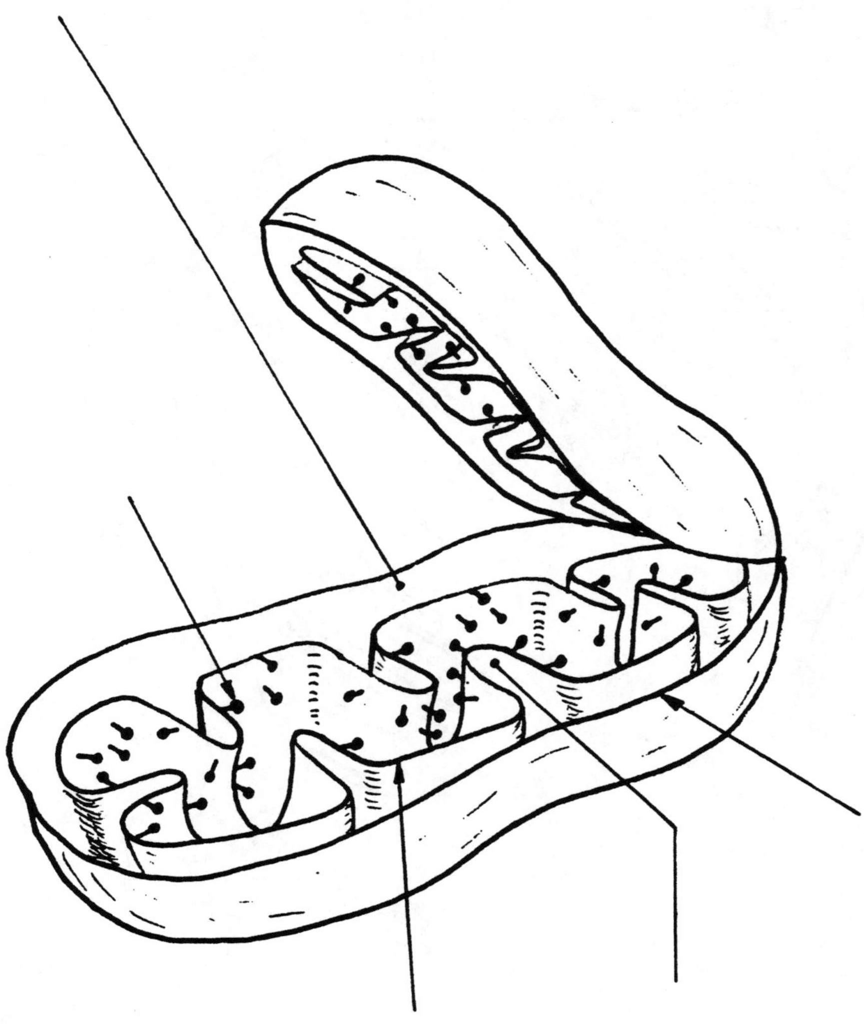

2). The **ENDOPLASMIC RETICULUM (ER)**

a). Types

1)) Smooth ER

2)) Rough ER

b). **FUNCTIONS** of the ER

1)). Smooth ER

2)). Rough ER

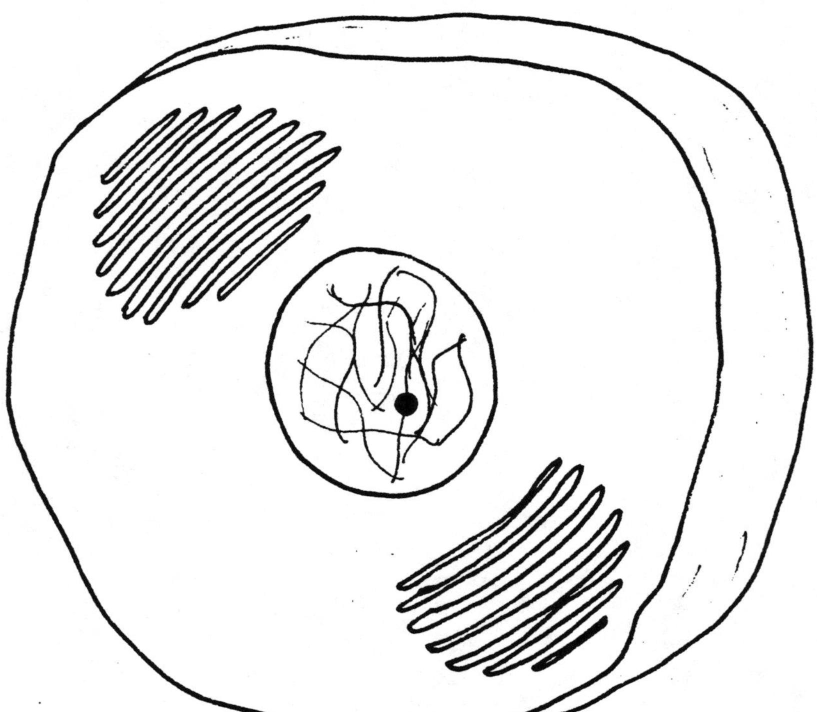

3). The **GOLGI APPARATUS (BODY)**

a). **FUNCTION** =

b). Structure

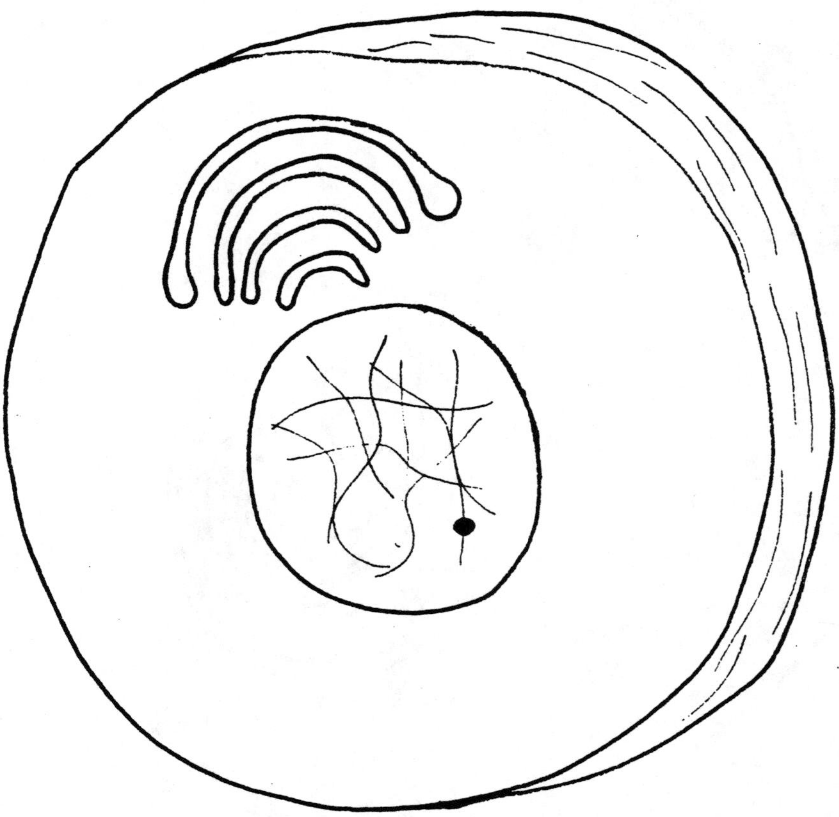

4). The **LYSOSOME**

a). Derivation of term:

LYSO = to break open

SOME = from the term soma, meaning body

b). Phagocytosis and lysosome function

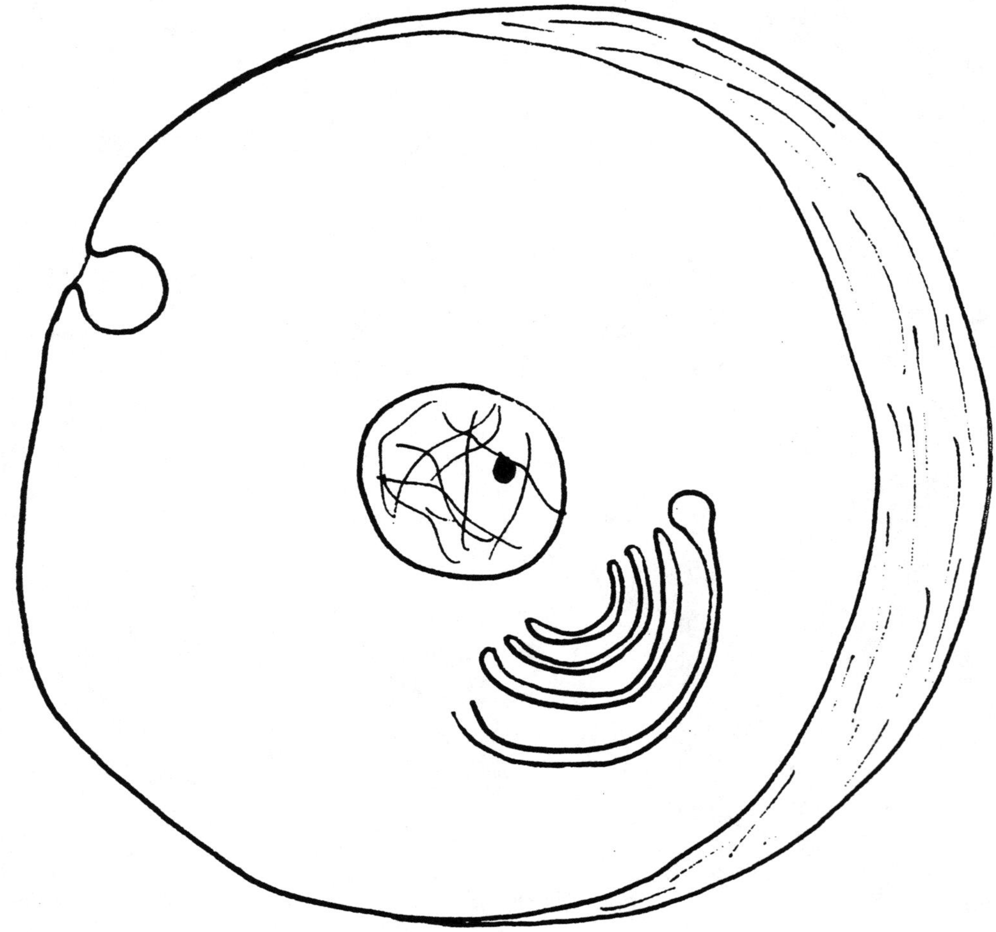

c. **NONMEMBRANOUS ORGANELLES**

1). Nonmembranous organelles in the cytoplasm

a). The **RIBOSOME**

1)). There are thousands in each cell

2)). **FUNCTION** =

b). Free and bound ribosomes

2). The **CENTROSOME and CENTRIOLES**

a). Centrosome =

b). Centrioles

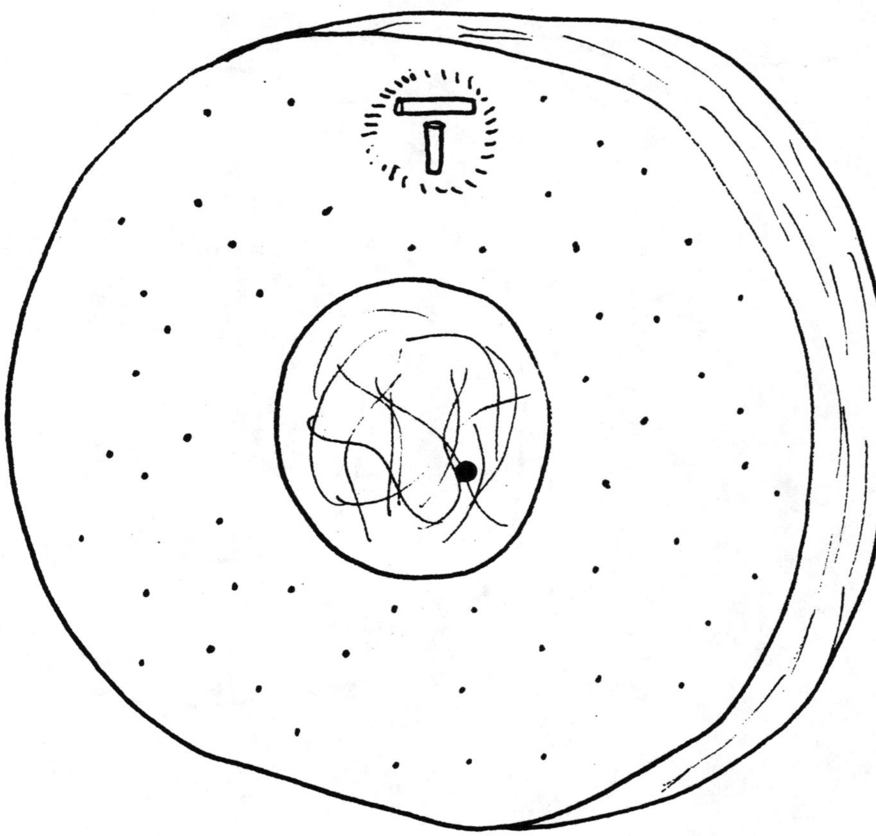

2) Nonmembranous organelles in the nucleus

a). The **NUCLEOLUS**

b). **CHROMATIN and CHROMOSOMES**

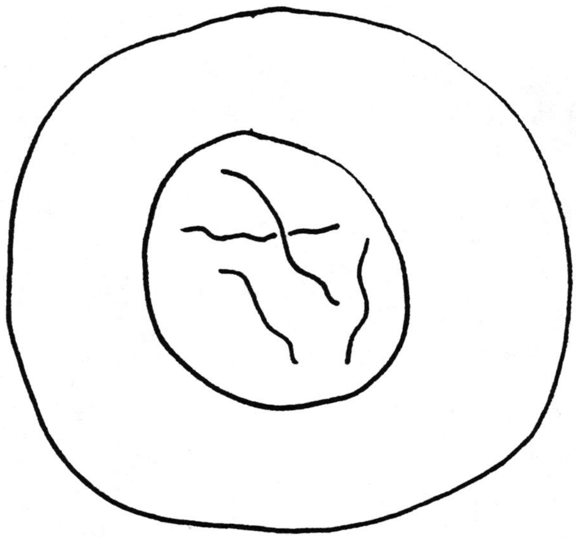

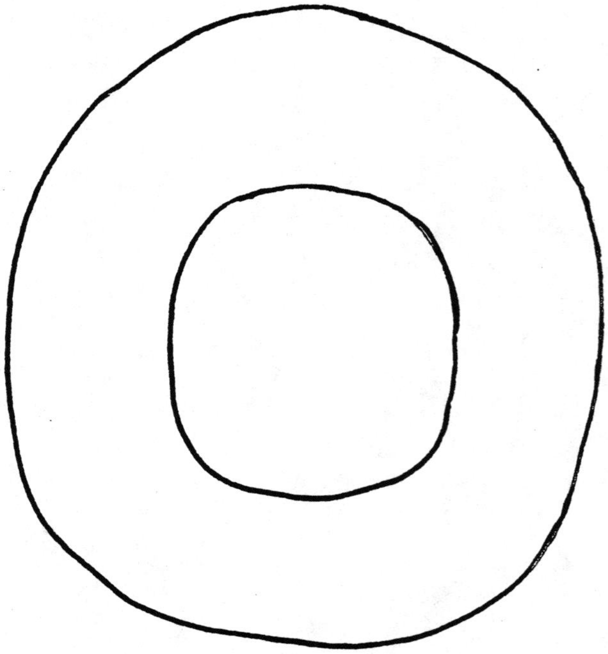

20. List some of the functions that cells do regardless of their types.

21. What is the full name of the compound that the cell makes for use as source of chemical energy ? What is its abbreviation ?

22. What is the energy in ATP used for by the cell ?

23. What organelle in the cell makes ATP ?

24. Define an organelle.

25. What are the membranous organelles and where in the cell is each located (cytoplasm, nucleus).

26. What organelle uses oxygen ?

27. What are the nonmembranous organelles and where in the cell are they located ?

28. What does semipermeable transport mean ?

29. List the cytoplasmic organelles and state whether each is membranous or nonmembranous

30. Are there any membranous organelles in the nucleus ?

31. List the organelles in the nucleus.

32. What is the major function of the mitochondrion ?

33. Where in the mitochondrion does the Krebs citric acid cycle occur ?

34. Where in the mitochondrion does the electron transport occur ?

35. What is the relationship between the number of mitochondria in a cell and its energy needs ?

36. What is the ER ? What is the anatomical difference between the smooth and rough ER ?

37. What are the functional differneces between the smooth and rough ERs ?

38. What is the Golgi apparatus (body) ?

39. What is the function of the Golgi apparatus ?

40. What does the prefix "***Lyso***" refer to ?

41. What does the term "***some*** " refer to ?

42. Describe the relationship between the Golgi apparatus, the lysosome, and phagocytosis.

43. What is the function of the ribosome ?

44. What is the difference between free and bound ribosomes ?

45. What do you think would be the relationship between the number of ribosomes and the number of mitochondria in a cell ?

46. What is the relationship between the centrosome and centrioles ?

47. What two types of cells have no centrioles ?

48. What cellular function do the centrioles participate in ?

49. Where is the nucleolus located ?

50. What is the function of the nucleolus ?

51. What is the difference between a molecule of chromatin and a chromosome ?

52. What is the fluid mosaic model for membrane structure ?

53. Draw a mitochondrion and label its basic structures and areas.

54. What specific structure in the mitochondrion forms (makes) ATP ?

G. MITOSIS (NUCLEAR DIVISION), and CYTOKINESIS (CYTOPLASMIC DIVISION)

1. The purpose and aim of the process of mitosis =

2. The structure of a chromosome

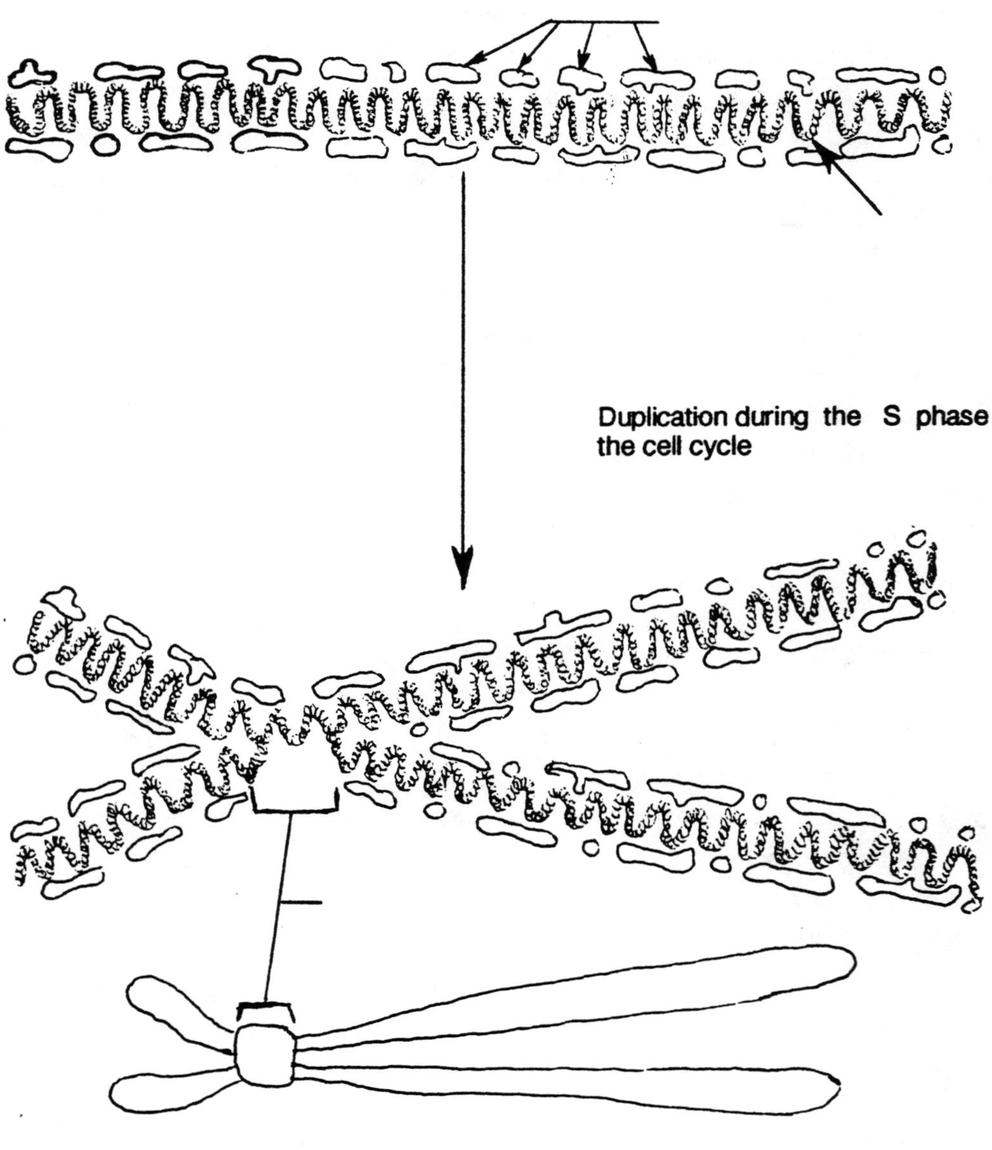

3. The cell cycle and the place of mitosis within this cycle

a. <u>General introduction</u> - All cells will pass through predictable stages during its life. For example, all cell grow to attain a certain size. Most cells will duplicate their genetic material and undergo mitosis. Two cell types, nerve cells and muscles cells, grow to attain their mature size, but do not duplicate their genetic material and therefore do not undergo mitosis.

b. The <u>cell cycle</u>

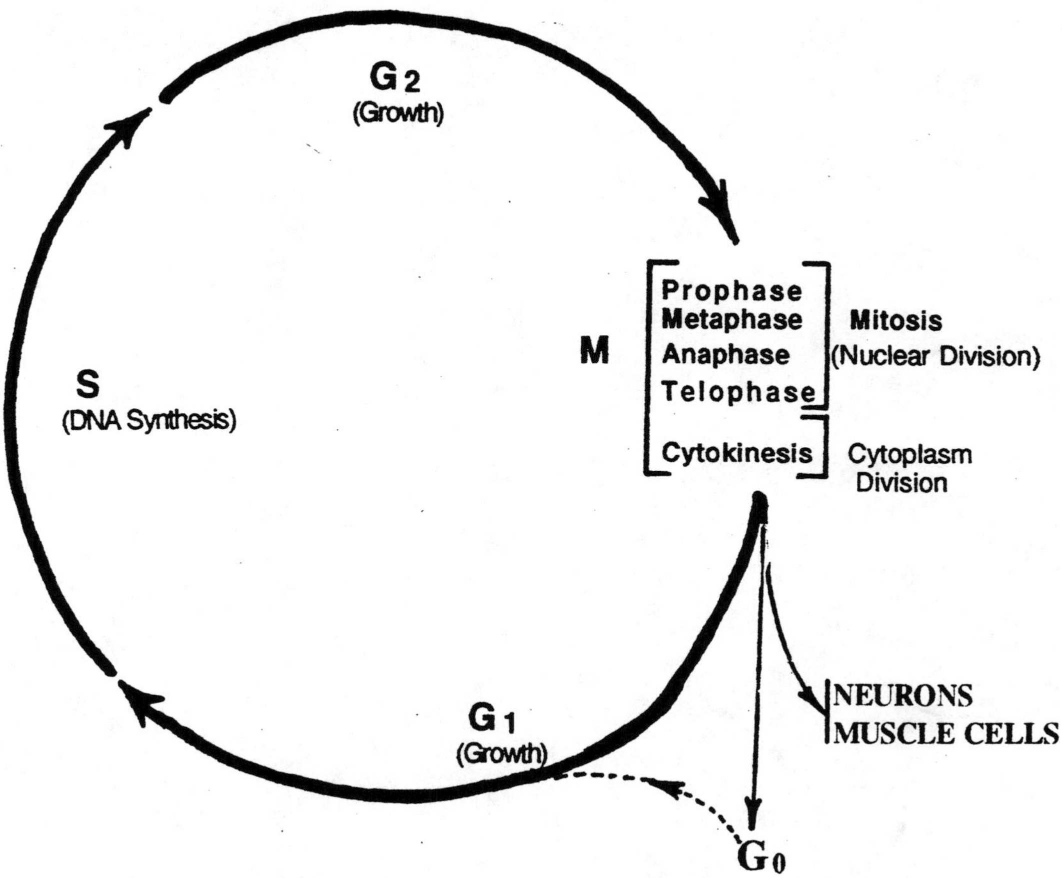

$G_1 + S + G_2$ = Interphase

G_0 = Temporary nondividing state

c. What is mitosis

4. Interphase stage of the cell cycle

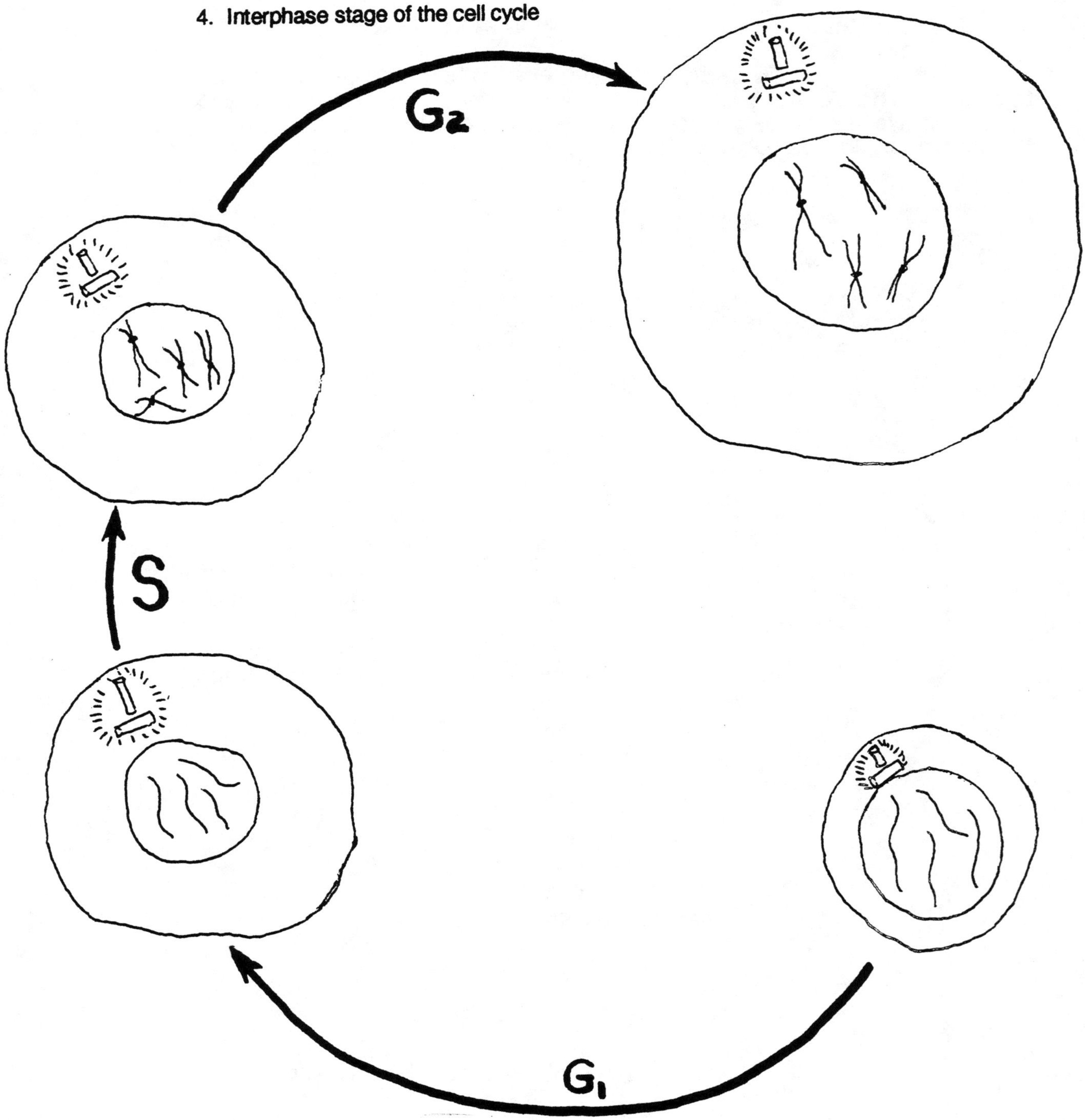

5. Stages in mitotic phase of the cell cycle (Prophase, Metaphase, Anaphase, and Telophase)

a. **PROPHASE**

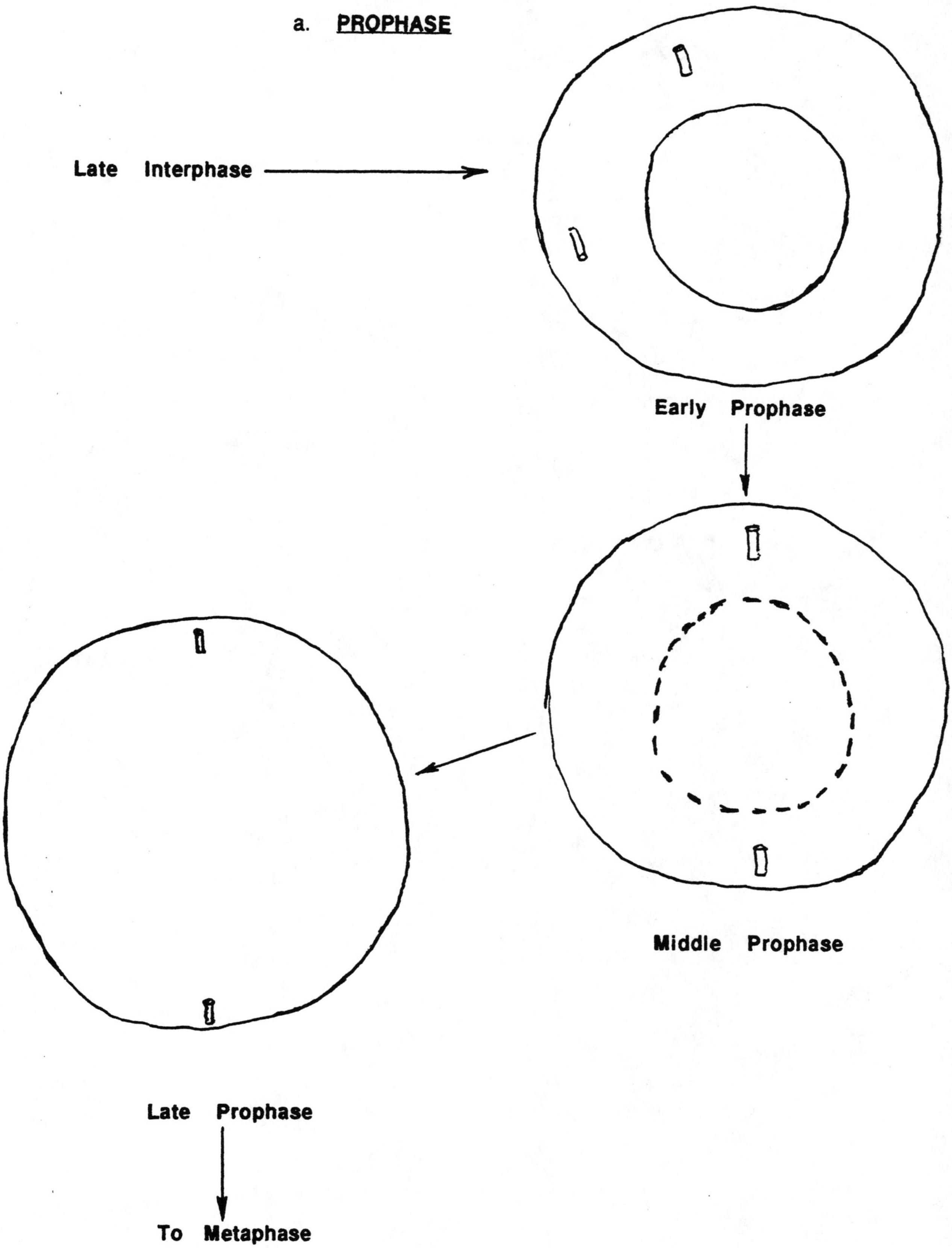

b. METAPHASE

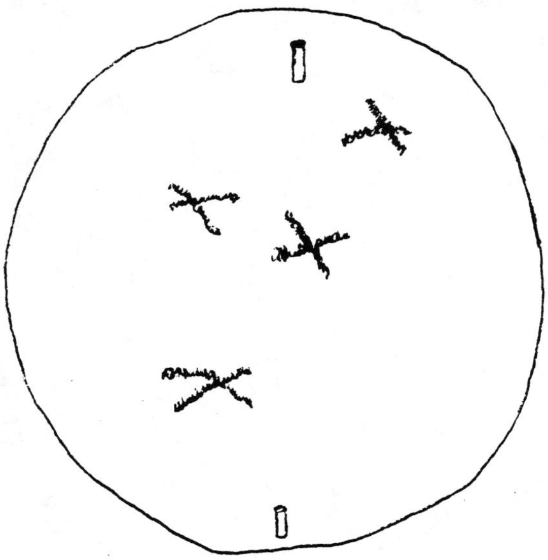

Late Prophase

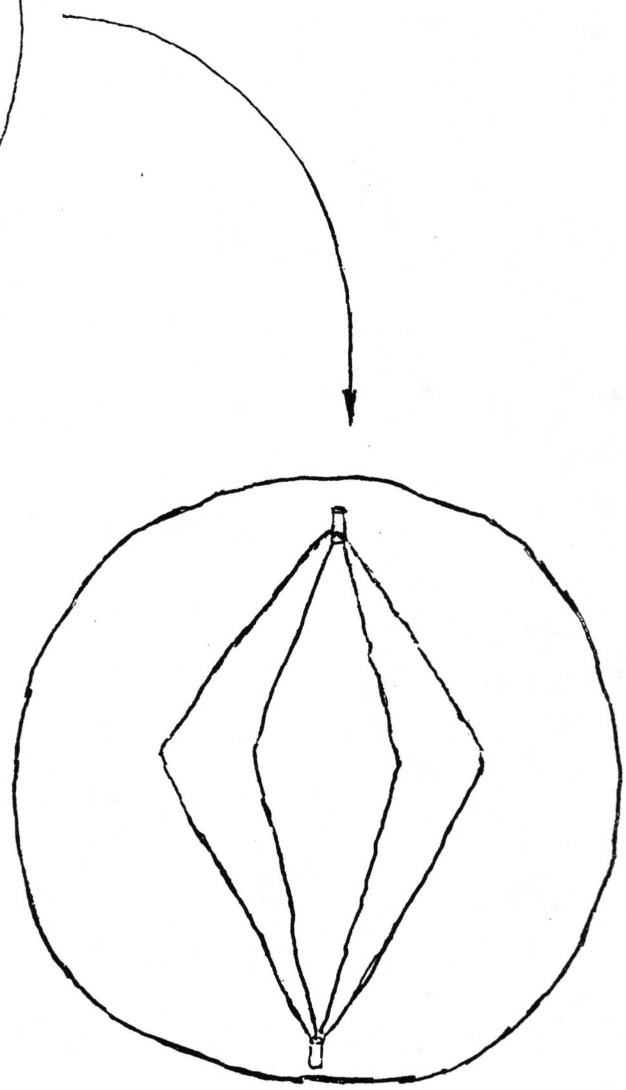

Metaphase

↓

To Anaphase

c. **ANAPHASE**

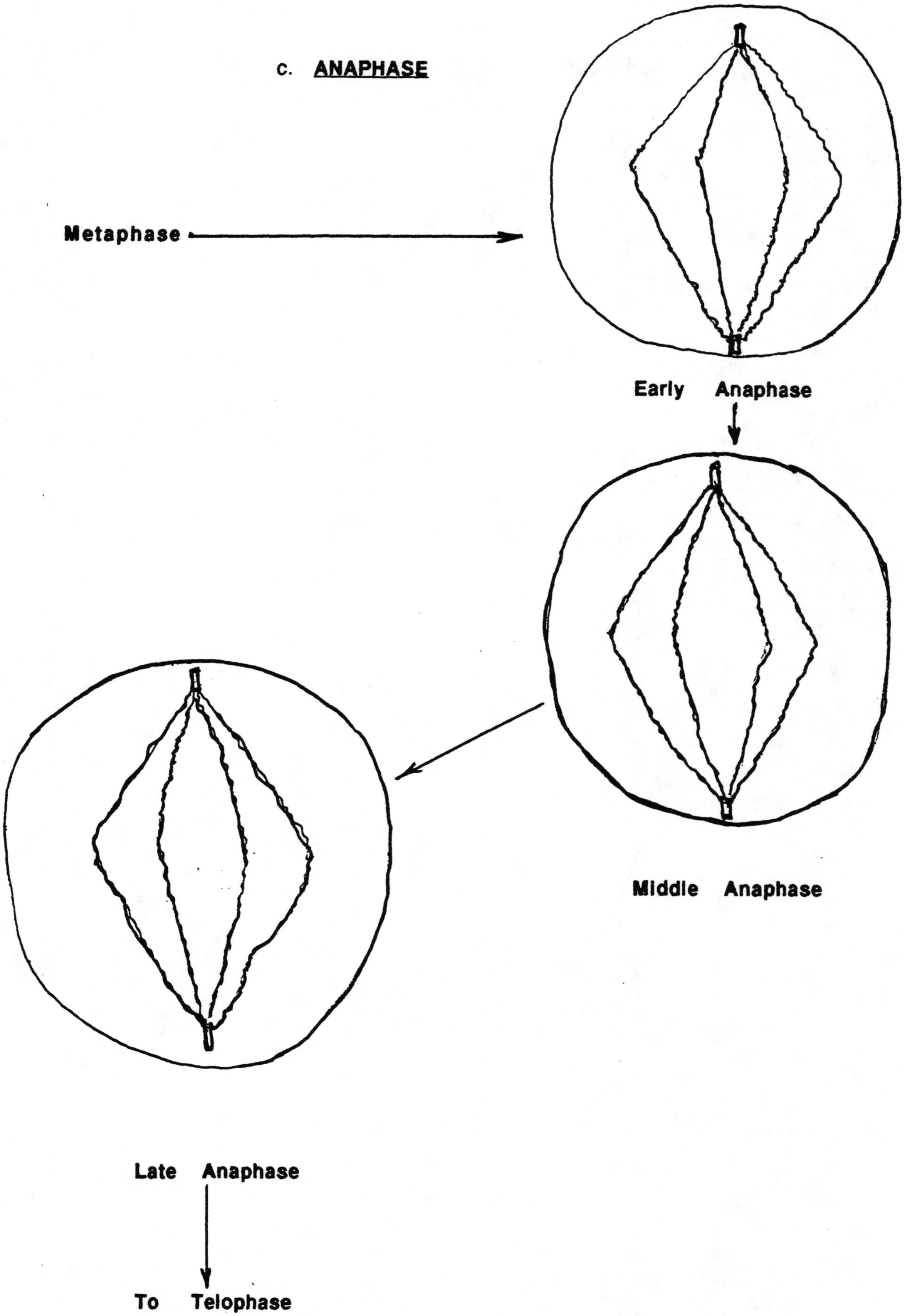
Metaphase
Early Anaphase
Middle Anaphase
Late Anaphase
To Telophase

d. **TELOPHASE**

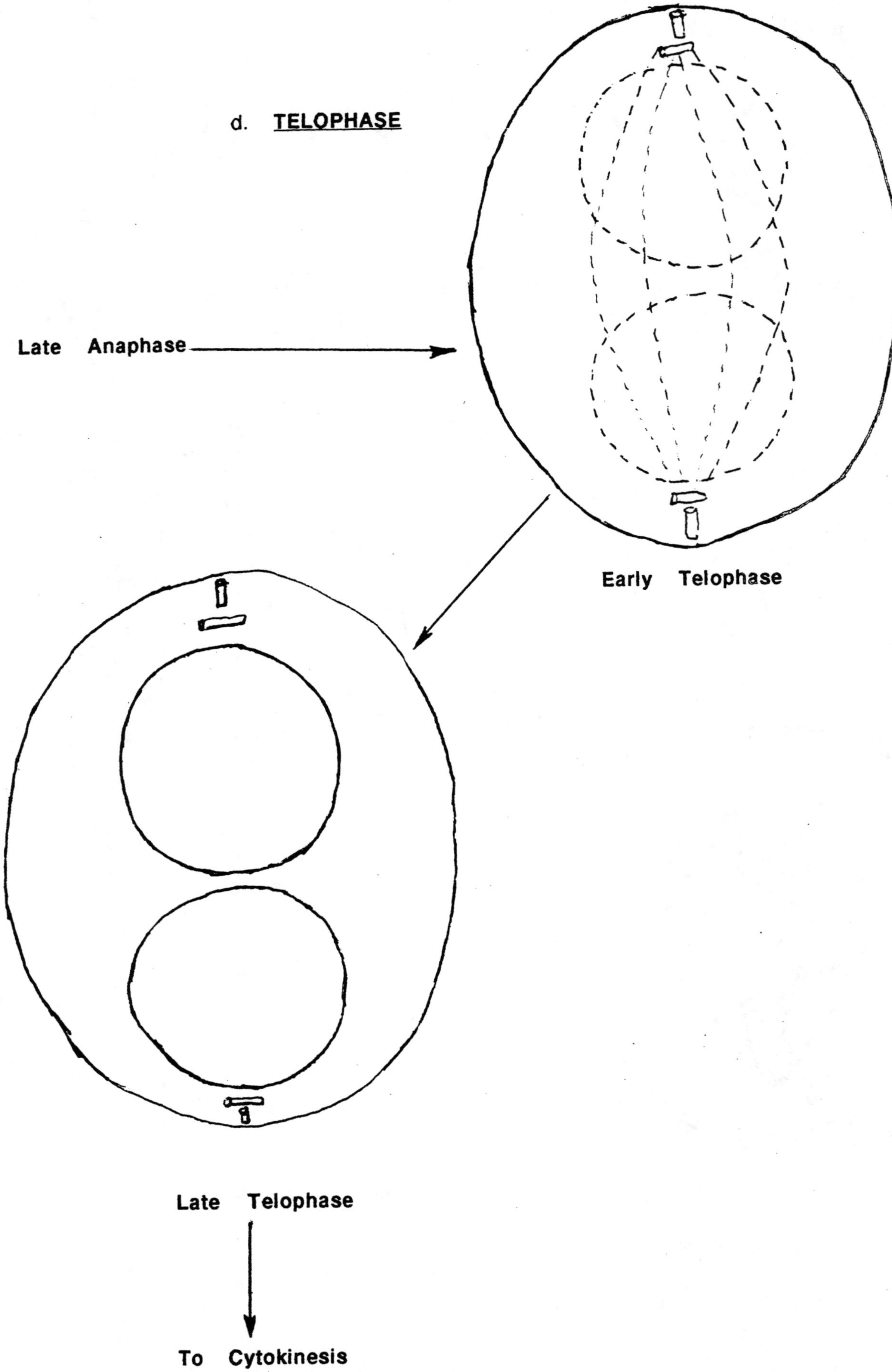
Late Anaphase
Early Telophase
Late Telophase
To Cytokinesis

6. **Cytokinesis**

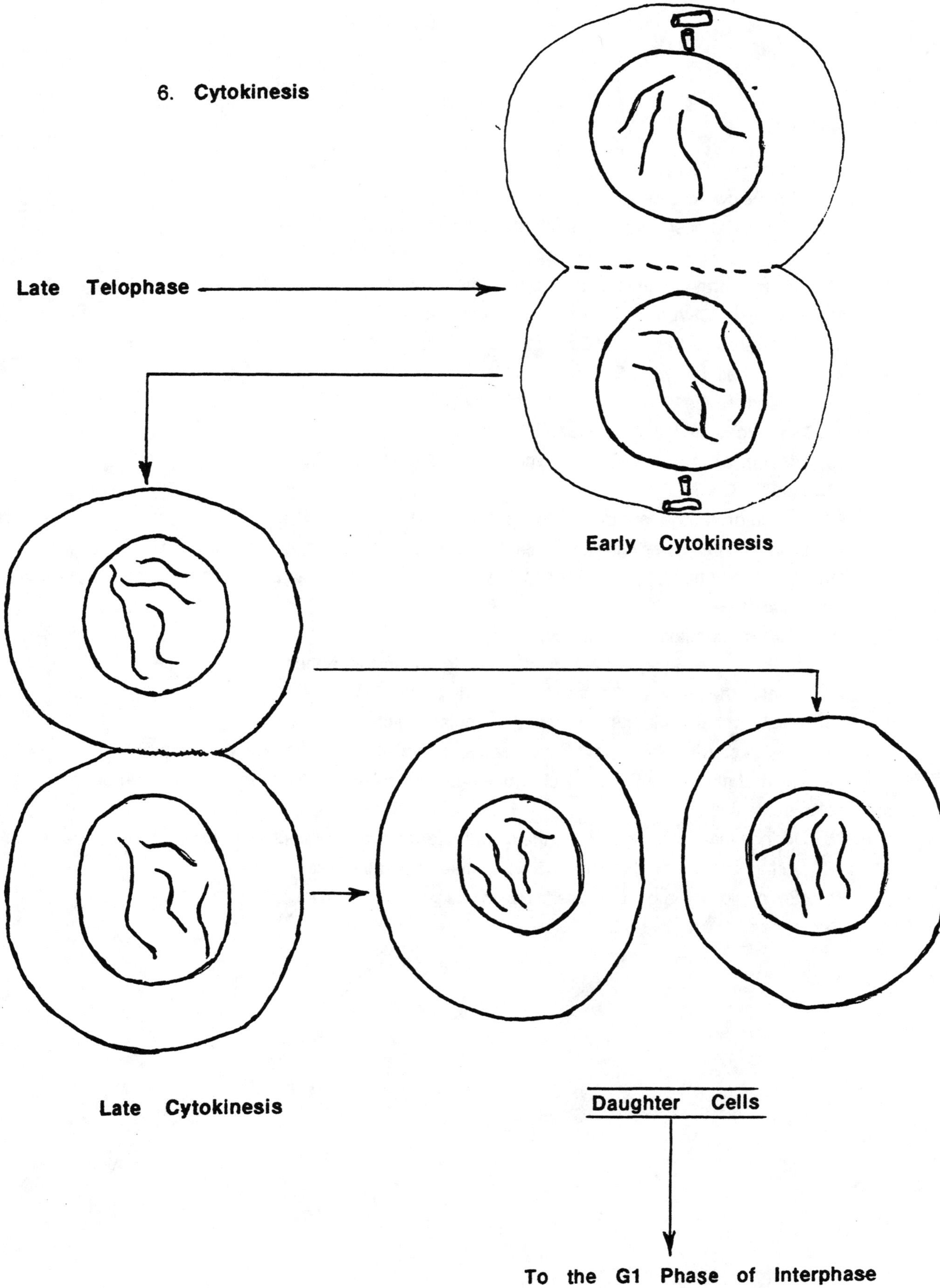

55. What is mitosis ?
56. What is cytokinesis ?
57. What is the diference between mitosis and cytokinesis ?
58. What is the purpose and aim of the process of mitosis ?
59. Describe the structure of a chromosome.
60. How many DNA molecules are in a chromosome ?
61. List, in order, the four phases (stages) of the cell cycle.
62. What does the letter G stand for in the cell cycle ?
63. What does the letter S stand for in the cell cucle ?
64. What does the letter M stand for in the cell cycle ?
65. What three stages of the cell cycle make up interphase ?
66. What is G_0 ?
67. List, in order, the four phases of mitosis.
68. What two cell types can not complete the M phase (stage) of the cell cycle ?
69. During prophase, what happens to the centrioles, the nuclear membrane, and the chromatin molecules ?
70. What is the mitotic spindle apparatus ?
71. In which mitotic stage is the mitotic spindle apparatus formed ?
72. What is the purpose of the mitotic spindle apparatus ?
73. What are the equitorial region and the polar regions of the cell ?
74. In what stage of mitosis do the chromosomes align along the equitorial region ?
75. In what mitotic stage do the chromosomes separate and begin to move toward opposite poles of the cell ?
76. During what mitotic stage are the nuclei reformed in the daughter cells ?
77. During what stage does the mitotic spindle apparatus disappear ?
78. You could say that telophase is the opposite of which mitotic stage - Prophase, Metaphase, or Anaphase ?

III. HISTOLOGY

A. Definition and types of tissues

1. Definition =

2. Types of tissues

a.

b.

c.

d.

e.

B. Epithelial tissues

1. Characteristics of epithelial tissues

a.

b.

c.

2. Diagram of these characteristics of epithelial tissues using the intestinal tract as an example:

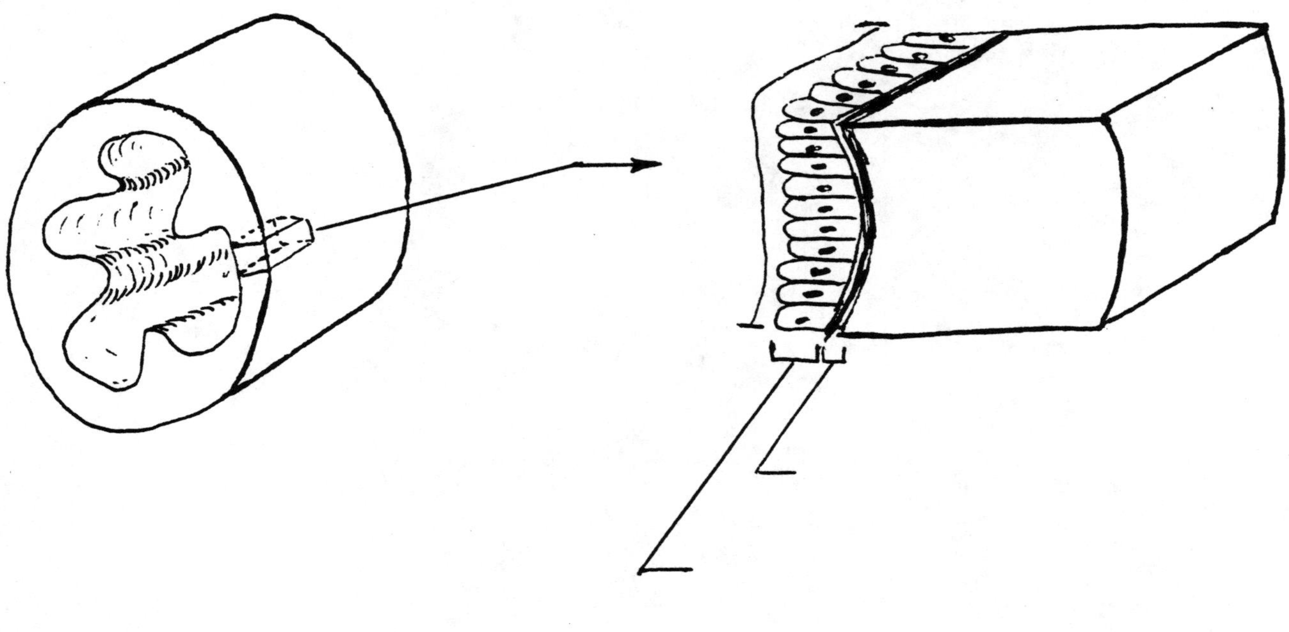

3. Classification of epithelial tissues is based on the **shape of the cells** and the **number of cell layers** between the basement membrane and the free surface

a. Shapes that epithelial tissue cells can have

1) FLAT =

2) CUBED =

3) TALL (ELONGATED) =

b. Number of cell layers between basement membrane and free surface

1) If only one cell layer, then it is called :

2) If two or more cell layers, then it is called :

c. A few combinations to show how these two schemes are used in conjunction to classify epithelial tissues

1) SImple types

a)

b)

c)

2) Stratified types

a)

b)

c)

3) Pseudostratified columnar

4. **SIMPLE SQUAMOUS EPITHELIUM** and its locations

5. **SIMPLE CUBOIDAL EPITHELIUM** and its locations

6. **SIMPLE COLUMNAR EPITHELIUM** and its locations

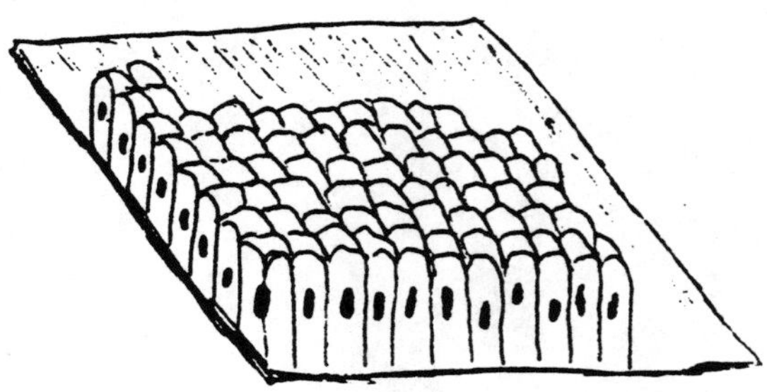

7. **PSEUDOSTRATIFIED COLUMNAR EPITHELIUM** and its locations

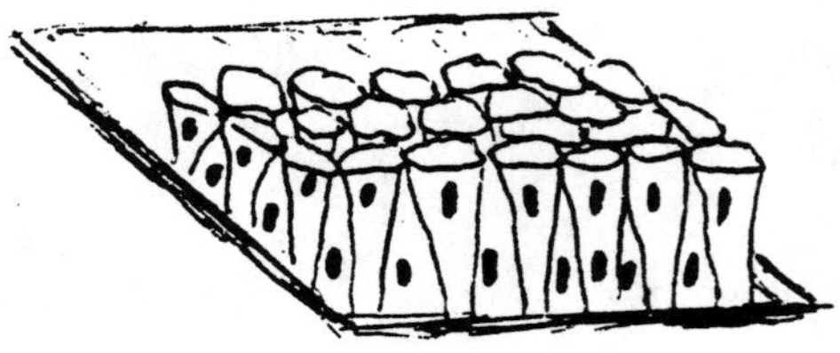

8. **STRATIFIED SQUAMOUS EPITHELIUM** and its locations

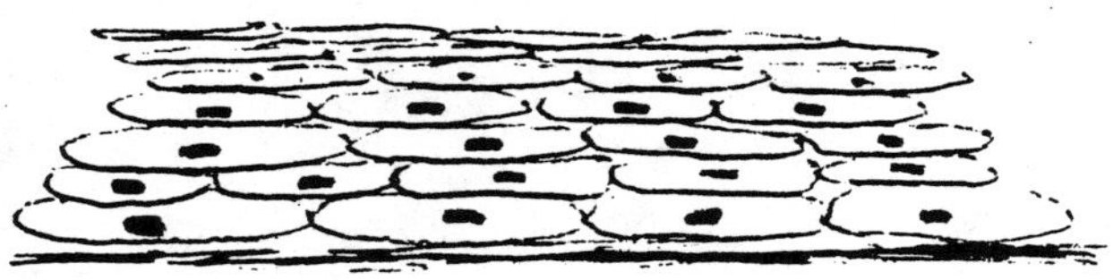

9. **STRATIFIED CUBOIDAL EPITHELIUM** and its locations

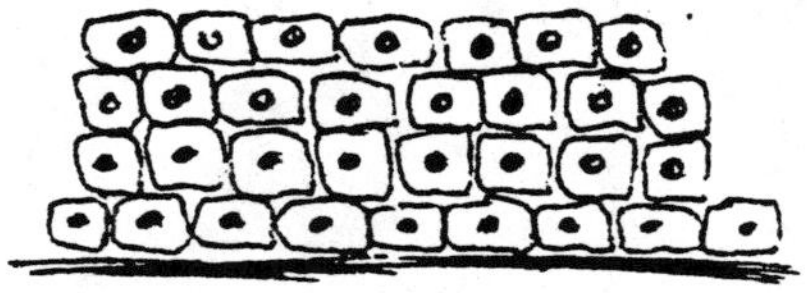

10. **STRATIFIED COLUMNAR EPITHELIUM** and its locations

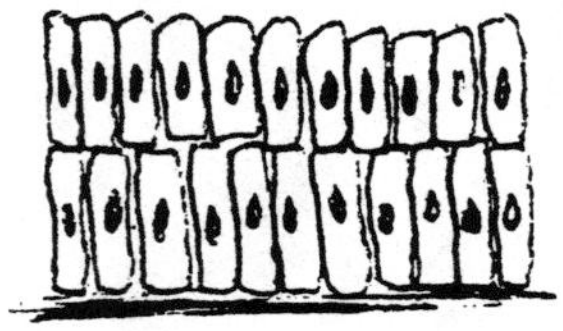

11. **TRANSITIONAL EPITHELIUM** and its locations

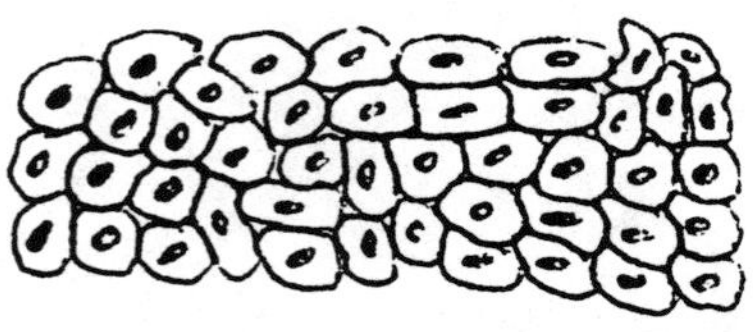

C. Connective tissues

1. Embryonic connective tissue

2. Adult connective tissues and their classification

   a. Functions of connective tissues
      1) Support of cells of other tissues
      2) Form protective sheaths and coverings around organs
      3) Involved in storage, repair, and transport

   b. Basic structure = Consists of:
      1) Cells
      2) Ground substance ┐
      3) Extracellular fibers ┘ — MATRIX
      4) Extracellular fluid

   c. Diagram of generalized connective tissue

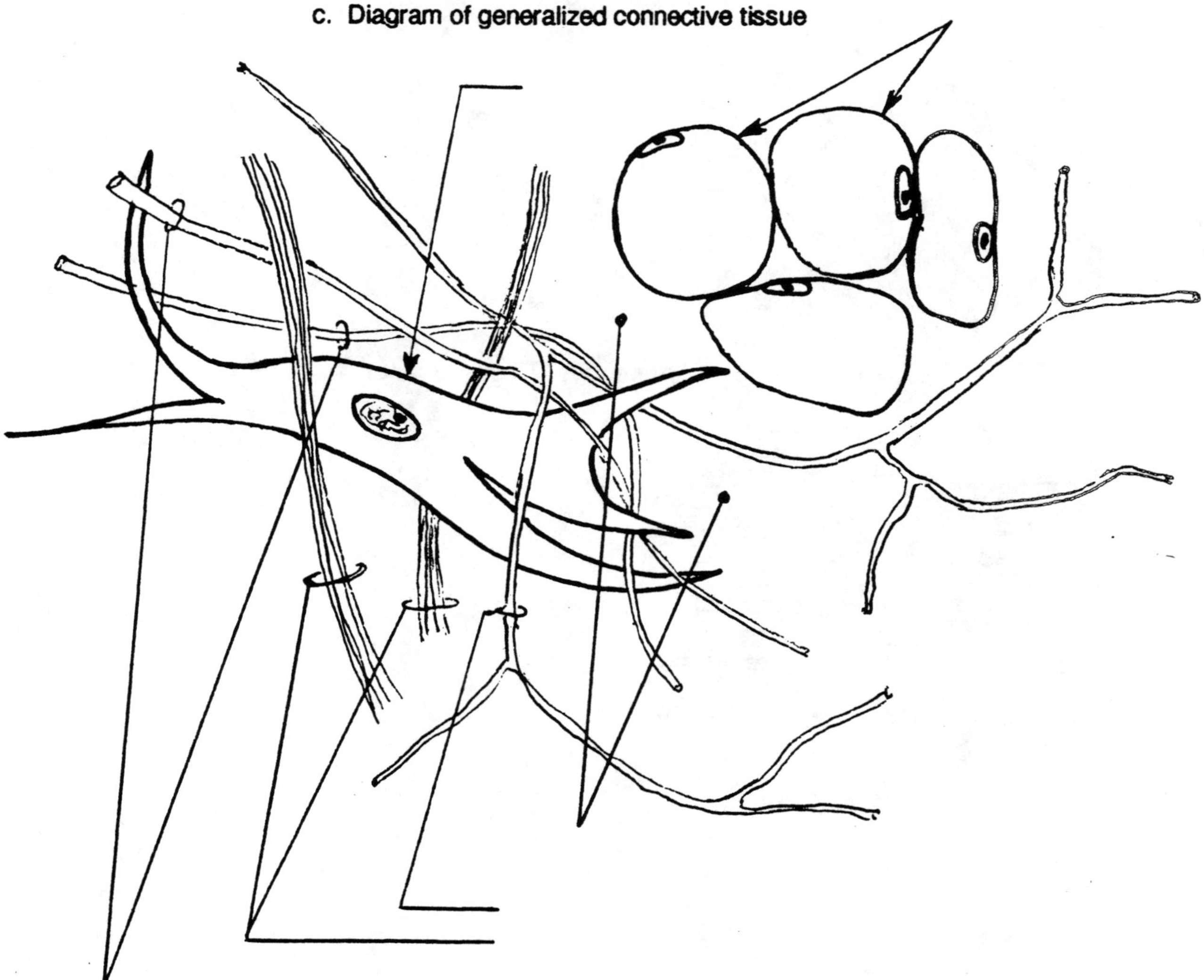

c. Types of adult connective tissues
  1) Connective tissue proper
    a) Loose connective tissue
    b) Dense connective tissue
    c) Elastic tissue
    d) Reticular tissue
    e) Adipose (Fat) tissue

  2) Cartilage

  3) Bone

  4) Vascular (Blood)

3. **CONNECTIVE TISSUE PROPER**

a. **Loose (Areolar) Connective Tissue Proper**

b. **Dense Connective Tissue Proper**

c. **Elastic Connective Tissue Proper**

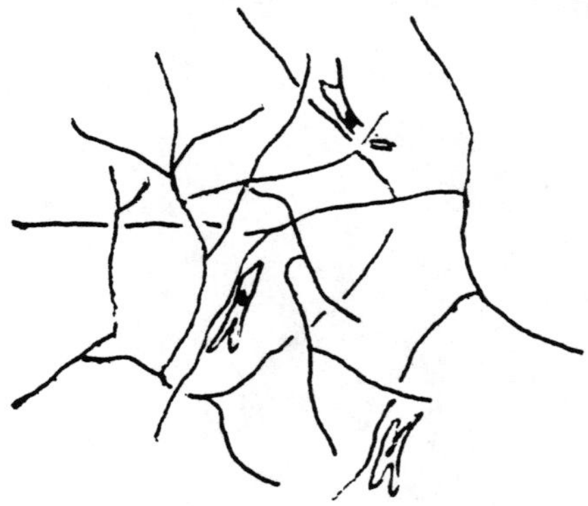

d. **Reticular Tissue**

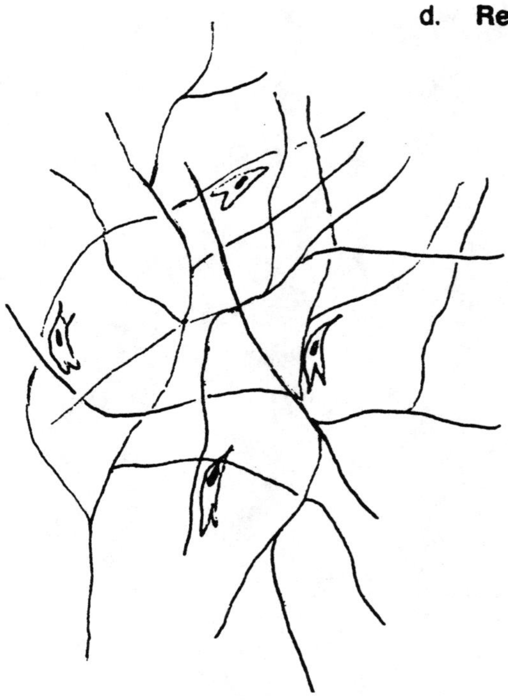

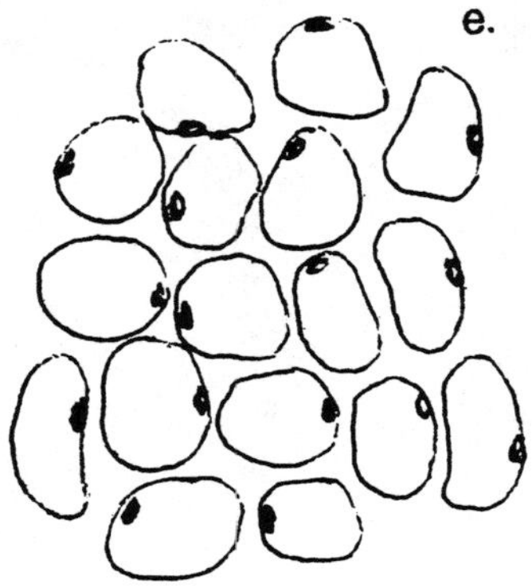

e. **Adipose Tissue**

4. **CARTILAGE**

a. Basic cell type of cartilage and its location

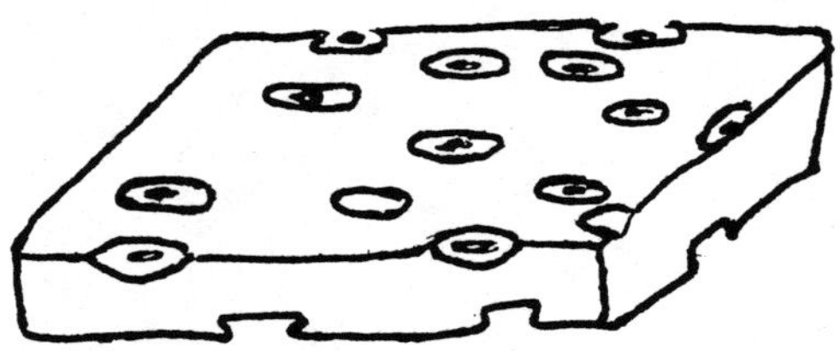

b. Types of cartilage

1)

2)

3)

5. **BONE**

6. **VASCULAR TISSUE**

D. **NERVOUS TISSUE**

E. MUSCULAR TISSUE

1. Types of muscle

    a. Voluntary = Skeletal muscle = Striated muscle

    b. Involuntary
        1) Smooth
        2) Cardiac

2. Function and location of skeletal muscle tissue

3. Function and location of some smooth muscle tissues

4. Function and location of cardiac muscle tissue

79. Define a tissue.
80. What are the four, or five types of tissues ?
81. What are the three characteristics of epithelial tissues ?
82. What are two criteria used to classify epithelial tissues ?
83. What are the three shapes that epithelial cells can have and what are their names ?
84. What do the terms simple and stratified refer to ?
85. Classify the types of epithelial tissues.
86. Give some locations of the following tissues : Simple squamous; Simple cuboidal; Simple columnar; Pseudostratified columnar; Stratified squamous; Stratified cuboidal; Stratified columnar ; Transitional.
87. What is embryonic connective tissue ?
88. List the three functions of adult connective tissues.
89. Diagram and label the structures of generalized connective tissue.
90. What two structures make up the matrix of connective tissues ?
91. What are the four types of adult connective tissues ?
92. What are the five types of connective tissue proper ?
93. What seems to be some differences between loose and dense connective tissues ?
94. What are some similarities between loose and dense connective tissues ?
95. Give some examples of loose and dense connective tissues.
96. Elastic connective tissue is found in structures that have to undergo stretching, such as the stomach and arteries ?
97. What are some locations of reticular connective tissue and what is the major function of this connective tissue ?
98. What is the basic cell of loose, dense, elastic, and reticular connective tissues ?
99. In what type of connective tissue proper do collagenous fibers predominate ? Elastic fibers? Reticular fibers ?
100. Which of the connective tissues proper does not have the fibroblast as its basic cell type and apparently makes no extracellular fibers ?
101. What is the function of adipose tissue ?
102. What is the basic cell type of adipose tissue called ?
103. What is the basic cell type of cartilage termed ?
104. What is a lacuna, and is cartilage the only tissue in which lacunae are found ?
105. What is located in a lacuna in cartilage ?
106. What are the three types of cartilage ?
107. What is the flexible form of connective tissue called ?
108. What is the rigid form of connective tissue called ?
109. What is the basic cell type found in bone ?
110. What is the function of an osteoblast ? Of an osteoclast ? Of an osteocyte ?
111. In what is an osteocyte located ?
112. What is a lacuna, and is bone the only tissue in which lacunae are found ?
113. What is vascular tissue ?
114. What is the basic cell type of nervous tissue ?
115. Classify the types of muscle tissue.
116. What are the two involuntary muscles ?
117. What are some additional names for voluntary muscle ?
118. What is the function and the general location of skeletal muscles ?
119. What is the function and the general location of smooth muscle ?
120. What is the function of cardiac muscle and where is it located ?

THE REPRODUCTIVE SYSTEMS

I. General Introduction

An organism, such as a bacterium or a yeast cell, is made up of only a single cell and each is considered a **UNICELLULAR ORGANISM.** The Escherichia coli bacteria that live in your large intestine are unicellular organisms. You, on the other hand, are made up of trillions of cells, and therefore are a **MULTICELLULAR ORGANISM**.

Reproduction among unicellular organisms is usually simple and is usually accomplished by mitosis and cytokinesis, without the exchange or unification of genetic material from another organism. This type of simple reproduction is termed **ASEXUAL REPRODUCTION**. Asexual reproduction does not lead to genetic variety in the offspring, and indeed they all look alike.

Not so with us multicellular organisms. The individuals of a multicellular species, such as Homo sapiens ("wise man"), can readily be distinguished. Each of us humans, except identical twins, are genetically unique (different genotypes), which leads to noticeable variability in physical characteristics (different phenotypes). The different genotypes and phenotypes seen among the individuals of a multicellular species are created by **SEXUAL REPRODUCTION**, a wild and crazy procedure by which the genetic material of one individual is united with that of another individual.

In considering asexual and sexual reproductions you might want to ask yourself if it ever happens that a human organism is initiated and develops by the process of mitosis and cytokinesis (asexual reproduction), and not from the fertilized egg ? The answer is yes. It has occurred millions of times during human history, and more than likely, there are a few of you in this course who were started in this manner. **HOW CAN THIS BE ! ?**

When we look inside a typical "**somatic**" human cell with a microscope we discover that the nucleus contains 46 chromosomes. This number of chromosomes is referred to as the **DIPLOID** number of chromosomes. There are some human cells, however, that do not contain 46 chromosomes. For example, mature erythrocytes (red blood cells) lack a nucleus and therefore have no chromosomes; skeletal muscle cells have multiple nuclei and therefore have multiples of 46 chromosomes; and sex cells (**gametes**), sperm and egg, each contain 23 chromosomes. Gametes have one half the chromosome number as somatic cells and are referred to as being **HAPLOID**. Each chromosome contains a single long DNA molecule. Therefore, in most human cells there are 46 DNA molecules, but in sperm and egg cells there are only 23 DNA molecules, and in mature erythrocytes there are no DNA molecules.

Each human DNA molecule contains about 2,000 genes along its length and so overall there are 46 x 2,000 = 100,000 genes in the human genotype. These genes store genetic information about protein structure and functions. From these various functions come such noticeable physical traits as eye color, skin color, height, sex, hair color, etc. We also inherit tendencies, and capacities about how certain environmental factors, both physical and mental, affect our lives. The interaction between our genetic makeup and the environment is extremely complicated and certainly beyond the range of this course.

There is absolutely nothing magical about having 46 chromosomes. It is not the number 46 that makes us human for there are undoubtedly many other species on earth whose cells contain 46 chromosomes. The uniqueness of a species derives not from the chromosome number, but from the genes along its DNA molecules.

Among the 46 chromosomes in a human cell, 44 of them are called **somatic chromosomes,** and two of them are termed sex chromosomes. There are two types of **sex chromosomes, X** and **Y,** and 22 types of somatic chromosomes, numbered 1 - 22. More on these types later.

Now we arrive at a real problem. In sexual reproduction where chromosomes from one individual are united with chromosomes from another individual, how is it that the chromosome number is kept constant ? That is, Mom and Dad have cells with 46 chromosomes. During fertilization, in which chromosomes in Dad's sperm is united with the chromosomes in Mom's egg cell, why doesn't the resulting offspring wind up with cells having 92 chromosomes ? How is it that the cells of the human offspring contain the same chromosome number as either Mom, or Dad ? The answer is that in the testes of Dad, and the ovaries of Mom there is a process, **MEIOSIS**, that results in the production of sex cells (gametes), sperm in males, ova (eggs) in females with one half the chromosome number (23) of normal body cells (46). Formation of gametes is termed **GAMETOGENESIS.** Gametogenesis is of two types: 1) Gametogenesis that involves the production of sperm is called **SPERMATOGENESIS** ; and 2) Gametogenesis that involves the production of the ovum (egg) is termed **OOGENESIS.** So, in humans, the sperm contains the 22 somatic chromosomes and one sex chromosome (either X, or Y), and the ovum contains the 22 somatic chromosomes and one sex chromosome, the X chromosome (human ova do not contain the Y chromosome).

Fertilization of the ovum with its 23 chromosomes by the sperm with its 23 chromosomes results in a human zygote with 46 chromosomes. The 22 somatic chromosomes from the sperm are homologous with the 22 somatic chromosomes in the egg. The sex chromosomes are a slightly different matter. If the sperm that fertilizes the egg contains an X sex chromosome, then the zygote is XX for the sex chromosomes and develops into a female. If, however, the fertilizing sperm contains a Y chromosome, then the resulting zygote is XY for the sex chromosomes and develops into a male. It therefore appears that it's Dad who determines the gender of the offspring.

Sexual intercourse is the primary and probably the preferred method used by us humans to get sperm and egg together for fertilization. In this we are not unique, for many other sexually reproducing animals accomplish fertilization the very same way. But when we take a closer look at things we have got to stop and wonder just what sexual intercourse and the resultant orgasms mean for us humans. The human female seems to be unique for several reasons. For example, the females of our nearest relatives (apes, chimpanzees, other primates) have very distinct estrous cycles where overt signs and signals are given when ovulation (release of the egg from the ovary so that fertilization can occur) occurs. These signals and signs are of course noticed by males and from this a great deal of sexual activity ensues. Sexual activity therefore is not uniform, but waxes and wanes with the ovulatory cycle of the female. In the human female, however , there is no overt sign or signal that ovulation has occured ; ovulation is hidden from the human male, and also from the human female. Moreover, the human female seems to be the only female among all the mammals that can experience orgasm. And we have to certainly consider, too, how often an individual, woman, or man, desires / wants/ needs to have sexual intercourse, but does not want fertilization or pregnancy to be the result. This is not true for other animals, just us. All other animals use sexual intercourse to accomplish fertilization and pregnancy only. So, we might be led with this information to consider that for us humans reproduction is a secondary result of our sexual activity. What do you think is the primary reason for sexual activity for us ?

Whether wanted, or not, sometimes fertilization and pregnancy results from sexual activity. This is where Dad kind of fades out and Mom takes on the burden of reproduction. Implantation and growth occur in the uterus of Mom; childbirth (parturition) is Mom's job, and so is nourishment following birth (mammary glands). The term mammary has an interesting derivation. The Greeks and early Romans noted that one of the first enunciations made by babies is "mamma". This is an almost universal phenomonon and in almost all societies one of the first things a baby speaks is "mamma". The Greeks thought that when babies uttered "mamma" that they were hungry and wanted to be fed, hence the term for breast - **MAMMAry** gland.

1. What is a unicellular organism ? A multicellular organism ?
2. What type of reproduction do unicellular organisms undergo?
3. What type of reproduction do multicellular undergo ?
4. The species name for humans is Homo sapiens. What does Homo sapiens mean ?
5. How is it that some humans are initiated by mitosis and cytokinesis (asexual reproduction) and not from the fertilized egg ?
6. What do the terms diploid and haploid mean ?
7. What are somatic cells ? What are gametes ?
8. How many DNA molecules are in a chromosome ?
9. What is the diploid number of chromosomes in Homo sapiens ? What is the haploid number ?
10. Where are the genes located and for what are they responsible ?
11. How many somatic chromosomes are in a somatic cell of humans ?
12. How many sex chromosomes are there in a somatic cell ?
13. How many somatic chromosomes are there in a gamete ?
14. How many sex chromosomes are there in a gamete ?
15. What is the name of the process that creates gametes ?
16. What are the two types of gametogenesis ?

17. Where does oogenesis occur ?
18. Where does spermatogenesis occur ?
19. Who determines the gender of the offspring, Mom, or Dad ?
20. What is the primary method of getting egg and sperm together ?
21. What do you think is the primary function of sexual activity among humans ?

II. Meiosis

A. The result of Meiosis = Creation of a group of cells called gametes that contain one half the chromosome number (haploid number) of somatic cells.

B. **Gametogenesis**

1. **SPERMATOGENESIS**
2. **OOGENESIS**

III. The Male Reproductive System

A. Components of the male reproductive system and their general functions

1. TESTES

2. PENIS

3. COMPLEX TUBULE SYSTEM

   a . Seminiferous tubules

   b. Epididymus

   c. Vas (ductus) deferens

   d. Urethra

4. Accessory reproductive organs and their functions

   a. Bulbourethral gland -

   b. Prostate gland -

   c. Seminal vesicles -

B. Structure of the male reproductive system

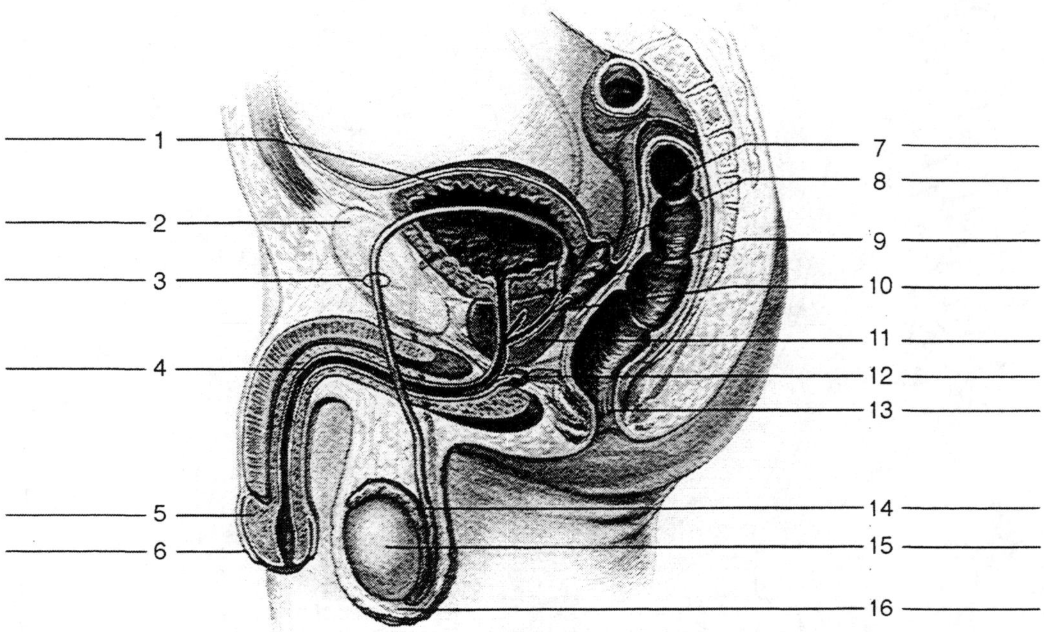

C. The testes

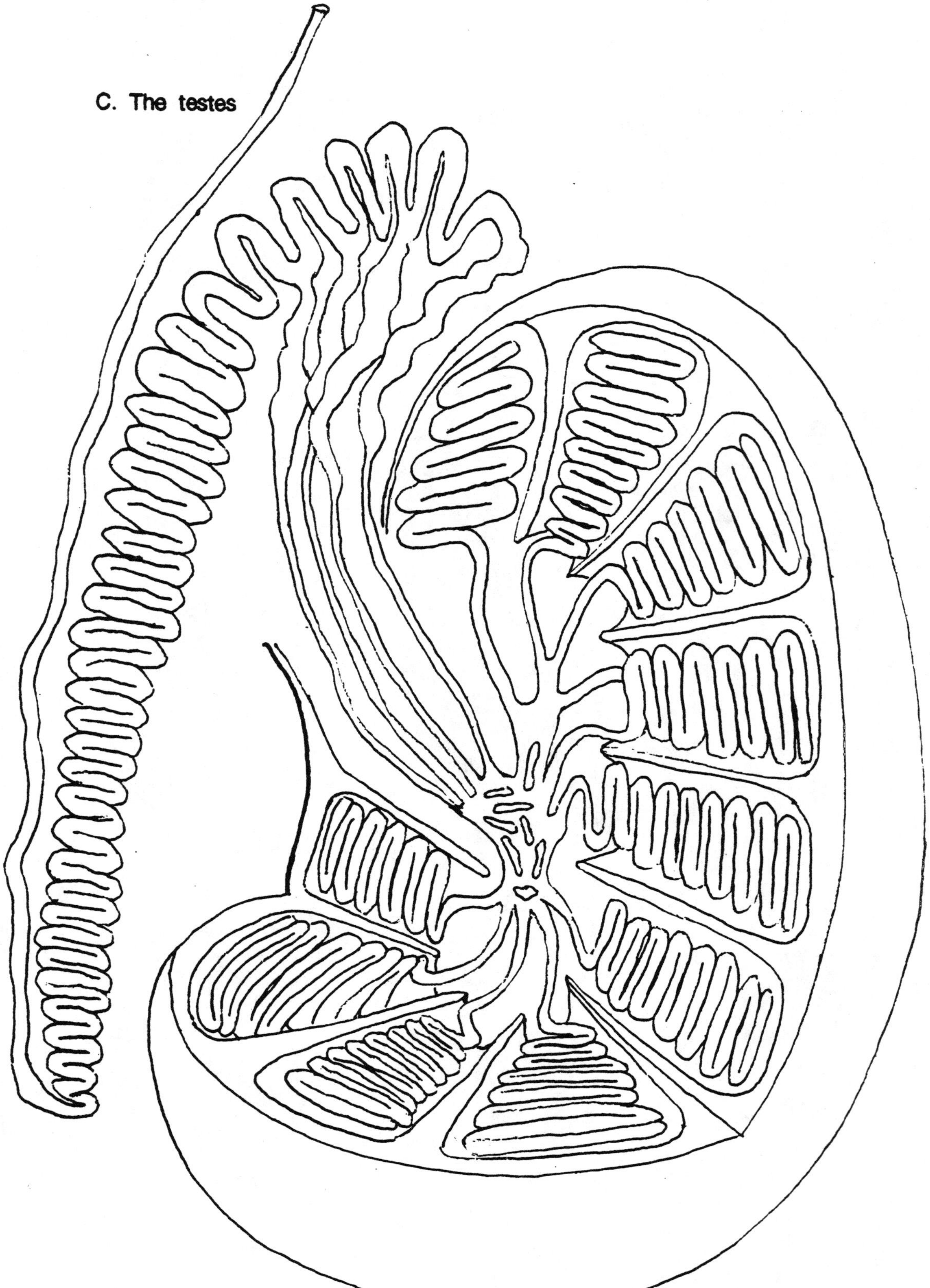

D. Comparative anatomy of the penis and clitoris

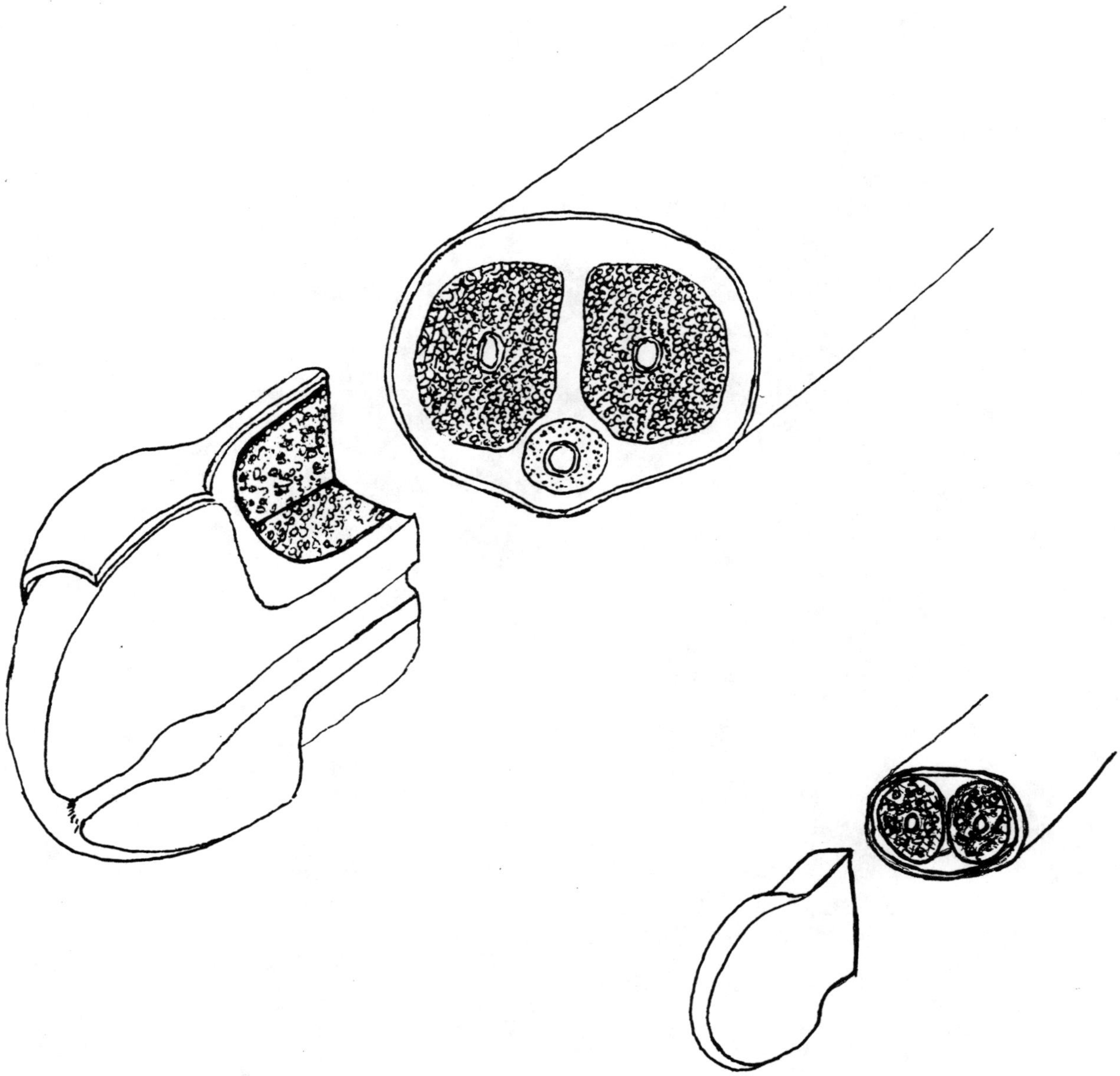

E. Accessory reproductive organs, and semen and its contents

F. The anatomical reasons for penile erection

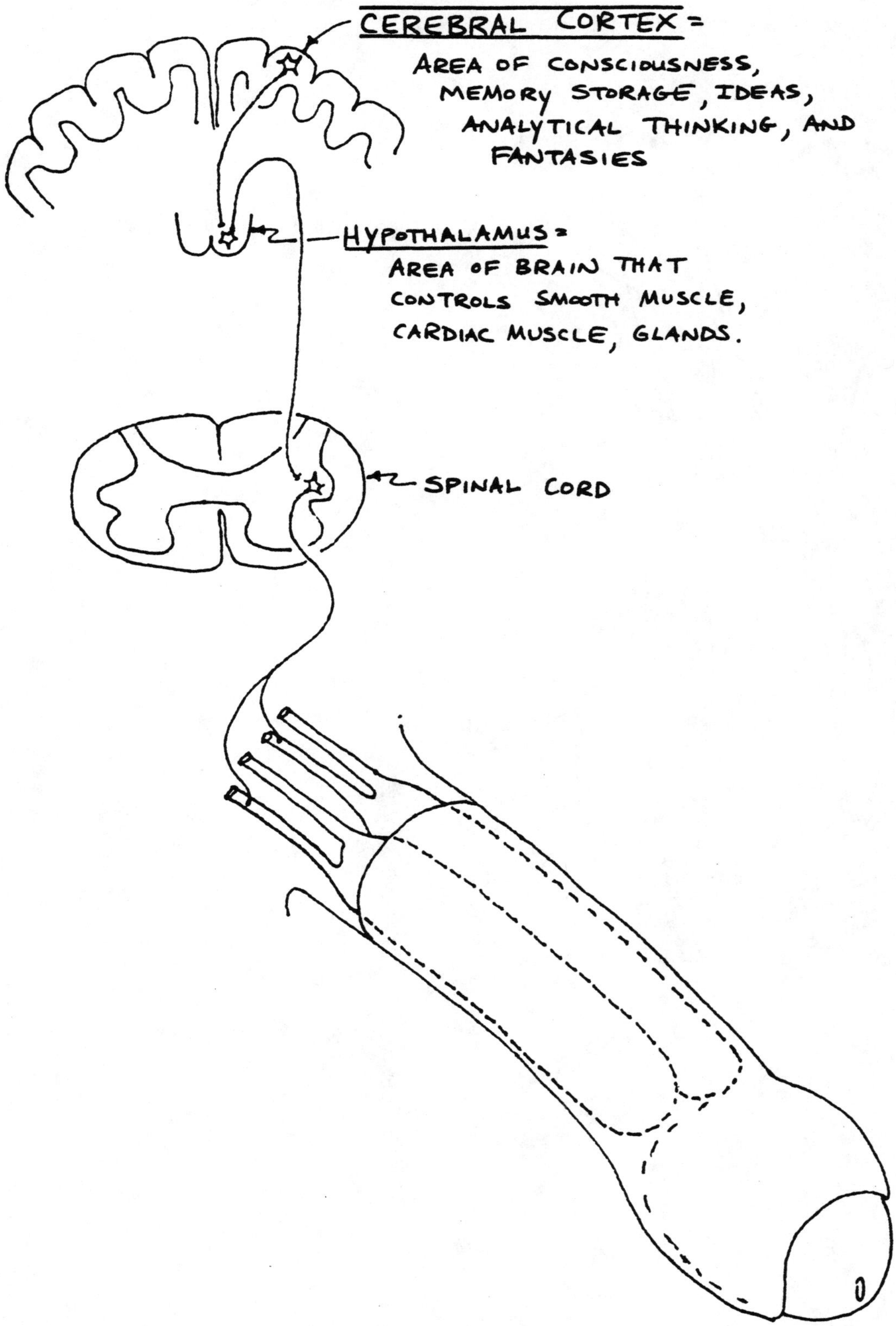

G. Vasectomy and vasovasotomy

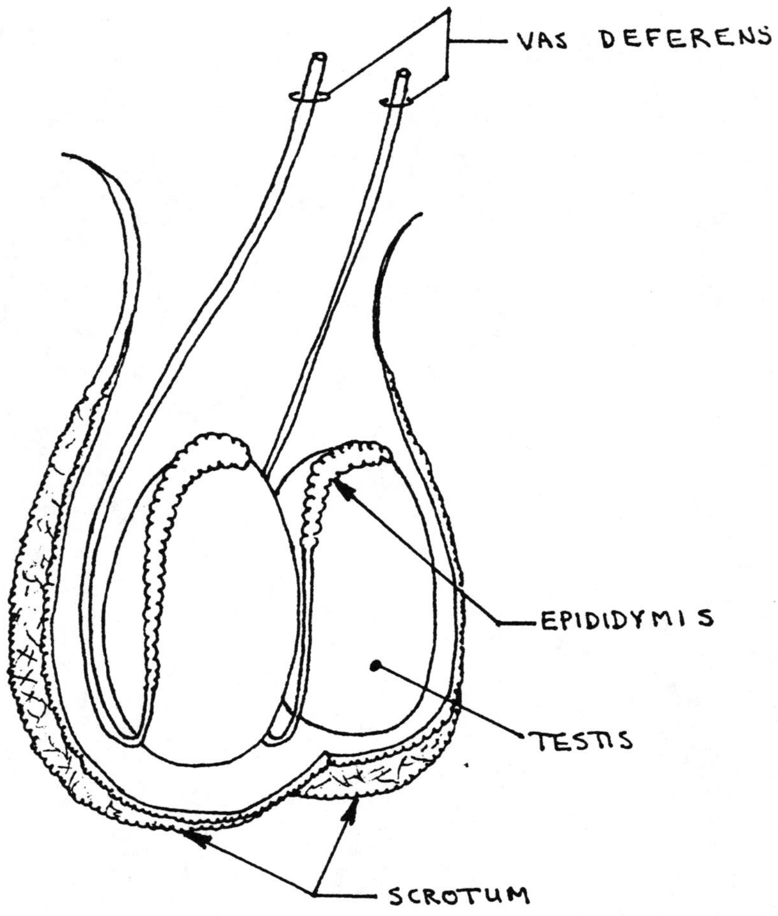

22. What is the result of the process of meiosis ?
23. What does the term diploid mean ?
24. What does the term haploid mean ?
25. Starting with a diploid cell, go through the process of meiosis to produce four haploid cells.
26. Does the process of mitosis occur anywhere during the process of meiosis ?
27. List the components of the male reproductive system and their functions.
28. What are the three accessory reproductive organs and what are their functions ?
29. What are the three sections of the male urethra ?
30. Where do the testes develop in the male before birth ?
31. Where are the testes found in the adult male ?
32. In order for the testes to get from the abdominal cavity into the scrotum, the testes must pass through small openings in the lower abdominal wall. What is the name of these openings ?
33. Why is the scrotum wrinkled and why does its skin contain numerous sweat glands ?
34. What process in the male is hampered by body temperature ?
35. Where are sperm stored until they are summonded ?
36. What do you think would happen to the prostatic urethra if the prostate gland becomes infected, or if the gland becomes cancerous, and enlarges ?
37. Where, specifically, does spermatogenesis occur ?
38. Is the epididymus inside, or outside the testis ?
39. What does the term "rete" mean ?
40. The seminiferous tubules run together to form a network of tubules called what ?
41. The tubes that emerge fron the rete testes are the what ?
42. What do the straight tubules become ?
43. What do the efferent ductules become ?
44. What does the epididymus become ?
45. When the vas (ductus) deferens and the duct of the seminal vesicle merge they form a tubule known as the what ?
46. In what gland is the ejaculatory duct found (located) ?
47. The ejaculatory duct connects to what portion of the male urethra ?
48. The following structures are in alphabetical order : **EJACULATORY DUCT ; EPIDIDYMUS ; EFFERENT DUCTULES; MEMBRANOUS URETHRA; PENILE URETHRA ; PROSTATIC URETHRA ; RETE TESTES ; SEMINIFEROUS TUBULES ; STRAIGHT TUBULES ; VAS (DUCTUS) DEFERENS.** These are the structures that spermatozoa must pass through to get to the outside of the male body. Starting with the place where sperm are made, place these structures in the order that the sperm would take to get to the outside of the body.
49. Name the spongy blood sinuses in the penis.
50. Which of the blood sinses is missing in the clitoris ?
51. Which blood sinus becomes the glans penis ?
52. Which blood sinus becomes the glans clitoris ?
53. The corpus spongiosum is in the penis, but not the clitoris. What other structure does the penis have, but the clitoris does not ?
54. What is the name of the fold of skin that partially covers the glans penis and glans clitoris ?
55. What are the three main components of semen ?
56. Approximately how many spermatazoa are stored in each epididymus before ejaculation ?
57. Describe the anatomical reasons that lead to penile erection.
58. What is a vasectomy ?
59. What is a vasovasotomy ?
60. Which is the more simpler of the two surgical procedures - vasectomy, or vasovasotomy ?

IV. The Female Reproductive System

A. Divisions of the female reproductive system and their components

1. External female reproductive structures

a.

b.

c.

d.

2. Internal female reproductive structures

a.

b.

c.

d.

B. General structures of the external and internal female reproductive systems

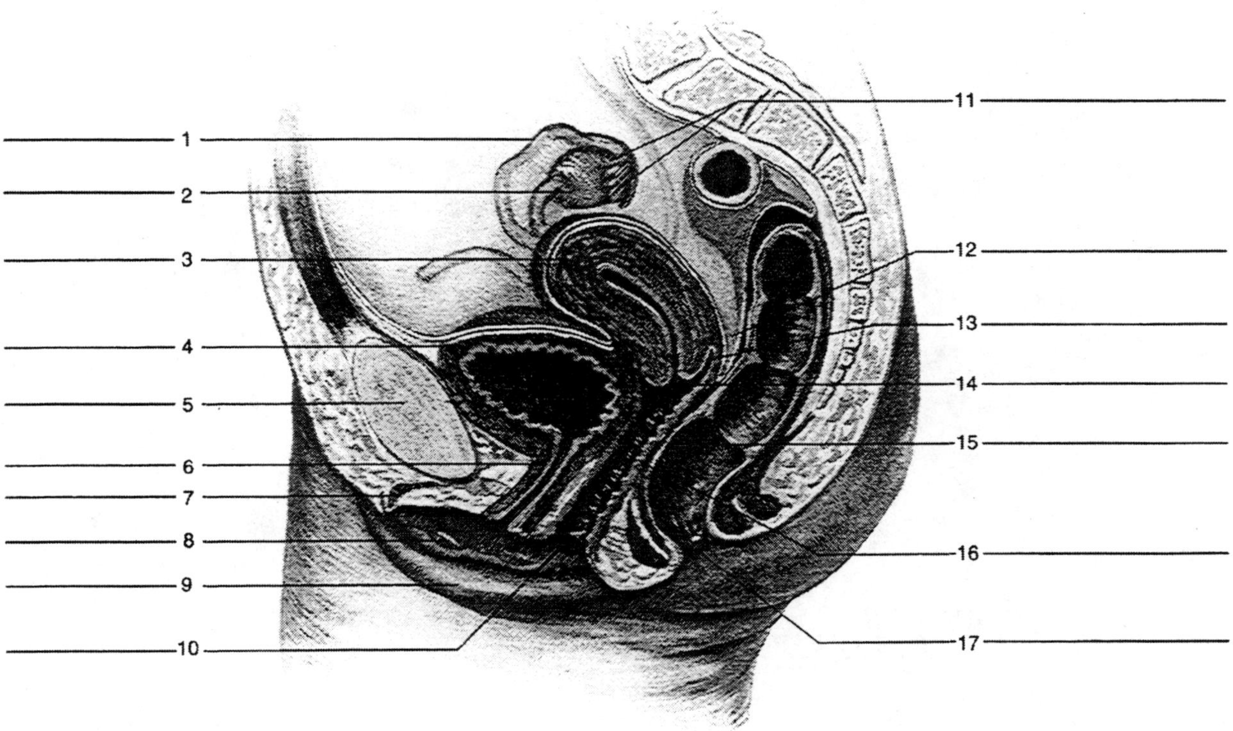

C. Anatomically homologous structures between female and male reproductive systems

D. External structures of the female reproductive system

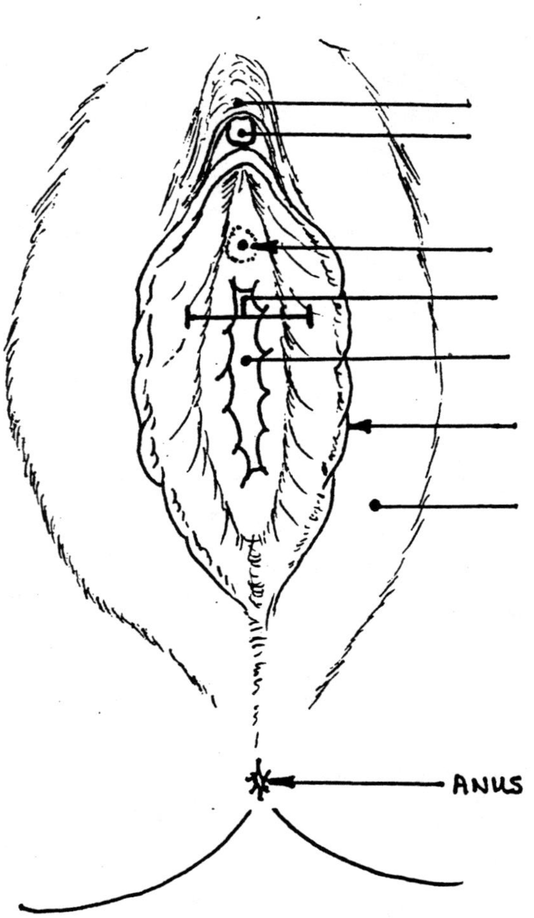

E. Internal Female reproductive structures

1. The **VAGINA, UTERUS, FALLOPIAN (UTERINE ; OVARIAN) TUBES**

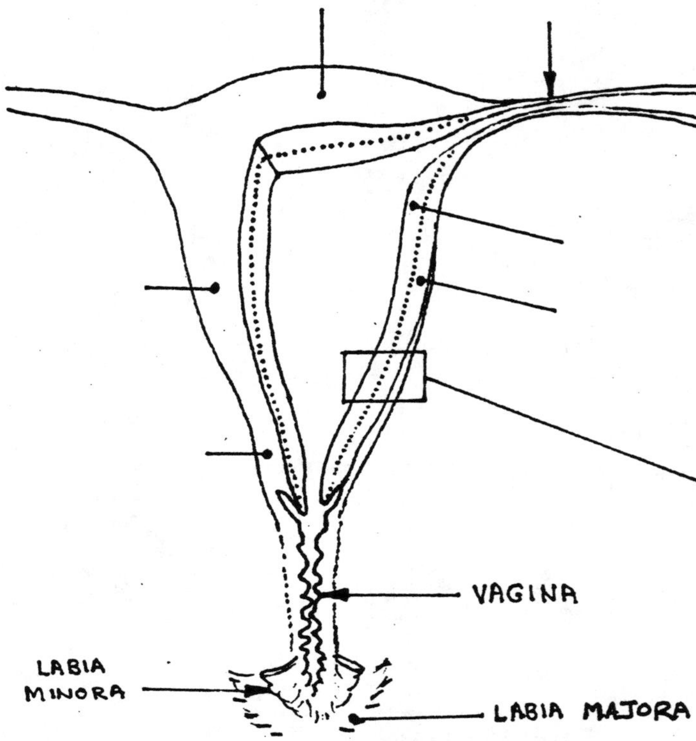

2. The wall of the uterus

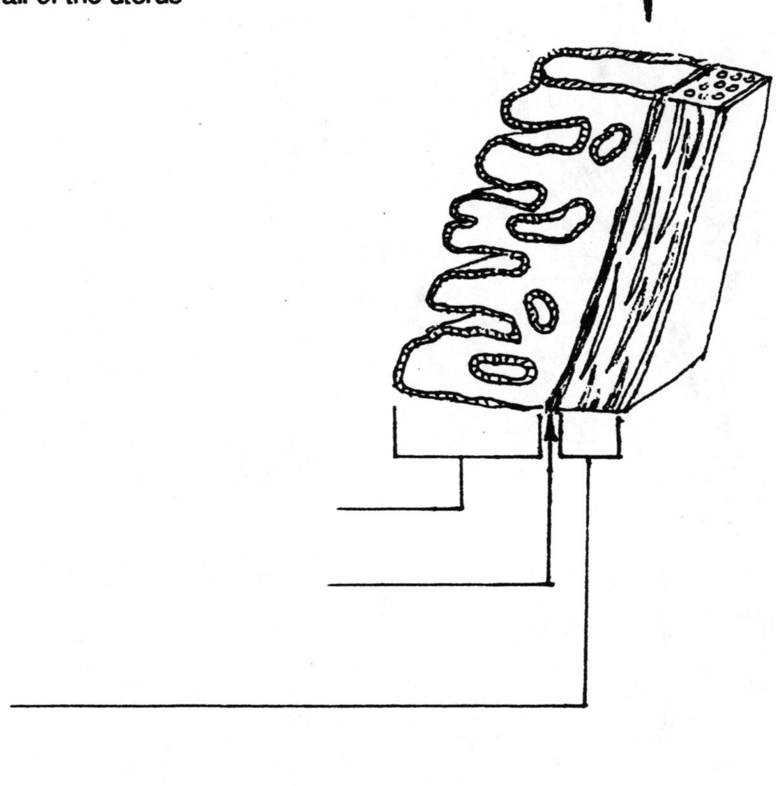

F. Fertilization of the ovum, early development, and implantation of the blastocyst

1. Fertilization

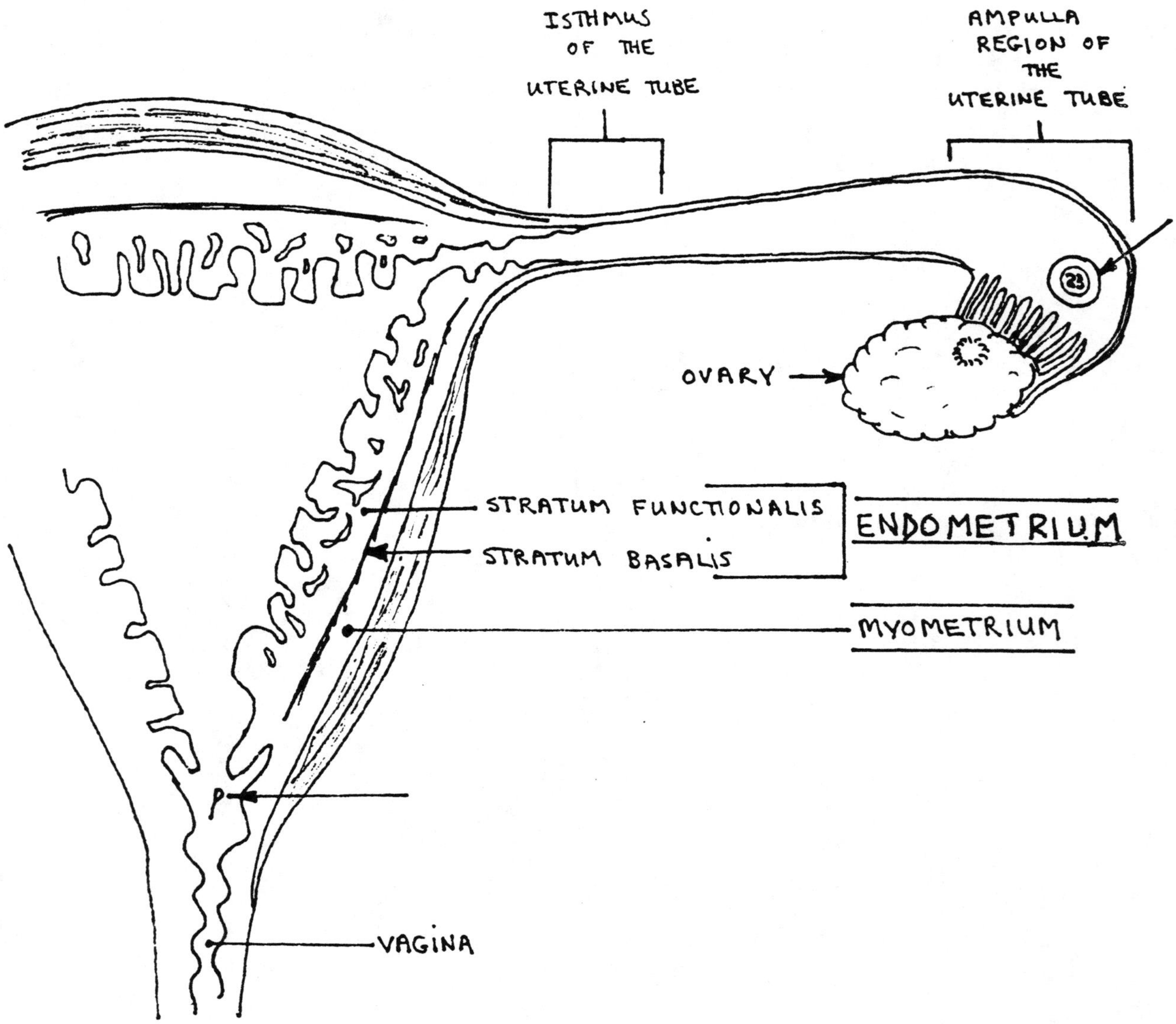

2. Early development and implantation of the blastocyst

a. Development of the blastocyst from the zygote occurs as the individual moves down the fallopian tube. The time required for the formation of the blastocyst is from 7 to 9 days. Normally, by the time the blastocyst is formed it has made its migration into the uterine cavity.

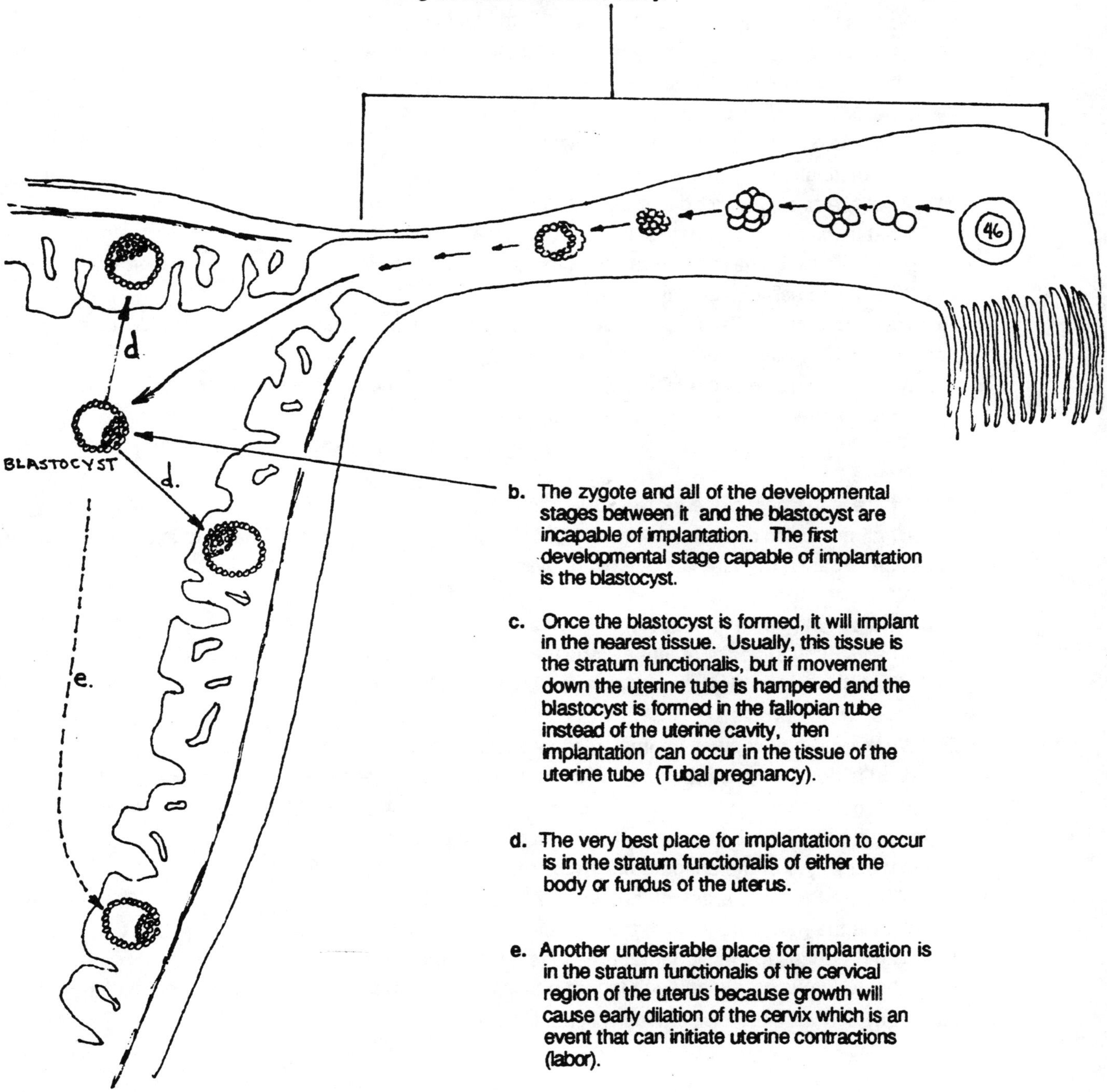

b. The zygote and all of the developmental stages between it and the blastocyst are incapable of implantation. The first developmental stage capable of implantation is the blastocyst.

c. Once the blastocyst is formed, it will implant in the nearest tissue. Usually, this tissue is the stratum functionalis, but if movement down the uterine tube is hampered and the blastocyst is formed in the fallopian tube instead of the uterine cavity, then implantation can occur in the tissue of the uterine tube (Tubal pregnancy).

d. The very best place for implantation to occur is in the stratum functionalis of either the body or fundus of the uterus.

e. Another undesirable place for implantation is in the stratum functionalis of the cervical region of the uterus because growth will cause early dilation of the cervix which is an event that can initiate uterine contractions (labor).

61. What are the external female reproductive system structures / areas ?
62. What are the internal female reproductive system structures ?
63. What is the clinical significance of the rectouterine pouch and the posterior fornix ?
64. What is the male anatomical equivalent of the glans clitoris ?
65. What is the male anatomical equivalent of the labia minora ?
66. Wha t is the male anatomical equivalent of the labia majoral ?
67. What is the male anatomical equivalent of the vagina and uterus ?
68. What is the male anatomical equivalents of the ovaries ?
69. What male structure do you think might be the anatomical equivalent of the fallopian tubes ?
70. The vestibule is an area that is between what structures ?
71. What are two other good and acceptable names for the fallopian tubes ?
72. What is the name of the muscle layer of the wall of the uterus ?
73. Is the muscle in the uterine wall smooth, cardiac, or skeletal ? Voluntary, or involuntary ?
74. What is the name of the inner layer of the uterine wall ?
75. The endometrium itself can be further divided into an inner and an outer layer . Whay is the name of the inner layer that faces the uterine cavity ?
76. What is the name of the other endometrial layer that is next to the myometrium ?
77. Which of the endometrioal layer is shed at menses ?
78. Which layer is not shed at menses and is responsible, for regenerating the stratum functionalis ?
79. Where is the best place for fertilization to occur ?
80. What structures capture and direct into the ampulla region of the fallopian tube the egg that has been released from the surface of the ovary ?
81. What is the term for a fertilized egg ?
82. Is the zygote capable of implantation ?
83. What is the first developmental stage that is capable of implantation ?
84. About how many days does it take for the processes of mitosis and cytokinesis to produce a blastocyst from a zygote ?
85. What are the best place for implantation of the blastocyst to occur ?
86. What is not a very desirable region of the uterus for the blastocyst to implant ? Why ?
87. In consideration of the answers to questions 82, 83, 84, 85, and 86 why is the ampulla region of the uterine tube the best place for fertilization to occur ?
88. In which layer of the endometrium does the blastocyst implant ?
89. What are the three areas of the uterus ?
90. What is the narrow regions of the ovarian tubes called ?
91. Who produces the most gametes in a lifetime, women or men ?
92. If one ovum is produced each month between the ages of 13 and 50 , then <u>approximately</u> how many ova will this individual make in her lifetime ?

SYSTEMS INVOLVED IN SUPPORT AND MOVEMENT

I. General Introduction

There are two systems that are involved in support and movement and these are the **SKELETAL SYSTEM** and the **MUSCULAR SYSTEM.** So closely do these two systems interact that many texts and anatomists will combine them into one system called the **MUSCULOSKELETAL SYSTEM**.

The adult human skeletal system consists of 206 bones, give or take a few. This isn't quite as bad as it sounds because, remember, we are bilaterally symmetrical, and therefore most of our 206 bones occur as pairs. So, take heart, there are only about slightly over 100 names to learn. For examples, there are right and left tibias, right and left femurs, right and left scapulae, right and left pubic bones, right and left parietals, right and left pisiforms, right and left clavicles, etc. Not all bones are paired. The vertebrae are not, the frontal and occipital bones are not, and many of the bones of the face (sphenoid, ethmoid, vomer) are not, and although the mandible begins as two separate bones they eventually fuse to form a single bone which hold your lower teeth.

In looking at the human skeleton, it is obvious that our 206 bones have unique shapes. Nevertheless, our bones can be organized into categories based on their shapes, or location. Bones that are based on their shape are **long bones**, **short bones**, **flat bones**, and **irregular bones**. Bones that are based on their location are the **sesamoid bones,** which are located in tendons. We are born with at least four sesamoid bones, our two kneecaps (patella), and two pisiforms. In general, long bones are located in the extremities (" arms and legs "), short bones are found in the wrist and ankle, flat bones are found in the skull and ribs, and irregular bones include such bones as the vertebrae and some of the bones of the face. In addition, there are **Wormian (sutural) bones**, which are additional small flat bones in the skull. These bones are named not for their shape or location, but for the17th century Danish anatomist, Olaus Worm.

Bone is the rigid form of connective tissue. It is rigid because Calcium and Phosphorus become deposited as a form of Calcium Phosphate among the collagen fibers. The bones of your skeleton are therefore a tremendous **RESERVOIR OF CALCIUM AND PHOSPHORUS** for the body. In addition to being a source of calcium, the bones of your skeleton perform other functions. Obviously, the bones of your skeleton provide **SUPPORT and FORM**. Also, muscles known as skeletal (voluntary) muscles are attached to our bones and contracting these muscles cause **MOVEMENT**. Some bones participate in the **PROTECTION OF SOFT ORGANS**. As examples, the flat bones of the skull protect the brain, the vertebrae protect the spinal cord, the ribs protect the lungs, liver, stomach, and spleen, the sternum is a good protector of the heart, and the bones of the pelvis protect such organs as the urinary bladder, and the uterus. In addition to being a storage place for calcium and phosphorus, another less obvious function of the skeletal system includes providing a place, in the medullary cavities, for **BLOOD CELL FORMATION** in children.

Although our bones have different shapes, they nonetheless all have something in common and that is that each of the 206 bones contains two kinds of bone known as **compact** (dense), and **spongy** (cancellous). As shown below, using the femur as an example, compact bone surrounds spongy bone. And as a general rule, the more weight a bone supports, the thicker the compact bone component.

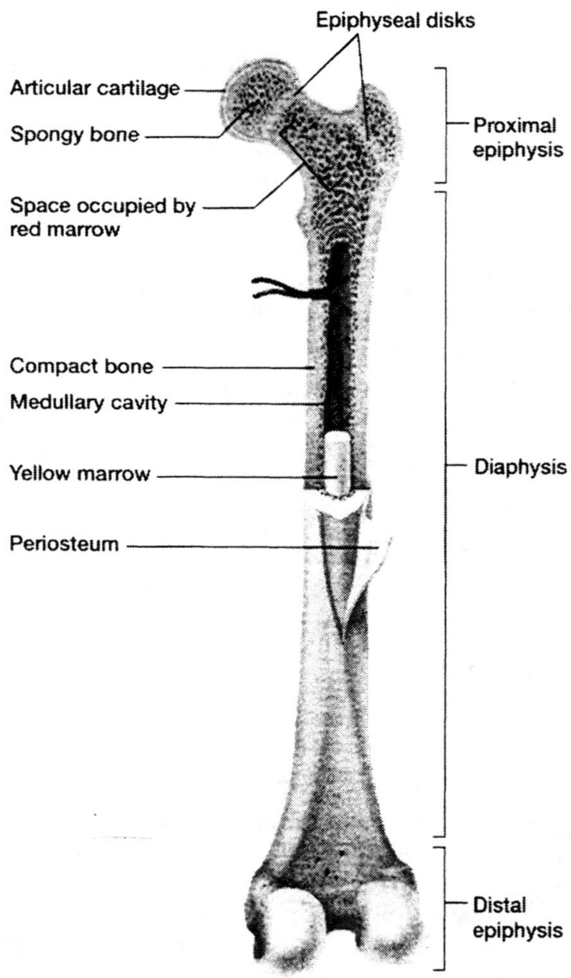

Femur

The human skeleton can be divided conveniently into an **AXIAL DIVISION (SKELETON)** and an **APPENDICULAR DIVISION (SKELETON)**. The term axial refers to an axis of rotation, which in the human body is the vertebral column. The term appendicular refers to our "arms and leg ", better known as the appendages, or limbs, or extremities. The axial division contains the vertebrae, bones of the skull and face, ribs, sternum, and the hyoid. Altogether, there are 80 bones that make up the axial skeleton of the human. The appendicular division includes the scapulae plus the bones in the two upper limbs, and the bones of the pelvis (minus the sacrum which is group of vertebrae and therefore belong in the axial skeleton) plus the bones in the two lower limbs. There are 126 bones in the appendicular skeleton.

You can readily see and understand that the skeleton is not totally rigid. There are places where bones move relative to one another. These places are where two or more bones meet one another and are called **ARTICULATIONS**, or **JOINTS**. You already know from your own experience that not all joints are equal. There are differences not only in the types of movements among your joints, but also in the degree of freedom of movements, and the ranges of motion. For instance, your knee can basically have only back and forth movements (flexion and extension), but your shoulder can have flexion, extension, abduction, adduction, and circumduction movements. The hip joint is the same type of joint as the shoulder joint (ball and socket), and can indeed have the same types of movements. Yet, you know, and can readily demonstrate, that the range of motion is greater for each of these movements at your shoulder compared to the hip. But with joints there is always a tradeoff : An increase in the ranges of motion at a joint is accompanied by a decrease in the stability of the joint, and for this reason, your shoulder is a less stable joint than your hip.

Just because two bones meet and form a joint does not mean that they can move relative to one another . Therefore, you have some **IMMOVEABLE JOINTS** in your body, notably the sutures in your skull and face. The epiphyseal plate is another immoveable joint. It is also the most numerous joint in children, but is totally missing in the adult. There are **SLIGHTLY MOVEABLE JOINTS**, too. The intervertebral joints, and the pubic symphysis are two good examples. We do not notice immoveable and slightly moveable joints because there isn't enough movement to notice, but we do notice **FREELY MOVEABLE JOINTS**, joints such as the shoulder, elbow, wrist, interphalangeals (fingers and toes), hip, knee, ankle, to name a few. Immoveable joints are termed **SYNARTHROTIC JOINTS**, slightly moveable joints are called **AMPHIARTHROTIC JOINTS**, and freely moveable joints are termed **DIARTHROTIC JOINTS**.

Besides classifcation on the basis of their degree of freedom of movement (immoveable, slightly moveable, of freely moveable) joints can be categorized on the basis of what is **between** the adjacent bones that make up the joint. If cartilage is between two bones, then it is simply classified as a **CARTILAGINOUS JOINT**. The epiphyseal plate, the intervertebral disks, and classified as a **CARTILAGINOUS JOINT**. The epiphyseal plate, the intervertebral disks, and the pubic symphysis are examples of cartilaginous joints. If tough fibrous material is between the bones, then it is a **FIBROUS JOINT** and the suture is a good example of a fibrous joint. A much more complicated joint than either the cartilaginous or fibrous joint is the **SYNOVIAL JOINT**. What is between bones in this type of joint is a thick and slippery fluid called synovial fluid.

The reason that you can have movements at amphiarthrotic and diarthrotic joints is because you have muscles attached to your bones. There are basically two types of muscles in your body, **VOLUNTARY MUSCLE** and **INVOLUNTARY MUSCLE**. Voluntary muscles are muscles that you can contract at will, or voluntarily. The majority of these muscles are attached to your bones and for that reason voluntary muscles are also called **SKELETAL MUSCLES**. Voluntary muscles are the ones that you use to move your skeleton around in space. They are located in a peripheral position in the body and in fact you can see many of the muscles contracting just beneath the skin. There are two types of involuntary muscles, **SMOOTH MUSCLE**, and **CARDIAC MUSCLE**. You can not contract these two types of muscle voluntarily and that is why, of course, they are called involuntary muscles. Both of these involuntary muscles are located internally. Smooth muscles are located in the walls of blood vessels, inside the eye, and in the walls of hollow organs such as the esophagus, stomach, intestines, uterus, gall bladder, and urinary bladder. Cardiac muscle is confined to the wall of the heart where it is often referred to as the myocardium.

There are about 650 skeletal muscles in your body. Once again, however, we luck out because most of these, like bones, are paired, and so we have only about 300 names to contend with. The way that skeletal muscles are arranged in the body is that at the ends of each muscle is a tendon. These tendons are the way that muscles are attached to bones. The tendon at one end attaches to one bone, and the tendon at the other end crosses an amphiarthrotic or diarthrotic joint and attaches to another bone.

When a given skeletal muscle contracts it normally happens that one of the bones that its tendon is attached to will move and the other bone to which it is attached will remain stable and not move. The place on the bone that does not move when that muscle contracts is termed the **ORIGIN**. The place on the bone that does move when that muscle contracts is termed the **INSERTION.** Each of the 650 muscles in your body has an origin and an insertion.

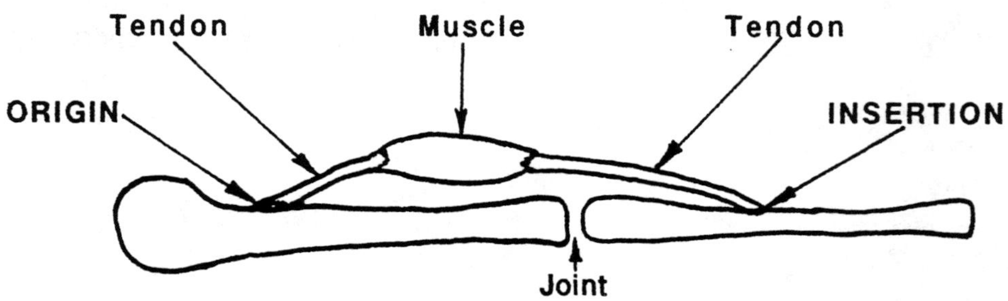

There are several types of movements that can occur at your diarthrotic joints. Some common movements include **FLEXION**, **EXTENSION**, **ABDUCTION**, **ADDUCTION**, and **CIRCUMDUCTION**. **Flexion** is defined as a decrease in the angle between two bones. **Extension** is the opposite of flexion and is defined as an increase in the angle between two bones. We encounter trouble right away with these two definitions when we consider the shoulder and hip joints. For these two joints flexion is defined as moving the humerus or the femur in a forward direction, and extension is defined as the movement of the humerus or femur in a backward direction. **ABduction** is the movement of away from the midline, while **ADduction** is the movement toward a midline. **Circumduction** is a movement in which one end of the moving part is fixed and the other end moves in a circular manner. There are two rotational movements associated with the hand, **PRONATION** and **SUPINATION**. **Pronation** is a movement in which the hand is rotated from the palms facing upward to palms facing downward, and **supination** is the opposite movement. These two movements are actually the result of a rotation movement of the radius that occus at the elbow.

1. What are the two systems that are involved in support and movement ?
2. How many bones are in the human skeleton ?
3. Where does the term Wormian come from ?
4. Besides Wormian bones, what are the five categories of bones based on shape and location ?
5. Think about your lab work with the skeletal system, and about your lecture notes, and classify each of the following bones as to whether each is long (L), short (S), flat (F), irregular (I), or sesamoid (SM): **Femur** ____, **Parietals** _____, **Ribs** ____, **Sphenoid** ____, **Radius** ____, **Patella** ____, **Atlas** ____, **Metacarpal III** ____, **Pisiform** ____, **T12** ____, **Frontal** ____, **Mandible** ____, **Calcaneus** ____, **Humerus** ____, **Body of sternum** ____, **Ulna** ____, **Maxilla** ____, **Capitate** ____, **Fibula** ____, **Occipital** ____, **Ilium** ____, **Axis** ____, **Talus** ____, **Ulna** ____, **Temporal** ____, **Vertbra prominens** ____, **Metatarsal IV** ____, **Scapula** ____, **Pubis** ____, **Sacrum** ____, **Metacarpal I** ____, **Navicular** ____, **L3** ____, **Coccyx** ____, **Proximal phalanx II** ____, **Zygomatic** ____, **Middle phalanx V** ____.
6. What are the five functions of bone ?
7. Name the two kinds of bone found in all 206 bones.
8. What are the two divisions of the human skeleton ?
9. Which division, axial or appendicular, contains the most bones ?
10. For each of the bones listed in question 5, state which skeletal division each is in .
11. Define, in a general way, what an articulation (joint) is.
12. Think about the various joints in your body. Which has the most types of movements ? The least ? Which joints have wide ranges of motion ? Which do not ?
13. What is the general relationship between range of motion and stability of a joint ?
14. What are the three categories that joints can fall into based on their degree of freedom of movement ?
15. What are the more complicated, scientific terms for these three classes of joints ?
16. What are the three categories that joints fall into based on the material between the bones ?
17. Give some examples of syarthrotic, amphiarthrotic, and diarthrotic joints.
18. Give some examples of cartilaginous, fibrous, and synovial joints.

19. Give an example of a fibrous synarthrotic joint.
20. Give an example of a cartilaginous synarthrotic joint.
21. Give two examples of a cartilaginous amphiarthrotic joint.
22. Give several examples of synovial diarthrotic joints.
23. Classify the types of muscles in the body.
24. Where is voluntary muscle found ?
25. Why is voluntary muscle called voluntary ?
26. To what are voluntary muscles attached ?
27. Where are some locations of smooth muscles ?
28. Where is cardiac muscle located ?
29. What is another name for cardiac muscle ?
30. About how many skeletal muscles are there in the human body ?
31. What is the origin of a skeletal muscle ?
32. What is the insertion of a skeletal muscle ?
33. Skeletal muscles are attached to bones by what ?
34. What is the definition of flexion ? Of extension ? Of abduction ? Of adduction ? Of circumduction ?
35. What is flexion at the shoulder ? At the hip ?
36. What is extension at the shoulder ? At the hip ?
38. Describe the movements pronation and supination.
39. What is the general relationship between the amount of compact bone in a bone and the amount of weight that it has to support ?

II. The Skeletal System

A. Axial and appendicular divisions (skeletons)

1. What does the term axial refer to and what are some of the bones in the axial skeleton ?

2. What does the term appendicular refer to and what are some of the bones in the appendicular skeleton ?

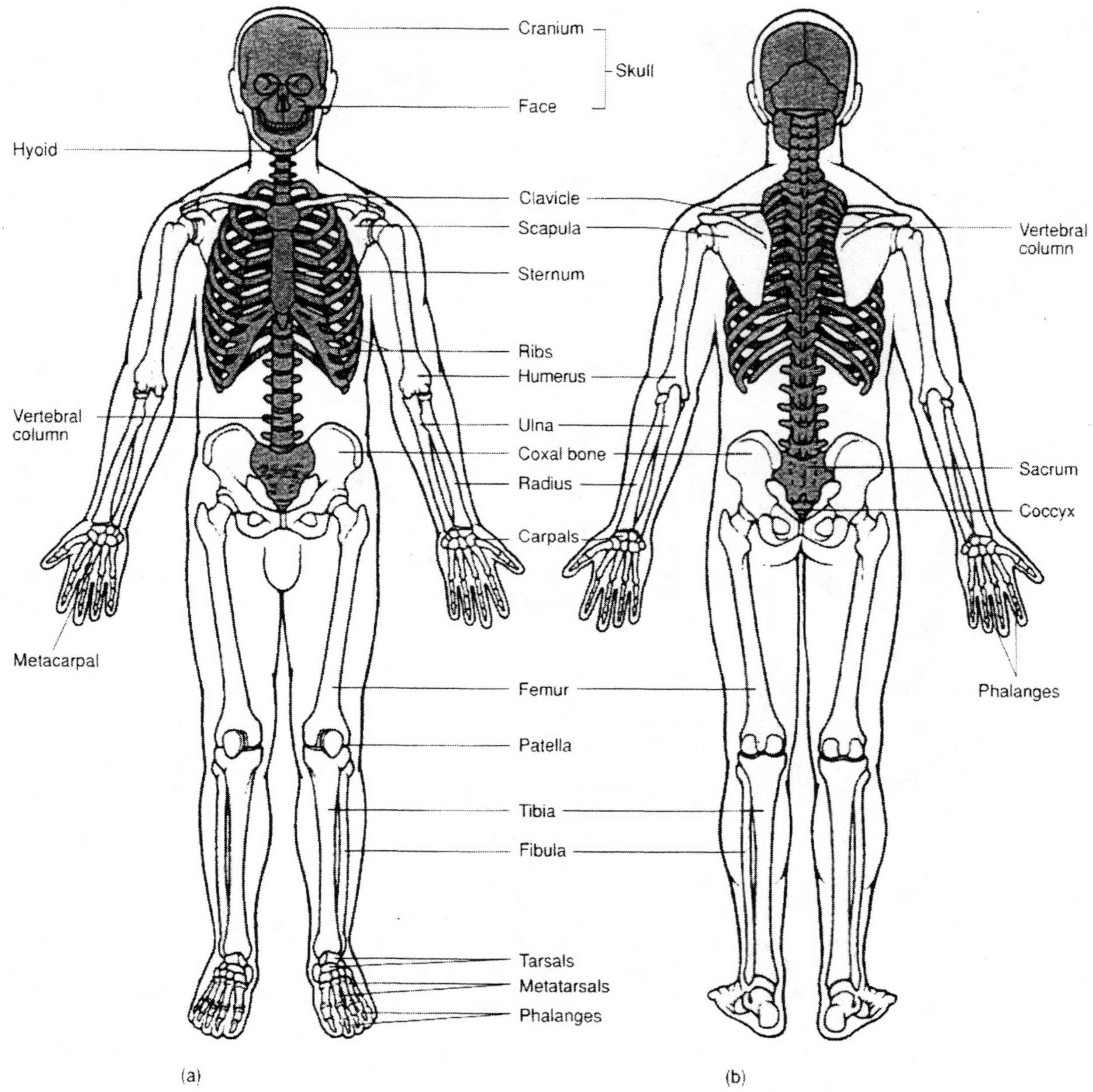
Cranium
Skull
Face
Hyoid
Clavicle
Scapula
Sternum
Ribs
Humerus
Vertebral column
Ulna
Coxal bone
Radius
Carpals
Metacarpal
Femur
Patella
Tibia
Fibula
Tarsals
Metatarsals
Phalanges
Vertebral column
Sacrum
Coccyx
Phalanges
(a)
(b)

B. Some common problems that occur within the axial skeleton:

1. Development of the shapes in the vertebral column and problems

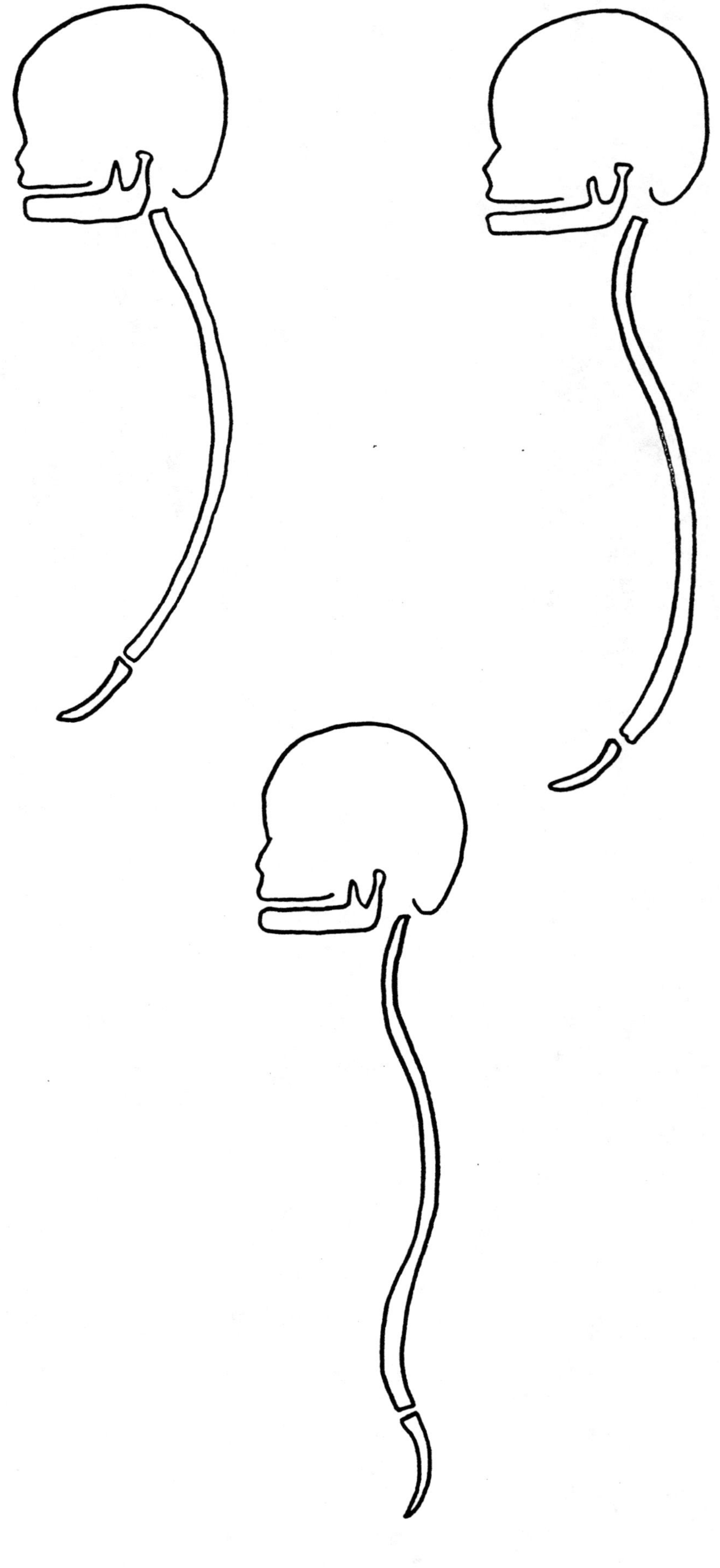

4. Laminectomy

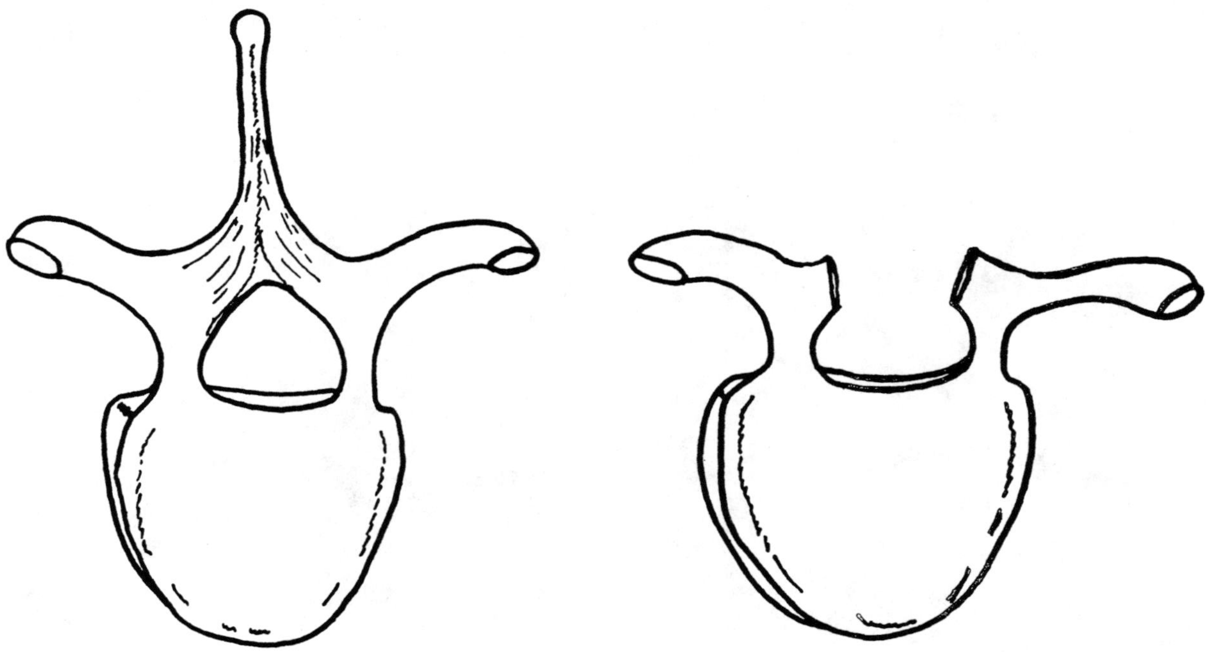

5. Spina bifida

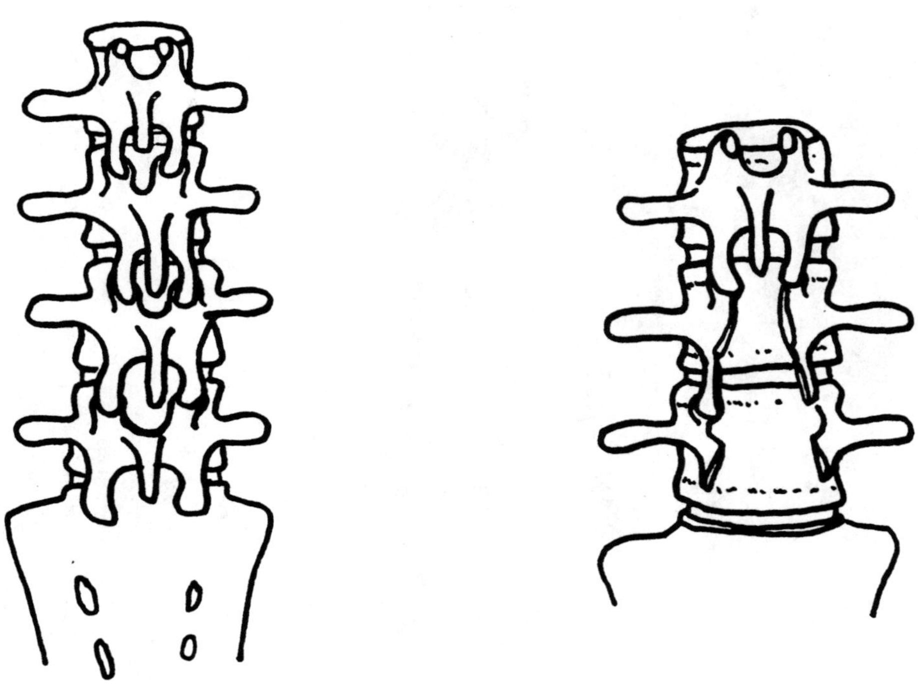

C. Some problems that occur within the appendicular skeleton

1. Colles' fracture

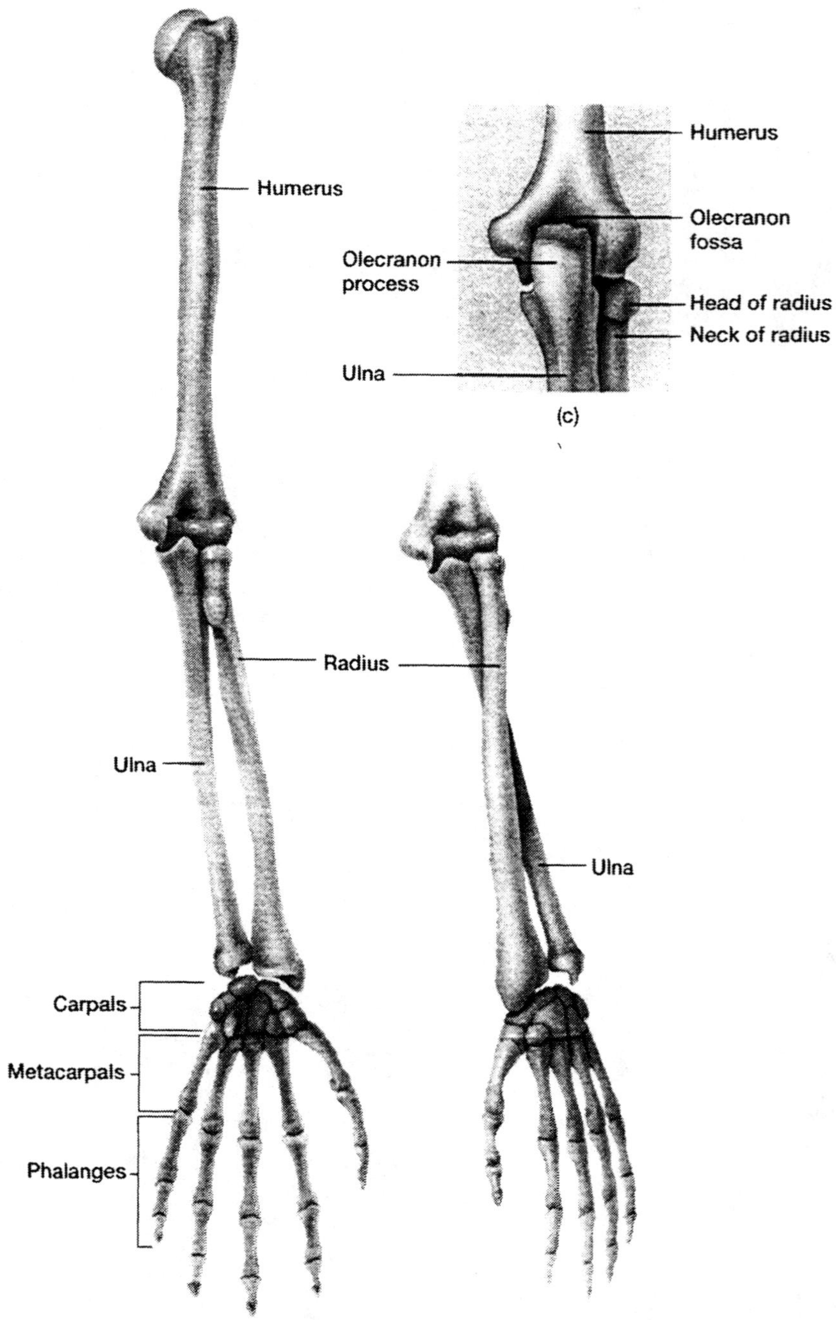

(a) Hand supinated (b) Hand pronated

2. "Breaking the hip "

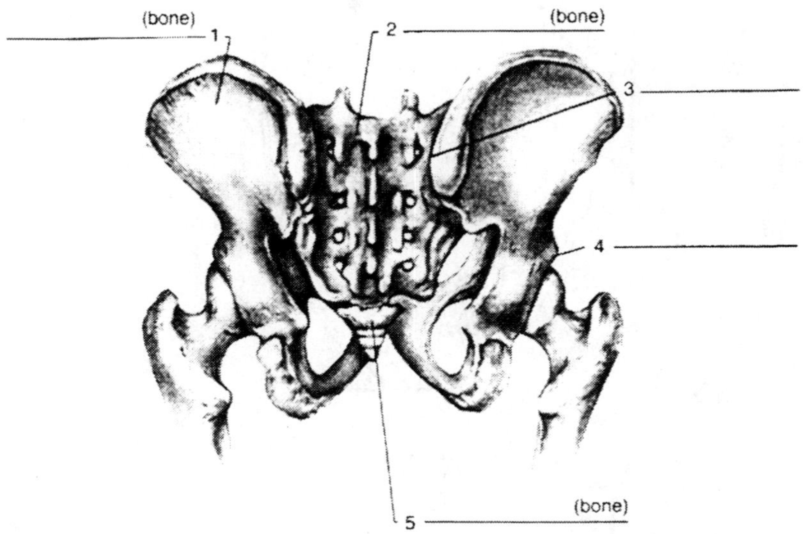

3. Potts Fracture

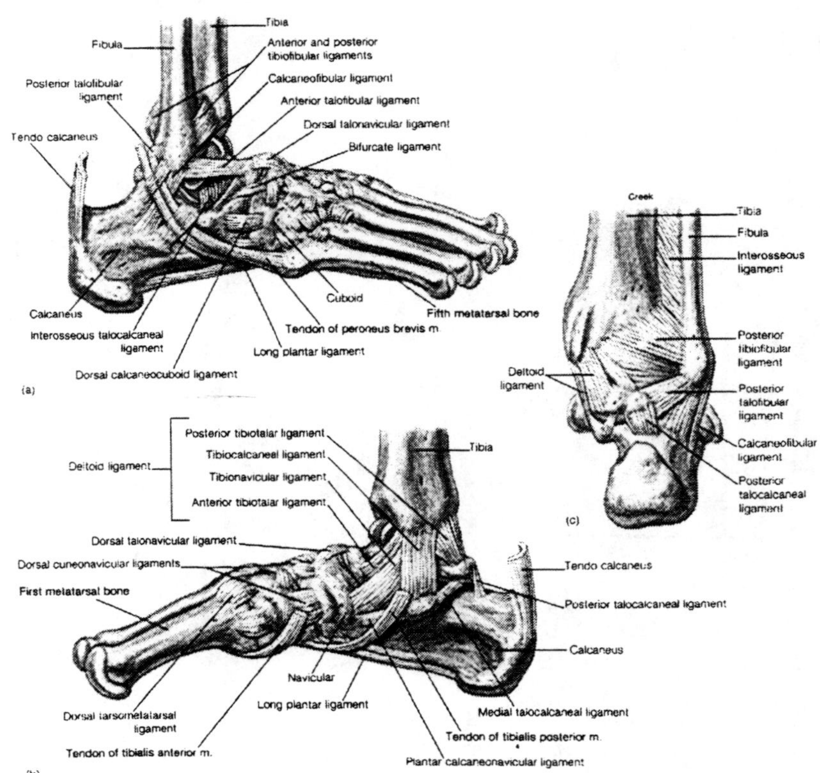

4. Carpal tunnel syndrome

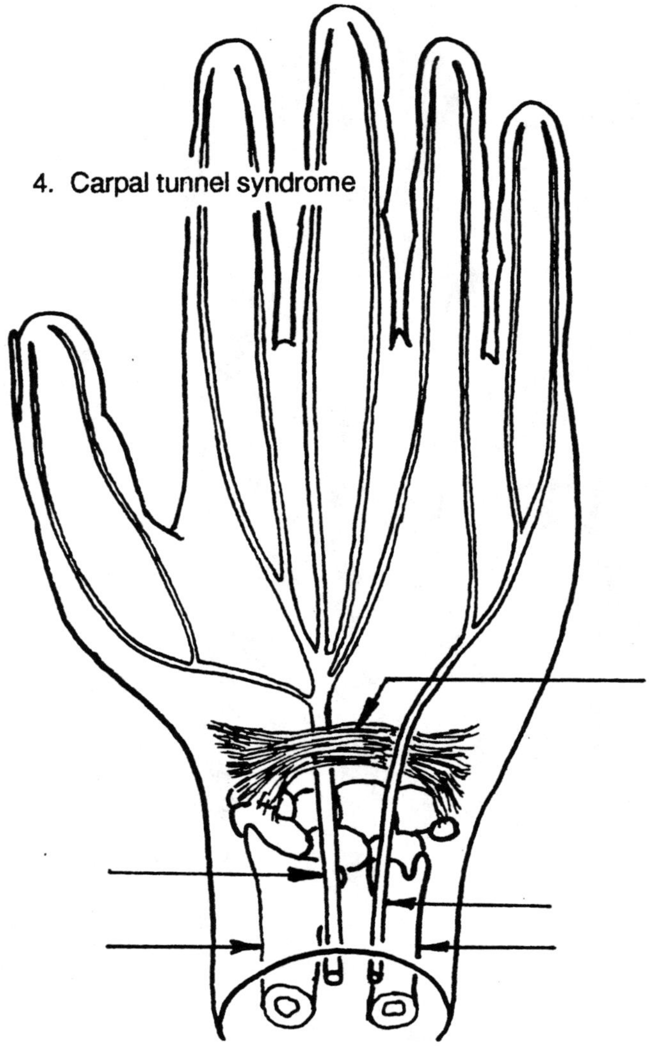

5. Subluxation (dislocation)of radius at the elbow

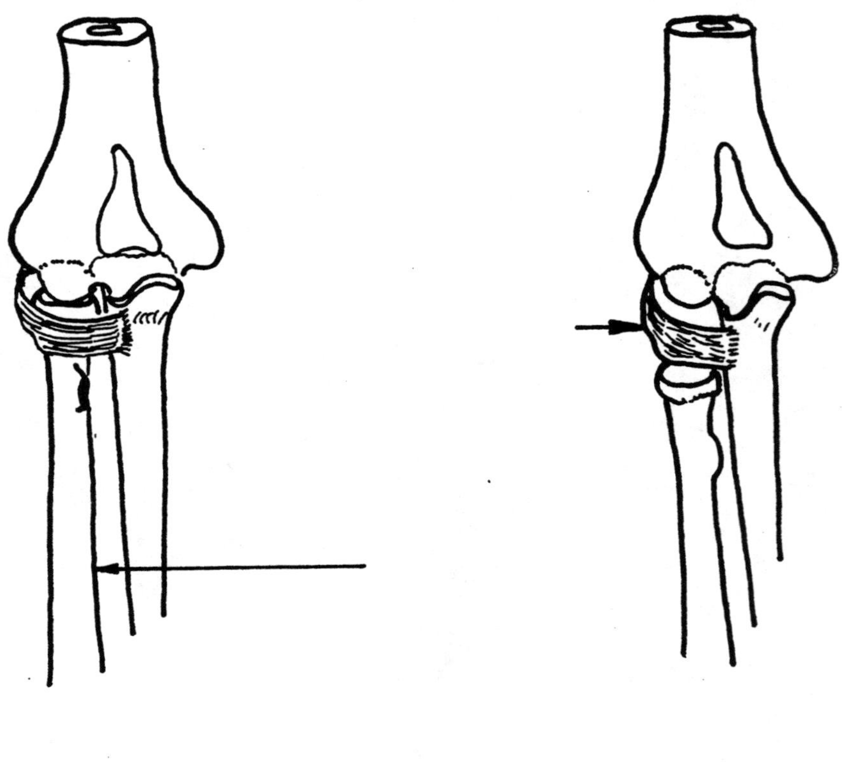

40. What does the term axial refer to in the human body ?
41. What does the term appendicular refer to in the human body ?
42. You will need to draw on your laboratory knowledge to answer the following questions. For each of the folowing bones and bone markings, state whether each is in either the axial, or the appendicular division: **Humerus___; Foramen ovale___; Patella___; Hyoid___; Frontal___; Infraorbital foramen___; Greater trochanter___; Exteral auditory meatus___; Lateral malleolus___; Sphenoid___; Infraorbital fossa___; Optic foramen___; occipital___; Greater tuberosity (tubercle)___; Talus___; Capitate___; Internal auditory meatus___; Vomer___; Medial epicondyle___; Supraorbital foramen___; Mental foramen___; Lesser tuberosity (tubercle)___; Mandible ___; Atlas___; Ilium___; Jugular foramen ___; Temporal ___; Foramen magnum___; Anatomical neck___; Head of femur___; Ulna___; Mandibular foramen___; Lesser trochanter___; Trochlea ___; Medial malleolus___; Deltoid tuberosity___; Maxilla___; Sacrum___; Surgical neck___; Cuneiforms I, II, III___; Pisiform___; Occipital condyles___; Transverse processes___; Vertebra prominens___; Sternum___; L3 ___; Pubis___; Obturator foramen___; Mandibular fossa___; zygomatic___; Coccyx___; Capitate___; Metatarsal V___; Tibia___; Radius___; Ribs___; Navicular___; Clavicles___; Anterior superior iliac spine___; Olecranon process___; Intertubercular (bicipital) groove___; Carotid canal___; Vertebral canal___; Acromion___; Intervertebral foramen ___; Inferior angle___; Laminae___; Coracoid process___; Infraspinatous fossa___; Intercondylar fossa___; Coronoid process___; Scaphoid___; Proximal phalanx II ___; Lateral border ___; Olecranon fossa___; Neck of femur___; Medial (vertebral) border___; Hamate___; Pedicle___; Ischial tuberosity___; Iliac crest___; Mastoid process___; Trochlear notch ___; Capitulum___ ; Head of humerus ___; L4___.**
43. What is the orbit of the eye ?
44. What major facial sinus lies immediately below the orbit ?
45. What is the problem with having physical trauma near the orbit ?
46. What is a cleft palate and what bone is involved ?
47. With which vertebral curve were you born with ?
48. What event caused you to develop your cervical curve ?
49. What event caused you to develop your lumbar curve ?
50. Which vertebral curve gives us the most trouble as an adult ?
51. What is a laminectomy and why is it needed ?
52. What is spina bifida ?
53. What is Colles' fracture ?
54. What is "breaking the hip " in the elderly ?
55. What is the Potts fracture ?
56. What is the carpal tunnel syndrome ?
57. What is subluxation of the radius and why is it more common in children than in adults ?

III. ARTICULATIONS, AND MOVEMENTS AT DIARTHROTIC (SYNOVIAL) JOINTS

A. ARTICULATIONS

1. Definition of an articulation =

2. Classification of joints

a. Classification by function (Degree of freedom of movement)

1) **IMMOVEABLE =**

2) **SLIGHTLY MOVEABLE =**

3) **FREELY MOVEABLE =**

b. Classification by anatomical structure

1) Joints with no joint cavity

a)) **CARTILAGINOUS JOINTS**

b)) **FIBROUS JOINTS**

2) Joints with a joint cavity = **SYNOVIAL JOINTS**

a)) General structure of a synovial joint

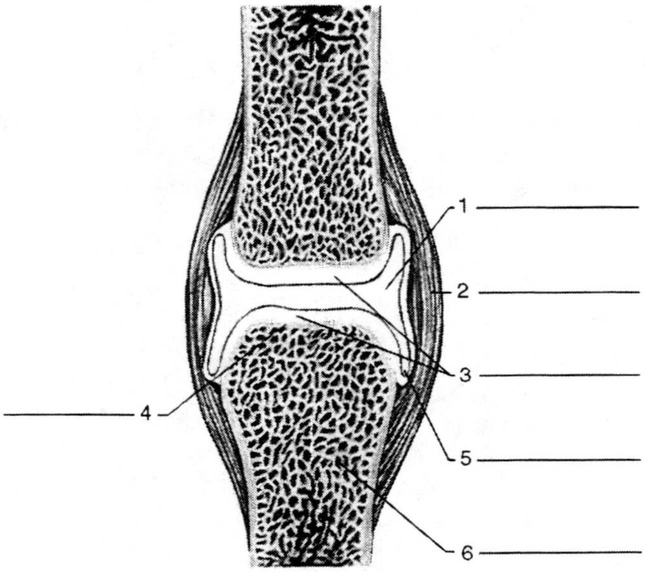

b)) Some examples of synovial joints

3. Specific joints and their function and importance

a) A **CARTILAGINOUS SYNARTHROTIC** joint

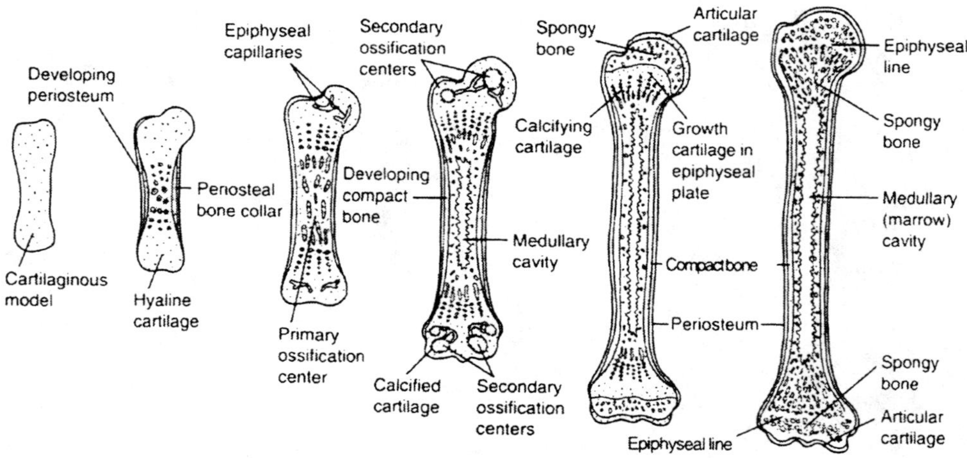

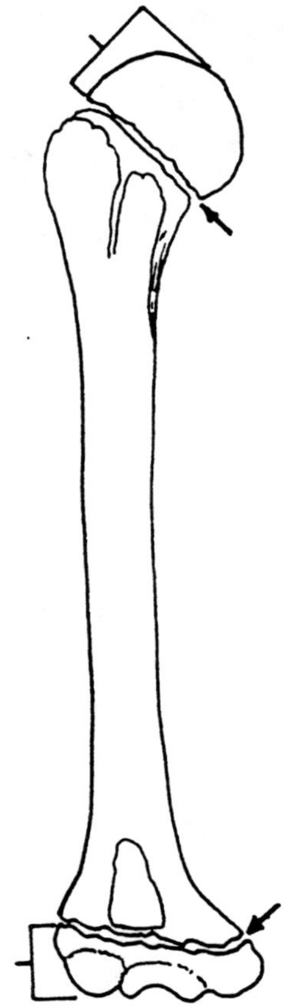

b) Some **CARTILAGINOUS AMPHIARTHROTIC** joints

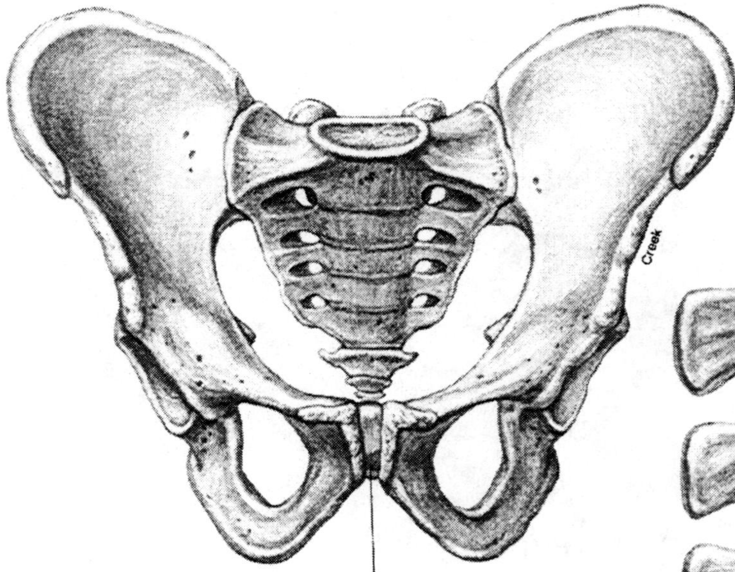

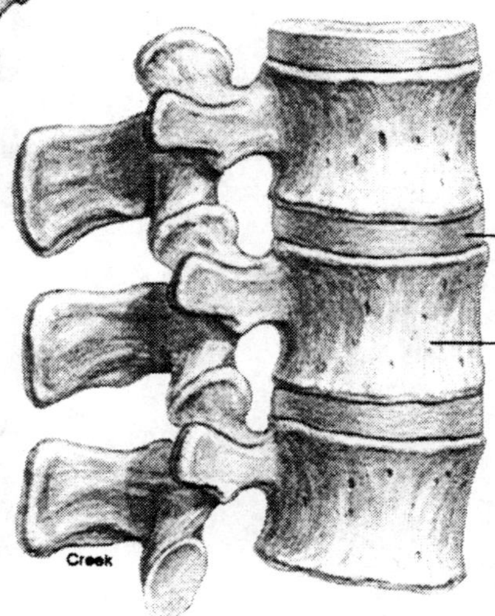

b) Some **CARTILAGINOUS AMPHIARTHROTIC** joints (continued)

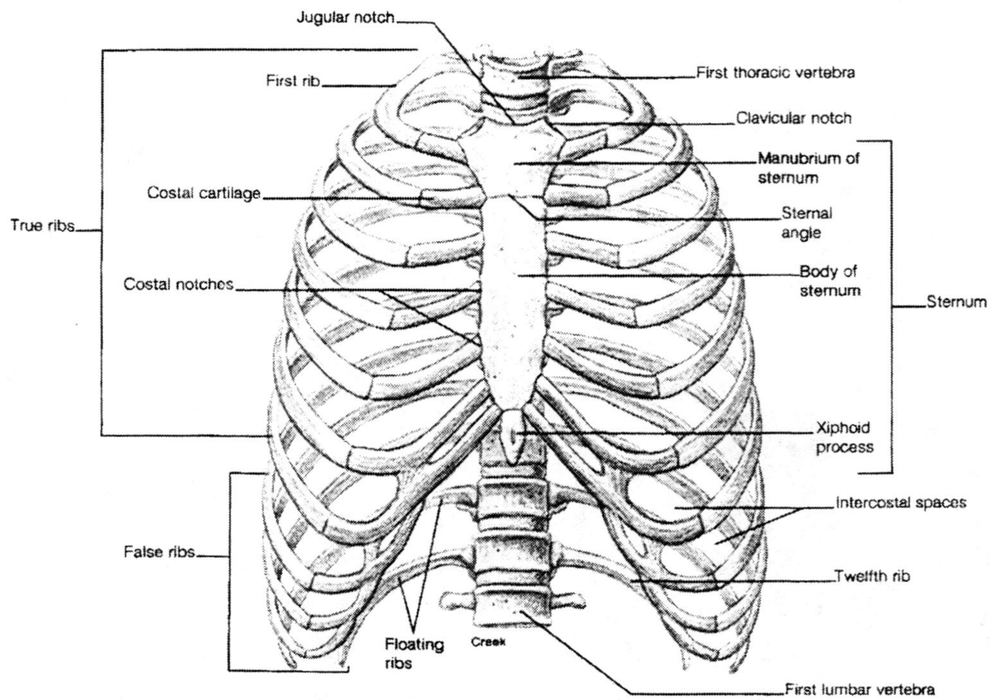

c. A **FIBROUS SYNARTHROTIC** joint

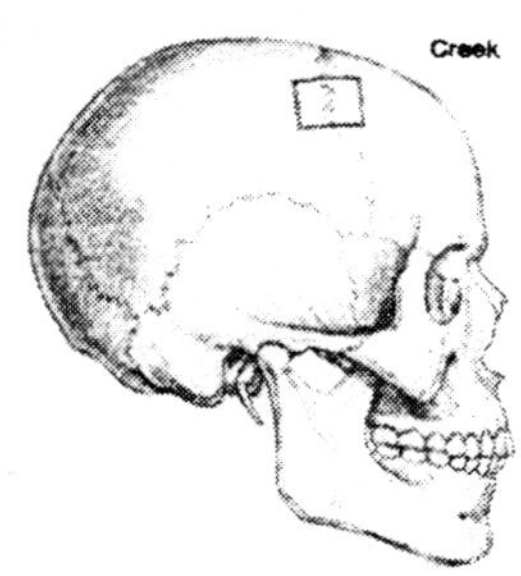

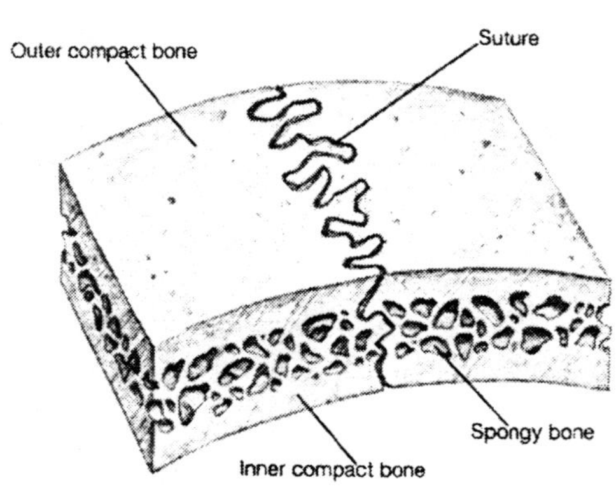

d. **SYNOVIAL JOINTS**

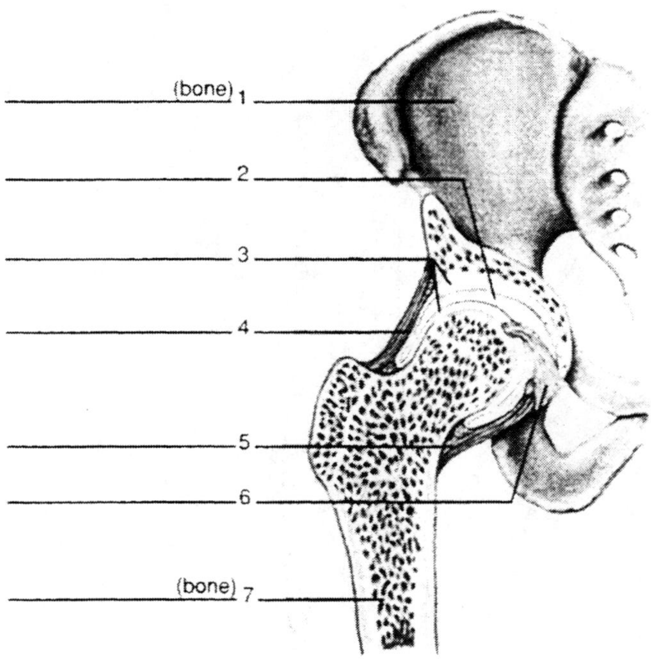

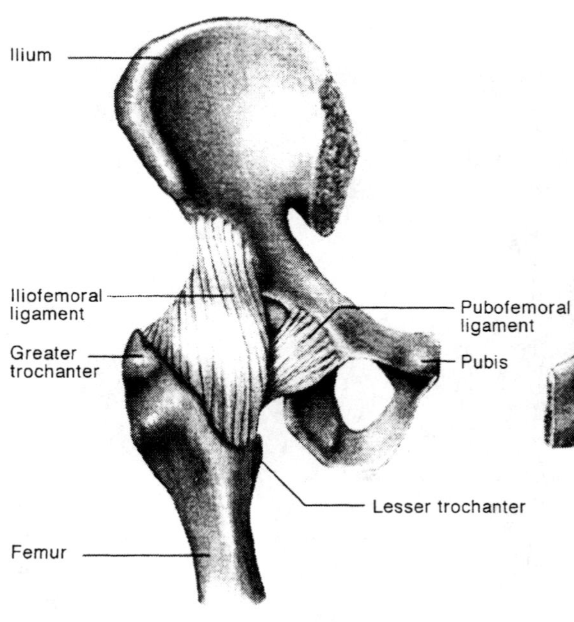

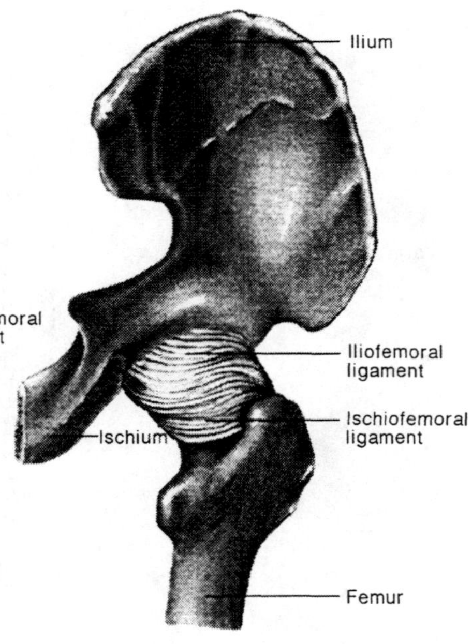

B. MOVEMENTS at synovial joints

1. **FLEXION / EXTENSION**

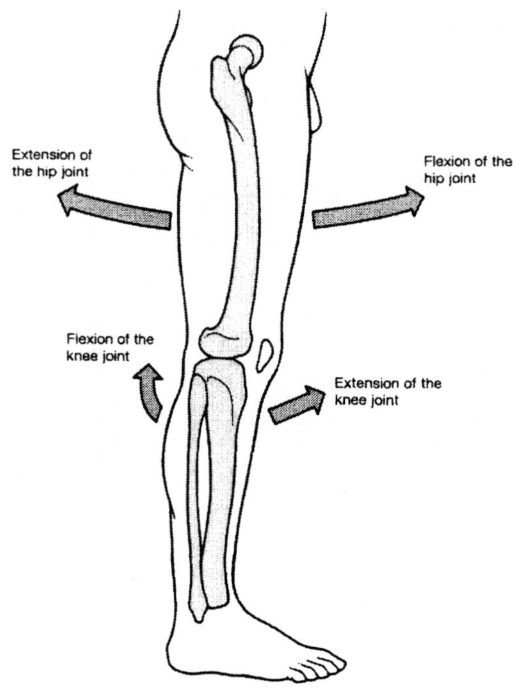

2. **ABDUCTION / ADDUCTION**

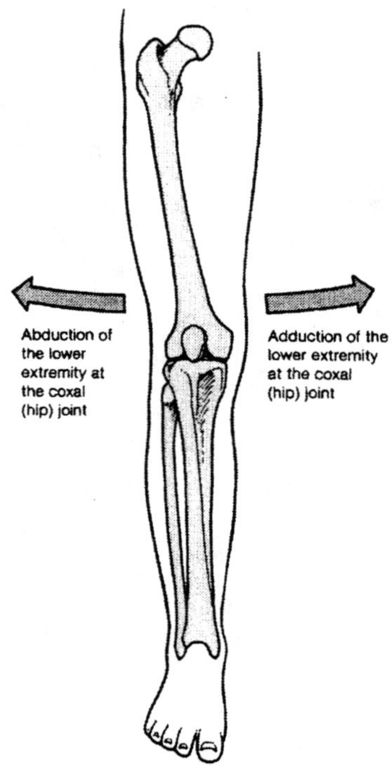

3. **PRONATION / SUPINATION**

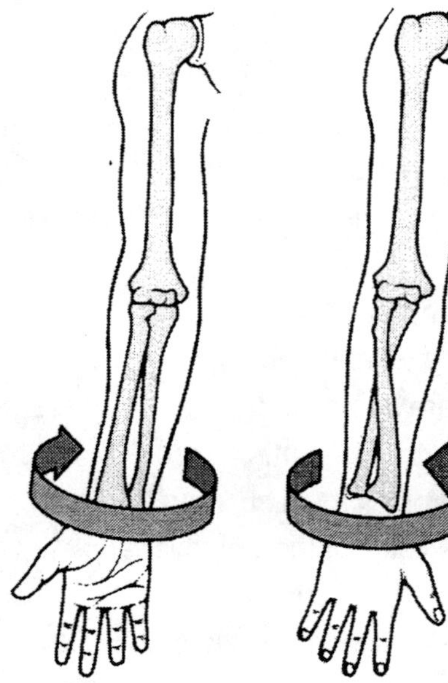

4. **DORSIFLEXION / PLANTAR FLEXION**

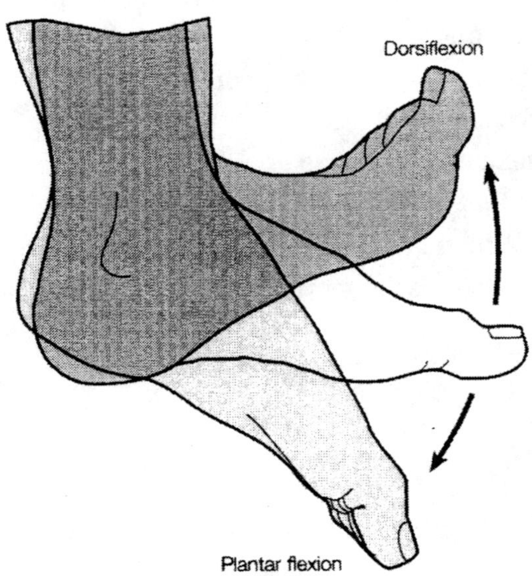

58. What is the definition of an articulation ?

59. On what two bases are joints classified ?

60. Write out the scheme used to classify joints based on the degree of freedom of movement.

61. Write out the scheme used to classify joints based on anatomical structure

62. What is another name for an immoveable joint ?

63. Give two examples of synarthrotic joints ? Which one is also a cartilaginous joint ? Which one is also a fibrous joint ?

64. Give another name for a slightly moveable joint.

65. Give two examples of an amphiarthrotic joint. Are these two joints also cartilaginous, fibrous, or synovial ?

66. Give some examples of a diarthrotic joint.

67. What is the material between bones in synovial joints ?

68. Besides synovial fluid, what are some other structures that are found with synovial joints ?

69. Name one cartilaginous synarthrotic joint.

70. Which joint is the most numerous joint in children ?

71. List some important characteristics of the epiphyseal plate.

72. Name two cartilaginous amphiarthrotic joint.

73. The pubic symphysis is a joint between what two bones ?

74. The intervertebral disks are located between what of the vertebrae ?

75. What is the name of the jelly-like material in the center of the intervertebral disk ?

76. What is a herniated disk ?

77. What is a meniscus and where is it found ?

78. What type of movement do the collateral ligaments of the knee prevent ?

79. On what basis are the trms anterior and posterior used in naming the cruciate ligaments ?

80. What movement at the knee does the anterior cruciate ligament limit ?

81. What movement at the knee does the posterior cruciate ligament limit ?

82. Define the terms flexion and extension.

83. What is the definition for flexion at the shoulder ? At the hip ?

84. Define the terms abduction and adduction.

85. Define the terms pronation and supination.

86. Which bone is the one that rotates during pronation / supination movements ?

87. What is dorsiflexion ? Why is this movement called dorsiflexion ?

88. What is plantar flexion ? Why is this movement called plantar flexion ?

IV. The Muscular System

A. Some definitions

1. Agonist =

2. Origin =

3. Insertion =

4. Synergist =

5. Antagonist =

6. Action =

B. How skeletal muscles are named

1. Shape =

2. Action =

3. Origin / Insertion =

4. Location =

C. Muscles of the Anterior and Posterior Thoracic region.

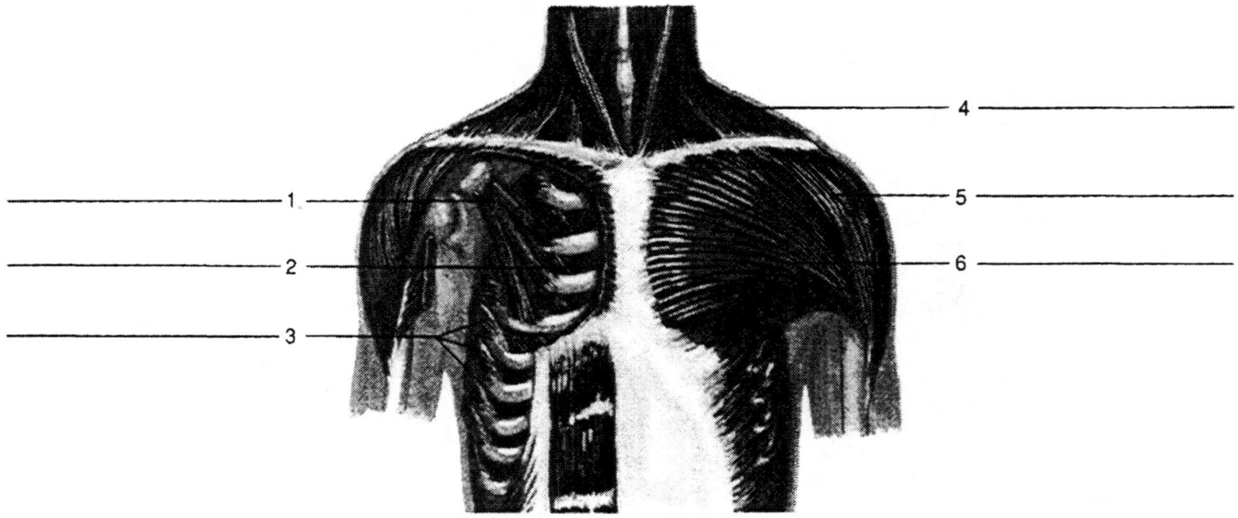

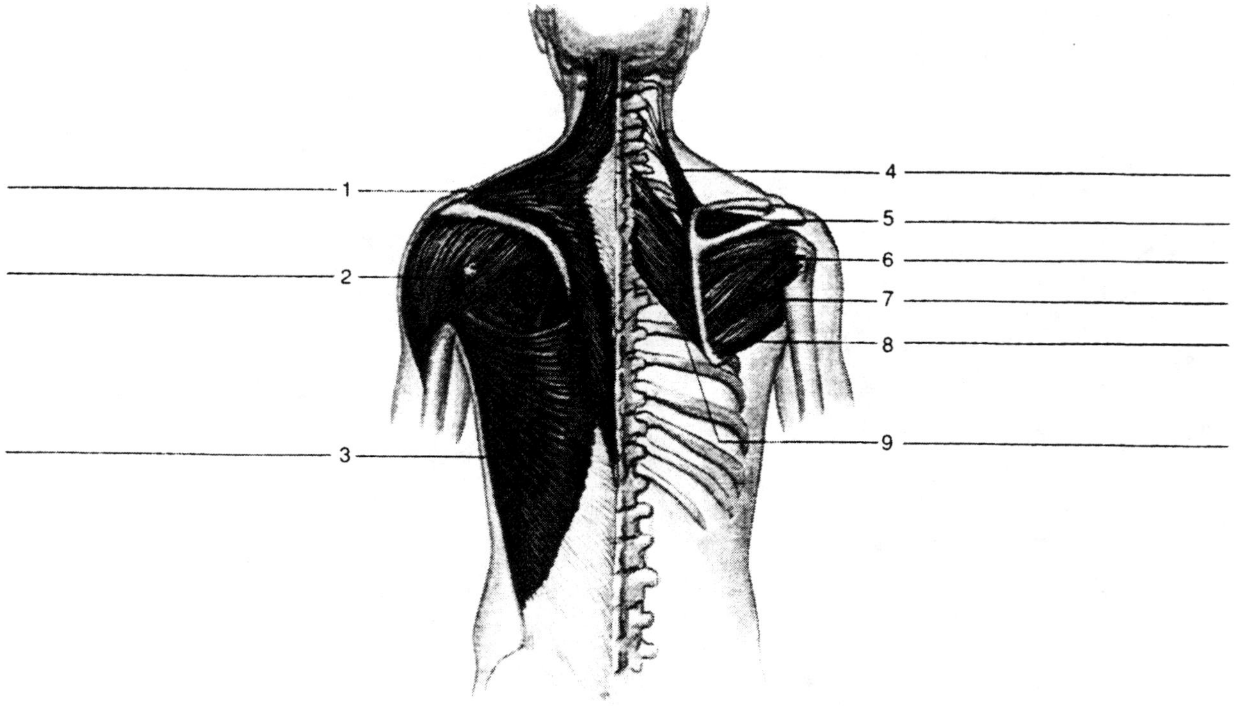

D. The pectoral girdle, regions of the upper limb, and muscle group locations

1. The pectoral girdle and regions of the upper limb

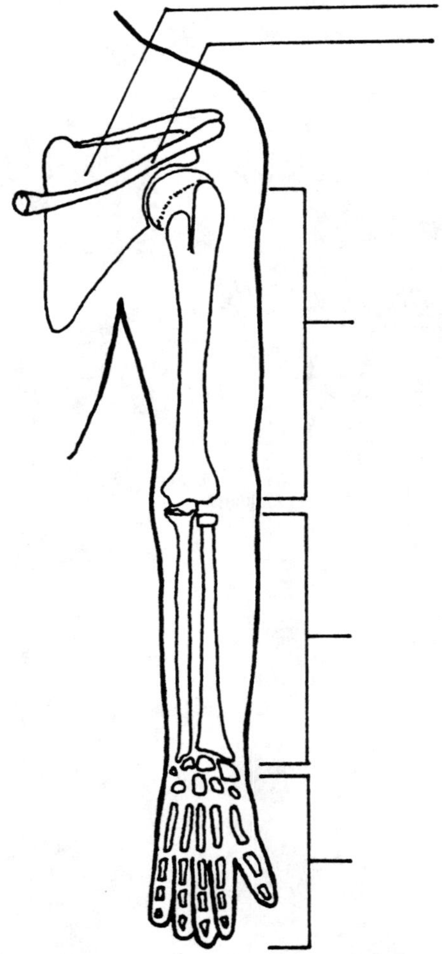

2. Muscle group locations in the upper limb

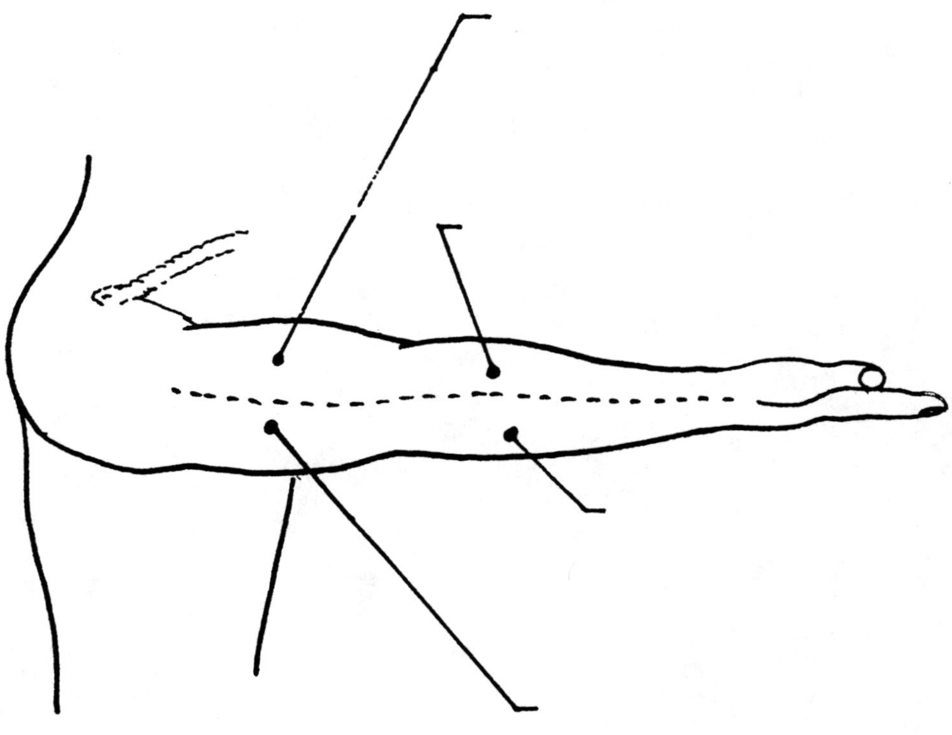

E. Major muscles that are the agonists for moving the humerus at the glenohumeral joint

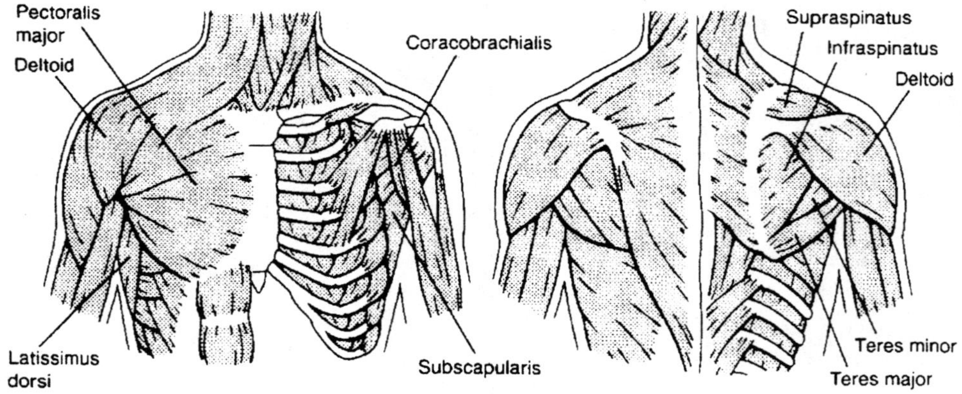

Anterior view **Posterior view**

F. Muscles that flex and extend the upper limb at the elbow joint

1. Names and locations of the agonists for flexion at the elbow

    a. **BICEPS BRACHII (BB)**
    b. **BRACHIALIS (B)**
    c. **BRACHIORADIALIS (BR)**

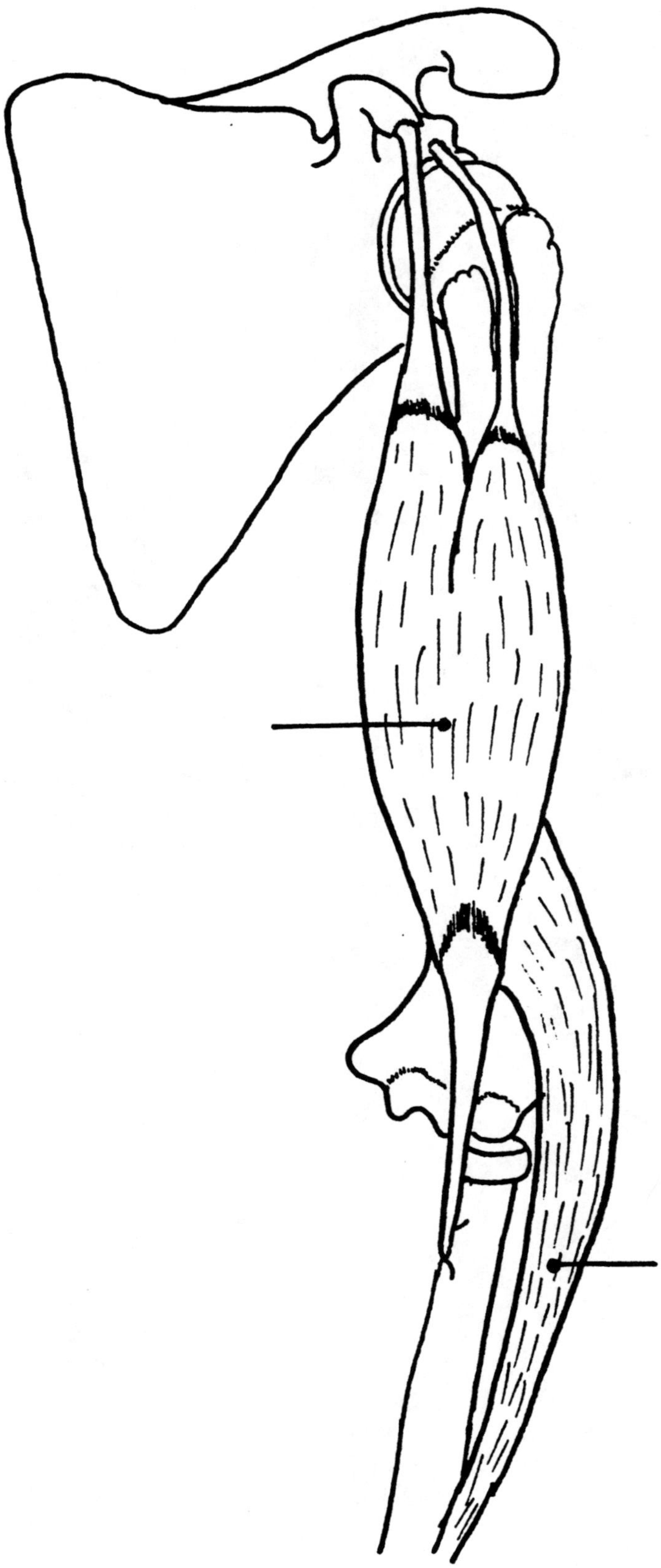

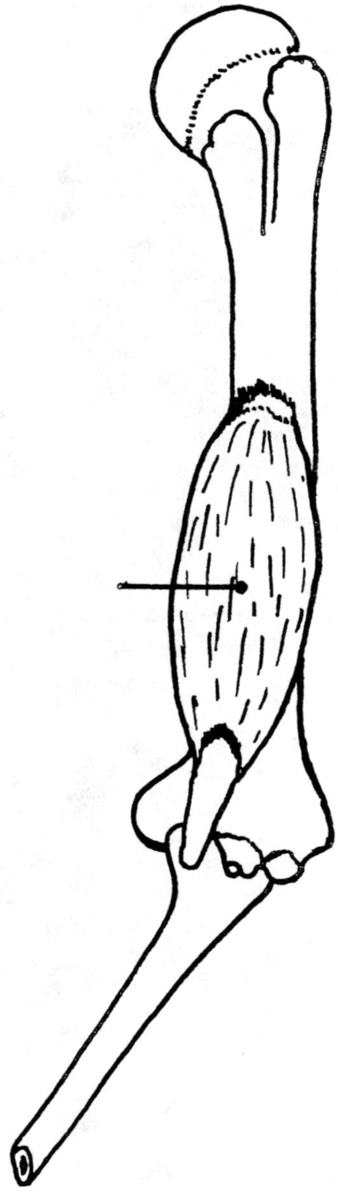

2. Name and location of the agonist for extension at the elbow

a. **TRICEPS BRACHII**

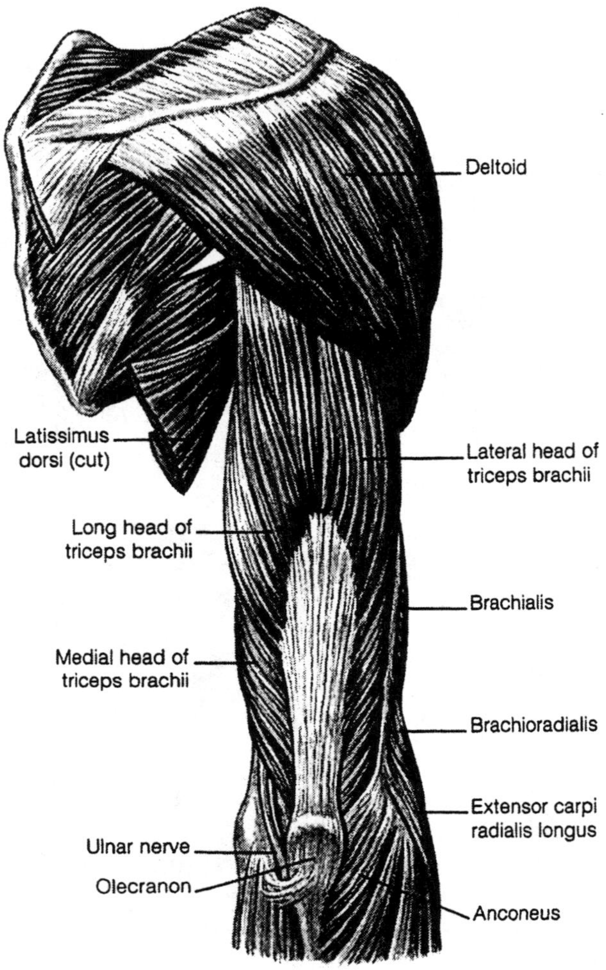

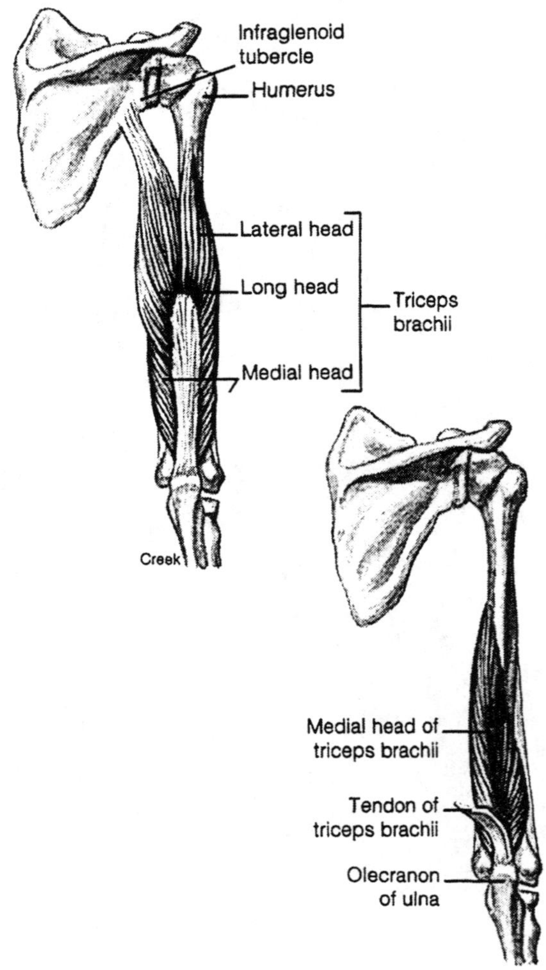

G. Muscles that flex and extend at the wrist (carpals)

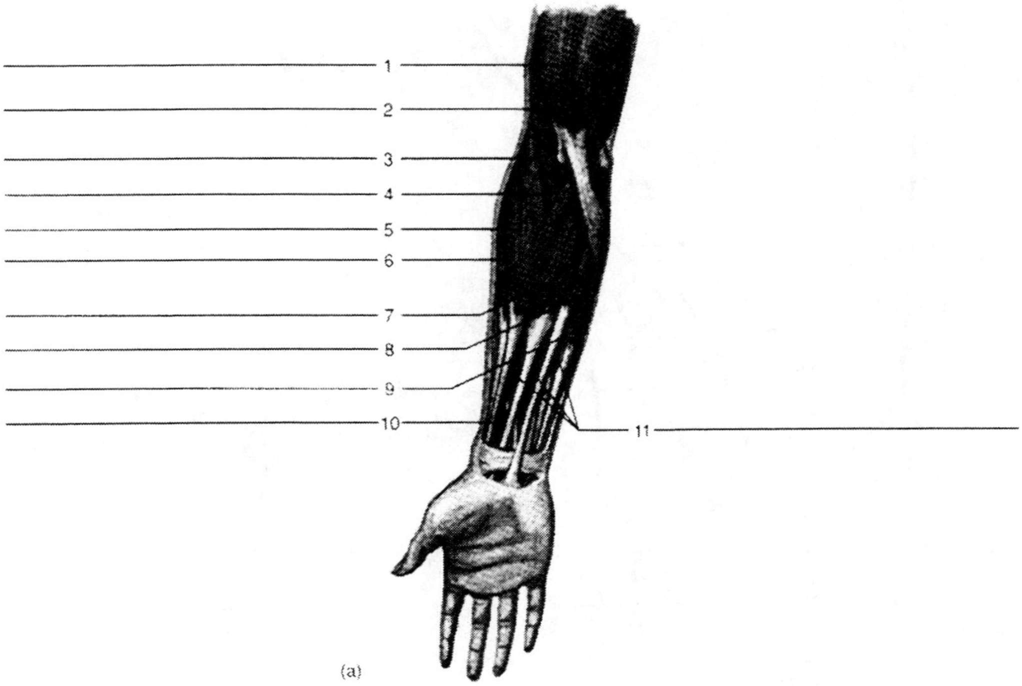

(a)

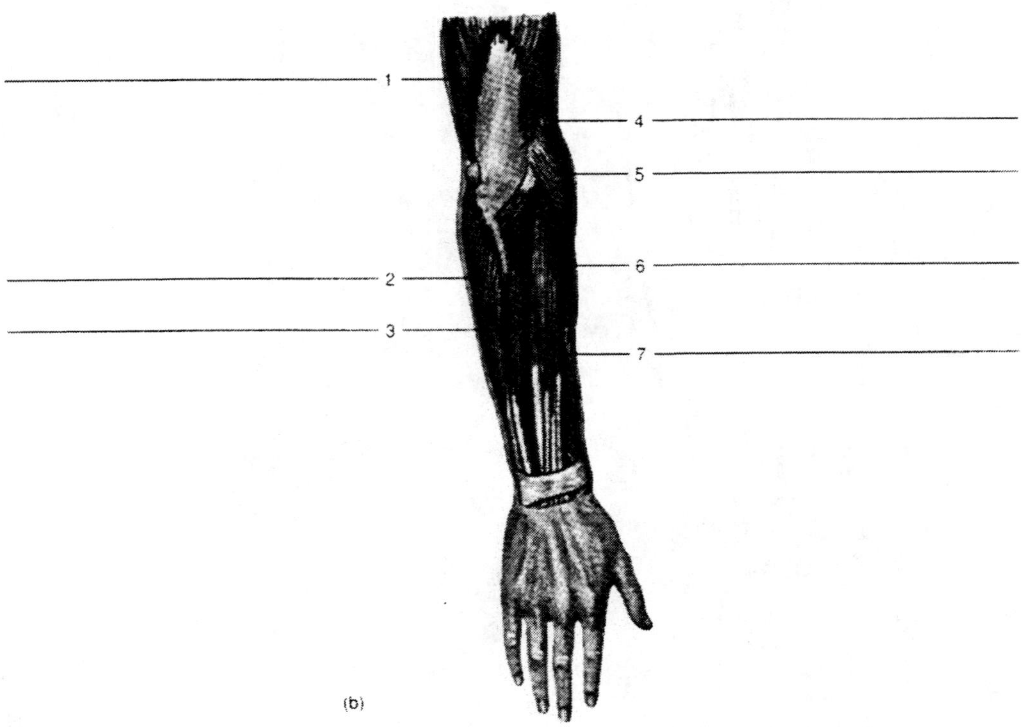

(b)

2. Names and locations of the agonists for flexion at the wrist (Carpiflexion)

Superficial Flexor Layer:

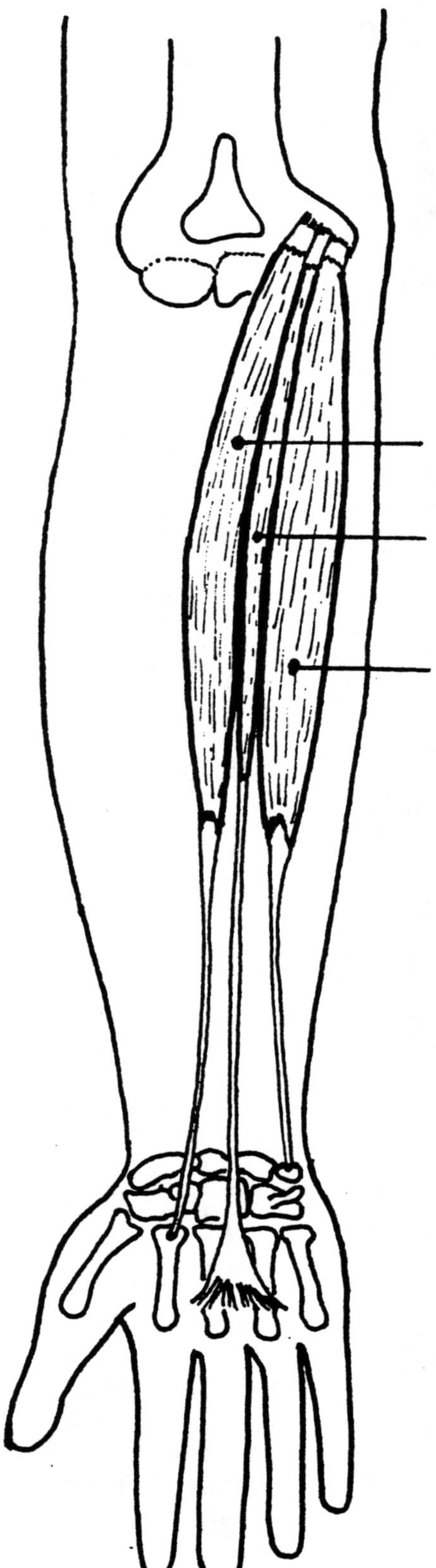

Intermediate Flexor Layer:

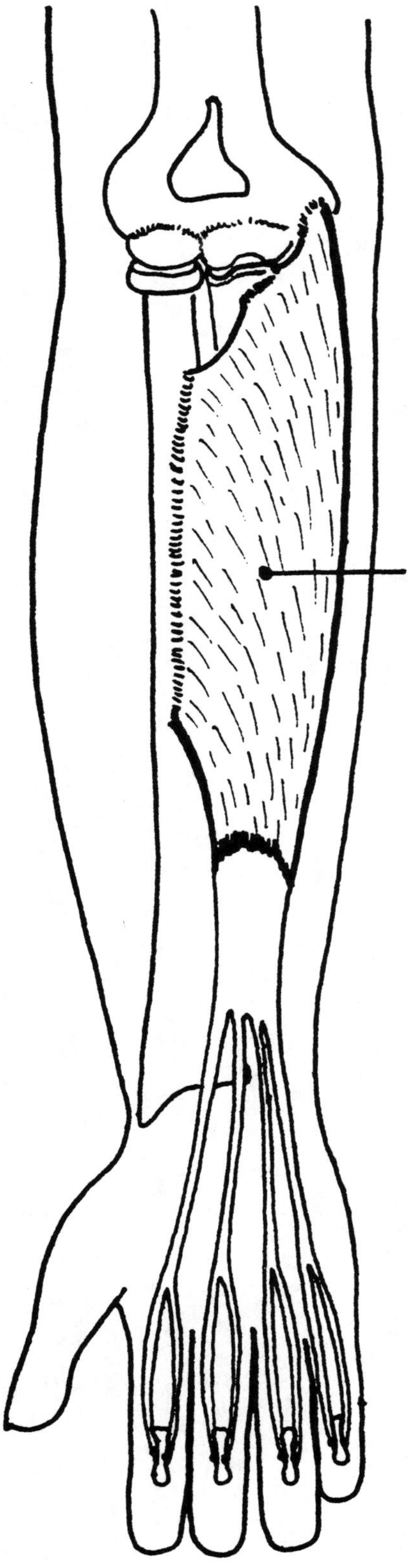

Deep Flexor Layer:

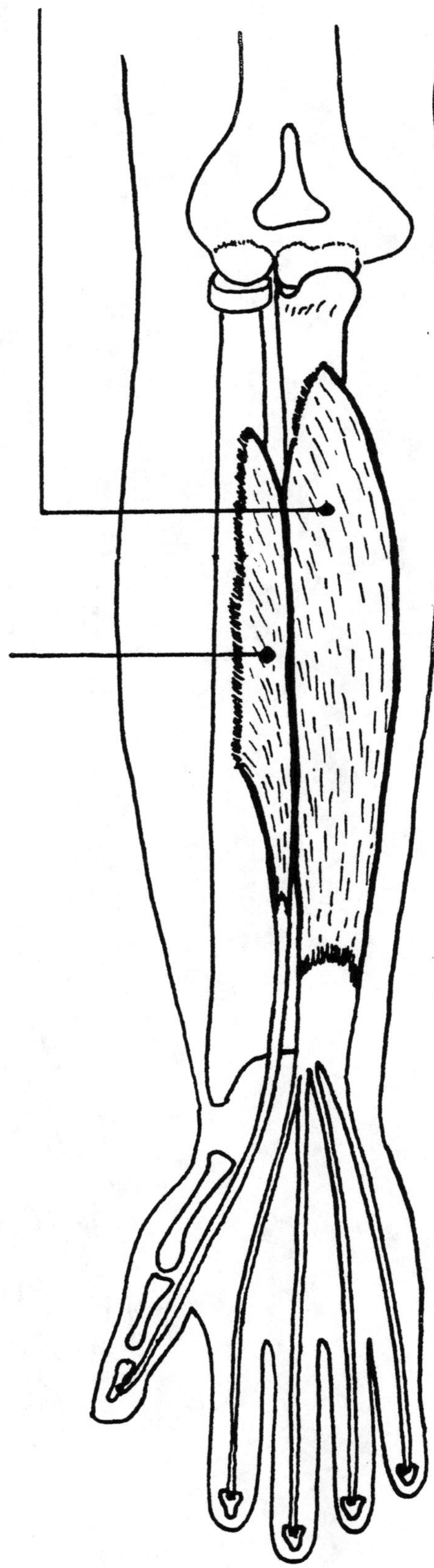

3. Names and locations of the agonists for extension at the wrist (Carpiextension)

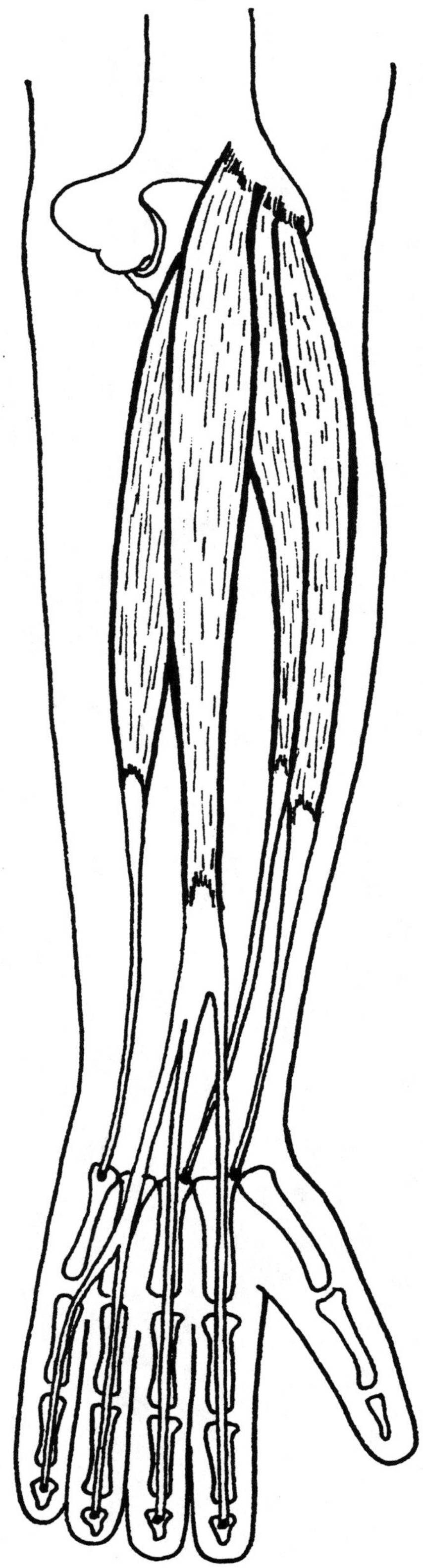

89. Define the term agonist.
90. Define the term origin.
91. Define the term Insertion.
92. Define the term synergist.
93. Define the term antagonist.
94. Define what the action of a muscle is.
95. State the ways that a muscle can be named and give some examples of each.
96. What are the muscles on the dorsum of the body ?
97. What is the two bones of the pectoral girdle ?
98. What is the arm?
99. What is the forearm ?
100. What is the manus ?

101. For each of the following bones or bone markings write an **A** if it is in the arm, an **F** if it is in the forearm, or an **M** if it is in the manus : **Capitate___; Humerus___; Pisiform ___; Metacarpal II ___; Proximal phalanx V ___; Ulna ___; Greater tuberosity___; Olecranon process___; Olecranon fossa___; Radial tuberosity___; Interphalangeal joints ___; Hamate ___; Medial epicondyle ___; Coracoid process ___; Scaphoid ___; Distal phalanx III ___; Trochlear notch ___; Coronoid fossa ___; Radius ___; Proximal phalanx I ___; coronoid fossa ___; Lesser tuberosity ___; Anatomical neck ___; Triquetrum ___; Intertubercular groove___.**

102. For each of the following muscle groups write an **AA** if it is located in the anterior arm, **PA** if in the posterior arm, **AF** if in the anterior forearm, or **PF** if in the posterior forearm: **Extensors at the wrist ____; Flexors at the elbow ____; Flexors of the fingers (digits) ____; Flexors at the wrist ____; Extensors of the fingers (digits) ____; Extensor at the elbow ____.**

103. What muscles are the agonists for flexion at the shoulder ?
104. What muscles are the agonists for abduction at the shoulder ?
105. What muscles are the agonists for adduction at the shoulder ?
106. What muscle is an agonist for extension at the shoulder ?
107. What are the three agonists for flexion at the elbow ? What are their locations ?
108. What is the agonist for extension at the elbow ? What is its location ?
109. Which of the three muscles, biceps brachii, brachiali, brachioradialis, can you see on yourself ?
110. What are the five agonists for flexion at the wrist ? Where are these muscles located ?
111. What are the two agonists for flexion of the fingers (digits) ? Where are they located ?
112. What are the agonists for extension at the wrist ? Where are they located ?

H. The pelvic girdle, regions of the lower limb, and muscle group locations

1. The pelvic girdle, and regions of the lower limb

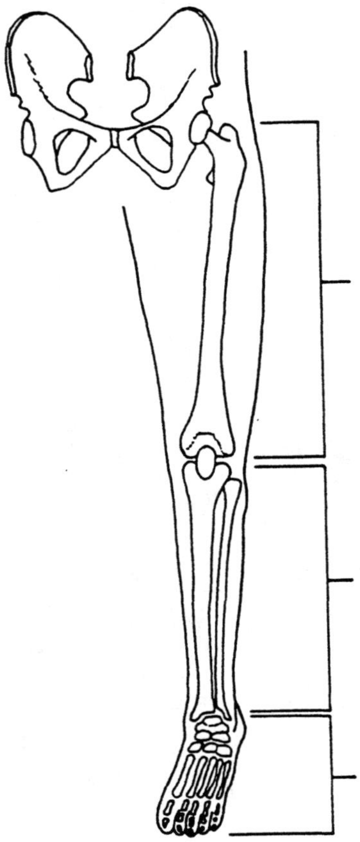

2. Muscle group locations for the lower limb

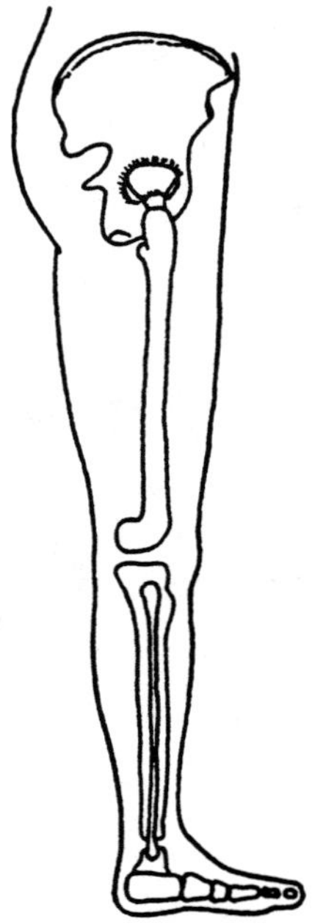

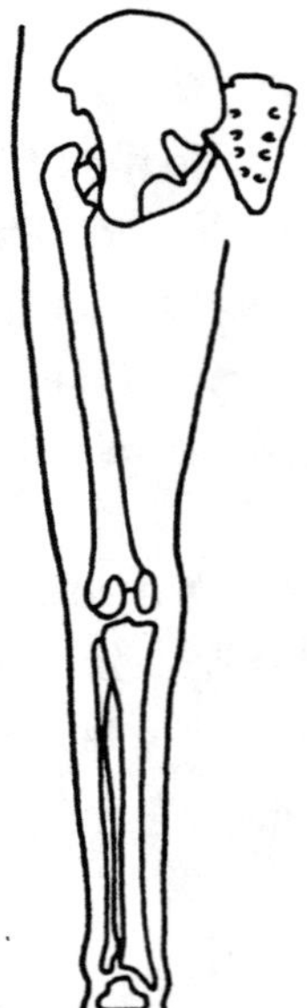

I. Muscles that are attached to and move the femur

1. Names and location of the agonists for flexion at the hip

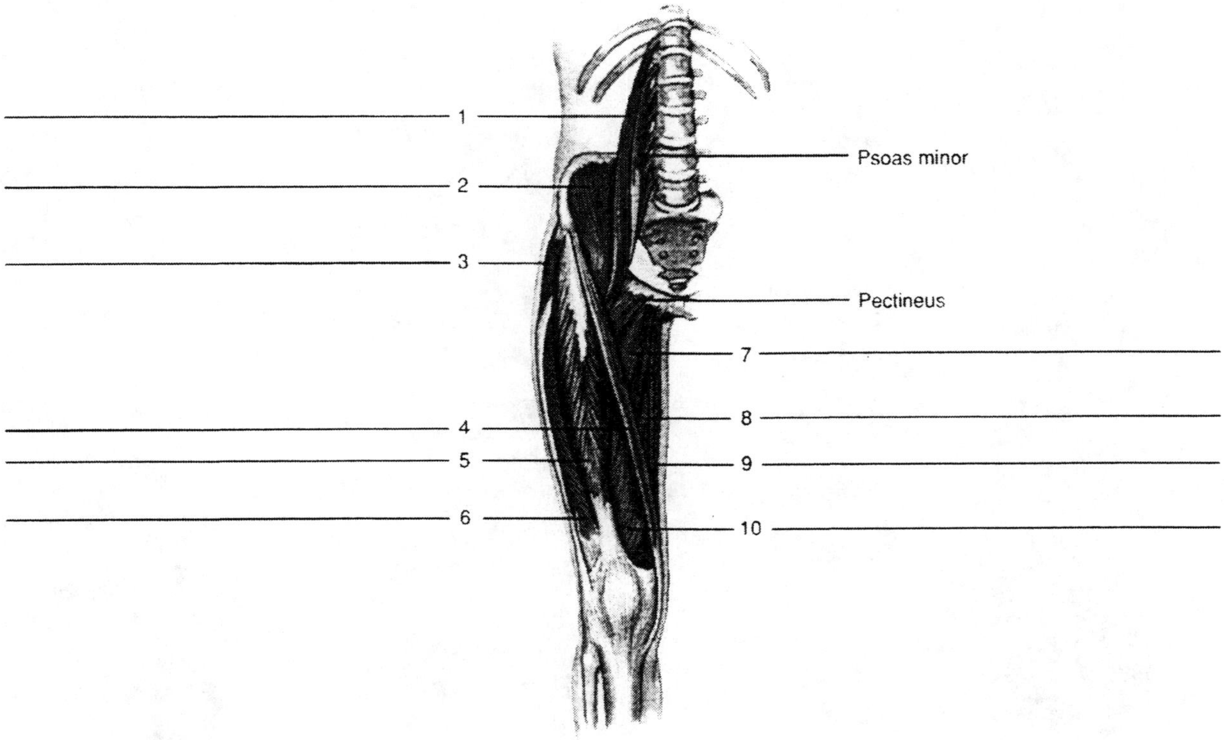

2. Names and locations of the agonists for extension at the hip

   a. **Gluteus maximus** =

   b. **HAMSTRINGS** =

      1) Semimembranosus
      2) Biceps femoris
      3) Semitendinosus

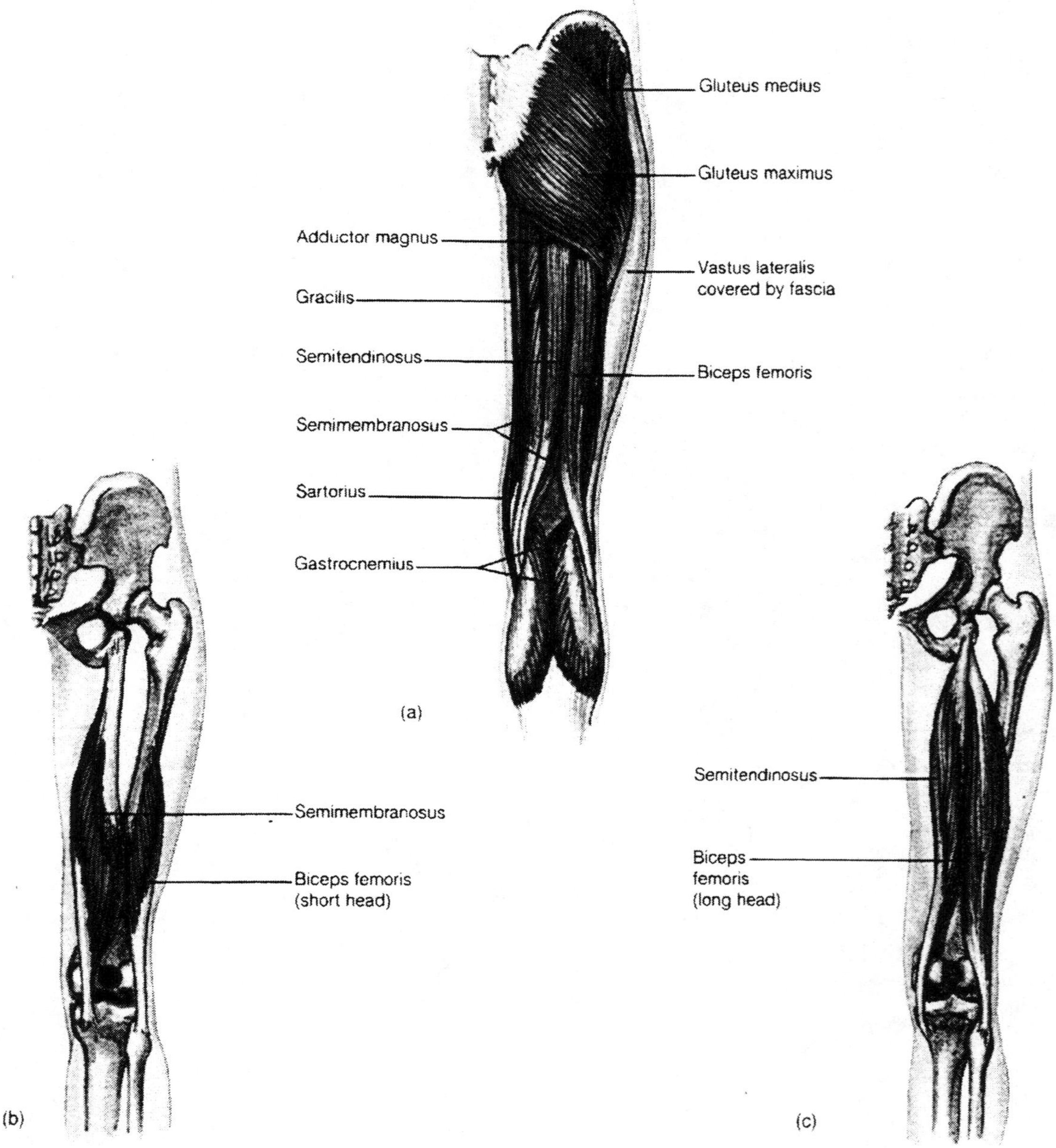

3. Names and locations of the agonists for abduction and adduction at the hip

a. **<u>AB</u>ductors** =

1) **Gluteus medius**
2) **Gluteus minimus**

b. **<u>AD</u>ductors** =

1) **Adductor magnus**
2) **Adductor longus**
3) **Adductor brevis**
4) **Gracilis**

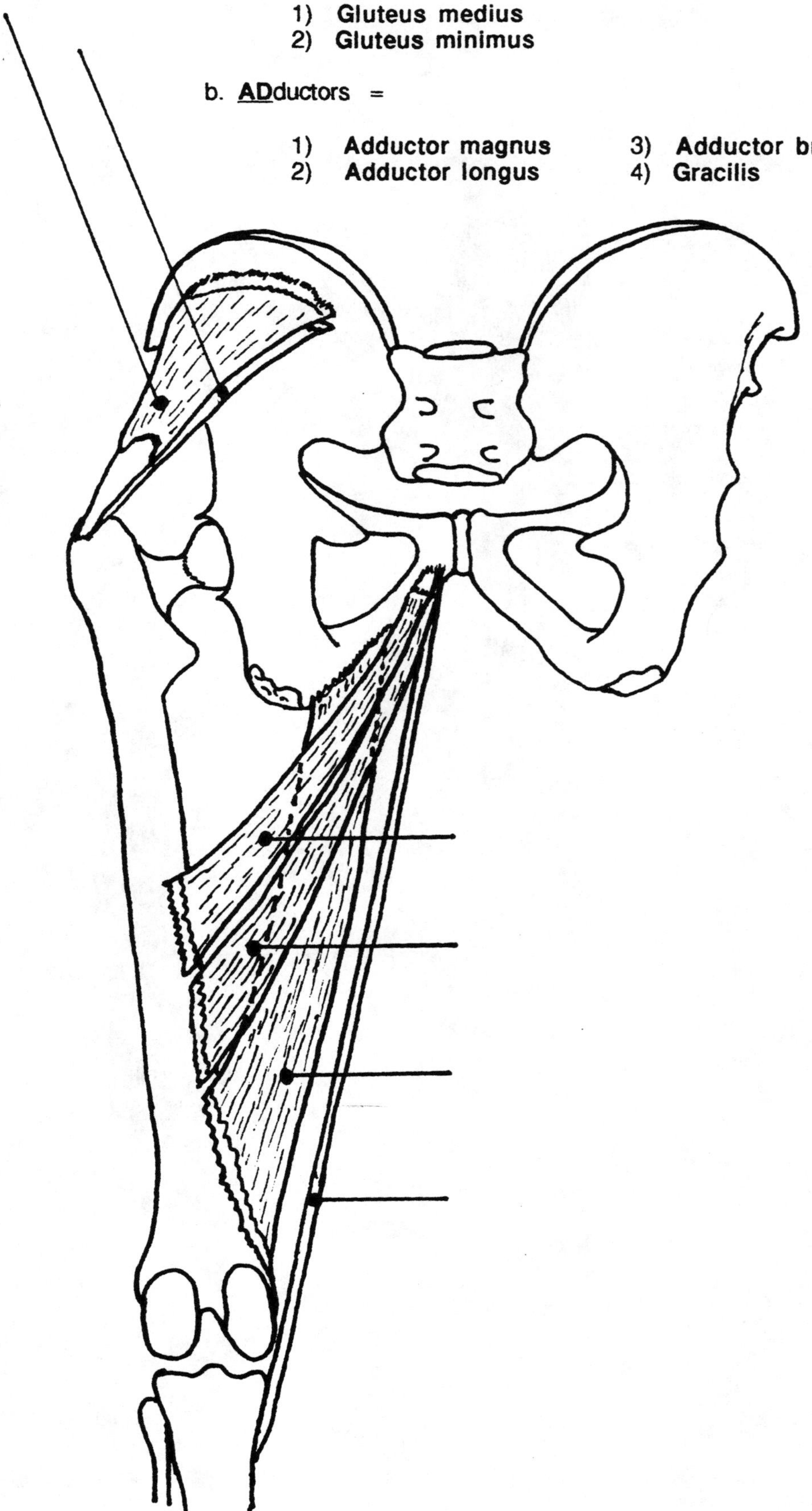

4. Names and locations of the agonists that flex and extend at the knee joint

   a. Flexors of the knee =

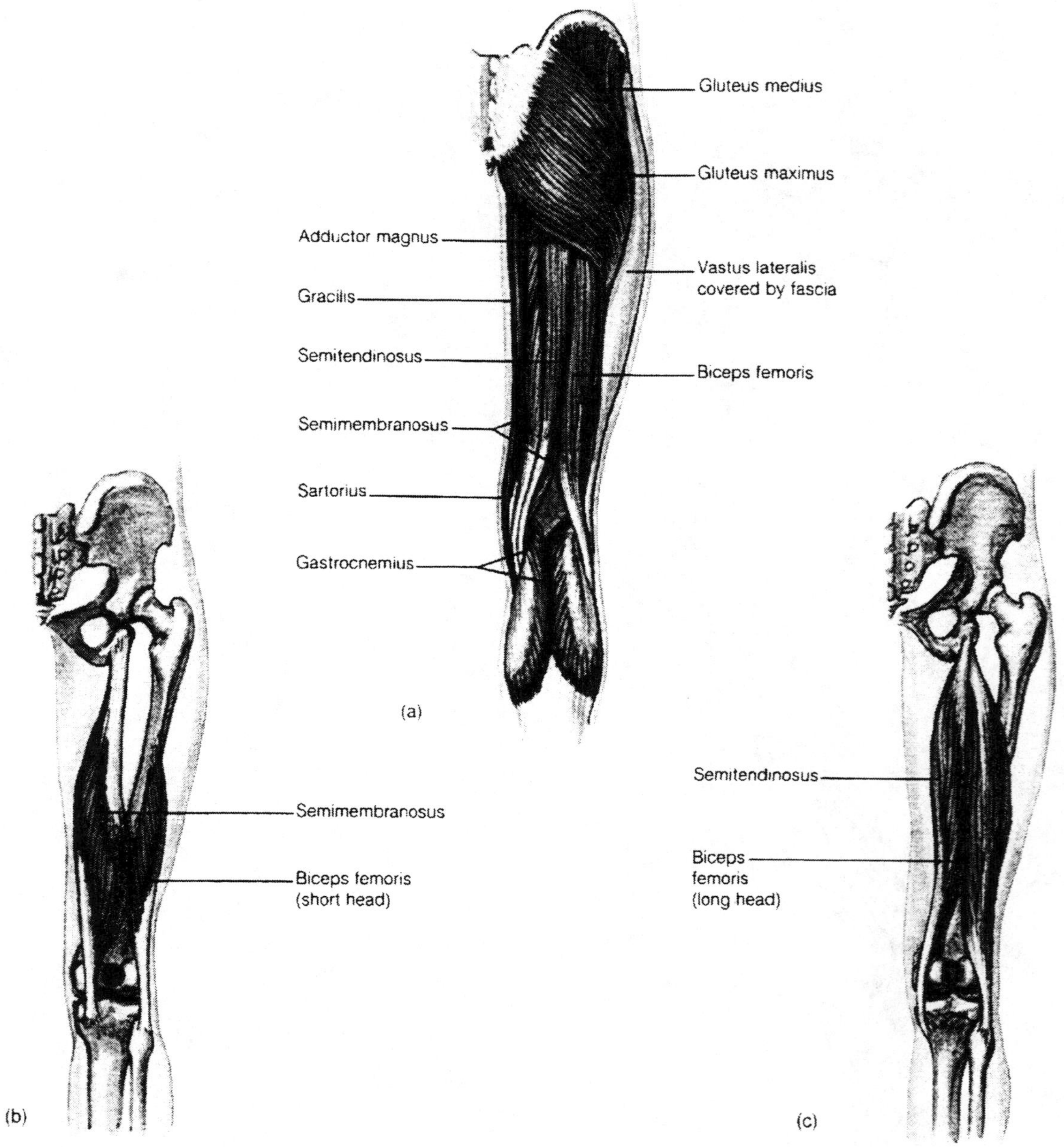

b. Extensors at the knee = **QUADRICEPS FEMORIS**

1) Location =

2) Names
   a) **Rectus femoris**
   b) **Vastus lateralis**
   c) **Vastus intermedius**
   d) **Vastus medialis**

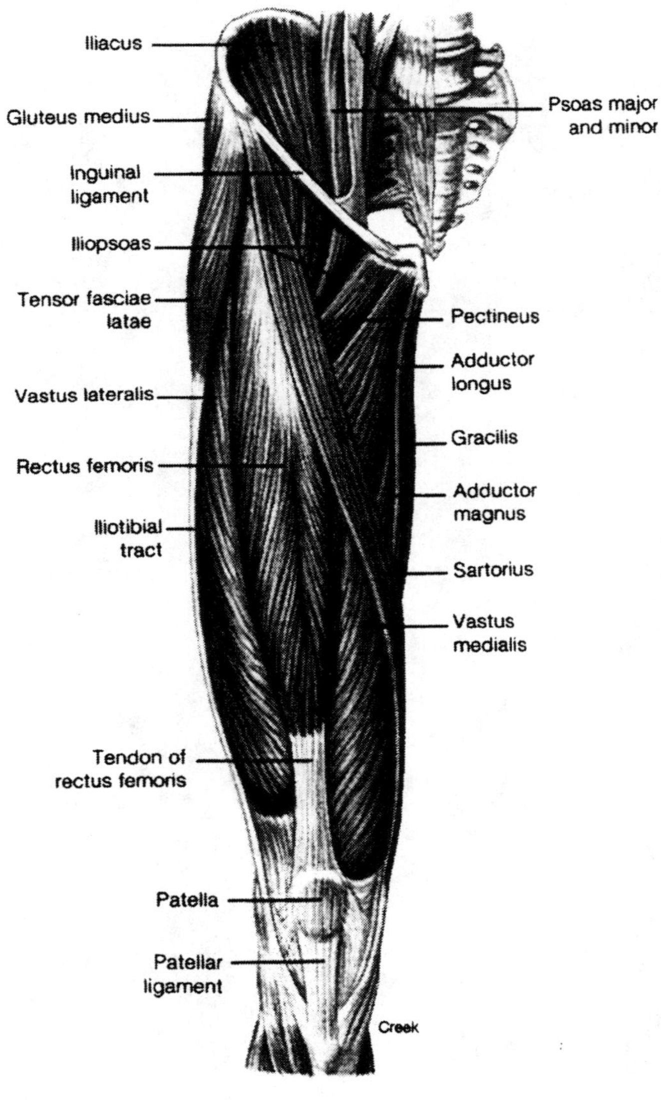

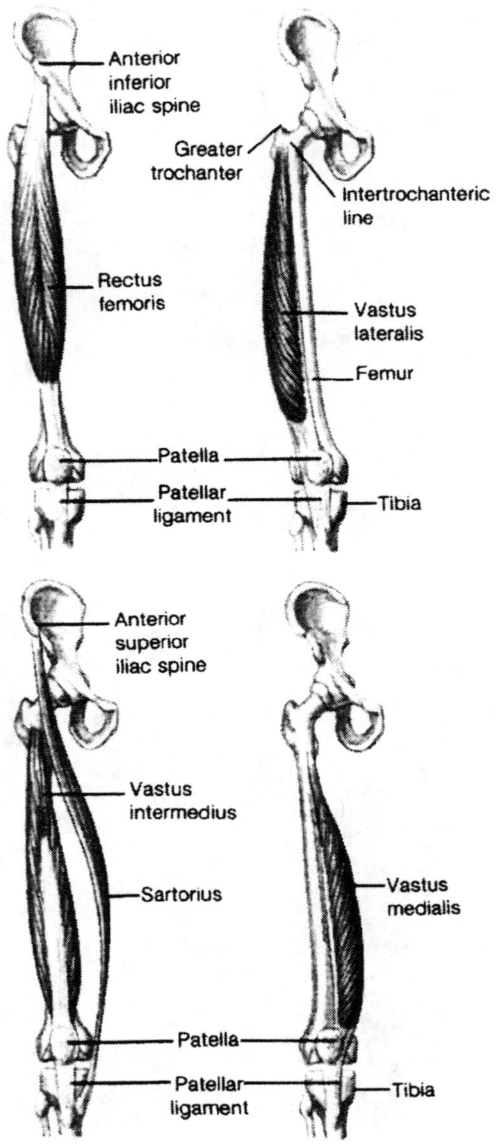

5. The names and locations of the agonists for dorsiflexion and plantar flexion

   a. The dorsiflexors

      1) Location =

      2) Names
         a) **Tibialis anterior**
         b) **Extensor digitorum longus**

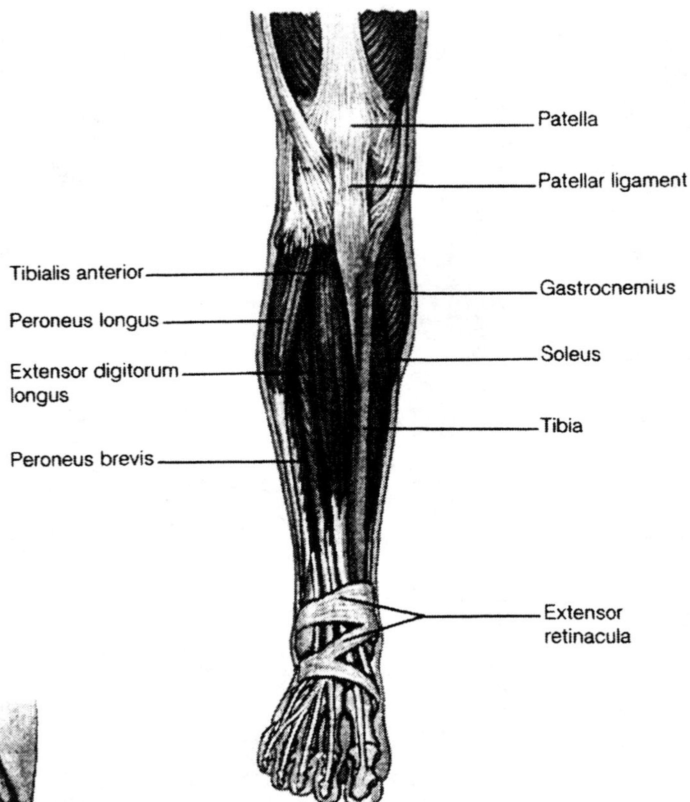

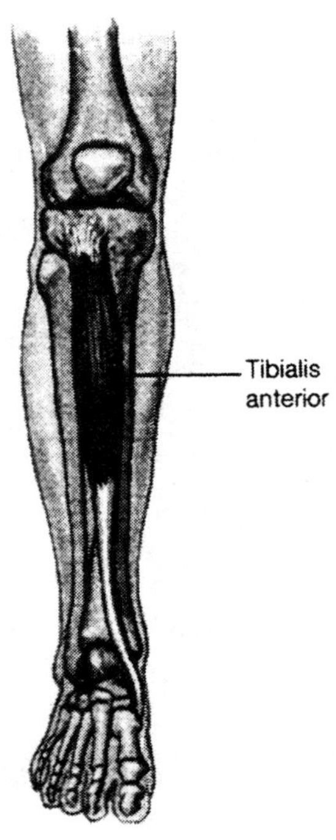

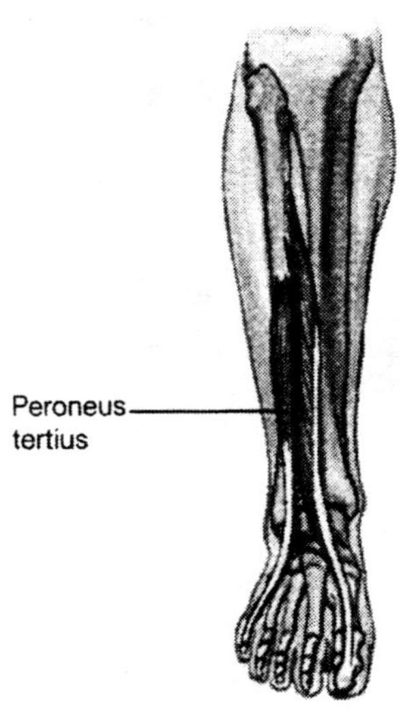

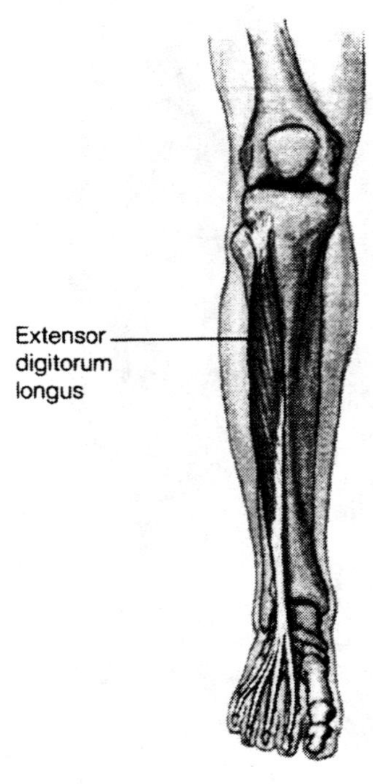

b. The plantar flexors

1) Location =

2) Names

a) **Gastrocnemius**

b) **Soleus**

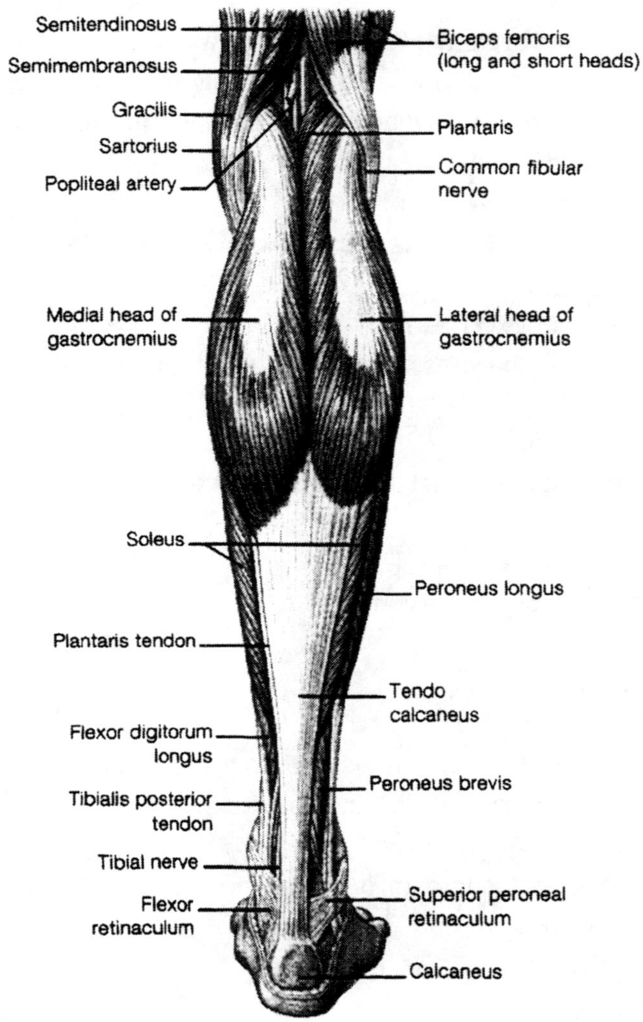

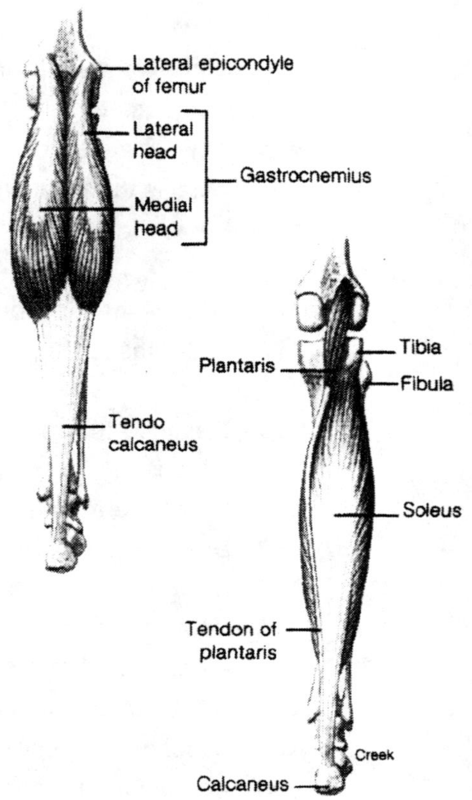

113. What are the bones of the pelvic girdle ?
114. What are the regions of the lower limb ?
115. What is the bone of the thigh ?
116. What are the two bones of the leg ?
117. What are the bones of the pes ?
118. The thigh is between what two joints ?
119. The leg is between what two joints ?
120. What is the location of the flexors at the hip ?
121. Name the two agonists for flexion at the hip.
122. What are the locations of the extensors at the hip ?
123. Name the agonist for extension at the hip that is located in the posterior pelvis
124. Name the three agonists for extension at the hip that are located in the posterior thigh. As a group, what are these three muscles called ?
125. What is the location of the flexors at the knee ?
126. Name the three agonists for flexion at the knee. As a group, what are these three muscles called ?
127. What is the location of the extensors at the knee ?
128. Name the four agonists for extension at the knee. As a group, what are these four muscles called ?
129. Describe the movement of dorsiflexion.
130. What is the location of the muscles (agonists) that perform the movement of dorsiflexion ?
131. Name the agonists for dorsiflexion.
132. Describe the movement of plantar flexion.
133. What is the location of the agonists for plantar flexion ?
134. Name the agonists for plantar flexion.
135. What does the term agonist mean ?
136. What kind of joint is both the shoulder and hip ?
137. Describe the similarities of bone content and structure between the arm and thigh. Between the forearm and leg. Between the manus (hand) and pes (foot).
138. In the leg, which bone is lateral, the tibia, or the fibula ?
139. In the anatomical position, which bone of the forearm is lateral, the radius, or the ulna ?
140. What type of joint is both the elbow and knee ?
141. How many carpals are there in each manus ?
142. How many tarsals are there in each pes ?
143. There are seven tarsals and eight carpals. Nevertheless, the number of bones in the lower limb = number of bones in the upper limb. How can this be ?

144 - 182 : For each of the following muscles use **AA** if it is located in the anterior arm, **PA** if in the posterior arm, **AF** if in the anterior forearm, **PF** if in the posterior forearm, **AP** if in the anterior pelvis, **PP** if in the posterior pelvis, **AT** if in the anterior thigh, **PT** if in the posterior thigh, **MT** if in the medial thigh, **AL** if in the anterior leg, **PL** if in the posterior leg, or a **D** if on the dorsum of the trunk:

Biceps brachii ____	**Psoas ____**	**Triceps brachii ____**
Hamstrings ______	**Soleus ____**	**Flexor carpi radialis ____**
Gastrocnemius____	**Trapezius ____**	**Gracilis ____**
Semimembranosus____	**Palmaris longus____**	**Teres major ____**
Iliacus ____	**Biceps femoris ____**	**Brachialis ____**
Flexor digitorum profundus____	**Teres minor ____**	**Vastus medialis ____**
Vastus lateralis ____	**Rectus femoris ____**	**Extensor digitorum ____**
Adductor longus ____	**Adductor magnus ____**	**Extensor carpi radialis longus ____**
Erector spinae ____	**Tibialis anterior ____**	**Quadriceps femoris ____**
Gluteus maximus ____	**Flexor carpi ulnaris____**	**Semimembranosus ____**
Extensor digitorum longus ____	**Gluteus medius ____**	**Gluteus minimus ____**
Extensor carpi radialis brevis ____	**Vastus intermedius ____**	**Adductor brevis ____**
	Flexor digitorum superficialis____	**Latissimus dorsi ____**
	Extensor carpi ulnaris ____	

183 - 220 : For each of the muscles listed mark an **A** if it is a flexor at the shoulder, **B** if it is an extensor at the shoulder, **C** if it is an abductor at the shoulder, **D** if it is an adductor at the shoulder, **E** if it is a flexor at the elbow, **AB** if it the extensor at the shoulder, **AC** if it is a flexor at the wrist only, **AD** if it is a flexor at the wrist and of the digits, **AE** if it is an extensor at the wrist only, **BC** if it is an extensor at the wrist and of the digits, **BD** if it a flexor at the hip, **BE** if it an extensor at the hip, **CD** if it is an adductor at the hip, **CE** if it is a flexor at the knee, **DE** if it is an extensor at the knee, **ABC** if it is a dorsiflexor, or **ABD** if it is a plantar flexor:

Biceps brachii____, Psoas____, Semitendinosus____, _____, Soleus _____, Extensor carpi radialis longus____, Gluteus maximus____, Palmaris longus ____, Rectus femoris ____, Flexor digitorum superficialis ____, Biceps femoris ____, ____, Extensor digitorum ____, Extensor digitorum longus ____ , Lateral deltoid ____, Tibialis anterior ____, Gluteus medius ____, Vastus medialis ____, Pectoralis major ____, Flexor carpi ulnaris ____, Triceps brachii ____, Gastrocnemius ____, Anterior deltoid ____, Vastus intermedius ____, Adductor magnus ___, Flexor digitorum profundus ____, Coracobrachialis___, Vastus lateralis ____, Brachialis ____, Extensor carpi radialis brevis ____, Posterior deltoid ___, Flexor carpi radialis ____, Adductor longus ____, Gracilis ____, Extensor carpi ulnaris ____, Semimembranosus ____, ____ , Iliacus ____, Hamstrings ____, ____, Quadriceps femoris ____

OPEN SYSTEMS THAT ALLOW FOR EXCHANGE OF SUBSTANCES BETWEEN THE ENVIRONMENT AND THE BLOOD

I. General Introduction

Let's go back in history to when you were a **zygote**. The zygote is a cell and as such it requires oxygen (O_2), nutrients (glucose, amino acids, fats, vitamins, minerals, etc.) , and there are chemical reactions (metabolism) going on within it that generate carbon dioxide (CO_2). CO_2 is a toxin (poison) to any cell and the cell must get rid of it. In the zygote, as in all other cells, CO_2 simply moves across the plasma membrane and out into the cell's environment. The environment of the zygote, you probably remember, is the space of the fallopian tube. It is also this environment from which O_2 can be directly obtained by the zygote. Nutrients are already present in the egg before fertilization and therefore the zygote has a sufficient amount of food.

What is true for the zygote is also true for the cells of the immediately subsequent developmental stages, and O_2 moves into the cell and CO_2 moves out of the cells by moving across the plasma membranes of the cells in the 2 - cell, 4 - cell, 8 - cell, and on up to the blastocyst. At the blastocyst stage, however, because of implantation, the environment abruptly changes and becomes the tissue and blood of the stratum functionalis of the endometrium. Now, the cells of the individual must obtain their O_2 from the surrounding tissue and blood, and CO_2 must be given up to this tissue. Nutrients, which were sufficiently present in the egg and zygote are almost depleted by the stage of implantation and then nutrients must also be obtained from the new environment in order to support the life, growth, and development of the cells of the developing individual. So, it is certainly true that the cells in the developing individual are no different in their basic needs than are the cells of the adult. This turns out to be an important consideration because of the changing environments that occur during development and the drastic change in environments that occurs at birth. These environments are important and we have to keep up with them.

For most of development between the zygote and the blastocyst the cells are in direct contact with their environment which does allow for direct exchange of O_2 and CO_2 between cells and environment. But think about your own body and notice that the vast, vast majority of your 100 trillion cells are not on your surface, but are located interiorly and can not exchange chemicals directly with the environment. After the formation of the blastocyst subsequent development begins to form an individual with cells pushed so far interiorly that they cannot exchange chemicals directly with the environment. Nevertheless, these interior cells are absolutely no different in their requirement for obtaining O_2 and nutrients and in getting rid of CO_2 than are those surface cells which are in direct contact with the environment. Blood vessels and blood cells therefore are developed in the early embryo so that O_2 and nutrients can be delivered to the interior cells, and so that CO_2 can be picked up when it moves out of the cells. These blood vessels and the blood flowing in them are collectively termed **fetal circulation**.

Now the questions arise as to where does the blood in fetal circulation get its O_2 and nutrients, and how does the fetal blood get rid of the CO_2 that comes into it from the cells ? After all, in order to continue to grow and divide, the cells of the embryo and fetus must continuously absorb and use the O_2 and nutrients, so how is it that more O_2 and nutrients get into fetal circulation ? Equally important is that CO_2 simply can not keep accumulating in the blood of the embryo and fetus. The CO_2 must be eliminated, so how is this accomplished ? The solution to these questions and problems is the **PLACENTA**, which is a joint structure developed by Mom and her embryo to allow easy exchange of chemicals between **maternal circulation** (Mom) and fetal circulation (embryo / fetus).

As development proceeds there are of course more and more cells produced by cell division, and the process of differentiation produces different types of cells (bone, muscle, nerve, epithelial cells, etc.). Tissues, organs, and systems eventually begin to form and undergo their developments. A **circulatory system** develops quite early and its function is to transport various chemicals to various parts of the fetus; **reproductive structures** develop but these truly will not be needed and used until several years following birth; the two systems of support and movement (**skeletal and muscular systems**) are developed also, but these, too, are not especially useful before birth, although in late pregnancy most Moms - To - Be can certainly feel the results of muscular contractions in the fetus' muscular system. The skeletal system is at first made mainly of cartilage, but over a period of years following birth the flexible cartilages that make up the "bones" are transformed into rigid bone and then the skeletal system truly becomes a system of support. The **nervous system** develops also, but it also is not essentially needed until after birth. There is little, perhaps nothing, for the developing embryo and fetus to see, or smell and taste while in the uterus, but nevertheless the **eyes, nose, and tongue** are developed before birth. Sound waves can travel very well through solid objects and so it could be that a fetus is able to detect loud sounds from outside the body. Therefore, the circulatory system is needed during embryonic and fetal development, but other systems / structures (reproductive, skeletal, muscular, nervous systems, the eye, ear, nose, and tongue), although they may be at least partially developed, are for functions that are not basically needed in the fetus, but are needed following birth (sexual activity, support, movement, coordination of bodily functions, seeing, hearing, smelling, tasting, etc.).

In addition to the above activities, the adult body has need to : **1)** Obtain O_2 for its cells (done by the **RESPIRATORY SYSTEM**); **2)** eliminate CO_2 from the blood (also done by the **RESPIRATORY SYSTEM**); **3)** obtain nutrients to support cellular metabolism, growth, and division (done by the **DIGESTIVE SYSTEM**); **4)** eliminate the toxin **UREA** from the body (done by the **URINARY SYSTEM**).

But when we look at the embryo and fetus we see that the embryo and fetus **also** have the same needs to obtain O_2, eliminate CO_2, obtain nutrients, and to eliminate urea. These are cellular functions and therefore are essential and necessary both before and after birth.

However, a problem is noticed right away because early in development the embryo / fetus changes its environment again by creating a membrane, called the **amniotic membrane**, around itself. What this membrane does is to form a sac around the embryo and fetus which becomes filled with a fluid, the **amniotic fluid.** This fluid becomes the immediate environment of the embryo and fetus. They are submerged in a fluid, but the adult is submerged in air. Two more different media would be difficult to imagine. There are O_2 and nutrients out there in the environment of the adult (air), but neither of these are found in the amniotic fluid. **Can you see the problems that being "under water" presents to the respiratory, digestive, and urinary systems ?** There is no way for the embryo / fetus to breathe and perform the O_2 / CO_2 exchange in the manner that we adults do, there is nothing in the amniotic fluid to eat, and the formation of urine to eliminate urea cannot be done from the fetal body (Think about it. If urine **could** be formed and eliminated from the fetal body, then it would have to be added into the already existing amniotic fluid. Let's face it, no one wants this to happen, and, in addition, on a more practical level, Mom would, over a nine month period, attain a truly enormous size !!!)

If you stop to think about these three systems in you as an adult, then you notice that each has an intimate contact with your circulating **blood**, and indeed these systems exchange O_2 / CO_2 with the **blood** and the alveoli of the lungs (**RESPIRATORY SYSTEM**), absorb nutrients from the intestinal tract into the **blood** (**DIGESTIVE SYSTEM**). and eliminate urea from the **blood** into the nephrons of the kidney (**URINARY SYSTEM**).

So, once more we are led to the question of if, as in the adult, O_2, nutrients, CO_2, and urea are necessarily in the fetal blood, then how did they get there ? The answer is that O_2 and nutrients enter fetal circulation from maternal circulation through the placenta. CO_2 enters fetal circulation from all cells of the fetus, and urea enters fetal circulation from the cells of the fetal liver. CO_2 and urea are eliminated from the blood of fetal circulation by passing from fetal blood to maternal blood through the placenta. These answers to the question bring us to this conclusion`:

OBVIOUSLY, FOR THE FETUS, THE PLACENTA BECOMES THE RESPIRATORY, DIGESTIVE, AND URINARY SYSTEMS.

1. What are the requirements of the zygote ?
2. How do oxygen and carbon dioxide get into the cell ?
3. What are the chemical reactions of the cell called ?
4. What is the environment of the zygote to blastocyst stages of development ?
5. What becomes the environment of the blastocyst ?
6. What is the problem of the interior cells of the body ?
7. How is this problem solved ?
8. What is fetal circulation ?
9. What must any cell, whether in the embryo, fetus, or adult have to do to remain alive, grow, and divide ?

10. How do the O_2 and nutrients in fetal blood and circulation get there ?

11. Where does the CO_2 in the fetal blood come from ?

12. How does CO_2 get out of fetal blood ?

13. What is the placenta, and what is its function ?

14. List the systems and structures that are at least partially developed during fetal development, but whose functions basically are not needed until after birth.

15. Correlate the systems and structures that you listed in question 14 with their functions / activities.

16. Can you see that the activities / functions that you correlated in question 15 really are those not needed until after birth ?

17. In addition to the functions / activities that you listed in question 15, what are four other activities pertaining to blood and cellular functions that the adult must perform ?

18. Correlate these four functions in question 17 with the three systems that are responsible for performing them.

19. Are these four functions also required in the fetal stages of development ?

20. What is the environment of the embryo and fetus ?

21. What is the name of the membrane that forms a sac around the embryo / fetus ?

22. What is the name of the fluid that the embryo / fetus is submerged in ?

23. What is the effect of being submerged in amniotic fluid on the functioning of the respiratory, digestive, and urinary systems in the fetus ?

24. Why can the respiratory system not function while the fetus is in a fluid ?

25. Why can the digestive system not function while the fetus is in fluid ?

26. Why can the urinary system not function while the fetus is in fluid ?

27. Can you see that the respiratory system, the digestive system, and the urinary system are involved in functions and activities that are occuring in both the fetus and the adult ?

28. With what do these three systems have intimate contact ?

29. What structure takes over the functioning of the respiratory, digestive, and urinary systems for the fetus ?

30. The placenta is a connector between what two circulations ?

31. Who makes the placenta ?

II. The Respiratory System

A. General information and the purpose of the respiratory system

The best place to start with the respiratory system is to simply state that all living cells have a critical and constant need for O_2 and also a critical and constant need to get rid of the CO_2 within them. Let's start with the problems that CO_2 present not only to the cells that make it, but to the entire body. CO_2 is generated inside all living cells as a perfectly normal byproduct of normal metabolism of the cell. Nevertheless, CO_2 is toxic to a cell and must be eliminated from the cell. This is easily done because CO_2 readily moves across the plasma membrane and out of the cell and into the blood flowing in capillaries around the cell .

Moving the CO_2 out of the cell and into the blood solves the toxicity problem for the cell, but in turn it creates a problem for the blood. Having too much CO_2 in the blood causes the blood to become more acid, a situation (respiratory acidosis) that is actually very life-threatening. Therefore, CO_2 must be removed from the blood as soon as possible (ASAP!) and this is accomplished by having all blood in the body eventually flow through the lungs and around the alveoli (air sacs). You must realize and thoroughly comprehend at this point that the transport of CO_2 from a muscle cell in the gastrocnemius, or from a nerve cell in the brain, or from any cell in any part of the body , to the capillaries around the alveoli is done by the circulatory system, not the respiratory system.

However, the CO_2 that is in the blood of the capillaries flowing around the alveoli moves out of the blood and into the alveoli of the lungs. The blood therefore loses its CO_2 to the respiratory system, which is good because with this loss respiratory acidosis is prevented.

If the individual is breathing and has taken in a breath of air, then then alveoli are full of O_2. In addition to losing its CO_2 into the alveoli, the blood picks up O_2 from the alveoli. This oxygenated blood flows on out of the lungs, back to the heart and then from there on out to the cells of the body. Once again, the transport of O_2 from the alveoli of the lungs out to the cells of the body is performed by the circulatory system, not the respiratory system.

So, what are the functions of the respiratory system ? There are two : **1)** To transport O_2 from the atmosphere to the alveoli of the lungs (done by the process of inhalation) so that it can be absorbed into the blood; **2)** To absorb the CO_2 from the blood and transport it from the alveoli out of the body and into the atmosphere (done by the process of exhalation).

The respiratory system is constructed precisely for the purpose of allowing movement of air from the atmosphere to the alveoli of the lungs (by using skeletal muscles to cause inspiration and the movement of air into the lungs) and of course allowing movement of air from the alveoli back into the atmosphere (by relaxing those muscles of inspiration to cause expiration). The difference in the air going in (inspired air)versus the air coming out (expired air) is that the inspired air contains O_2 and the expired air contains CO_2.

If we trace a molecule of $O2$ from the atmosphere to an alveolus of the lung, then we see that it passes through either the **nasal cavity** (nose) or the **oral cavity** (mouth), the **nasopharynx, oropharynx, laryngopharynx** ("voicebox"), **the trachea** ("windpipe"), **primary bronchus. secondary bronchus, tertiary bronchus, bronchiole, terminal bronchiole, respiratory bronchiole, alveolar duct,** and then finally into an **alveolus**.

These structures are listed on a one to one basis, but that is not the way it really is in the body. There is one each of a nasal cavity, oral cavity , nasopharynx, oropharynx, larynx, and trachea. After this, structures become plural. There are two primary bronchi, several secondary bronchi, several more tertiary bronchi, perhaps several thousands of bronchioles, many more terminal bronchioles, even more respiratory bronchioles, and then even more alveolar ducts. There are more than 300 million alveoli in each lung. It's a general priciple with our systems that contain tubules and tubules, including the respiratory system, that the tubes and tubules become more numerous and smaller.

Your respiratory system can conveniently be divided into an **UPPER RESPIRATORY TRACT** and a **LOWER RESPIRATORY TRACT**. The upper respiratory tract is composed of the **nasal** and **oral cavities, nasopharynx, oropharynx, laryngopharynx**, and the **larynx.** Structures in the lower respiratory tract include the **trachea, primary bronchi, secondary bronchi, tertiary bronchi, bronchioles, terminal bronchioles, respiratory bronchioles, alveolar ducts, alveoli,** and **other structures of the lungs,** including the **pleurae.**

There are of course several common problems that can occur in your respiratory system. These problems can be, once again, conveniently categorized as either upper respiratory tract problems, or lower respiratory tract problems. Some common upper respiratory tract problems include such nusances as rhinnitis ("runny nose"), nasal polyps, sore throat (pharyngitis), tonsilitis (tonsils are located in the pharynx), and laryngitis. Some common lower respiratory tract problems include bronchitis, asthma, and pleurisy (inflammation of the pleurae). Most of these listed problems are tissue inflammation in response to overwhelming bacterial infections.

B. The upper respiratory tract

1. Structures

a. Nasal cavity

b. Oral cavity

c. Pharynx

1) Upper region =

2) Middle region =

3) Lower region =

d. Larynx

1) Cartilages

a) Thyroid
b) Epiglottis
c) Cricoid
d) Arytenoids
e) Corniculates
f) Cuneiforms

2) True vocal cords
3) False vocal cords

2. General features

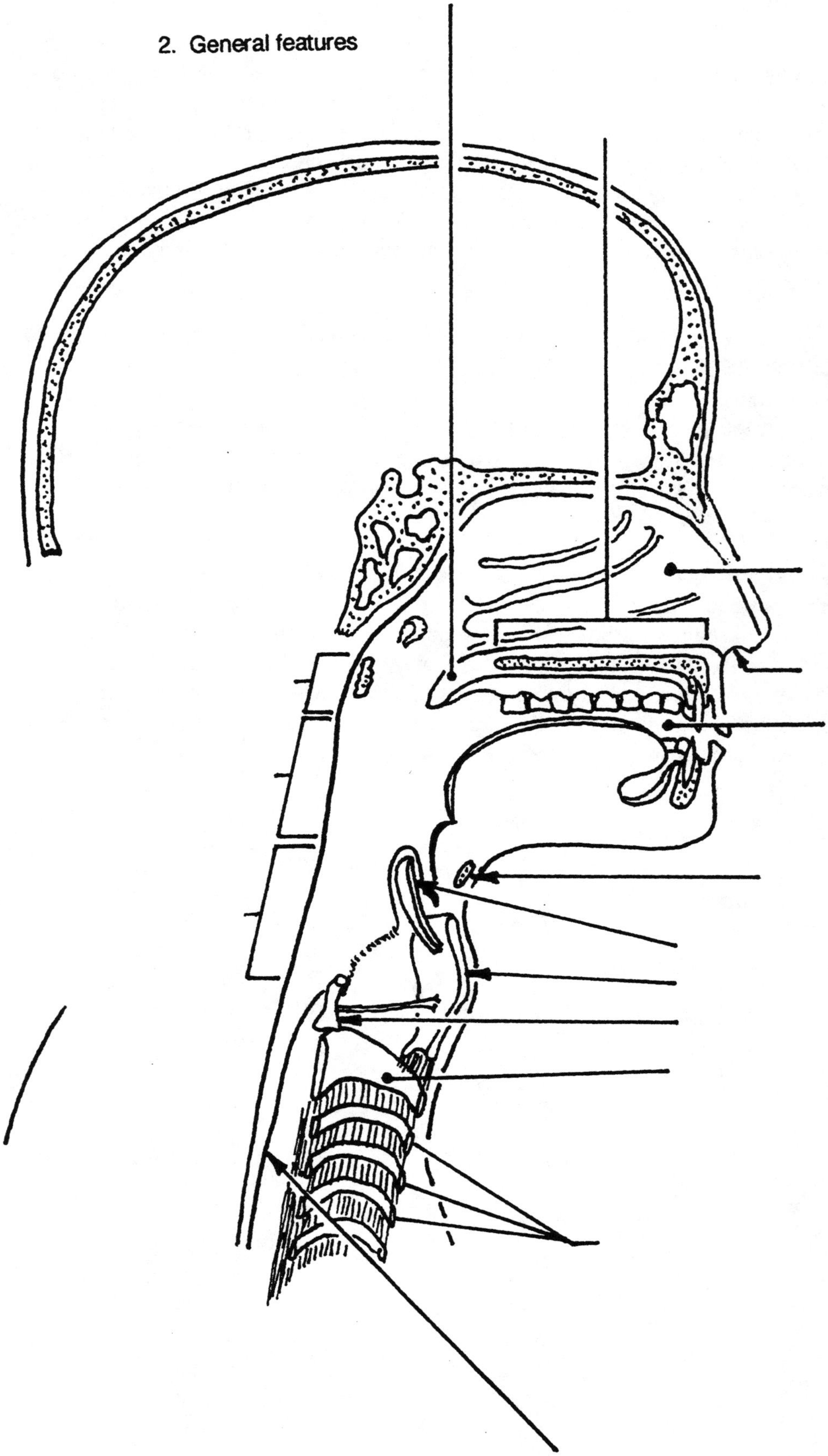

3. Paranasal sinuses

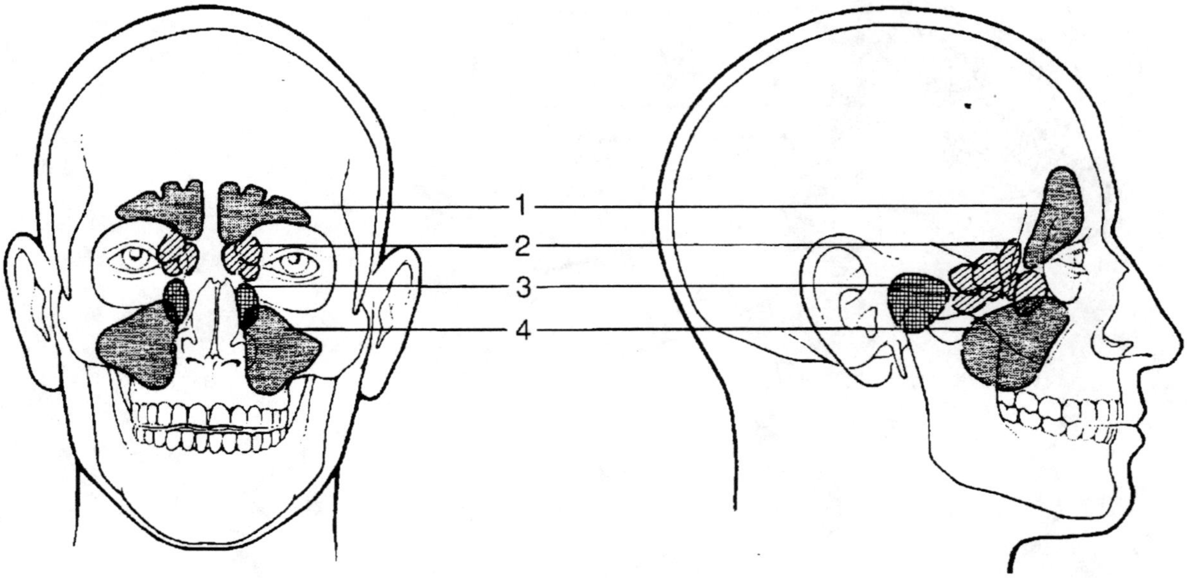

4. The larynx

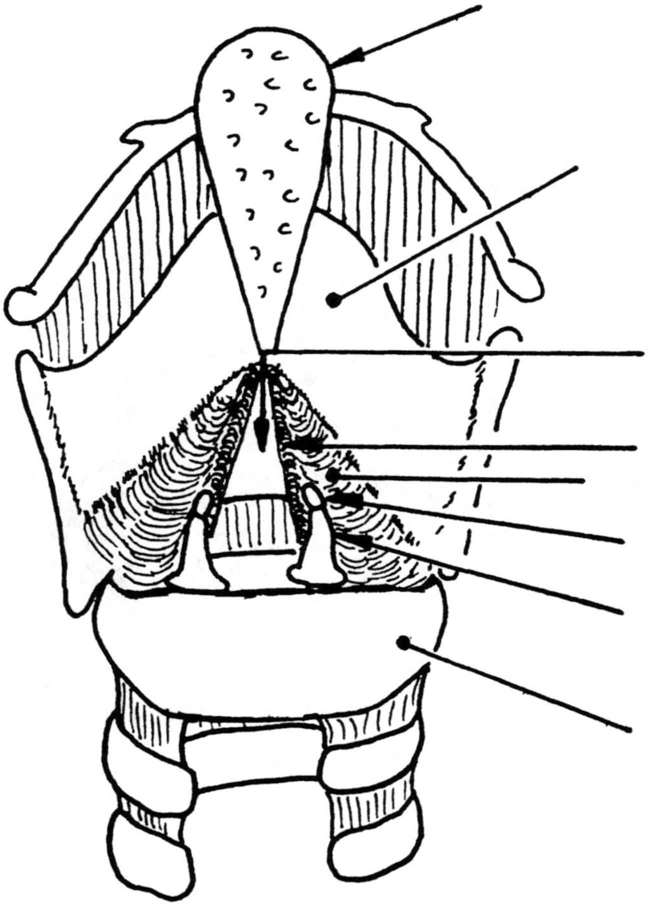

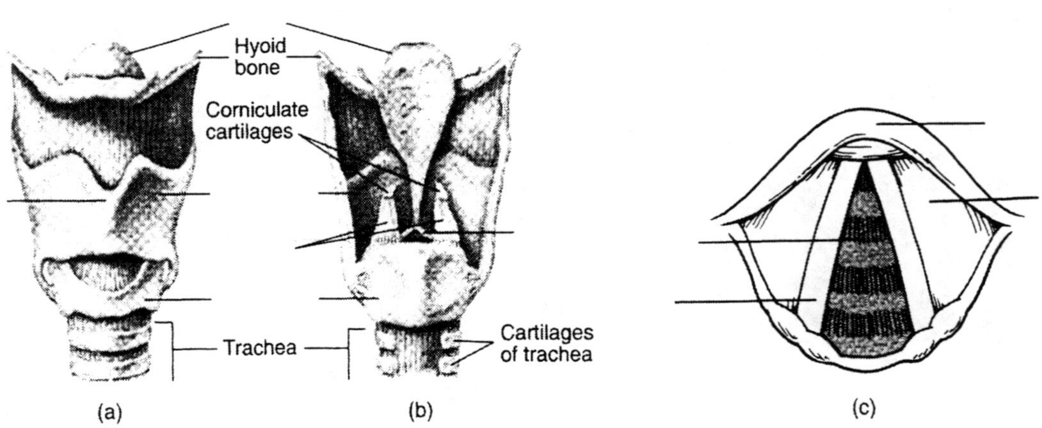

(a)

(b)

(c)

C. The lower respiratory tract

1. The trachea and bronchi

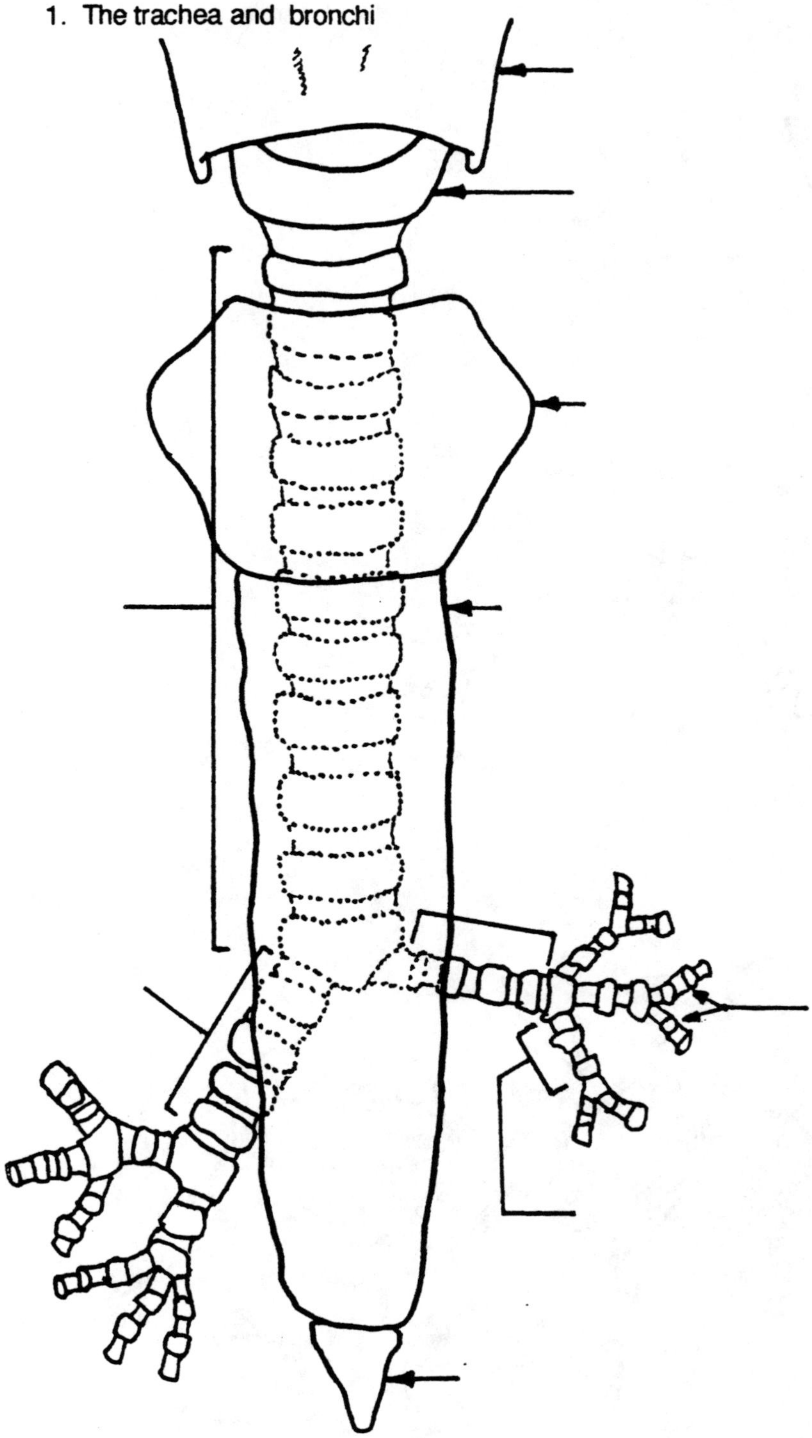

2. The bronchial tree

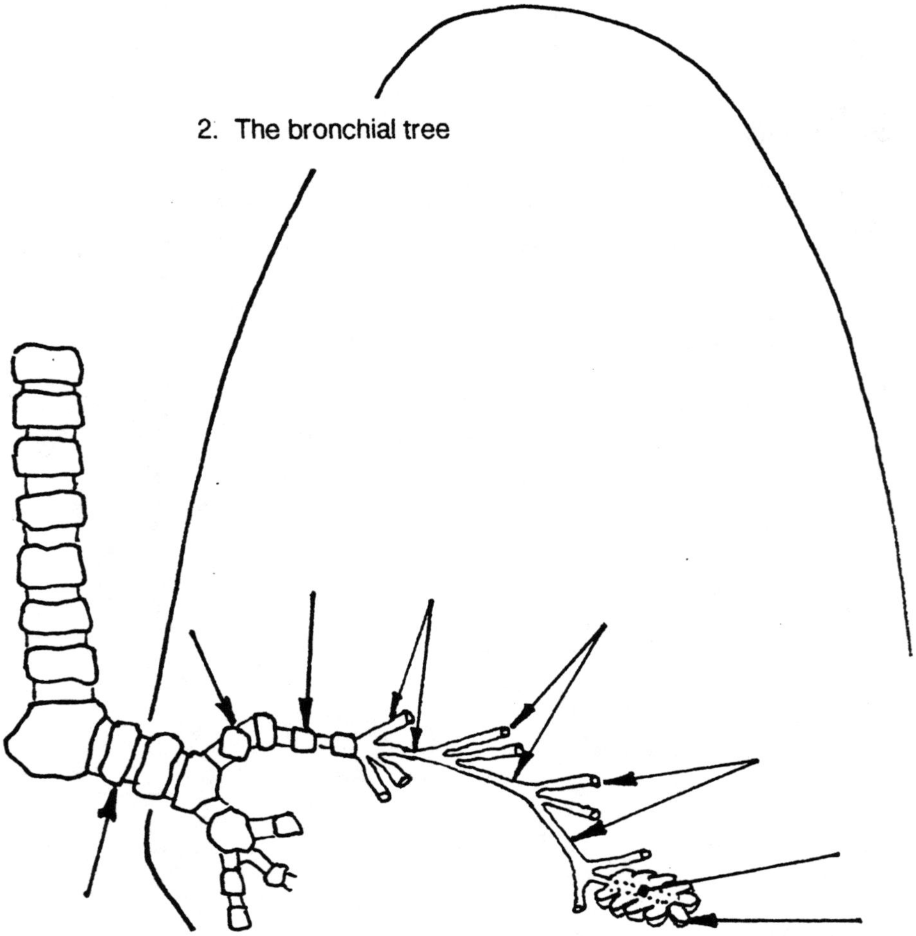

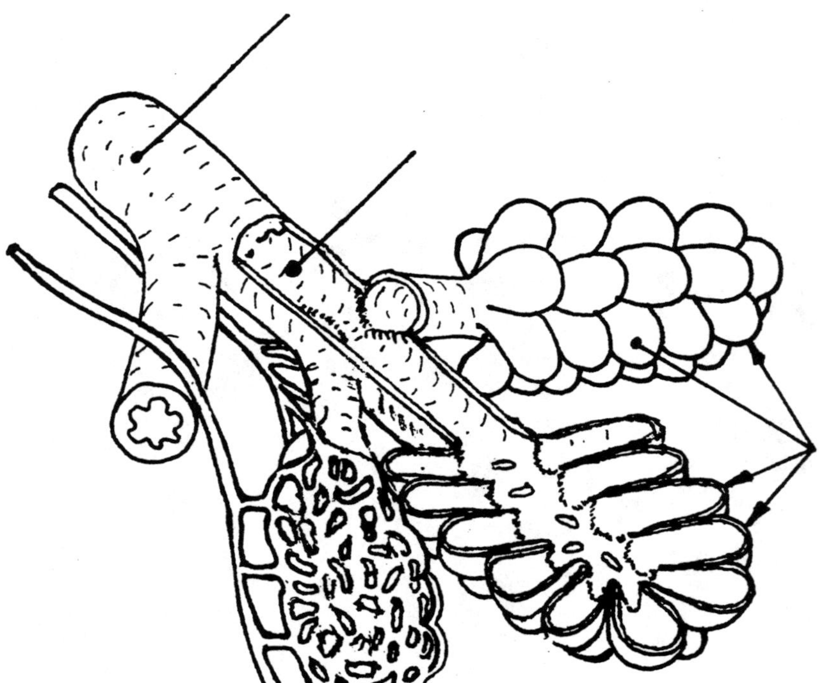

3. The gross structure of the lungs

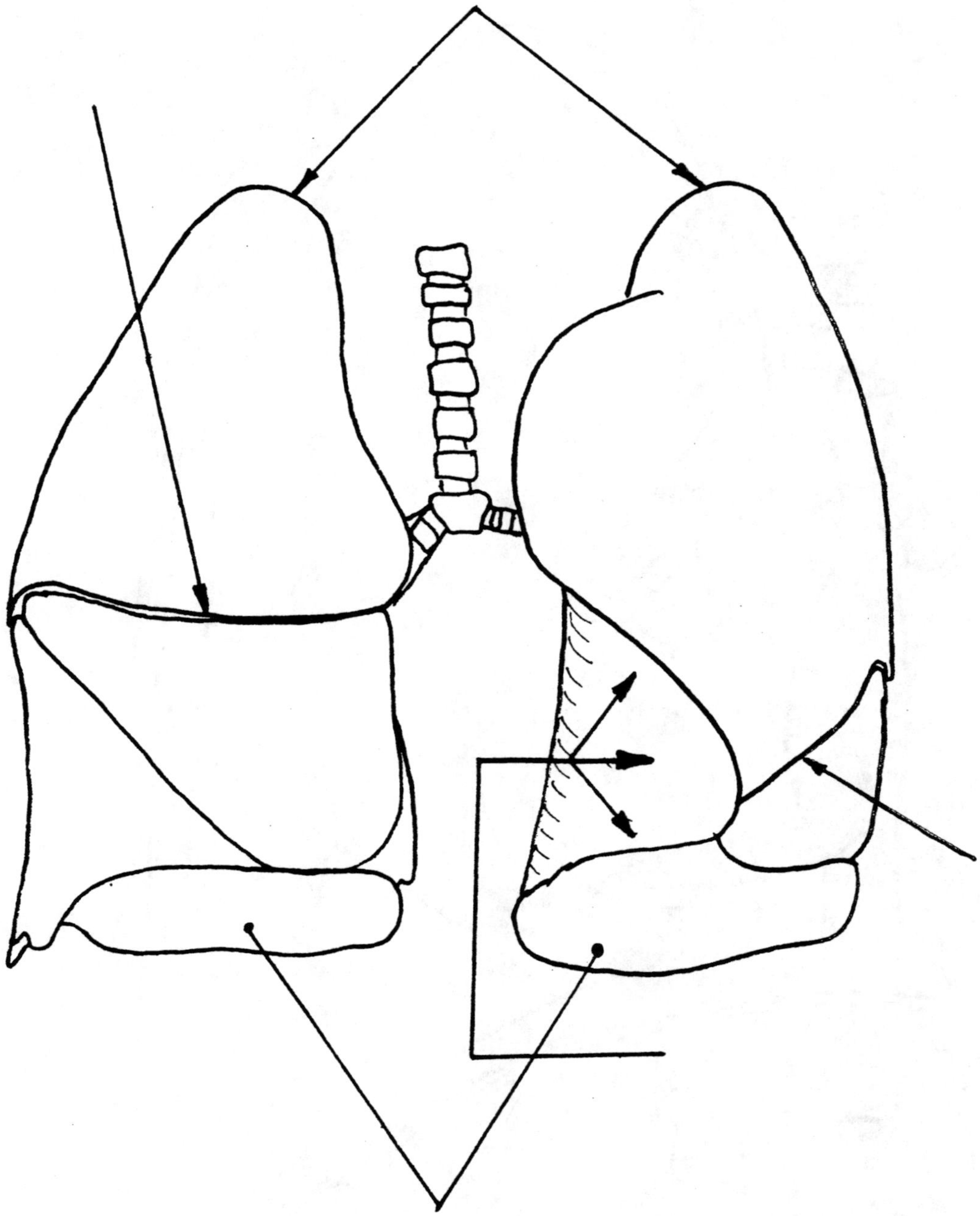

4. The parietal and visceral pleurae

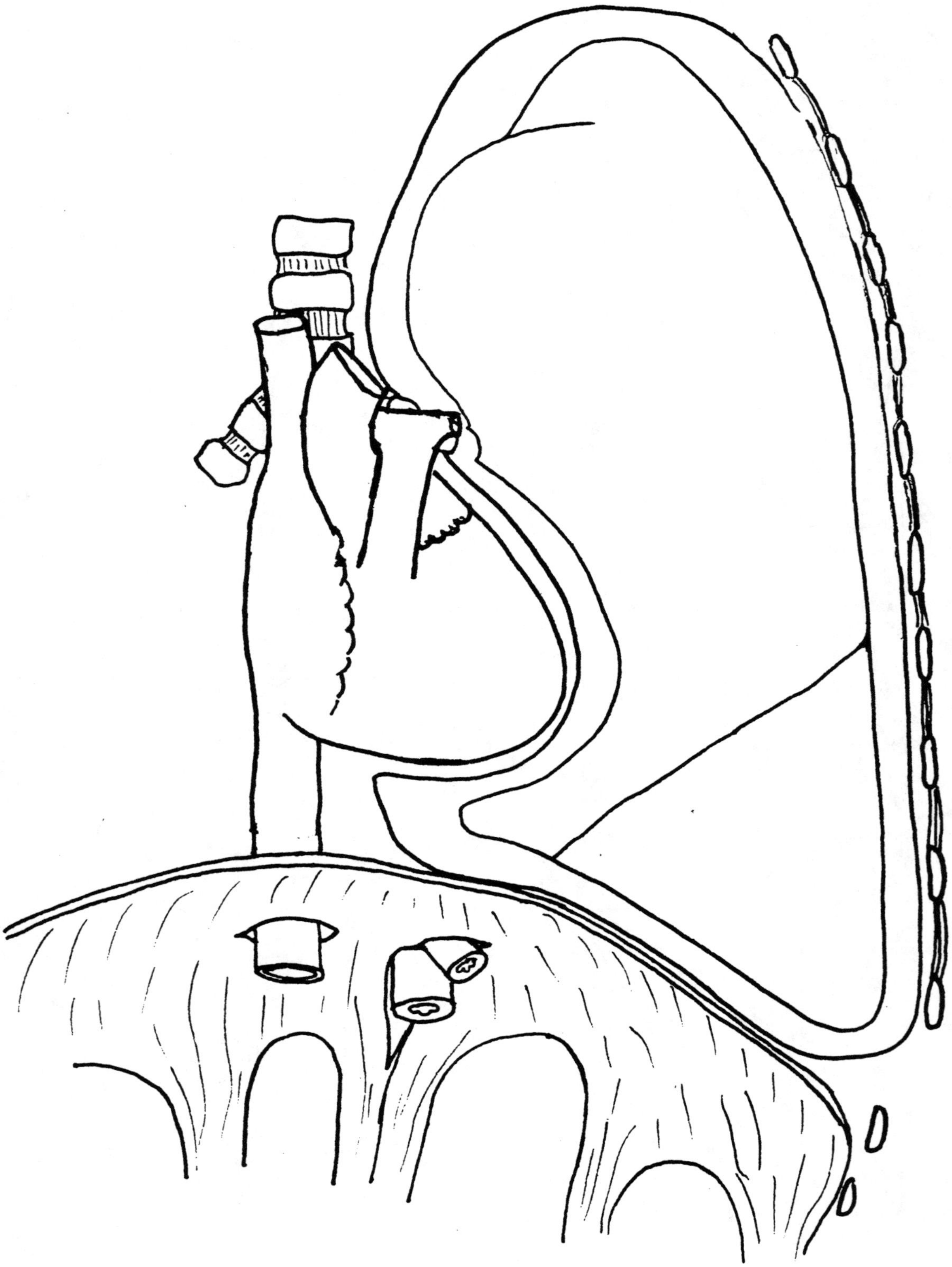

D. Muscles of inspiration and expiration

1. Muscles of inspiration

a. Contracting these muscles INCREASES the volume of the pleural cavities

b. Muscles involved in quiet breathing

1)

2)

c. Muscles involved in forced inspiration

1)

2)

3)

4)

2. Muscles involved in expiration

a. Normal quiet expiration requires only that the muscles of inspiration be relaxed

b. Muscles involved in forced expiration

1)

2)

32. As far as the respiratory system is concerned, what are the two critical and constant needs of all living cells ?
33. What is generated in all living cells as a byproduct of normal metabolism ?
34. How does CO_2 get out of the cell ?
35. Into what does CO_2 go when it passes across the plasma membrane of the cell ?
36. Why does CO_2 have to be removed from the cell ?
37. What problem does having too much CO_2 in the blood cause for blood ?
38. What is the name of this problem ?
39. How is CO_2 removed from the blood ?
40. What does removing CO_2 from the blood prevent ?
41. How does O_2 get into the blood ?
42. Which system transports CO_2 from the cells to the alveoli, and transports O_2 from the alveoli to the cells ?

43. What are the two functions of the respiratory system ?
44. Trace a molecule of O_2 from the atmosphere to an alveolus.
45. List in order the structures of the respiratory system according to their numbers.
46. About how many alveoli are there in each lung ?
47. What is the general principle concerning tubule size and numbers ?
48. What are the general structures of the upper respiratory tract ?
49. What are the general structures of the lower respiratory tract ?
50. What are some common problems that occur in the upper respiratory tract ?
51. What are some common problems in the lower respiratory tract ?
52. List the cartilages of the larynx.
53. Which cartilages are paired ? Which are unpaired ?
54. Which of the laryngeal cartilages can you feel on yourself ?
55. Where are tonsils located ?
56. Where is the ostium (opening) of the auditory tube located ?
57. What are the cartilages that sit on top of the arytenoids ?
58. The vocal cords are connected to what laryngeal cartilages ?
59. The opening between vocal cords is called what ?
60. What does the term epiglottis mean ?
61. What are C-rings ?
62. In which of the respiratory structures are C-rings located ?
63. Which structures of the bronchial tree do not have C-rings ?
64. At what level of the sternum does the sternum end ?
65. What is the name of the branch point of the trachea ?
66. What does the trachea branch into ?
67. What are some differences between the right and left primary bronchi ?
68. If someone aspirated an object and it got past the glottis and went down the trachea, then which primary bronchus would the object most likely become lodged in ? Why ?
69. What are the cavities in which the lungs are located ?
70. What are the pleurae ?
71. What is the difference between the parietal and visceral pleurae ?
72. What are the similarities in the gross structures of the right and left lungs ?
73. What are some differences in the gross structures of the right and left lungs ?
74. What is the purpose of the cardiac notch on the left lung ?
75. For each of the following structures write a **U** if located in the upper respiratory tract, or an **L** if in the lower respiratory tract: **Alveolar ducts ____; Nasal cavity ____; Parietal pleura____; Oropharynx____; Terminal bronchioles____; Glottis____; Epiglottis____; C-rings____; Cuneiforms____; Base____; Thyroid cartilage____; Laryngopharynx____; Oral cavity____; Apex____; Carina____; Arytenoids____; Middle lobe____; Alveoli____; Corniculates____; True vocal cords____; Horizontal and oblique fissures____; Visceral pleurae____; Bronchioles____; Nasopharynx_____; Cricoid____; Secondary bronchi____; Tertiary bronchi____; Respiratory bronchioles___.**

III. The Digestive System

A. General information and the purpose of the digestive system

Being hungry and in need of calories is a common experience. Physiologically, the process of being hungry is complicated and the basic reasons for the sensation of hunger are not completely understood. In our society, perhaps in all, we try to "satisfy our hunger" three / four times a day. Yet, being the social animals that we are, we humans attach and associate all sorts of other non-caloric activities and meanings to the activity of caloric intake (eating). Few of us enjoy eating alone, and so we have romantic dinners ("wine and dine"), business luncheons ("power lunches"), dinner parties, wine and cheese parties, birthday parties, banquets and receptions for all sorts of emotional occasions (retirements, marriages, various honors, etc.), beer and drinks with friends and co-workers (there are a lot of calories in alcohol), lunch with friends, etc. Restaurants and fast-food establishments abound and do a great business. They don't exist for nothing; they exist because we like to satisfy some aspects of our non-caloric needs along with our caloric needs. And it is interesting to note that one of the first things parents notice and lament when their children get older is that the family rarely "has a meal together anymore". With few exceptions, our social dining is for happy occasions, and we have to stop and wonder what the connection is between food and some of our emotional needs, and to our needs to be with others. So, as with sexual activity, there seems to be more to the act of getting caloric nourishment than simply satisfying the physiological needs of our bodies.

The emotional and social aspects of eating aside, let's admit that your body has a nutritional / caloric/ physiological need for food. The reason for your physiological need is due to the fact that the 100 trillion or so cells in your body have used up the the nutrients that you gave them during the previous meal (or snack). The nutrients in the food you eat are necessary for your cells to make energy (ATP), to grow, to divide. Without nutrients, your cells will eventually die. How the cells in your body let you become consciously aware that they are in need of more nutrients (that is, that you are hungry) is an interesting physiologically phenomonon. There are numerous theories for how it occurs, but none actually offers a satisfactory explanation.

The <u>nutrients</u> that your cells need and can absorb from the blood and use are simple carbohydrates (sugars), such as **glucose, amino acids, fatty acids, vitamins, and some minerals.** However, the nutrients that you take into your body from the environment are in the form of **complex carbohydrates** (starch and cellulose from plants, and glycogen from animals), **proteins** (linear forms of amino acids linked together), and **triglycerides** (a form of lipid, or fat, that contains three fatty acids). These forms of nutrients cannot be used by the cells of your body, and so these forms must be broken down into glucose, amino acids, and fatty acids. The glucose is used by your cells as a source of energy to make **ATP**. The amino acids are reconnected to make protein molecules that go into the structure of membranes, cellular organelles, or to become enzymes, and the fatty acids are used by your cells to make membranes, and also as a source of energy.

Once again we must recognize that the delivery of usable nutrients (glucose, amino acids, fatty acids) to your cells is done by your circulatory system, <u>**NOT**</u> your digestive system.

Your digestive system has two important functions to perform : **1)** To enzymatically break down complex carbohydrates into glucose, proteins into amino acids, and triglycerides into fatty acids; **2)** To provide an anatomically intimate contact with the circulatory system (blood and lymphatic circulations) so that glucose, amino acids, and fatty acids can be absorbed out of the digestive system and into the circulatory system.

Anatomically, the digestive system is composed of a long (about 25 feet) tube, called the **ALIMENTARY CANAL**, through which food passes while the nutrients are enzymatically degraded by **ENZYMES** which are made and secreted by some of the **ACCESSORY DIGESTIVE ORGANS.** Let's take a quick trip through this tube, from one end to the other.

Food enters the **oral cavity** where it is chewed by teeth and moistened by saliva from the salivary glands. Partial digestion of complex carbohydrates (starch and glycogen only; you do not have the enzymes to degrade cellulose) occur in the mouth thanks to the enzyme amylase, which is in saliva. Teeth and the salivary glands are the accessory digestive organs of the oral cavity. The swallowing mechanism, rapidly and with much force, pushes food backward into the **oropharynx, laryngopharynx, and into the esophagus**, the muscular tube between the laryngopharynx and the stomach. Food and drink will remain in your **stomach** for 2 - 4 hours and during this period mechanical action by the smooth muscle of the stomach, secretion of hydrochloric acid (HCl), and partial digestion of proteins by the enzyme pepsin create a liquid form of food called chyme. Afterwards, chyme passes from the stomach into the **duodenum of the small intestine** where digestion of sugars, proteins and triglycerides is completed. The enzymes needed for digestion are made by the pancreas. Bile, which is not an enzyme, but is needed to solubilize fats in the duodenum so that they can be absorbed, is made continuously by the liver and stored in the gall bladder between meals. Therefore, the liver, gall bladder, and pancreas are the accessory digestive organs of the duodenum. Water soluble nutrients (glucose, amino acids) are absorbed across the wall of the villi of the duodenum, jejunum, and ileum and into the blood where they are transported directly to the liver. Water insoluble nutrients, such as the fatty acids, are absorbed into the lymphatic circulation. From your **ileum** of the small intestine, undigested ingredients, such as the cellulose in your diet, pass into your large intestine, first into the **cecum**, then the **ascending colon**, the **transverse colon**, the **descending colon**, and then the **sigmoid colon**. The large intestine has the function of feces formation. Also some of the fluid in your large intestine is reabsorbed back into your body. There are many situations that will interfere with the correct amount of water reabsorption, including bacterial and viral infections of the large intestine, some drugs, your diet, etc. Not reabsorbing an adequate amount of water (fluid) will lead to diarrhea, while reabsorbing too much leads to constipation. From the sigmoid colon, feces moves out of large intestine and into the **rectum**, then into the **anal canal**, and finally out of the body through the **anus**.

B. General anatomy of the digestive system

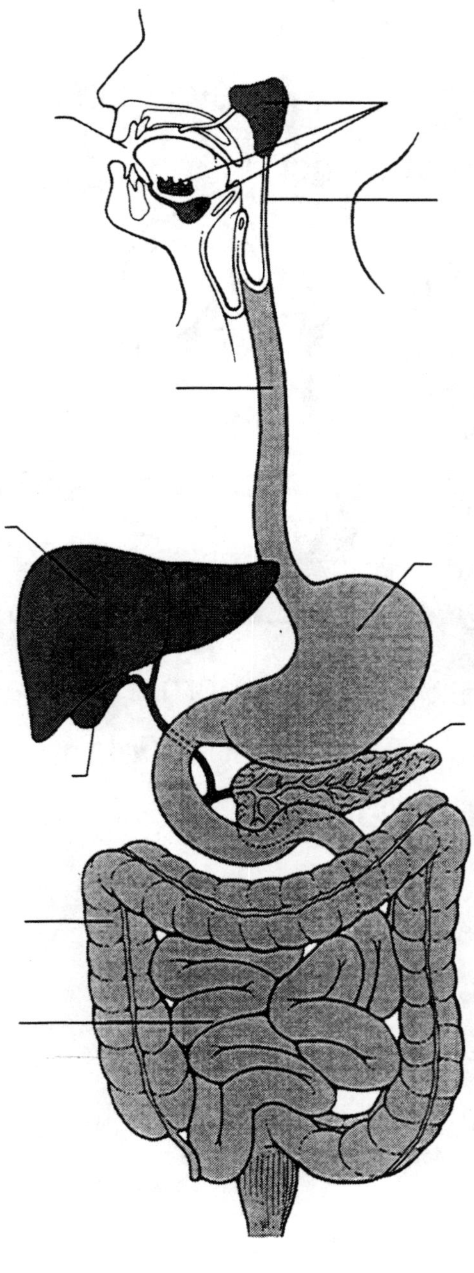

C. The oral cavity and its accesory digestive organs

1. Salivary glands

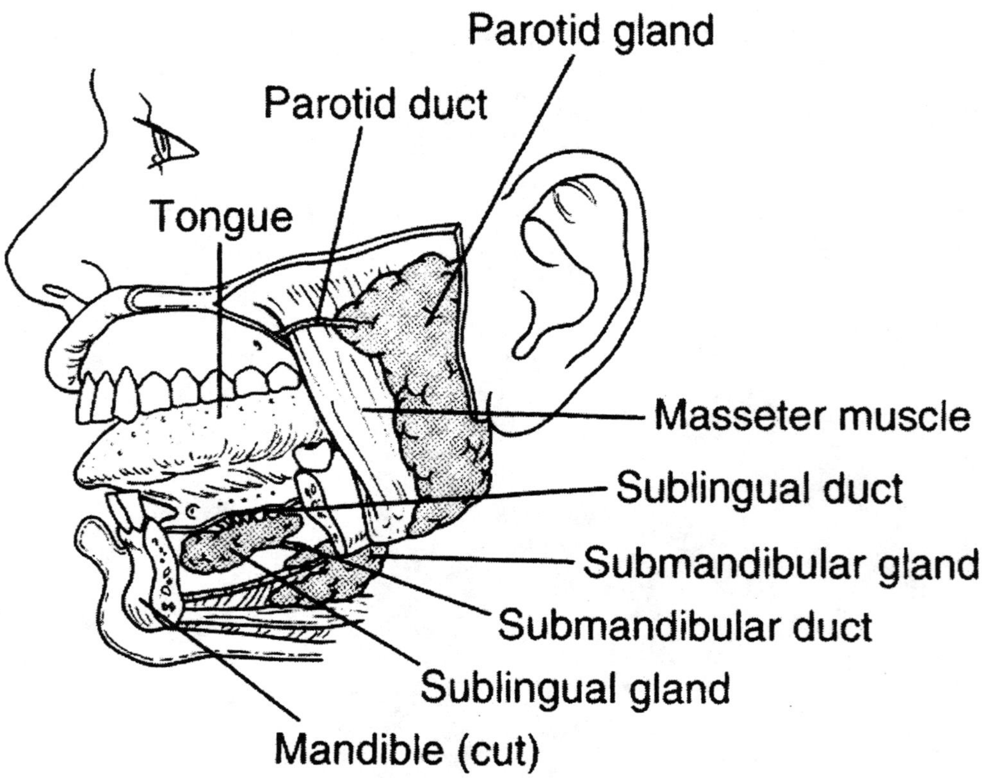

2. Teeth
    a. Function of teeth =

    b. Sets of teeth =

    c. Structure of a typical tooth located in the mandible

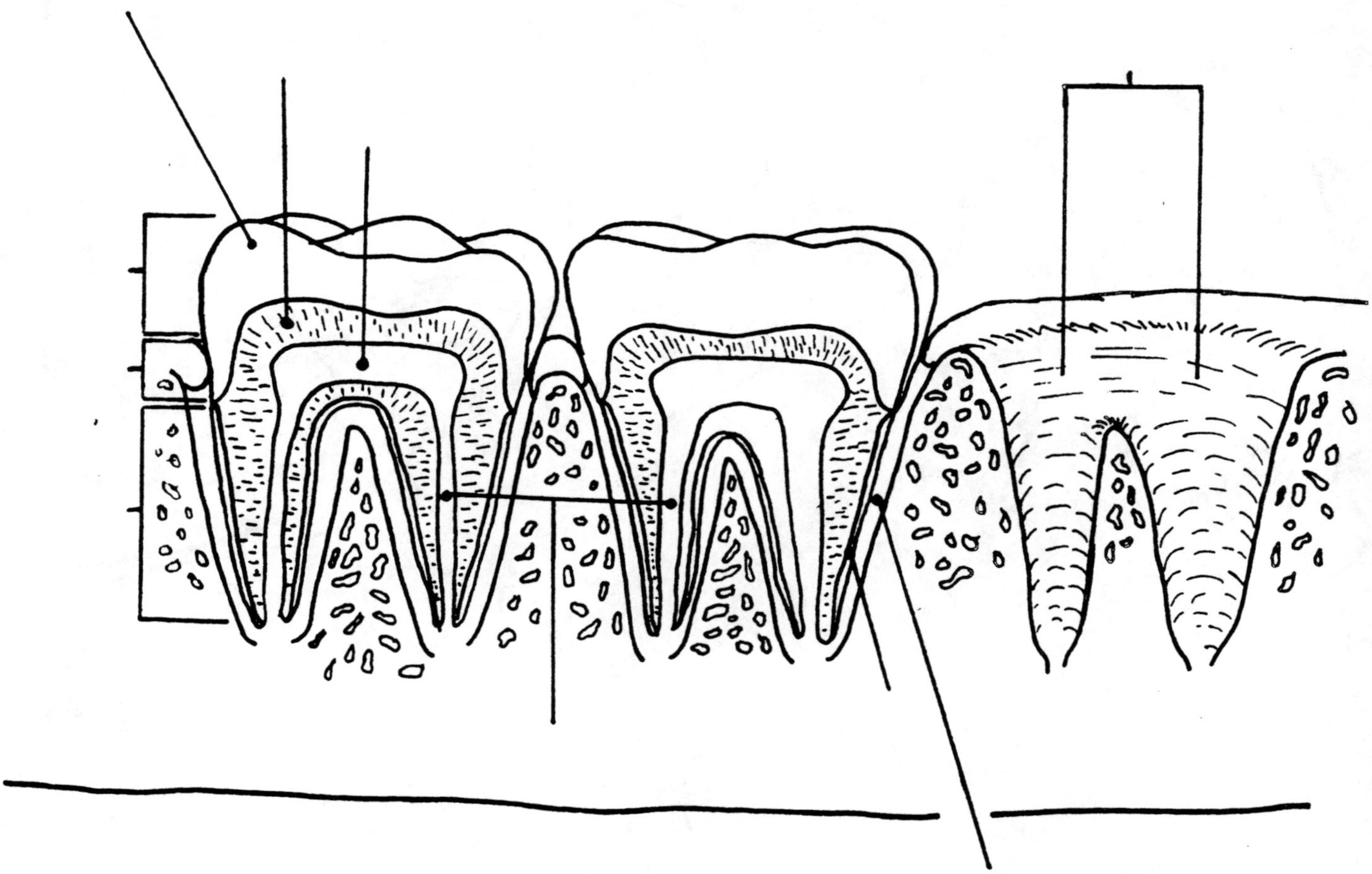

D. The esophagus and the mechanism of swallowing

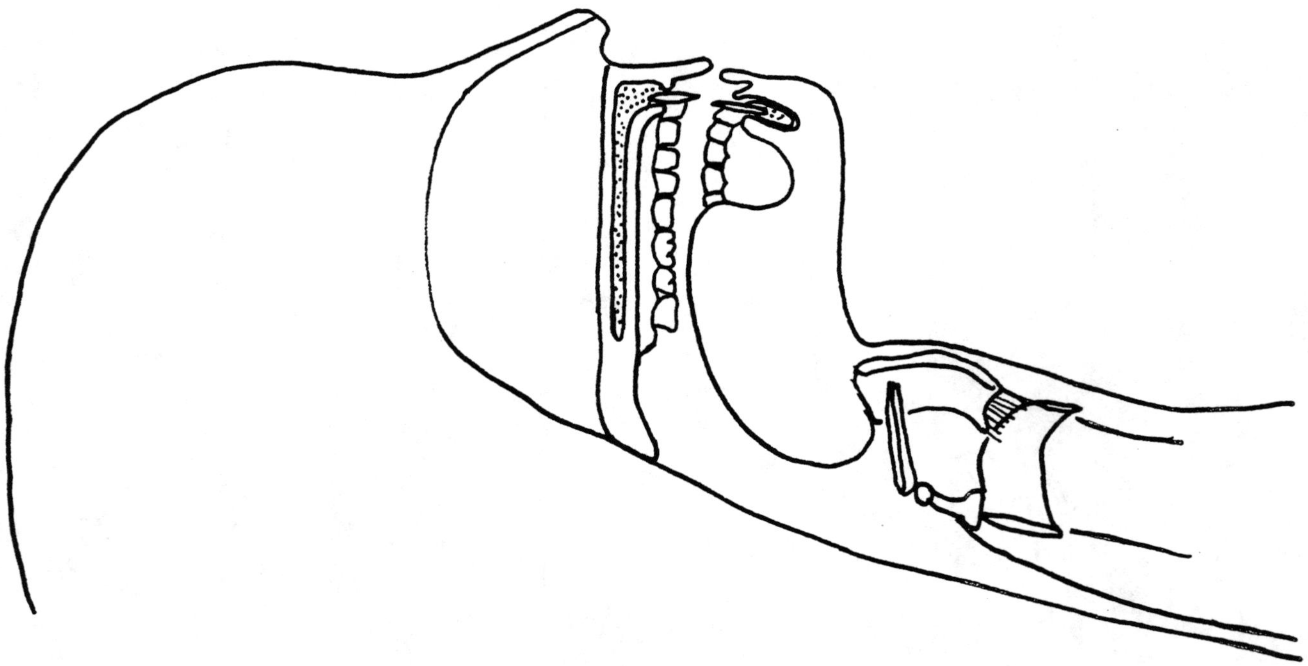

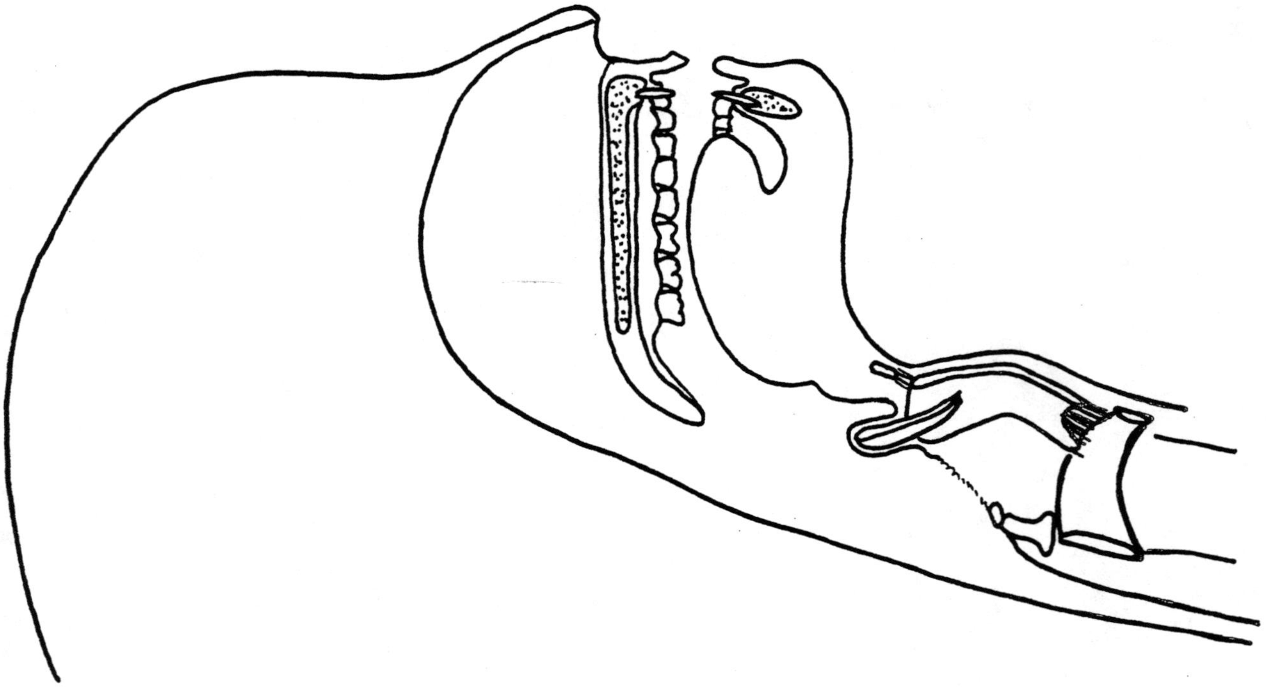

E. The stomach

1. Structure of the stomach

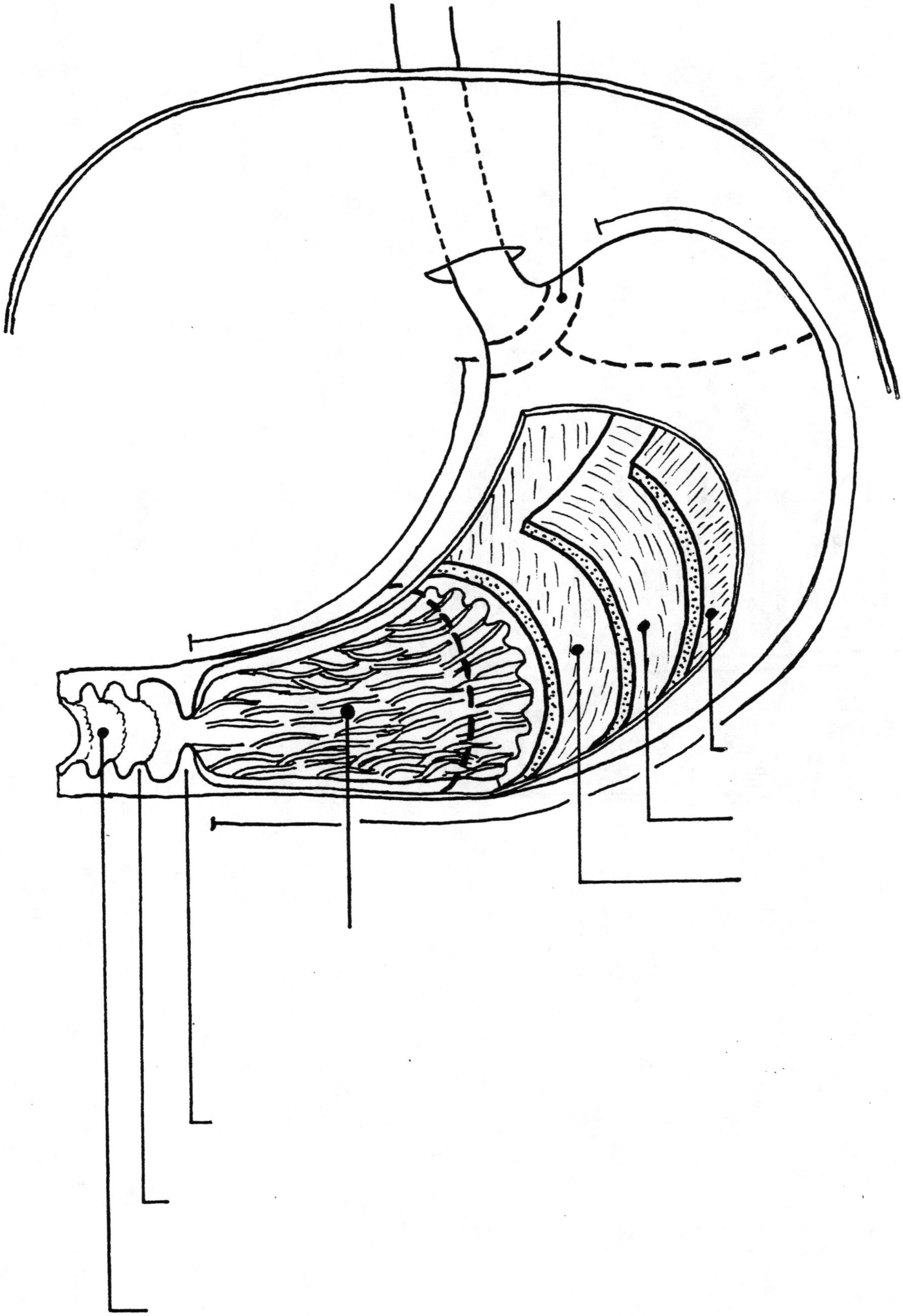

F. The wall of the stomach and digestive events occuring in the stomach

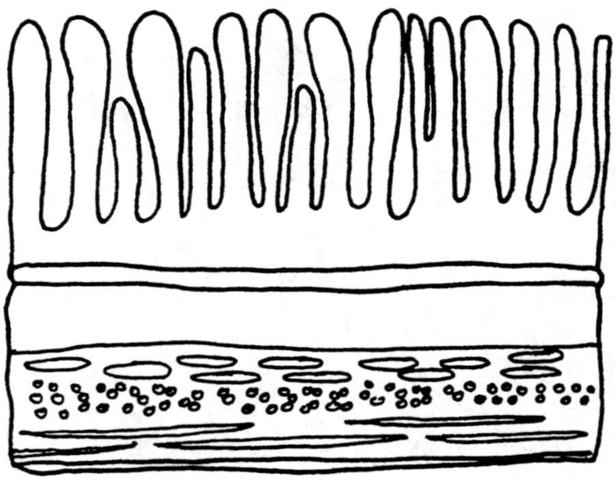

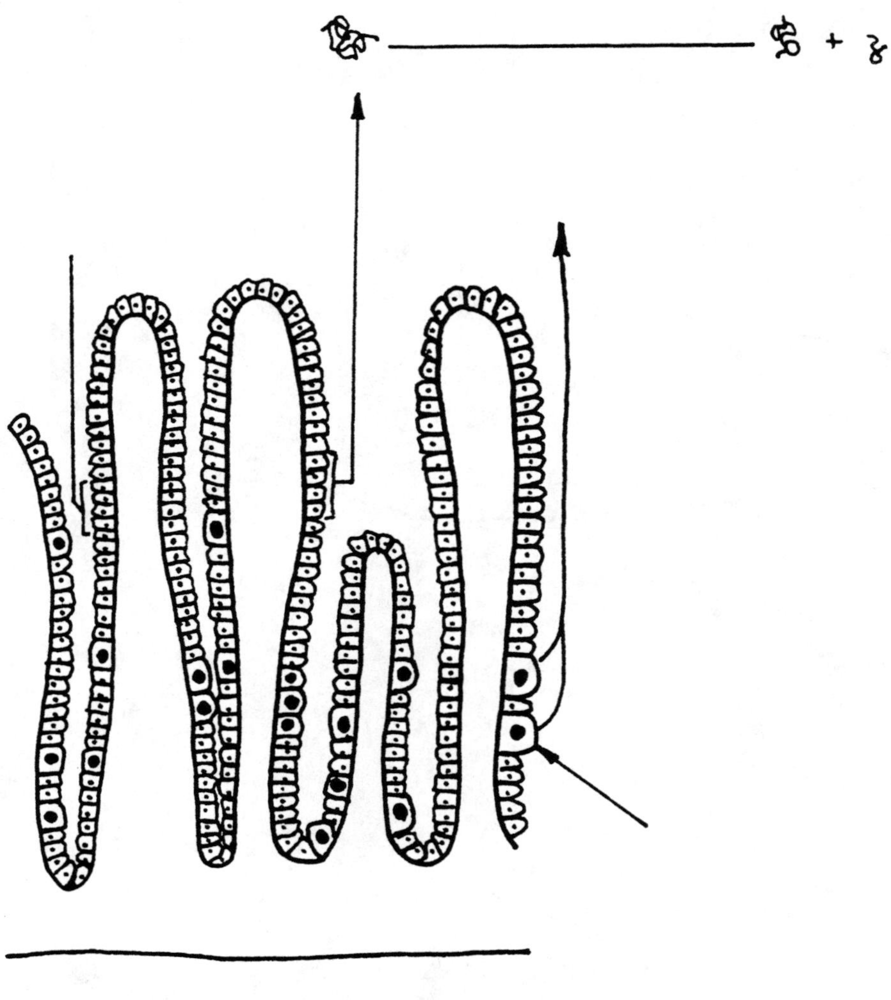

G. The small intestine

1. Areas of the small intestine

   a.

   b.

   c.

2. The four areas of the DUODENUM

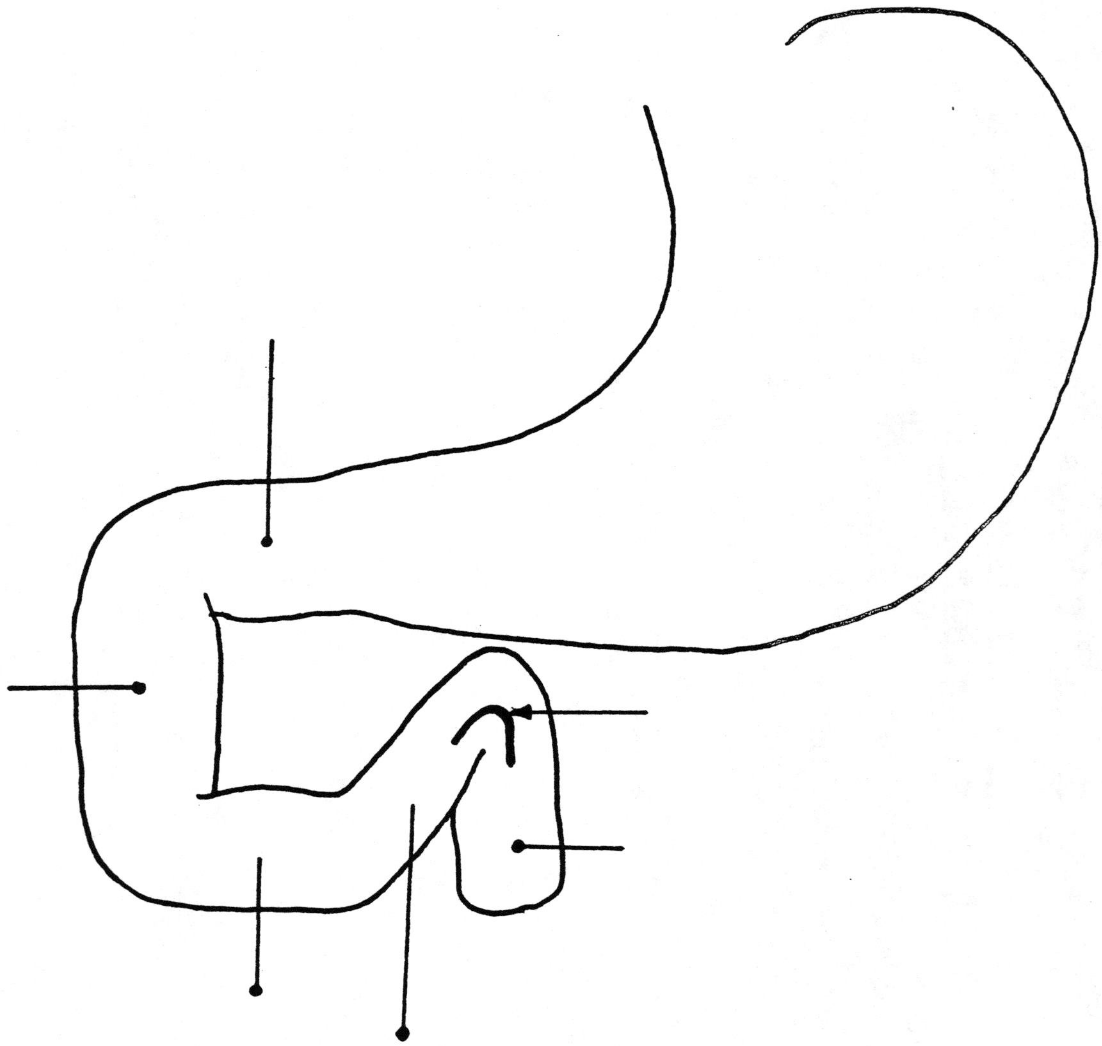

3. Plicae circulares and the villi
   a. These are found only in the duodenum, jejunum, and ileum
   b. PLICAE CIRCULARES

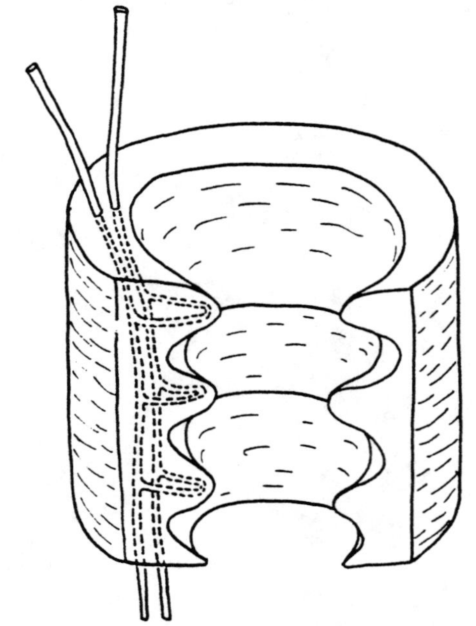

   c. VILLI

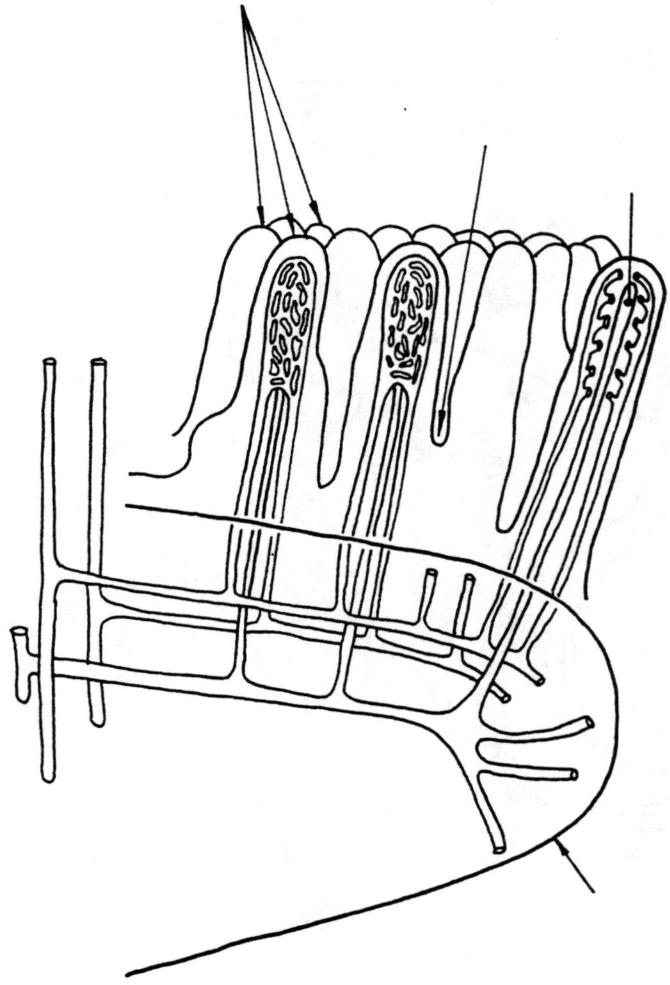

4. Digestive events and the fate of absorbed nutrients

   a. Digestive events

   b. Fate of nutrients after absorption from the small intestine

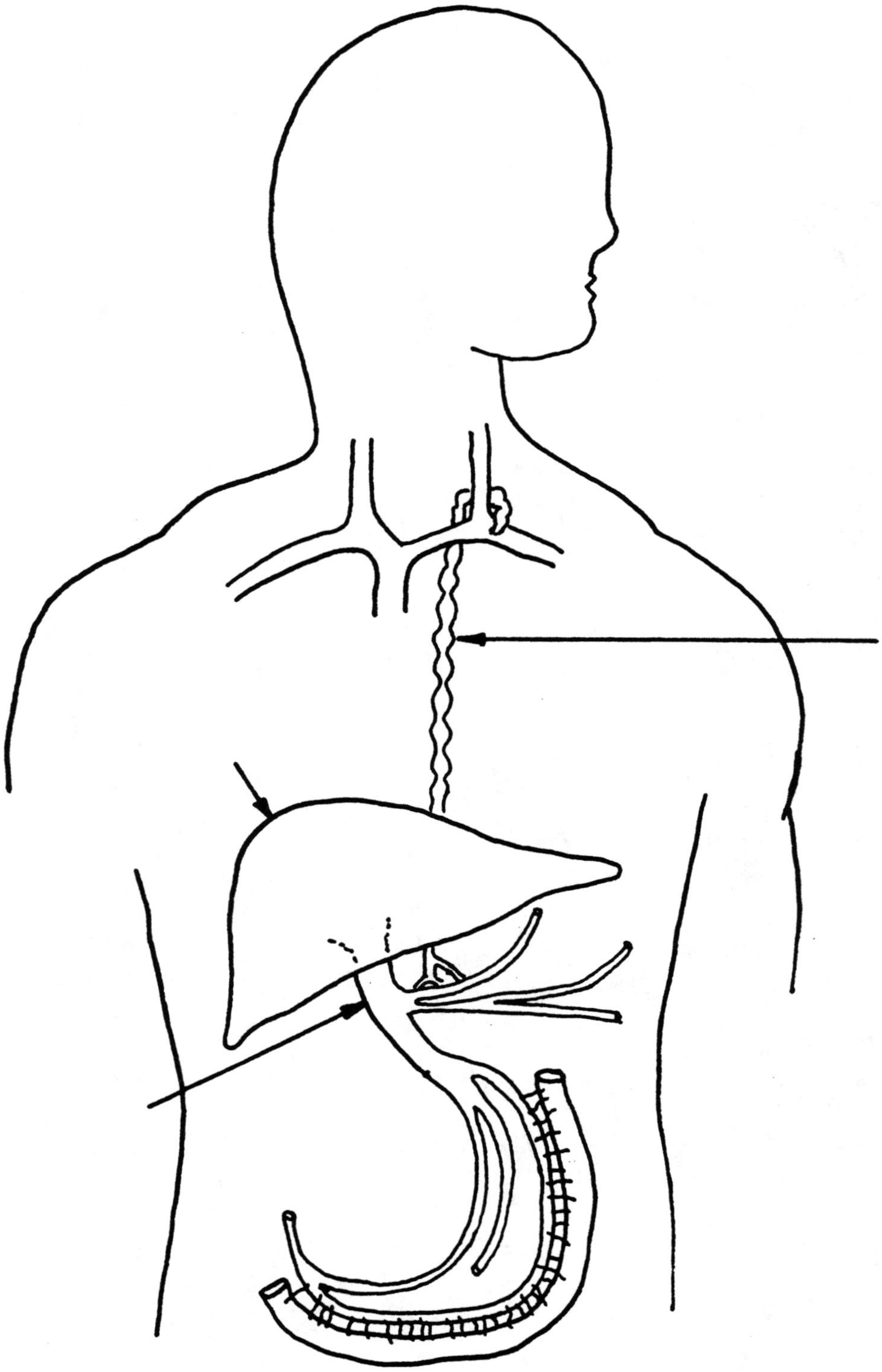

5. The pancreas, liver, gall bladder, and the biliary system as accessory digestive organs and structures associated with the duodenum

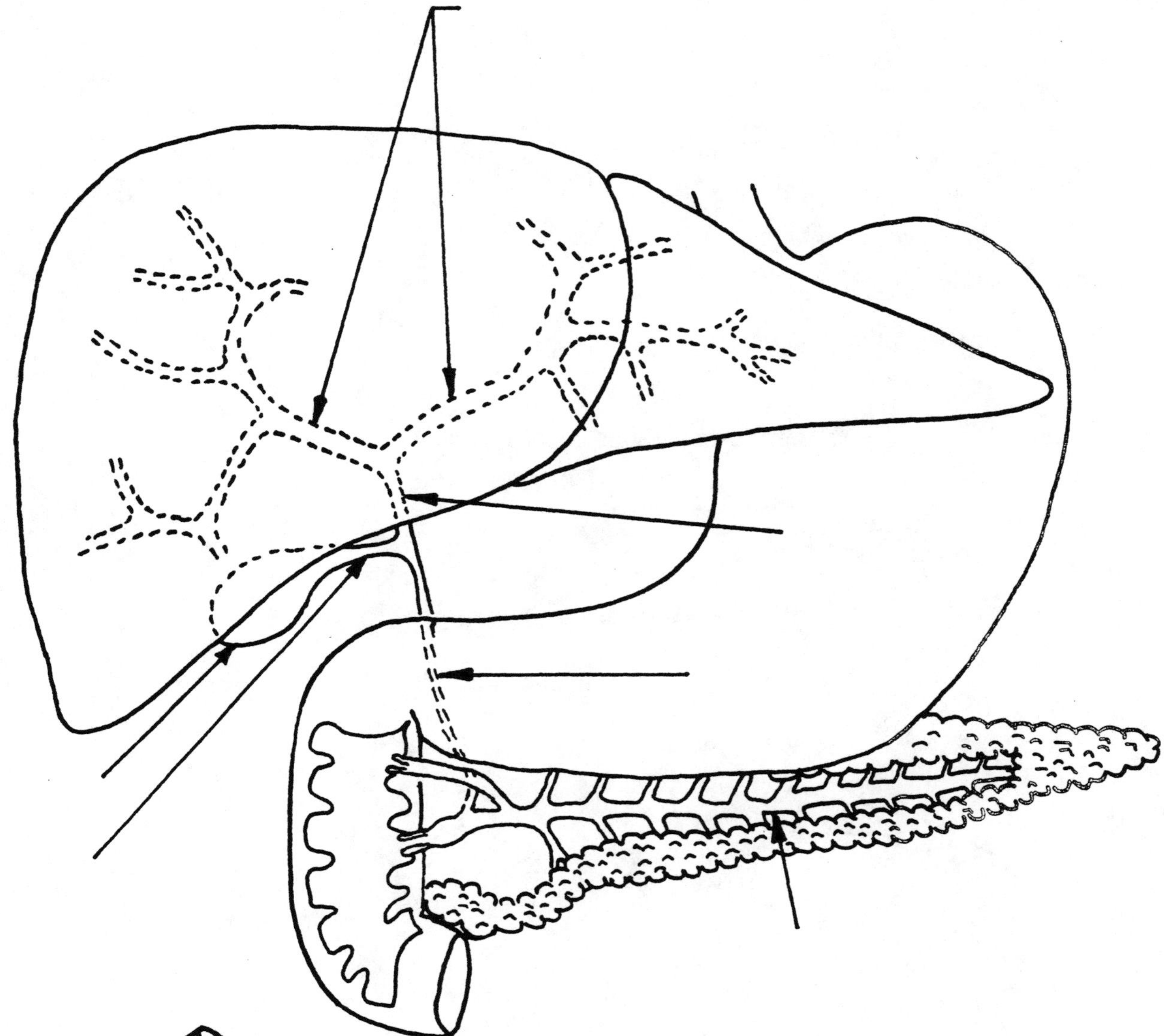

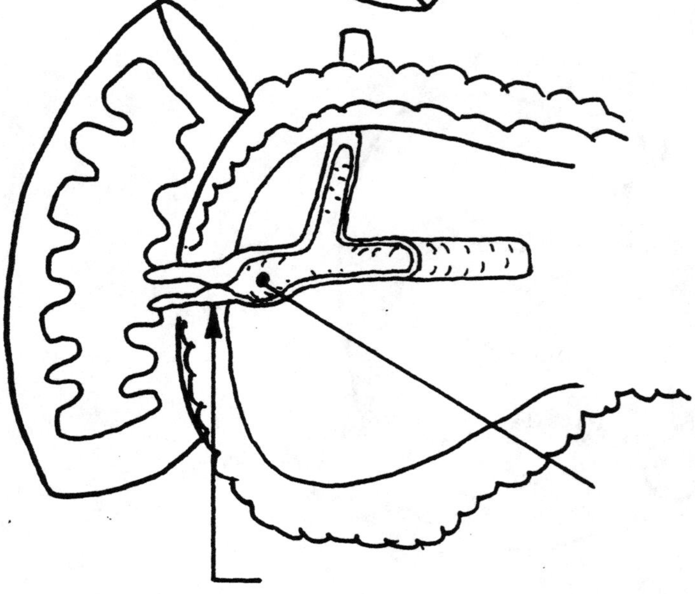

6. The liver

a. Some functions of the liver

1)

2)

3)

4)

5)

b. Gross external structure of the liver

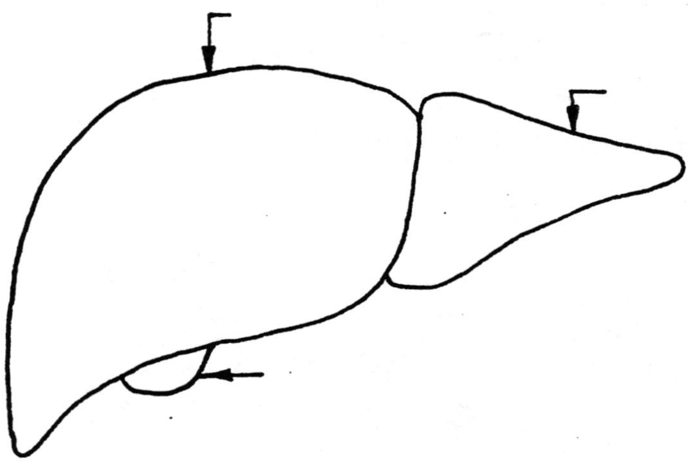

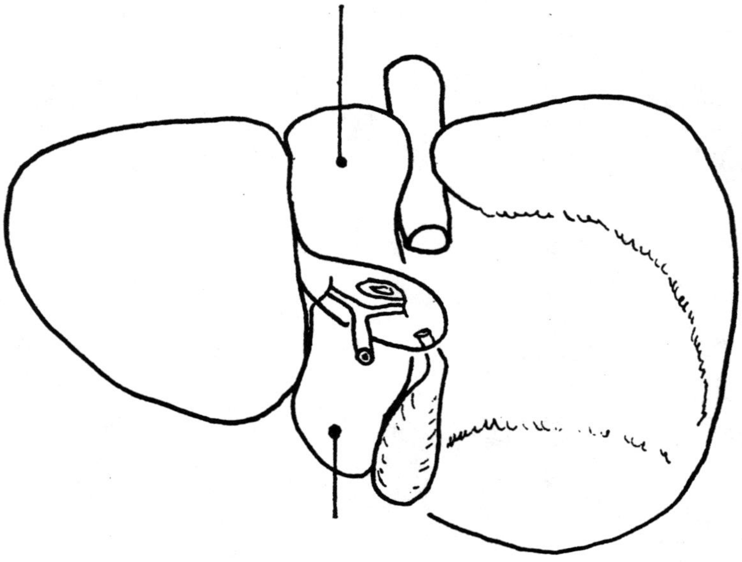

H. The structural features of the large intestine

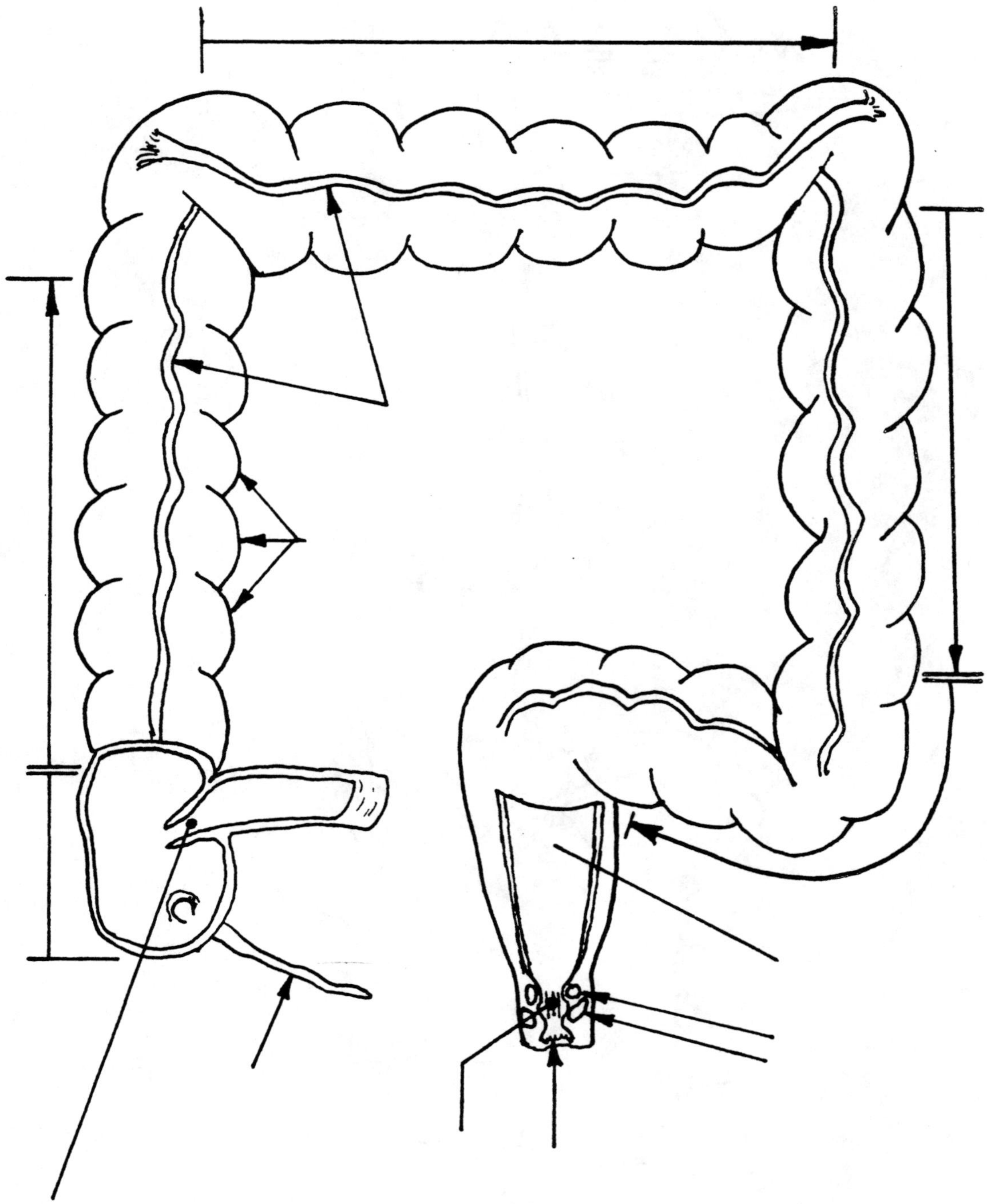

I. The function of the sphincters and valves in the digestive system

a. Overall function =

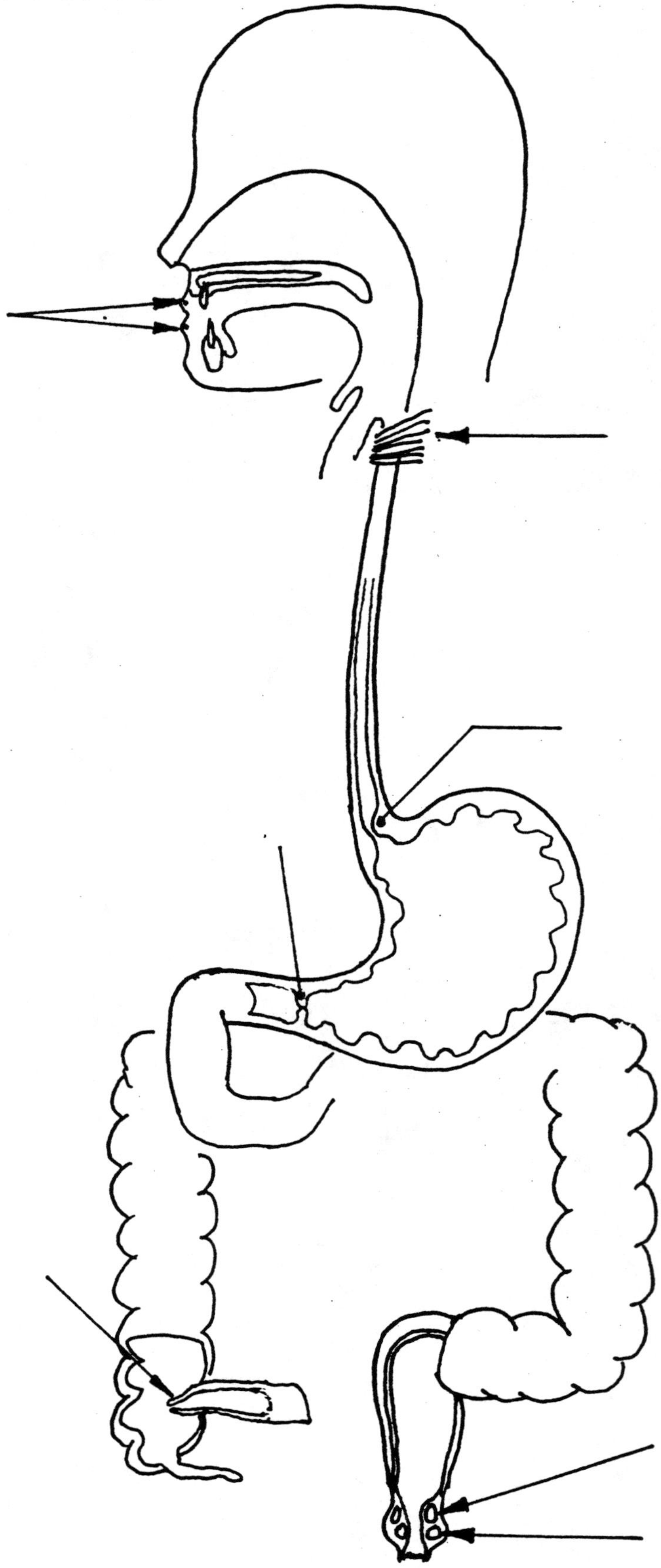

76. Why do your cells need nutrients ?
77. In what forms are the nutrients in your food ?
78. In what forms do your cells need nutrients ?
79. Which system delivers nutrients to the cells of your body ?
80. What are the two major functions of the digestive system ?
81. Anatomically, of what is the digestive system composed ?
82. Starting with the oral cavity, trace a molecule of cellulose through the digestive system.
83. What is another term for the oral cavity ?
84. What are the accessory digestive organs associated with the oral cavity ?
85. Name the three salivary glands.
86. What is the name of the enzyme that is in saliva ?
87. What does amylase partially degrade ?
88. Therefore, what is the digestive event that occurs in the oral cavity ?
89. What cranial nerve passes through the parotid gland ?
90. What is the purpose of the facial nerve ?
91. What will happen if this nerve is cut either by trauma, or during surgery on the parotid gland ?
92. Draw and label a typical tooth.
93. What is the purpose of teeth ?
94. Describe the process of swallowing. How is food pushed into the pharynx ? How are food / drink prevented from entering the nasal cavity ? How are food / drink prevented from entering the larynx ? Which is more life-threatening, having food / drink enter the nasal cavity, or enter the larynx ?
95. During swallowing where do you want food / drink to go ?
96. What does the esophagus connect ?
97. Are there any digestive events that occur in the esophagus ?
98. Name the four regions of the stomach.
99. What are the folds inside the stomach called, and in what other organs in the body are they found?
100. What is the function of rugae ?
101. Name the muscle layers of the stomach going from most superficial to deep.
102. Are these muscle layers made of smooth, skeletal, or cardiac muscle ?
103. Are these muscle layers made of involuntary, or voluntary muscle ?
104. What are two good names for the muscular sphincter at the end of the esophagus ?
105. Is this cardiac sphincter made of voluntary, or involuntary muscle ? Also, is it smooth, skeletal, of cardiac muscle ?
106. What is the purpose of the lower esophageal sphincter ?
107. What digestive event occurs in the stomach ?
108. What strong acid is made in the stomach ?
109. What cell of the inner stomach lining makes hydrochloric acid (HCl) ?
110. What is the purpose of having HCl in your stomach ?
111. How is HCl, a very strong and irritating acid, prevented from entering the esophagus ?
112. Getting HCl into your esophagus is termed acid reflux and is extremely painful. What is the common name of the pain brought about by acid reflux into the esosphagus ?
113. What is the name of the enzyme that partially digests proteins in the stomach ?
114. What is the name of the stomach cell that makes pepsinogen ?
115. Can pepsinogen digest proteins, even partially ?

116. How is pepsinogen converted to its active form, pepsin ?
117. Normally, you do not feel any pain from irritation by HCl in your stomach. HCl is really in your stomach, so why doesn't it irritate your stomach (You should know that HCl can actually eat holes through metal !!)
118. What cells of the stomach make and secrete the mucous that protects the stomach from the irritating effects of HCl.
119. Where else in the digestive system is HCl made ?
120. What is the name of the liquid form of food that is formed in the stomach following several hours of muscular churning, exposure to acid, and exposure to pepsin ?
121. What prevents stomach contents from leaving the stomach immediately and entering the duodenum ?
122. Is the pyloric sphincter smooth, cardiac, or skeletal muscle ?
123. Is the pyloric sphincter an involuntary, or voluntary muscle ?
124. What are the three sections (regions, areas) of the small intestine, and their order, between the pyloric sphincter and the ileocecal valve ?
125. What are the circular folds called that are found throughout the small intestine ?
126. What are the microscopic finger-like projections found on the surface of the plicae circulares called ?
127. What are found inside each villus ?
128. In going from the pyloric sphincter to the duodenojejunal flexure name the four areas of the duodenum.
129. What are the three accessory digestive organs associated with the duodenum ?
130. What are some functions of the liver ?
131. What substance does the liver make that has as digestive function ?
132. Bile is made constantly by the liver, but is needed in the duodenum only during a meal. So, where is bile stores in between meals ?
133. Is bile an enzyme ?
134. If bile is not an enzyme, then what is its function ?
135. Describe how bile gets from the liver to the gall bladder.
136. Describe how bile gets from the gall bladder to the duodenum.
137. Which two ducts merge to form the common hepatic duct ?
138. Which two ducts merge to form the common bile duct ?
139. Where are the digestive enzymes for the duodenum made ?
140. How do the digestive enzymes get from the pancreas into the duodenum ?
141. The common opening into the duodenum for the main pancreatic duct and the comon bile duct is denoted by what structure ?
142. What two hormones does the pancreas make ?
143. What is the accessory pancreatic duct ?
144. What are some digestive events that occur in the duodenum ?
145. What nutrients get absorbed across the villi into the blood ?
146. Where are these water-soluble nutrients directly transported to ?
147. Waha does the liver do with the glucose that comes in from the duodenum ?
148. What does the liver do with the amino acids transported in it from the duodenum ?
149. What nutrients get absorbed across the villi and into the central lacteals ?
150. Where do these water-insoluble nutrients that get absorbed into the central lacteals go ?
151. Put the three areas of the small intestine in order according to their increasing lengths.

152. What is the name of the muscular valve between ileum of the small intestine and the cecum of the large intestine ?
153. What is the function of the ileocecal valve ?
154. In what area of the abdominal cavity is the cecum located in ?
155. What is the name of the finger-like projection off of the cecum ?
156. What is the function of the appendix in us humans ?
157. In what abdominal area is the appendix located ?
158. What is the name of the section of large intestine that ascends out of the cecum ?
159. On which side of the abdominal cavity is the ascending colon ?
160. What would be two good names for the bend in the colon between the ascending and transverse colons ?
161. Why is one of these good names the hepatic flexure ?
162. What does the term flexure mean ?
163. The transverse colon is between what two colon areas ?
164. What are two good names for the flexure between the transverse colon and the ascending colon ?
165. Why is one of these good names the splenic flexure ?
166. On which side of the abdominal cavity is the descending colon located ?
167. What is the name of the large intestine that is the direct continuation of the descending colon ?
168. In what abdominal area is the sigmoid colon located ?
169. What does the sigmoid colon immediately become ?
170. What does the rectum narrow into ?
171. What is the opening to the outside of the body from the anal canal called ?
172. What is the name given to the three longitudinal bands of smooth muscle that go along the length of the cecum, ascending colon, transverse colon, descending colon, and sigmoid colon ?
173. What is the name of the pouches of the large intestine ?
174. What are two functions of the large intestine ?
175. What bones protect the stomach, liver, and spleen ?
176. Starting with the orbicularis oris, list the sphincters and valves in order and state whether each is made of voluntary, or involuntary muscle.
177. Trace a molecule of cellulose from the oral cavity to it passage through the anus. Be specific about all structures and digestive landmarks that are passed along the way.

IV. The Urinary System

A. General information and the purpose of the urinary system

All living cells make CO_2 which, we have seen, is eliminated from the body by the respiratory system. **UREA** , like CO_2, is toxic to the cells that make it and to the body in general. It has the following structure:

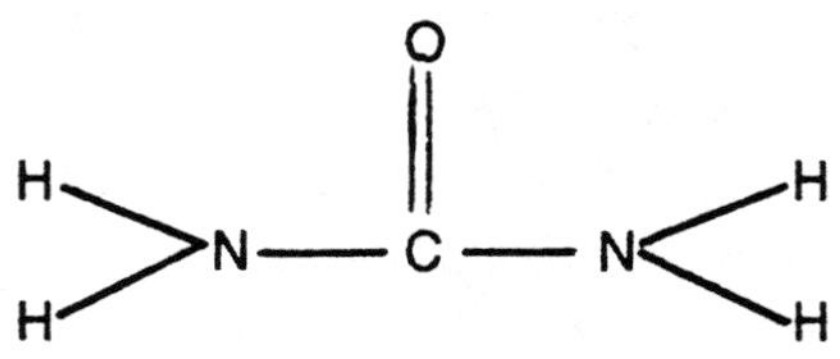

Urea is the normal metabolic byproduct of amino acid metabolism in **LIVER CELLS**.

As with CO_2, your liver cells dump their urea into the blood that is circulating around them, but although doing so relieves your liver cells of the toxic effects of urea, urea must then be eliminated from the blood and completely from the body. The removal of urea from the blood by filtration of the blood and from the body is one of the functions of the urinary system. In addition to removal of the waste product urea, the kidneys of the urinary system also participate in **1)** maintenance of the acid-base balance of the blood, **2)** regulation of the balance of some ions (sodium and potassium) in the blood, **3)** regulation of the level of some minerals (calcium, phosphorus) in the blood, and, most importantly, **4)** regulation of your daily, long-term blood pressure. Obviously, your kidneys are extremely important.

On a gross level, your urinary system is composed of two **KIDNEYS**, which perform the critical functions of the urinary system and in doing so make the fluid known as urine, two **URETERS**, which are the tubes that transport the formed urine from the kidneys to the **URINARY BLADDER** where it is stored until it can be conveniently voided to the outside of the body through a single tube called the **URETHRA**.

Your kidneys are indeed the functional structures of your urinary system, and it is in the renal cortex where your blood is filtered and urea is removed from the blood. Let's follow a molecule of urea from the main artery bringing blood into the kidney, the **renal artery**, all the way out to the circulatory structure, the **glomerulus**, located in the renal cortex, from which urea and nutrients are removed. A single **renal artery** enters a kidney and immediately branches into 15 - 20 smaller **interlobar arteries**. These arteries pass through the renal medulla region of the kidney out of the medulla / cortex junction where each branches into several smaller **arcuate arteries**. These arteries in turn give off numerous smaller **interlobular arteries** that penetrate out into the renal cortex; there are perhaps 50,000 - 100,000 of these small arteries. Several microscopic sized arteries, called **afferent arterioles**, branch off of each interlobular artery. The end of each one of the afferent arterioles branches into a small, arterial globular plexus called a **glomerulus**. There are about one million afferent arterioles in the cortex of the kidney and therefore there are about one million glomeruli in each kidney. So, once again we encounter the general principle of a tube and tubule system, blood vessels in this case, undergoing branching and in doing so becoming smaller and more numerous. We began with one vessel (renal artery) entering each one of your kidneys and eventually wound up out in each renal cortex with one million much, much smaller vessels (the afferent arterioles and glomeruli).

Practically everything, urea and nutrients, gets filtered from the blood flowing through your glomeruli into the surrounding Bowman's capsules. Therefore, the blood that flows into a glomerulus through its afferent arteriole contains urea and nutrients, but the blood that flows out of the glomerulus through its efferent arteriole contains neither.

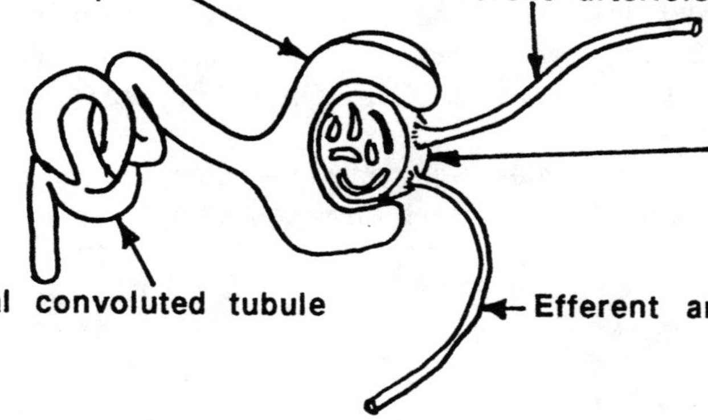

Located on the back of your Bowman's capsule (**BC**) is a single tubule, the <u>**proximal convoluted tubule (PCT)**</u>, into which the filtered urea and nutrients enter from **BC**. After the **PCT** straightens out the name is changed from the **PCT** to the <u>**loop of Henle (LOH)**</u>. It is still the same tubule, however ; it is just that the the name has been changed. **BC** and the **PCT** are both located in the renal cortex, but the **LOH** can dip down into the medulla region. Nevertheless, in passing through the entire **LOH**, urea passes through what is called the <u>**descending limb of the loop of Henle**</u>, and then the <u>**ascending limb of the loop of Henle.**</u> The <u>descending limb</u> dips into the medulla and the <u>ascending limb</u> comes back into the cortex. As nutrients pass through the **PCT** and the <u>descending limb</u> of the **LOH** they are reabsorbed back into the blood (circulatory system). Urea is not reabsorbed and continues on through the **LOH.**

Once back in the cortex the <u>ascending limb</u> of the **LOH** becomes convoluted, and, once again, although it is still the same tubule, the name is changed to the <u>**distal convoluted tubule (DCT).**</u> The **DCT** is connected to a tubule calle the <u>**collecting duct (tubule)**</u>. In fact, an important point, many **DCTs** are connected to a single **collecting tubule (duct).**

As urea is passing through the **PCT**, **LOH**, **DCT** and the **collecting duct**, various amounts of water are taken from the interstitial fluid of the kidneys and added to the urea. The amount of fluid taken into your tubules is dependent on how much fluid your body needs to retain or eliminate. The resultant fluid formed inside your tubules is called <u>urine</u>.

The urine (containing urea) that is in your **collecting tubules** flows on out into small spaces in your kidneys called the **minor calyces**, then into slightly larger spaces termed **major calyces**, and then on into a larger space called the **renal pelvis**. It is the renal pelvis that narrows and forms the tube known as the **ureter**, which transports the urine to the **urinary bladder** for storage until it can be, at your convenience, voided to the outside of the body through the **urethra**.

We will have to go backwards, against the direction of urine flow, to once again see that this tubes in this tubular system become smaller and more numerous. We begin with one urethra and urinary bladder, progress to two ureters and renal pelvices (one for each kidney, of course), several major and minor calyces, then perhaps 100,000 thousand collecting tubules in each kidney, then finally one million in each kidney of DCTs, LOHs, PCTs, and BCs.

When we take a close look at the tubule system we notice how simple it actually is, that between a glomerulus and a collecting duct there is really only a single tubule :

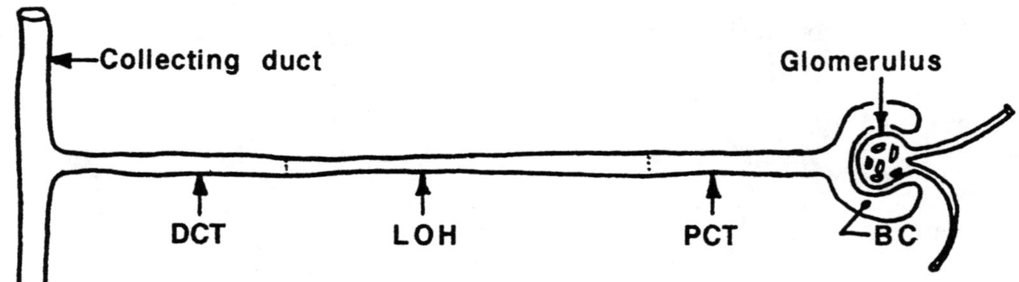

However, from what you have already read you know that this single tubule is not straight, but becomes twisted and convoluted along its length :

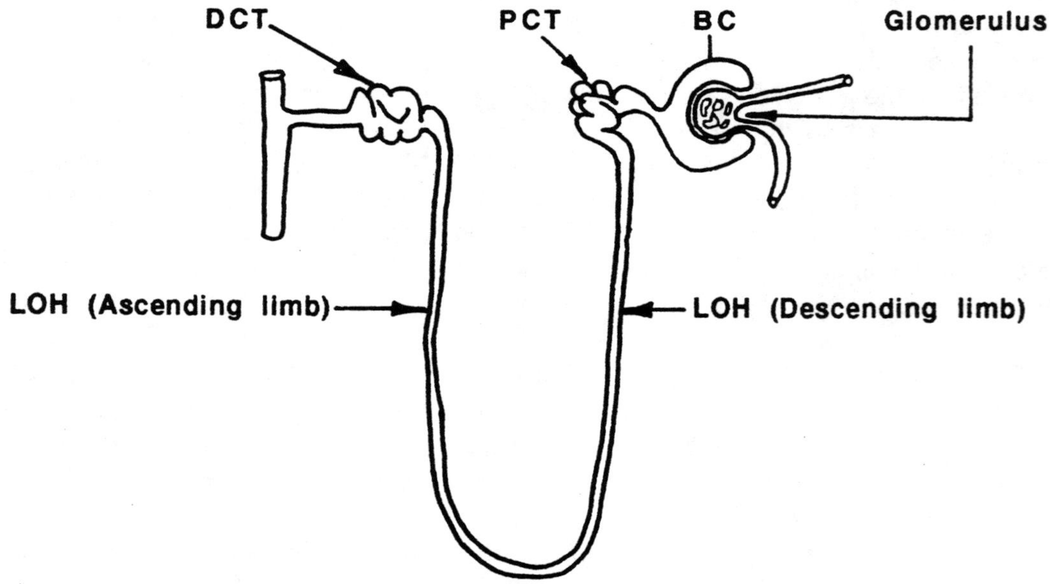

Nevertheless, it is still a single tubule and we are simply denoting, with names such as **BC, PCT, LOH, and DCT** its different regions. And with terms such as capsule, convoluted, and loop in these regional names you can tell that the different regions are named for anatomical reasons.

The <u>functional unit of the kidney</u> (and therefore the urinary system) is the **<u>NEPHRON</u>**. Anatomically, a nephron is based on the urinary structures that are on a one - to - one numerical correlation. These are found in that single tubule between the glomerulus and the collecting duct. Along this single tubule is <u>one</u> **BC**, <u>one</u> **PCT**, <u>one</u> **LOH**, <u>one</u> **DCT**. These four structures together form a filtration unit called a **NEPHRON**. The **collecting duct (tubule)** is <u>not</u> considered a part of the nephron because it is not on a one - to - one numerical correspondence with the other urinary structures (BC, PCT, LOH, DCT). The glomerulus is another matter for it is on a one - to - one correspondence because one BC surrounds one glomerulus. However, you most likely remember that the glomerulus really is part of the circulatory system. Nonetheless, the glomerulus forms such an intimate contact with the BC of the urinary system that many authors and texts will include the glomerulus as a component of the nephron.

We have described a single nephron and its components. How many nephrons do you have in each of your kidneys ?

B. Gross components of the urinary system

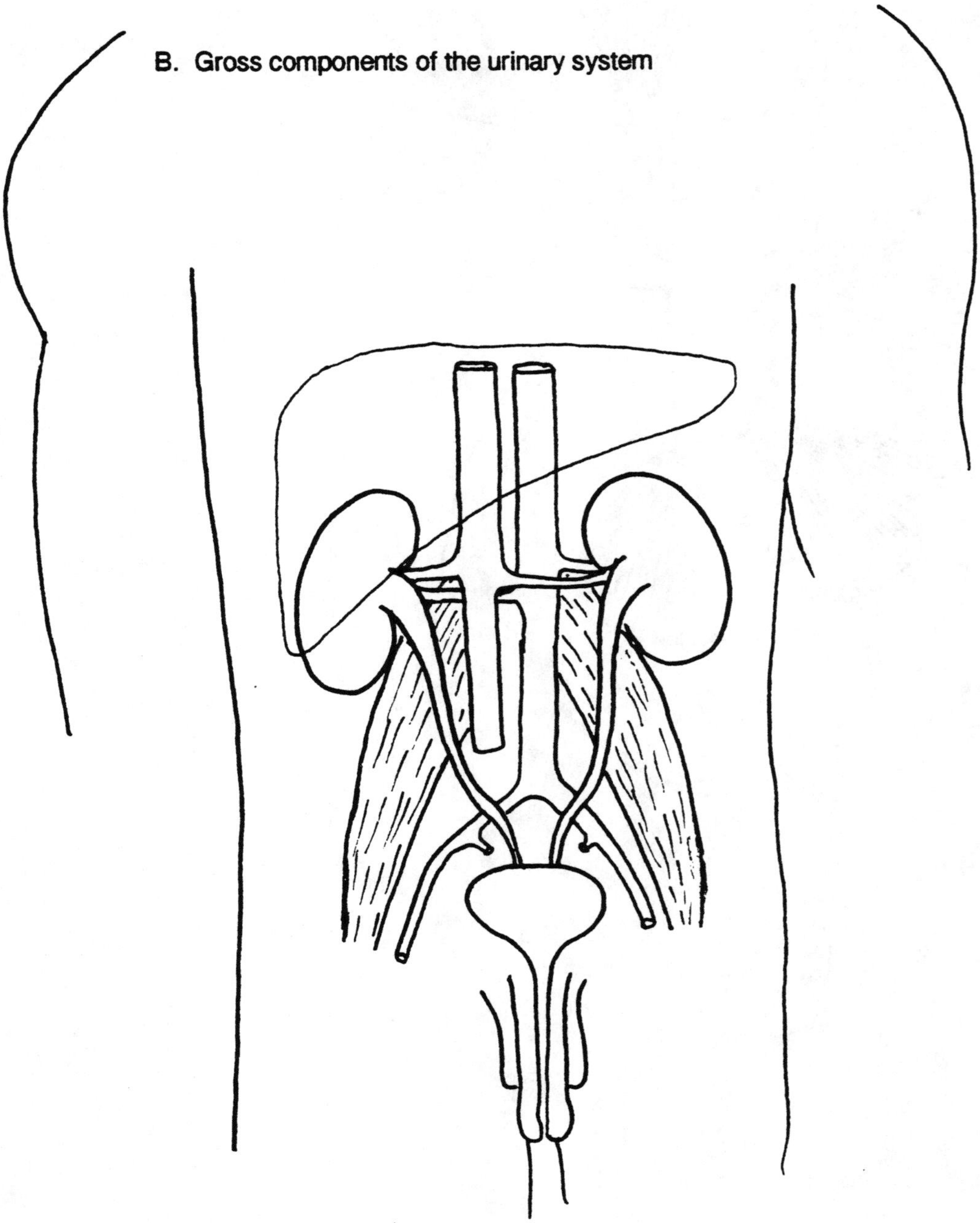

C. Gross structure of a kidney

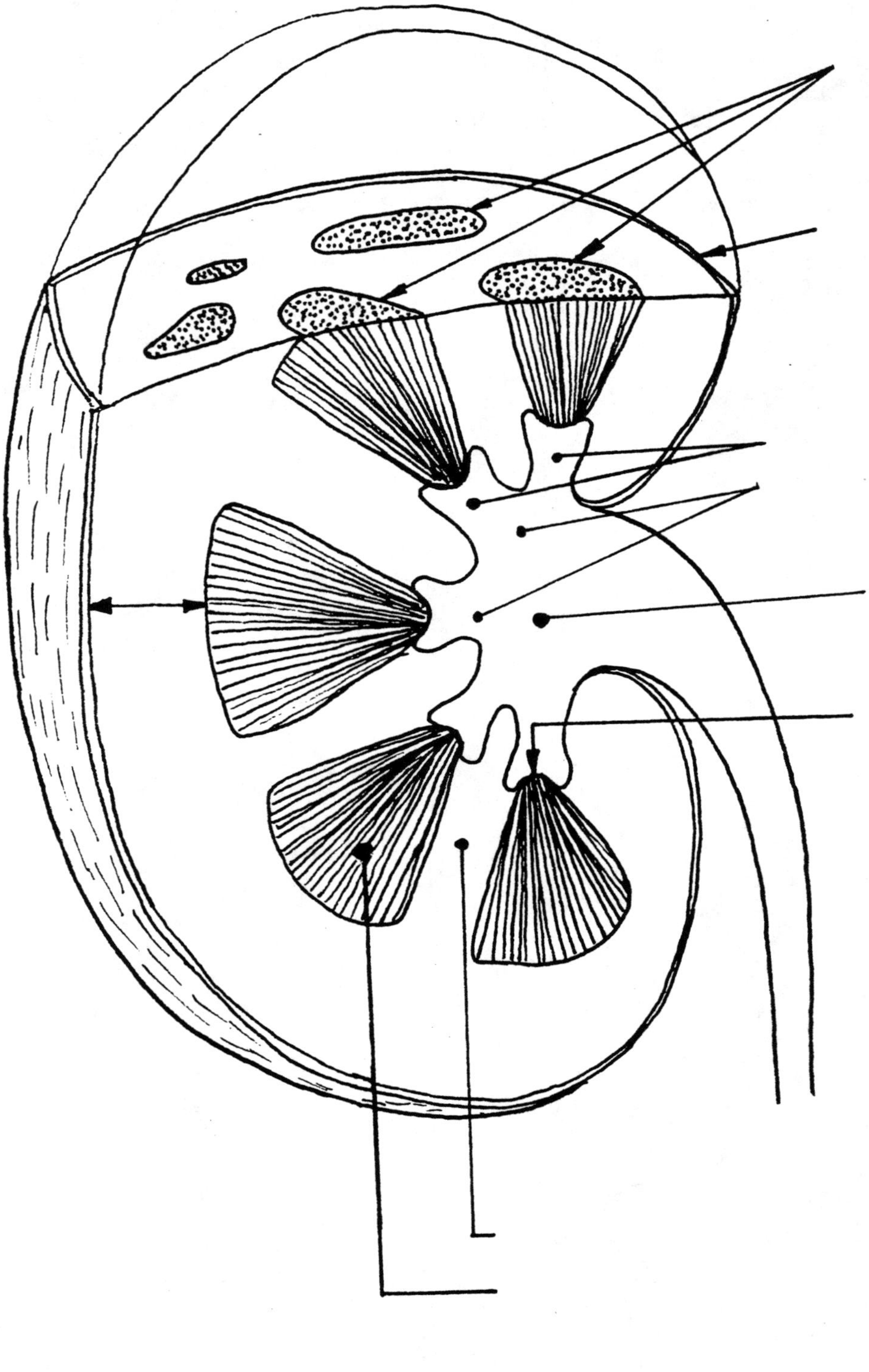

D. Arterial blood flow into the kidney

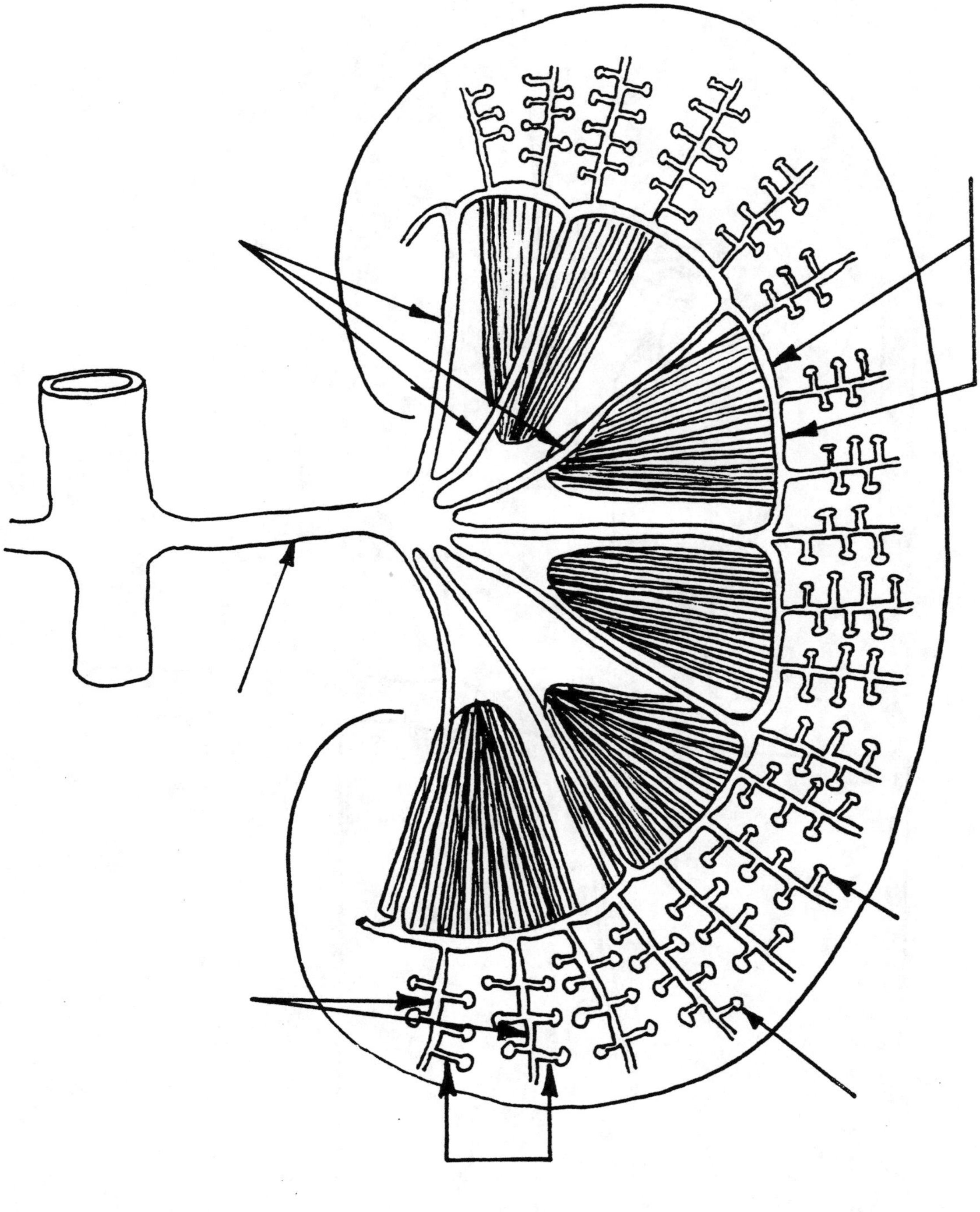

E. Filtration of the blood and the NEPHRON

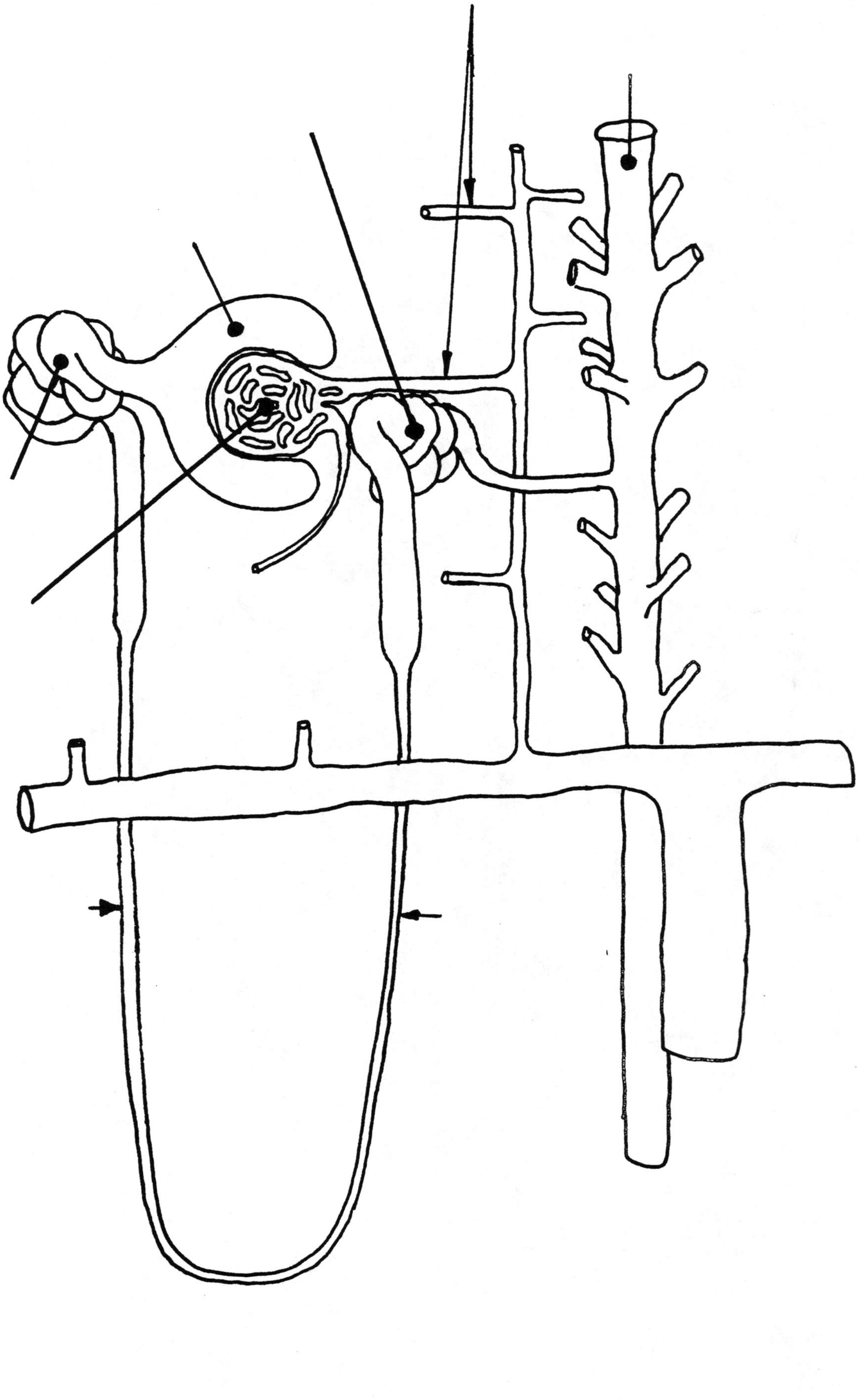

F. The juxtaglomerular apparatus

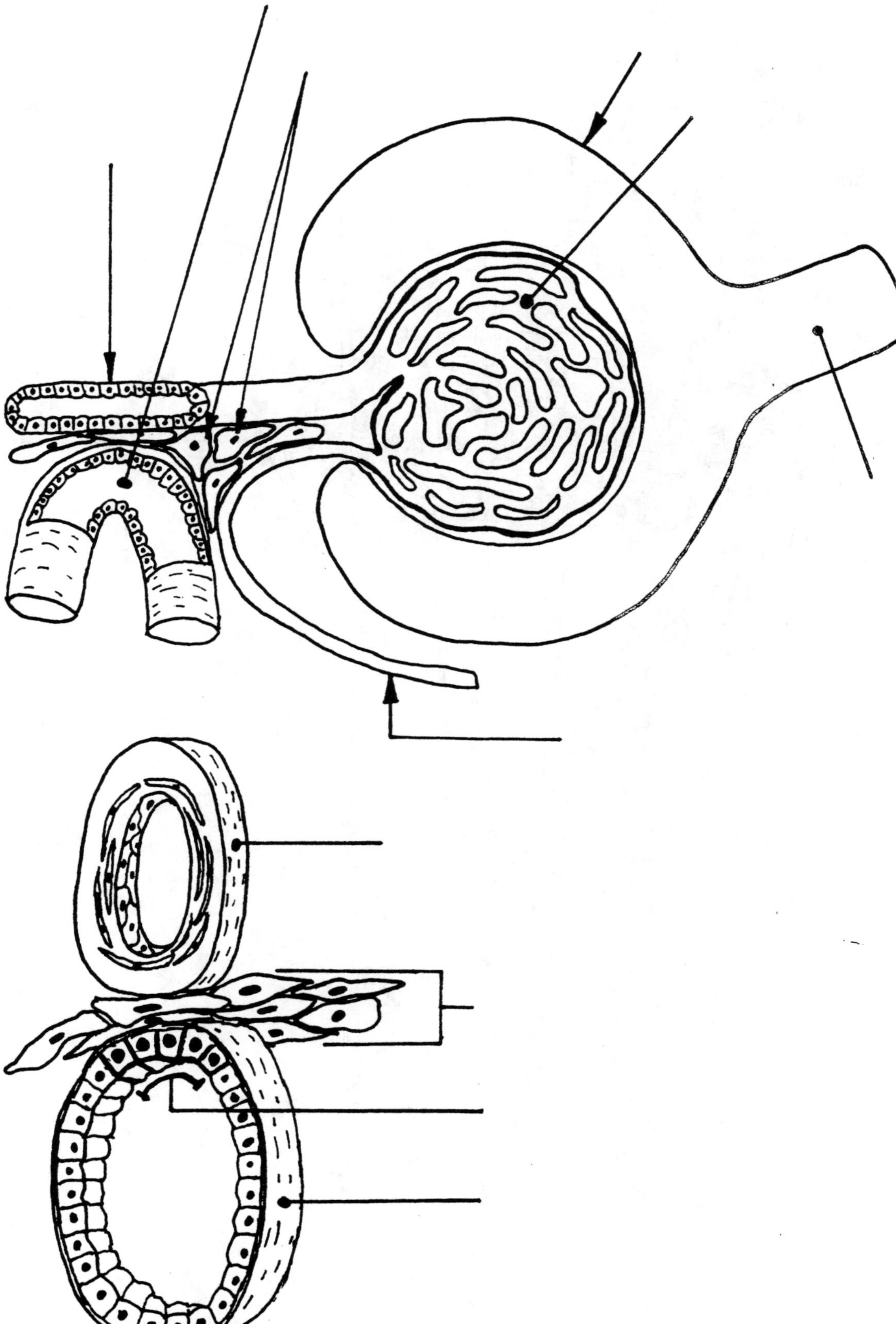

G. The structure of a renal pyramid

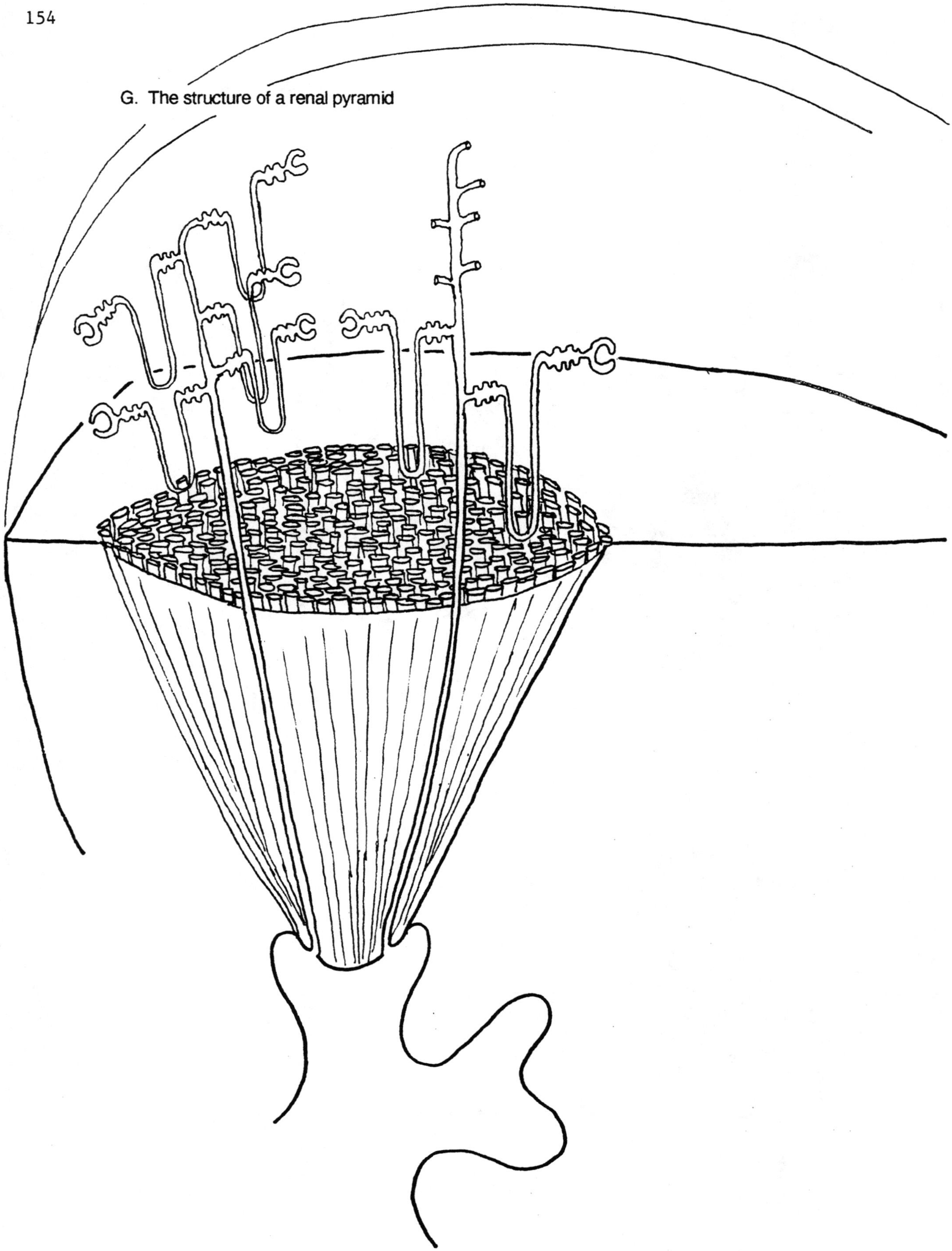

H. The structure of the urinary bladder

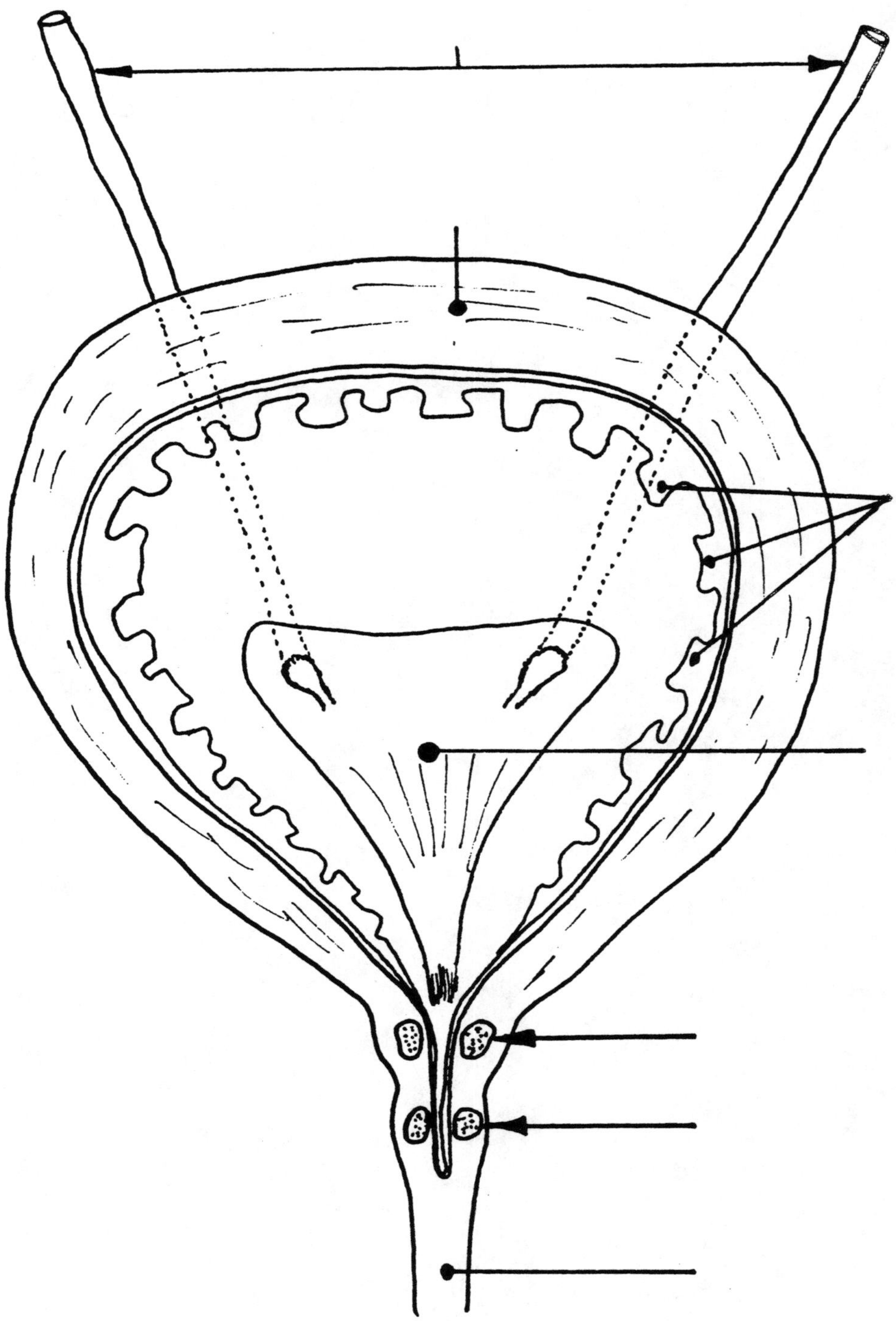

I. The difference in the female and male urethras and the clinical significance of this difference

178. In what organ is urea made ?
179. What are the gross structures in the urinary system ?
180. What are five functions of the kidneys ?
181. What is the function of the ureters ?
182. What is the function of the urinary bladder ?
183. What are folds inside the urinary bladder called ?
184. What is the function of these rugae ?
185. Through what tube is urine voided to the outside of the body ?
186. What is the name of the main artery of the kidney ?
187. A renal artery branches into what arteries ?
188. The interlobar arteries branch into what arteries ?
189. The arcuate arteries give off numerous smaller arteries that pass out into the cortex. What is the name of these arteries ?
190. What small arteries branch off from the interlobular arteries ?
191. What circulatory system structure is at the end of an afferent arteriole ?
192. About how many afferent arterioles are there in a kidney ?
193. Therefore, about how many glomeruli are there in a kidney ?
194. You start with one renal artery entering a kidney and wind up with about one million much smaller afferent arterioles and glomeruli out in the cortex. What general principle can you determine from this fact ?
195. What surrounds a glomerulus ?
196. Is Bowman's capsule a part of the circulatory, or the urinary system ?
197. What is filtered out of the glomerulus and into Bowman's capsule ?
198. What is the difference between the contents of urea and nutrients between the afferent arteriole and the efferent arteriole ?
199. Why does the efferent arteriole not contain urea and nutrients ?
200. List the components of a nephron. Make this list in order of components of going from the glomerulus to the collecting tubule.
201. Can you see that the nephron really is a <u>single</u> tube between the glomerulus and a collecting tubule ?
202. A renal pyramid is a bundle (collection) of what ?
203. What two structures make up the renal medulla region ?
204. The renal pelvis narrows to form a tube called the ---- ?
205. What is the juxtaglomerular apparatus, and what is its function ?
206. Trace a molecule of urea from the renal artery to the outside of the body.
207. Where exactly is urea filtered out of the circulatory and into the urinary system ?
208. Who has the shorter urethra, women or men ?
209. What is the clinical significance of women having shorter urethras than men ?

<u>210 - 225 :</u> For each of the following structures write an <u>**A**</u> if it is located in the cortex only, a <u>**B**</u> if located in the medulla only, a <u>**C**</u> if found in both, and a <u>**D**</u> if not found in either :

Collecting tubule (duct) ____, Bowman's capsule ____, Loop of Henle ____, Renal artery ____, Glomerulus ____, Interlobar artery ____, Interlobular artery ____, Afferent arteriole ____, Renal pelvis ____, Ureter ____, Efferent arteriole ____, Major calyces ____, Renal capsule ____, PCT ____, DCT ____, Minor calyces ____.

THE INTEGUMENTARY SYSTEM

I. GENERAL INTRODUCTION

It is often stated, and just as often accepted, that the skin is the largest organ in the body. Obviously, skin is an organ because it is composed of at least two types of tissues (epithelial and connective). However, if we accept at face value that a system consists of several organs, then we arrive once again at a problem similar to the skeletal system. Is the skin the only organ in the intergumentary system, and if so, then does that fact not violate the definition of a system, so that we should really call it the integumentary organ ? But how about the sweat glands and the sebaceous glands ? Are they organs ? If they are, then the integumentary system is at least a three organ system. What do you think ? Does either a sweat gland or a sebaceous gland qualify as an organ ? What is your evidence for what you think ?

Your environment is a contradictory place. Some of the organisms in it represent opportunities for nourishment (chickens, potatoes, beef cattle for us), reproductive opportunity (the opposite sex), and injury and / or death (bacteria. viruses, other animals, especially other humans). The events that occur out there in your environment may touch you, or emit "light" (electromagnetic radiation), cause compression and decompression of the molecules in the air (compressional waves, or "sound waves"), or emit chemicals that you can smell, or taste.

Your **Integumentary System** offers **PROTECTION** by being a barrier to bacteria entering your body, a barrier for prevention of fluids from leaving your body, and for absorbing the harmful ultraviolet (UV) radiation from the sun. Not only does your skin stop the UV light that comes in contact with it, but skin will use some of this UV light to **SYNTHESIZE** vitamin D, which is necessary for normal development of teeth and growth of bones. Your sweat glands are involved in **THERMOREGULATION** by being activated on a warm day to make and secrete a watery fluid (sweat) onto the surface of your skin which helps cool you with its evaporation and thereby maintain a normal body temperature. In addition, small amounts of some substances, such as urea and nitrogen, are **SECRETED** in sweat. Your integumentary system therefore functions in 1) **PROTECTION**, 2) **SYNTHESIS**, 3) **THERMOREGULATION**, and 4) **SECRETION**.

Many of the events in your environment will come into contact with you, which leads us to an additional important function of your integumentary system and that is for **SENSORY RECEPTION**. A **stimulus** that comes in contact with your skin can be fairly light to give your brain the **sensation** of **TOUCH**, or heavier to give your brain the sensation of **PRESSURE**. Intense touch and pressure can lead your brain to experience **PAIN**. In addition to touch, pressure, and pain, your skin can register the **TEMPERATURE** of your environment, and, as with touch and pressure, extremes in temperature can produce the experience of pain. There is an additional general sensation known as **PROPRIOCEPTION** ("muscle sense") that does not originate from your skin, but from your skeletal muscles and joints. The receptors in your muscles are the muscle spindle apparatuses which detect the stretching and unstretching of your muscles during movements. Proprioception is the sensation that allows you to know the position of your body, especially your limbs, without seeing them.

There are two reasons that you can experience these four sensations: The first reason is that you have **SENSORY RECEPTORS** that are located in the dermis of you skin. The receptors for pain and temperature are **free nerve endings**. those for touch are the **Meissner's corpuscles**, **Merkel's disks**, and **hair follicle plexuses**, and those for pressure are **Pacinian corpuscles**. These receptors are found in the skin all over the body. Lining the mouth , the surface of the tongue, and within teeth are receptors for touch, pressure, pain, and temperature. The receptor for pain and temperature is a sensory neuron, and those receptors for touch and pressure are attached to sensory neurons.

You should be aware that what happens to stimulate a receptor is **NOT A SENSATION**. The events in your environment create "stimuli" that activate your receptors. So, what happens at a receptor is a **STIMULUS**. The stimulus activates a receptor and then this message is conveyed to your brain over your sensory neurons. It is your brain that creates the **SENSATION**, and therefore the second reason that you can experience a sensation **is** your **BRAIN**.

Considering the five very important functions of your integumentary system, it is completely obvious that your integumentary system is important to your well - being. Of the five functions of your integumentary system, which one do you think is the most important ? The least important ? Or, are they all equally important ?

There is a way to easily estimate percent skin areas on the body. The method of estimation is called the **RULE OF NINES** and is based on areas that can be either nine percent, or some multiple of nine percent. The skin of each upper limb represents **nine percent**; the skin of each lower limb represents **18 percent**; the skin of the trunk (front plus back) represents **36 percent**; the skin of the face and skull (head) is **nine percent**. All of these areas should add up to 99 percent. The other **one percent** is the perineum, which includes mainly the external genitalia.

The amount of skin that you can lose the function of (due to burns, diseases, abrasions) and still survive is age dependent. There are burn centers at many hospitals in large and medium-sized cities that are of great importance to burn victims. However, if such a center is unavailable, then you can calculate the percent chances of survival from the following:

% Chances of Survival = 100 % - (% of Skin Loss + Age)

A 20 year old and a 60 year old each of whom have been burned over 30 % of his or her body do not have the same chances of survival. The 20 year old will have a 50 % chance, but the 60 year old person will have only a 10 % chance of survival.

You skin is composed of two major areas, an outer **EPIDERMIS**, and below this area, the **DERMIS**. The epidermis is composed of five **STRATA** (layers), and from superficial to deep these strata are 1) **STATUM CORNEUM**, 2) **STRATUM LUCIDUM** (found only in the skin of the palms and soles), 3) **STRATUM GRANULOSUM**, 4) **STRATUM SPINOSUM**, and 5) **STRATUM BASALE**. The statum spinosum and stratum basale are combined to form the **STRATUM GERMINITIVUM**.

The **DERMIS** is below the epidermis and contains your sweat glands, hair follicles, sebaceous glands, sensory receptors for pain, temperature, touch, and pressure, lymphatic vessels, and blood vessels.

Below your dermis is the **hypodermis**, but this fat containing layer **IS NOT** part of your skin.

1. Why may it be difficult to consider the integumentary system as a system ?
2. Why is your environment such a contradictory place for you as a living organism ?
3. As far as you are concerned, what can the events in your environment do to you ?
4. What are five important functions of your integumentaty system ?
5. What event in your environment does your skin use to make (synthesize) Vitamin D ?
6. What structure in your integumentary system is used for thermoregulation ?
7. What is the difference between a stimulus and a sensation ?
8. What is the sensation of proprioception, and what does it allow you to know ?
9. What is the receptor for pain and temperature ?
10. What are the receptors for touch ?
11. What is the receptor for pressure ?
12. What is the receptor for proprioception ?
13. Pain, temperature, touch, and pressure are grouped together and called what type of sensations ?
14. Which of the five general sensations does not originate from the skin ?
15. Explain the rule of nines.
16. What percent of total skin surface area is each one of your upper limbs ?
17. What percent of total skin surface area is each one of your lower limbs ?
18. What percent of total skin surface area is the trunk of your body ?
19. What percent of total skin surface is your head (including your face) ?
20. What percent of total skin surface area is your perineum ?
21. What is a critical factor in the survival chances of an individual following skin loss ?
22. What is another critical factor ?
23. Write the formula that describes the relationship between % survival chances and age and % skin loss.
24. What are the two major areas of skin ?
25. What are, from superficial to deep, the five strata of the epidermis ?
26. What are some structures found in the dermis ?
27. Is the hypodermis considered a part of skin ?

II. SKIN

A. Major areas of the skin

1.

2.

B. The EPIDERMIS

1. The strata (layers) of the epidermis

a.

b.

c.

d.

1)

2)

C. The DERMIS

1. Some characteristics of the dermis

a. Anatomical contents

1)

2)

3)

4)

5)

6)

b. Layers of the dermis

1) Papillary layer

2) Reticular layer

c. Langer's lines of cleavage

D. Diagram of the skin

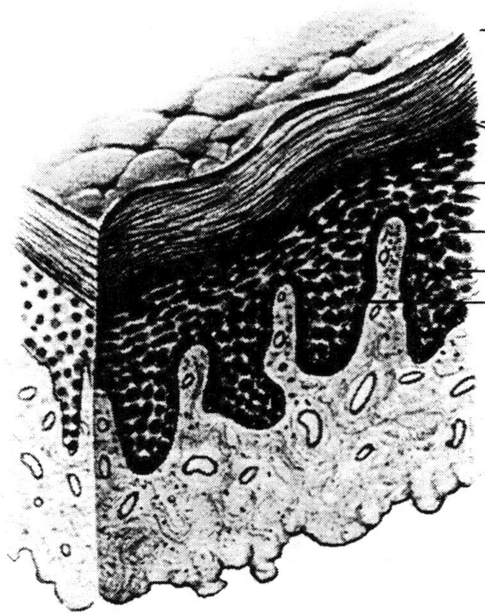

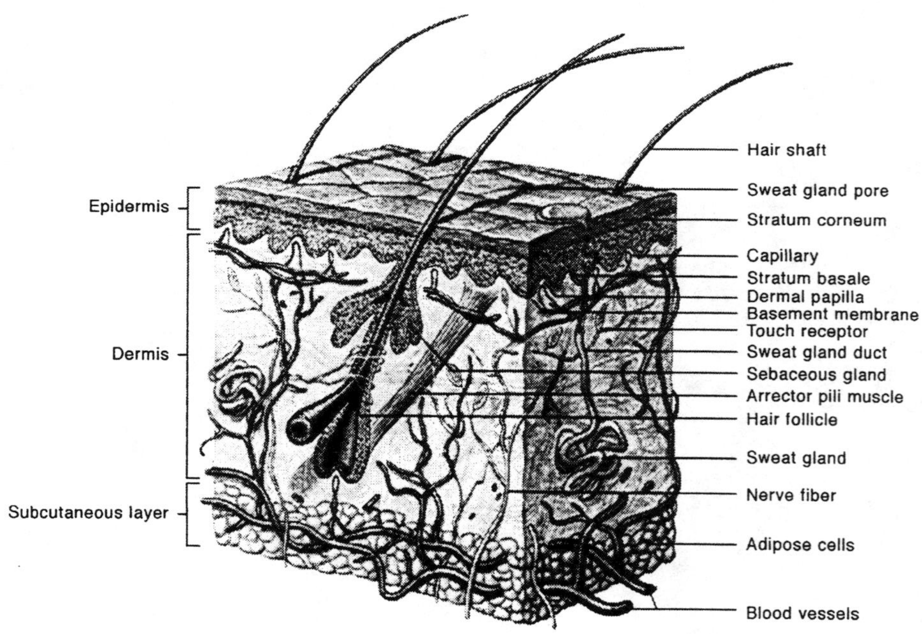

E. Functions of the skin

1.

2.

3.

4.

5.

F. Burns

1. First degree burns

2. Second degree burns

3. Third degree burns

G. Calculation of the percent survival chances of burn victims

HOW DO YOU ESTIMATE THE PERCENT OF TOTAL SKIN AREA LOST ?

H. The RULE OF NINES

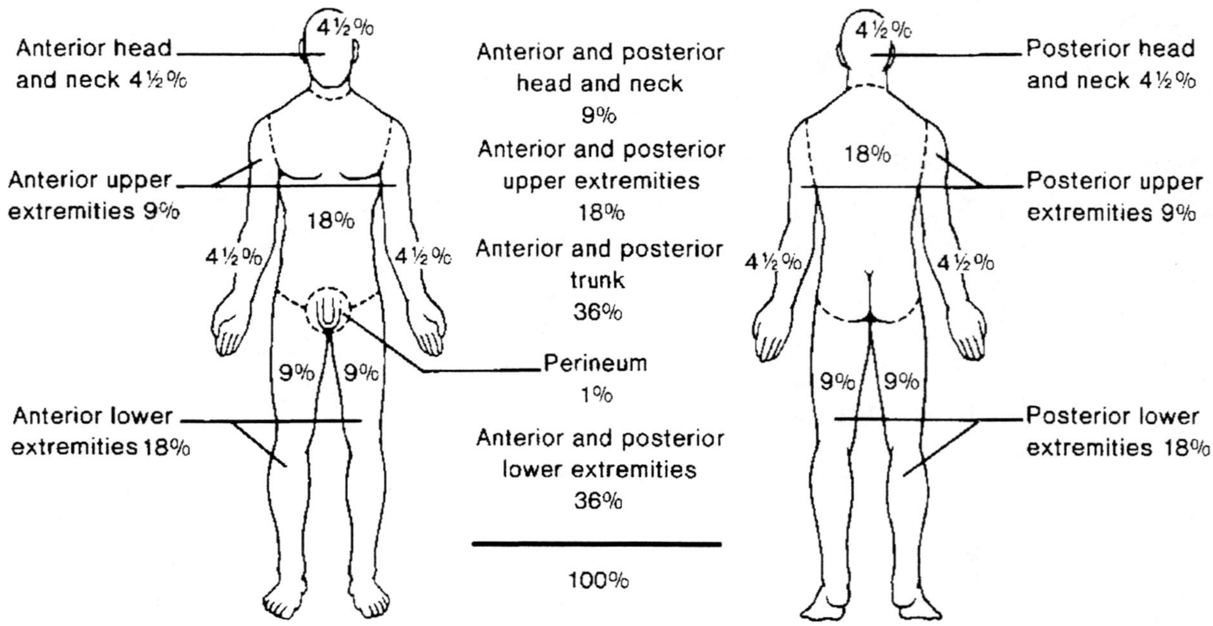

I. The **GENERAL SENSATIONS**

1. The human body is able to monitor and respond to changes in both the external and internal environments. It does so through a sensory reception network that receives (receptors), processes and interprets (analyzes) all types of sensory information. Humans are considered to be able to experience five general sensations. A greatdeal of information about your external environment is detected by sensory receptors in your skin. There are five (5) types of sensations that are referred to as "general sensations". Four of these you detect through your skin, but one (proprioception) is detected through joint pressure and muscle stretch:

a.

b.

c.

d.

e.

2. The receptors that are involved with each general sensation

a. Receptors for touch =

1) Meissner,s corpuscle

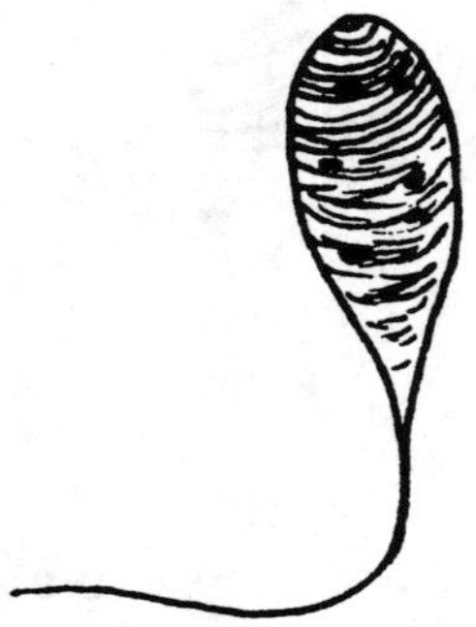

2) Merkel's disks

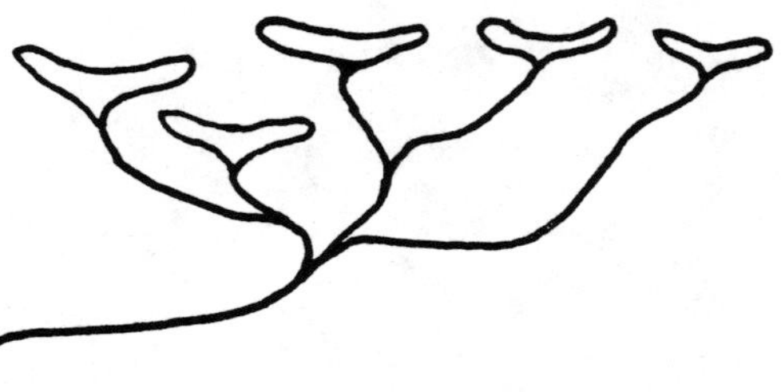

3) Hair follicle plexus

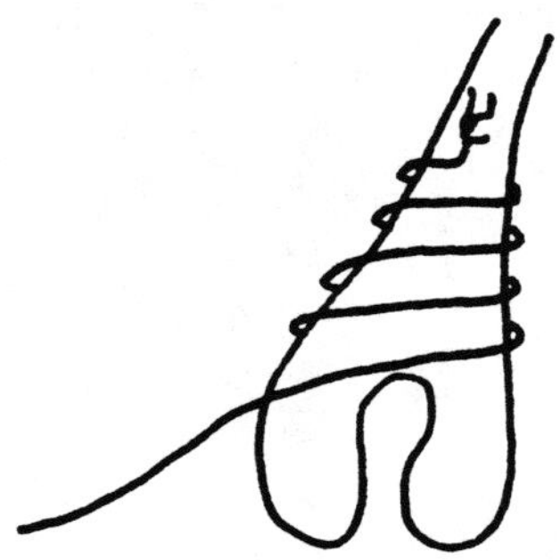

b. Receptor for pressure = Pacinian corpuscle

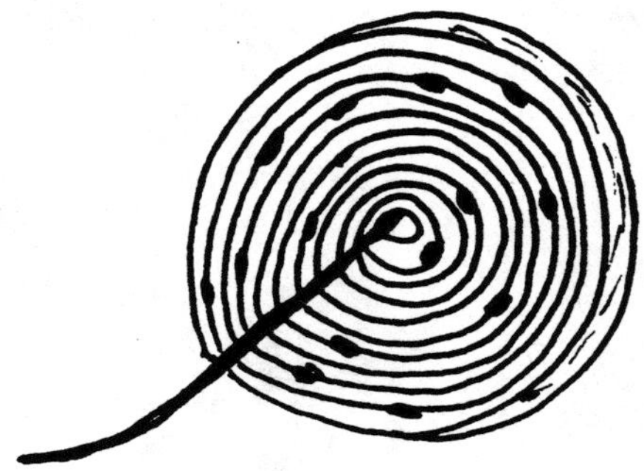

c. Receptor for pain and temperature = Free nerve endings of sensory nerves

d. Receptors for proprioception

1) Pacinian corpuscles in joints

2) Muscle spindle apparatus

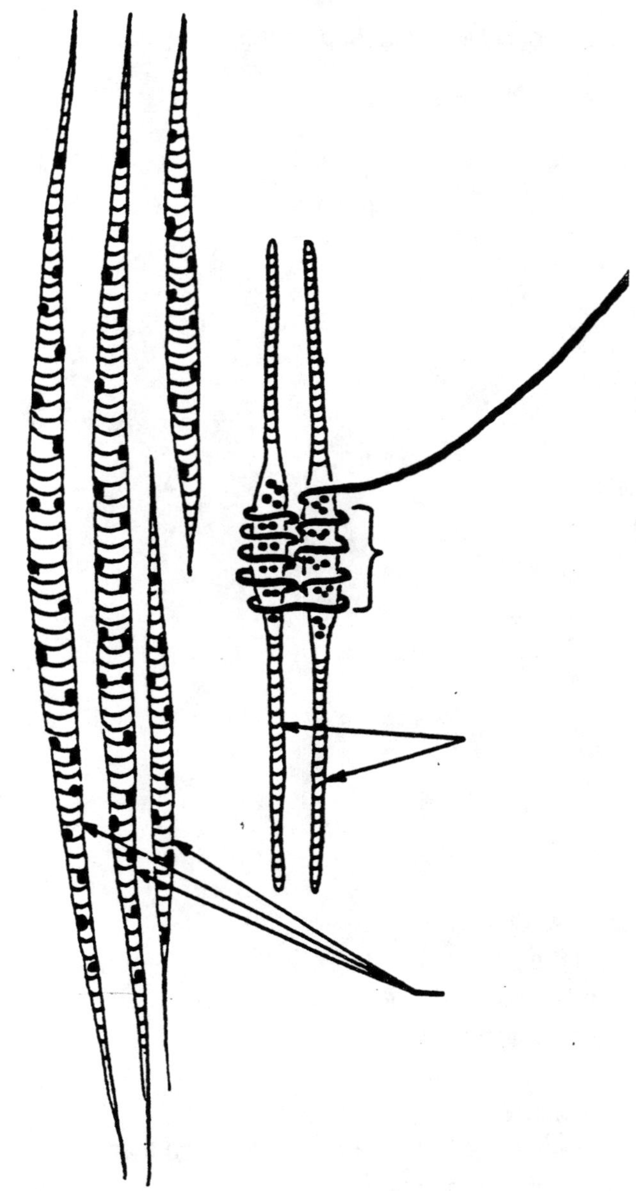

3. Classification of receptors based on their location

a.

b.

c.

d.

4. Classification of receptors based on the stimulus they respond to

a.

b.

c.

d.

e.

f.

28. What are the two major areas of the skin ?
29. What are the four strata of the epidermis ?
30. The stratum germinitivum can be divided into what two strata ?
31. What are the two layers of the dermis ?
32. What are Langer's lines ?
33. What are the functions of skin ?
34. What are the degrees of burns and the differences between them ?
35. What is the formula that describes the percent chances of survival in a non-burn center ?
36. In this formula what are the two important factors that determine the percent chances of survival after skin loss ?
37. What is the rule of nines ?

38. Give the percent of total skin area for each of the following body areas:

Right upper limb	________
Left upper limb	________
Right lower limb	________
Left lower limb	________
Trunk	________
Head / face	________
Perineum	________
TOTAL	________

39. What are the five general sensations ?
40. What is the difference between a stimulus and a sensation ?
41. What are the receptors for touch and where are they located ?
42. What is the receptor for pressure and where in the body are theses receptors located ?
43. What is the receptor for pain and temperature, and where in the body is this receptor located ?
44. What is proprioception ?
45. What are the receptors for proprioception and where are they located ?
46. Which one of the five general sensations does not originate from the skin ?
47. Classify receptors according to their location ?

THE SPECIAL SENSES

I. General Introduction

You obtain information about the events going on in your external and internal environments by a detection system made up of **RECEPTORS**. The receptors in your body are located in several places. We have already encountered the receptors in the skin which repond to stimuli that generate what we classify as the **GENERAL SENSATIONS** (pain, temperature, touch, pressure, and proprioception).

However, the skin and skeletal muscles are not the only place where sensations are located, and general sensations are not the only sensations that you can experience. You know for a fact from experience that you can see, hear, taste, smell, and keep your balance. These are the **SPECIAL SENSATIONS** and are 1) **VISION** (sight), 2) **AUDITION** (sound), 3) **GUSTATION** (taste), 4) **OLFACTION** (smell), and 5) **VESTIBULATION** (balance).

The receptors for your general sensations have a widespread, general distribution and are located in your skin, in the lining of your mouth and pharynx, your teeth, the surface of your tongue, and the lining of your nose. The receptors for your **special sensations** have specialized locations. Receptors for vision (**rods and cones**) are confined to the retina of the eye, those for audition (**hair cells**) are localized in the membranous cochlea of the inner ear, those for gustation (**taste buds**) are found on the surface of the tongue, those for olfaction (**olfactory neuroepithelium**) are in the lining of the nose, and the receptors for vestibulation (**hair cells**, but not the same ones involved with sound) are restricted to the saccule, utricle, and semicircular canals of the inner ear.

Each eye is located in a "socket" called the **ORBIT**. The structures of the eye that can be seen on yourself and others are the **SCLERA** (the outer tunic; "white of the eye"), the **CORNEA**, the **IRIS**, and the **PUPIL**. Associated structures of the eye that can be readily seen are the eyelids (**PALPEBRAE**), their lashes, and the eyebrows. Additionally, the **SUPERIOR PALPEBRAE** (upper eyelids) can be "elevated" (raised) to "open" your eyes, or "depressed (lowered) to "close" your eyes. With your eyes open the superior and inferior palpabrae create two angles at the corners where they meet. The medial angle is called the **MEDIAL CANTHUS** and the lateral angle is called the **LATERAL CANTHUS**. Just inside the medial canthus is a small, soft, mound of tissue, the **CARUNCLE**. In and around the caruncle are small openings through which the tears are drained from the surface of your eye. If we follow these tears through these openings , then we see that they enter two small tubules that lead into a small storage bag called the **LACRIMAL SAC**. From this bag tears drain into the nasal cavity through a tubule, the **NASOLACRIMAL DUCT**. In order to reach the nasal cavity this duct must pass through the **lacrimal foramen** of the skull.

Before discussing the internal structure of the eye, we should at least mention the very important functions of the lacrimal gland and tears. Each gland is located laterally and superiorly in the orbit, just above the eye. The function of the lacrimal glands is to make and secrete the tears that flow over the surface of the eyes. The tears are spread out over the anterior surface of you eye by "blinking" your upper eyelid, and in fact that is the reason for your blinking your eyes. This action also directs your tears medially and downward toward the caruncle where they can be removed.

The functions of your tears are several and critical. In their movement across your eye tears of course **1)** remove dirt and dust that has come in from the environment. **2)** They also contain a bacteriocidal agent that kills the bacteria on the surface of your eye, and **3)** provide oxygen to the outer cells of your cornea, which lack a blood supply.

Internally, the eye is divided by the **LENS** into two cavities. Posterior to the lens is the much larger **POSTERIOR CAVITY** , which is filled with a jelly-like fluid called vitreous humor, and for that reason this cavity is also termed the **VITREOUS CHAMBER**. Anterior to the lens is the smaller **ANTERIOR CAVITY**. This cavity is further divided into an **ANTERIOR AND POSTERIOR CHAMBERS** by the **IRIS**. The anterior cavity, and therefore the anterior and posterior chambers, are filled with a water-like fluid, the aqueous humor. Both of these fluids function to inflate the eye.

Inside your eye are three smooth muscles which together make up the **INTRINSIC EYE MUSCLES**. One is located in the **CILIARY BODY** and functions to change the shape of the lens during focusing. The other two are in the iris. These are the **SPHINCTER PUPILLAE** and the **DILATOR PUPILLAE**, which function to change the size of your pupil in order to regulate the amount of light entering the eye.

In addition to the tough, fibrous outer tunic (layer) of the eye, the **sclera**, there are two more. Immediately interior to the sclera is a layer (tunic) of blood vessels, the **CHOROID**, and just interior to the choroid is the layer (tunic) of photoreceptors (rods and cones). It is the retina that forms the optic nerve. There are neither rods nor cones at the small area where the retina forms the optic nerve and for this reason this area is called the **OPTIC DISK**,' or **BLIND SPOT**. Another area on the retina, the **FOVEA CENTRALIS** is unusual because it contains cones only (no rods). The fovea is also the place of most acute and detailed vision.

The **optic nerve** is the nerve of the eye. It carries visual sensory stimuli from your rods and cones to your brain. In order to reach your brain, each optic nerve must pass through an **optic foramen**.

Attached to the sclera of each of your eyes are six skeletal (voluntary) eye muscles. In addition there is one more skeletal muscle that does not attach to the eye, but enters the upper eyelid. These seven skeletal muscle outside the eye form a group called the **EXTRINSIC EYE MUSCLES**, whose functions are to move the eye and to elevate the upper eyelid.

Your ear is divided into three major areas: 1) **EXTERNAL EAR**, 2) **MIDDLE EAR**, and 3) **INNER EAR**. Your external ear is composed of the **PINNA**, and the **EXTERNAL AUDITORY CANAL**. The **TYMPANIC MEMBRANE** is at the end of the external auditory tube. On the other side of the tympanic membrane is a very small cavity called the **MIDDLE EAR CAVITY**. Within this cavity are the three smallest bones in your body, your **EAR OSSICLES** (**MALLEUS, INCUS and STAPES**). From the middle ear cavity there is a tube called the **AUDITORY TUBE** (or **EUSTACHIAN TUBE; PHARYNGOTYMPANIC TUBE**) that leads to the **nasopharynx**. The purpose of having this tube is allow you to equalize air pressure on either side of your tympanic membrane. However, bacteria in you nasopharynx can easily migrate through your auditory tube to infect and cause inflammation of your middle ear (**Otitis media**).

Anatomically, as well as functionally, our **inner ear** is slightly complicated. It consists of three interconnecting spaces in the petrous portion of your temporal bone. Collectively, these three spaces are called the **BONY LABYRINTH**. The three spaces of the bony labyrinth are the **VESTIBULE, BONY SEMICIRCULAR CANALS**, and the **BONY COCHLEA**. Within these three spaces are small, fragile, and membranous structures. Within the vestibule are the two small membranous and fragile bags, the **SACCULE** and **UTRICLE**. From the saccule emerges a single membranous tube, the **MEMBRANOUS COCHLEA**, that spirals around in the bony cochlea. From the urtricle emerges several membranous tubes, the **MEMBRANOUS SEMICIRCULAR CANALS** that are located and contained inside the bony semicircular canals. All of the membranous structures are together called the **MEMBRANOUS LABYRINTH.**

The fluid inside your membranous labyrinth structures and inflates them is called **ENDOLYMPH**. The fluid inside the bony labyrinth is called the **PERILYMPH**.

48. Of what is your detection system composed ?
49. What are the five general sensations ?
50. What are the five special sensations ?
51. On what basis are the sensations categorized as general and special ?
52. What are the receptors for vision and where are they located ?
53. What are the receptors for audition and where exactly are they located ?
54. What are the receptors for equilibrium and where exactly are they located ?
55. What are the receptors for gustation and where are they located ?
56. What are the receptors for olfaction and where are they located ?
57. What is the name of the depression in which the eye is located ?
58. List the structures of the eye that you can see on yourself in a mirror, or on someone else.
59. When you state that someone has green eyes, or blue eyes, or brown eyes, then what structure of the eye are you talking about ?
60. What is a canthus ?
61. What structure is just inside the medial canthus ?
62. Describe how tears get from your lacrimal glands to your nasal cavity.
63. What is the purpose of "blinking your eyes" ?
64. What is the term for eyelid ?
65. Which palpebra do you move when you open and close your eyes ?
66. What is the function of the lacrimal glands ?
67. What are the functions of the tears made and secreted by the lacrimal glands ?
68. What structure separates the anterior and posterior cavities of the eye ?
69. What is another name of the posterior cavity ?
70. Why is the posterior cavity also called the vitreous chamber ?
71. What are the two areas of the anterior cavity ?
72. What structure separates the anterior and posterior chambers ?
73. What is the name of the opening in the iris ?
74. What is the name of the fluid that inflates the chambers of the anterior cavity ?
75. What are the three intrinsic eye muscles ?
76. What type of muscle are the intrinsic eye muscles ?
77. Where inside the eye are the three intrinsic eyes located ?

78. What are the names of the two iris muscles ?
79. What is the function of the sphincter pupillae and the dilator pupillae ?
80. What is the function of the ciliary muscle ?
81. What are the three layers (tunics) of the eye ?
82. Which tunic is tough, fibrous, and opaque ?
83. Which tunic is the layer of blood vessels ?
84. Which tunic is the layer of photoreceptors ?
85. Which layer (tunic) forms the optic nerve ?
86. What foramen in the skull does the optic nerve pass through ?
87. What does the optic disk and the fovea centralis have in common ?
88. What are some differences between the optic disk and the fovea centralis ?
89. What is another name for the optic disk ?
90. What are the three major areas of the ear ?
91. What are the three structures of the external ear ?
92. Which of the external ear structures is a barrier between the external and middle ears?
93. What is in the middle ear cavity ?
94. What is the tube that connects the nasopharynx to the middle ear cavity ?
95. What is the purpose of the auditory tube ?
96. What is the clinical significance of having an auditory tube ?
97. What is another name for the auditory tube ?
98. In what portion of what bone of the skull is the inner ear located ?
99. Classify this bone.
100. What is the bony labyrinth ?
101. What are the three areas of the bony labyrinth ?
102. What two structures are in the vestibule ?
103. What tube emerges from the saccule ?
104. Into what space of the bony labyrinth does the membranous cochlea enter ?
105. What set of tubes emerge from the utricle ?
106. Into what spaces of the bony labyrinth do the membranous semicircular canals enter ?
107. What is thefunction of the membranous cochlea ?
108. What is the function of the saccule, utricle, and membranous semicircular canals ?
109. What structures make up the membranous labyrinth ?
110. What is the name of the fluid inside the membranous labyrinth structures ?
111. What is the name of the fluid in the bony labyrinth spaces ?

II. The EYE

A. Visible structures of the eye and structures associated with the eye

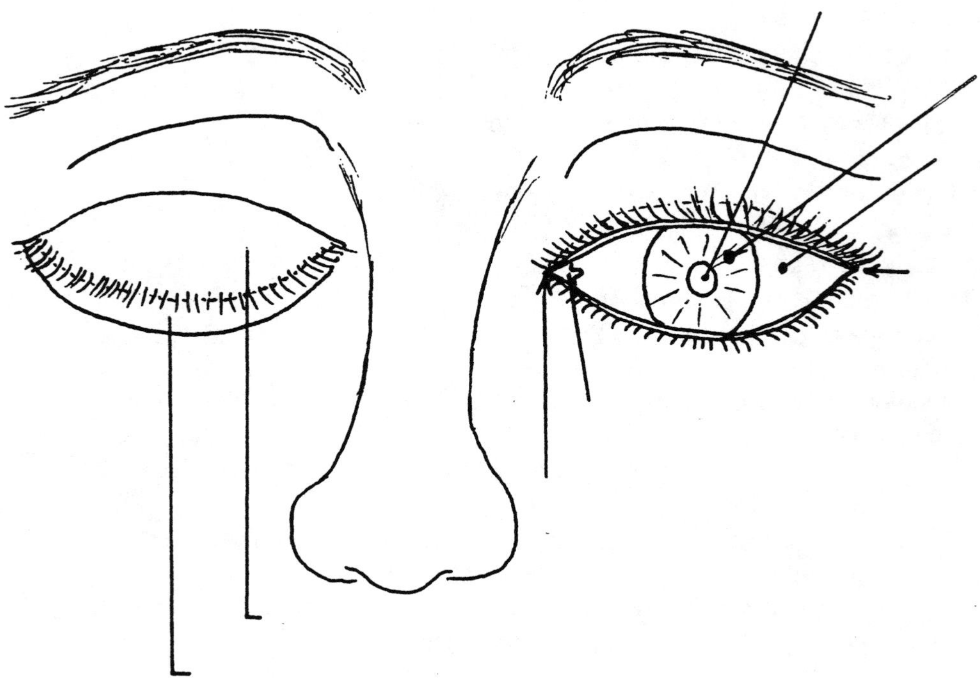

B. The lacrimal gland and the functions of tears

1. Location of the lacrimal gland and the direction of tear flow

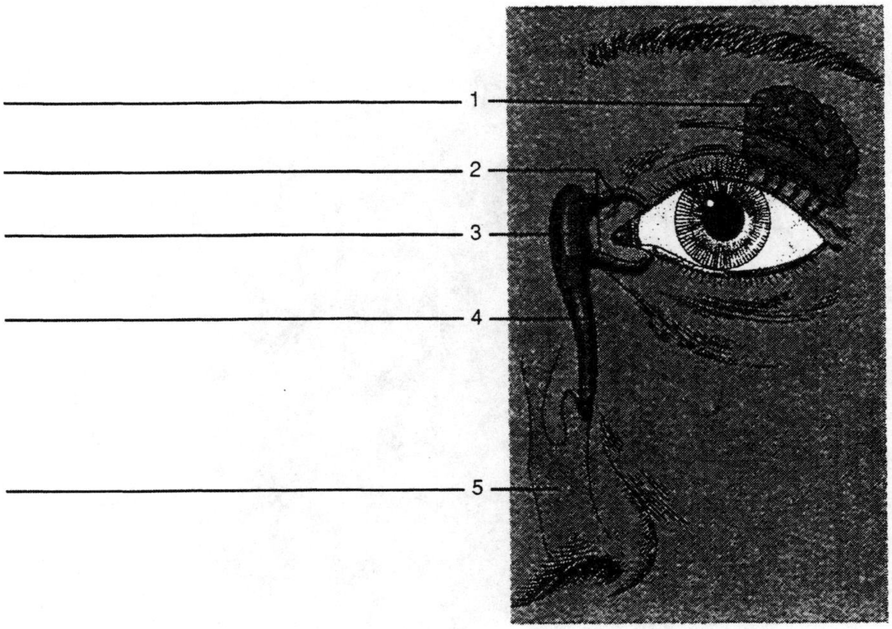

2. Functions of tears

a.

b.

c.

d.

C. Gross anatomy of the eye

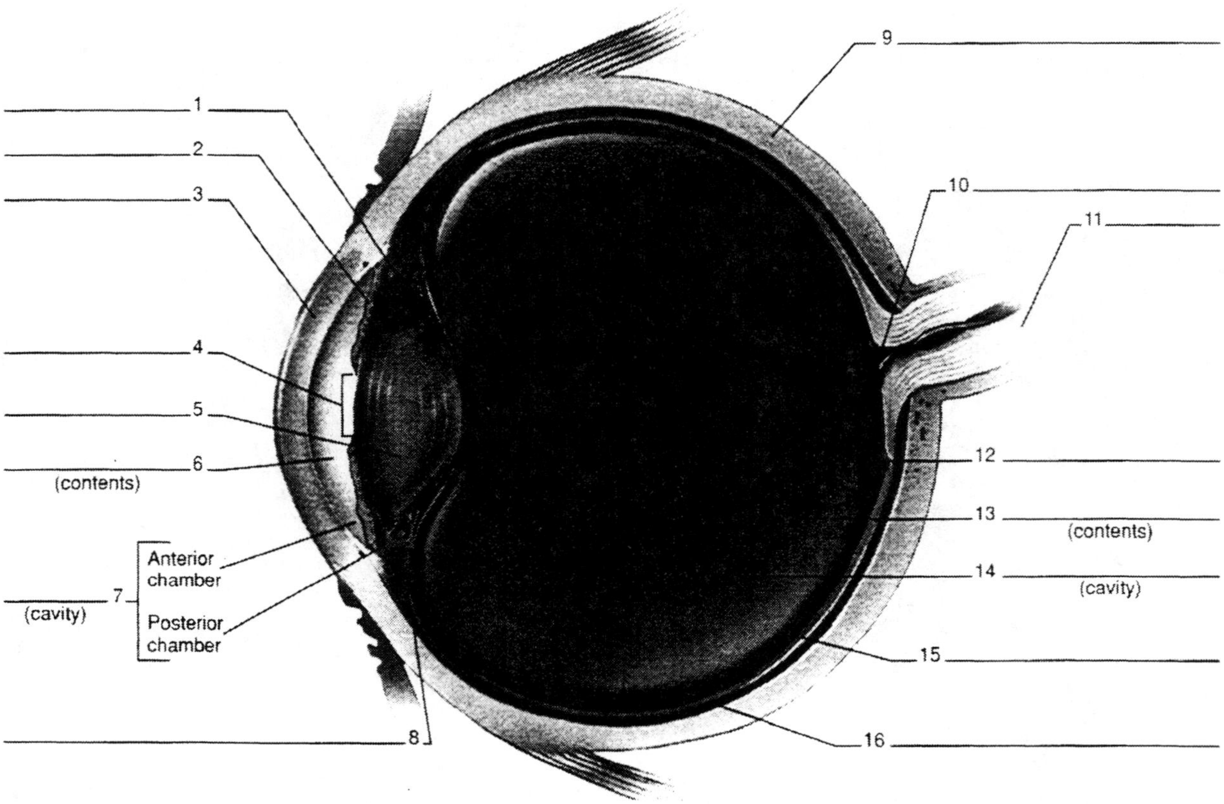

D. Aqueous humor production, the canal of Schlemm, and fluid pressure in the eye

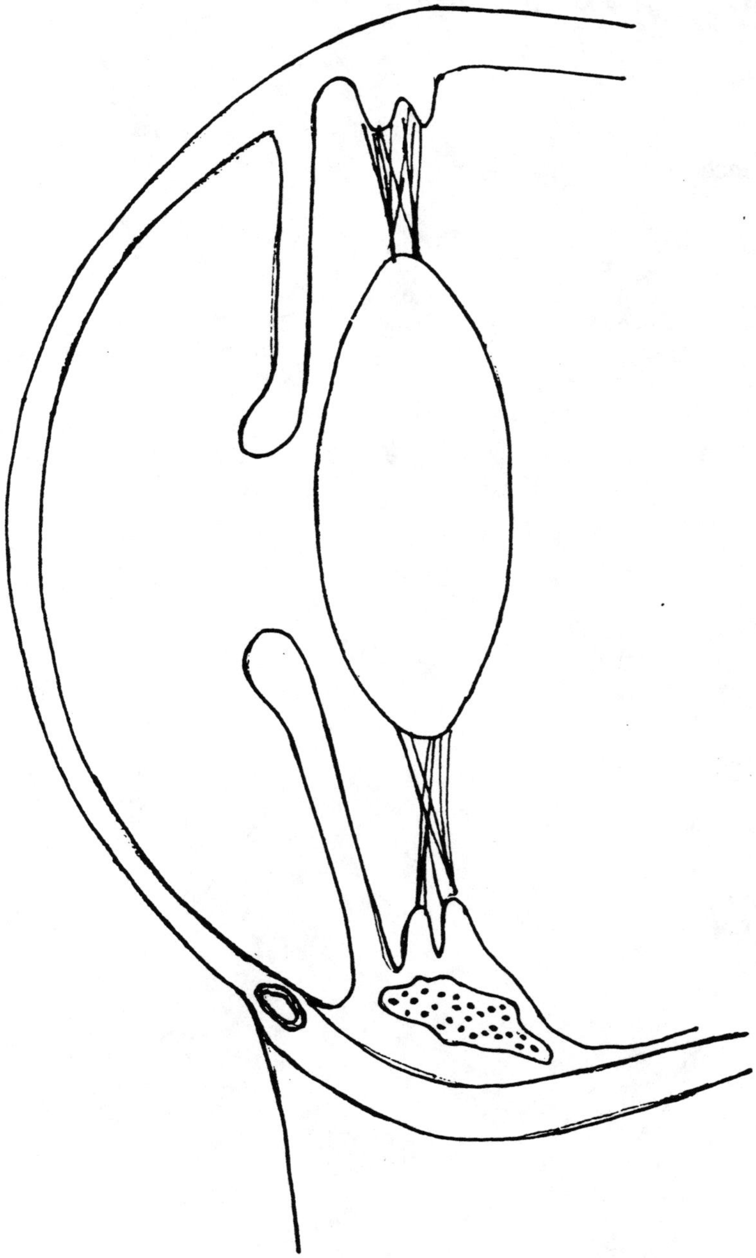

E. The three Intrinsic eye muscles

1. Type of muscle =

2. Intrinsic muscles located in the iris

   a.

   b.

3. The overall purpose of the iris muscles =

4. Functioning of the two iris muscles

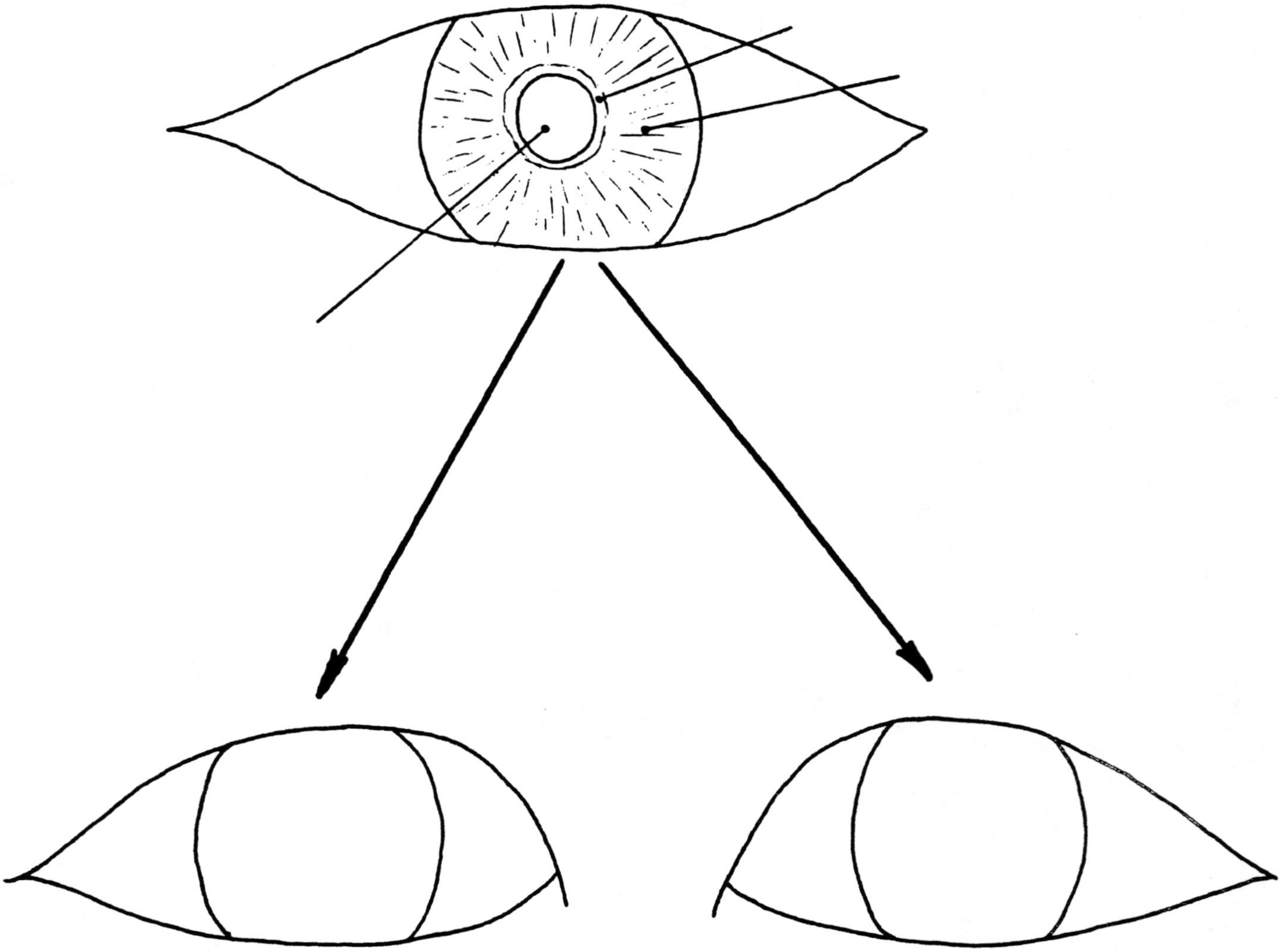

5. Functioning of the ciliary muscle and the suspensory ligament of the lens

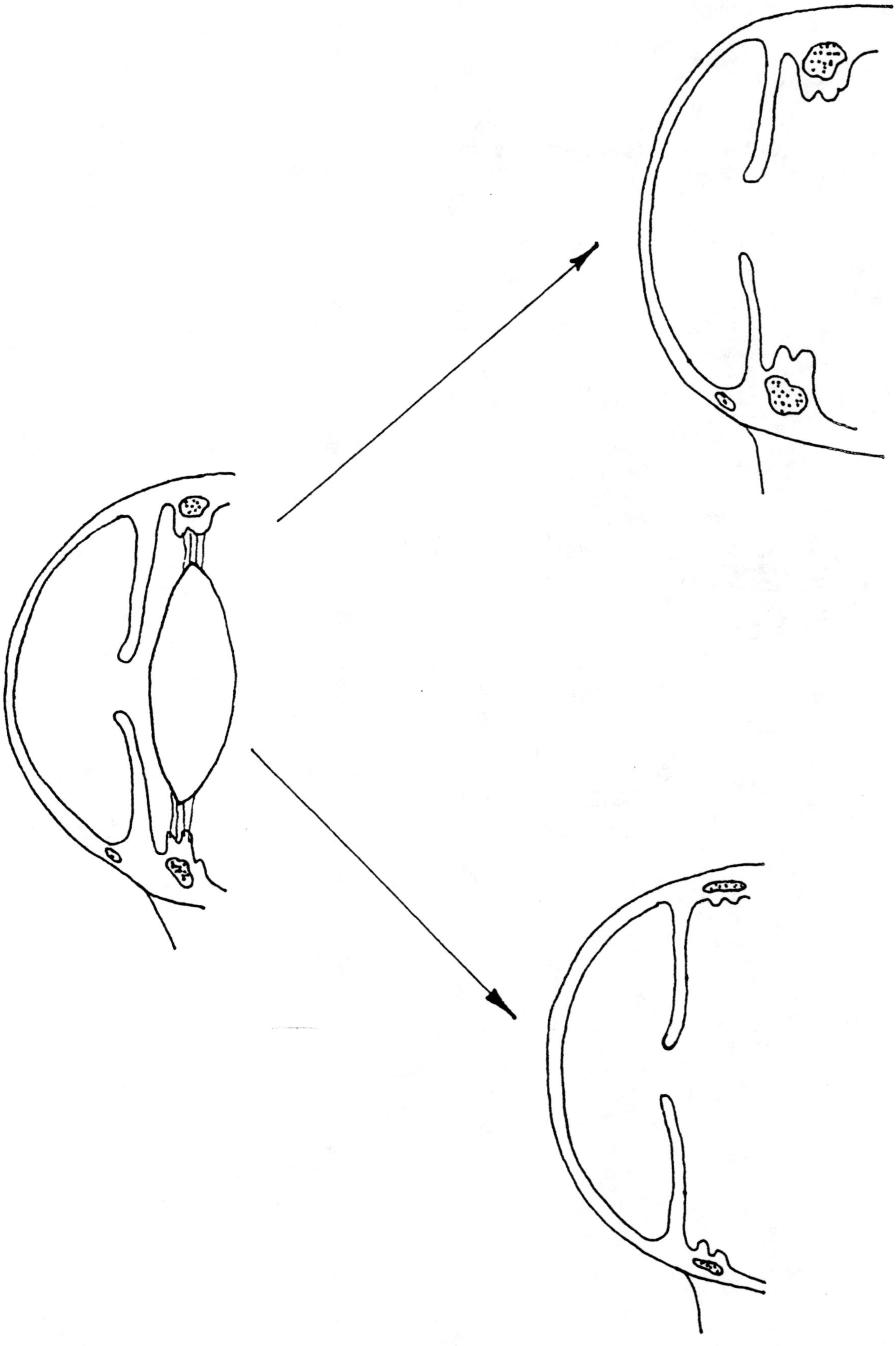

F. Extrinsic eye muscles

1. Type of muscle =

2. Functions

   a.

   b.

3. Diagrams of the extrinsic eye muscles

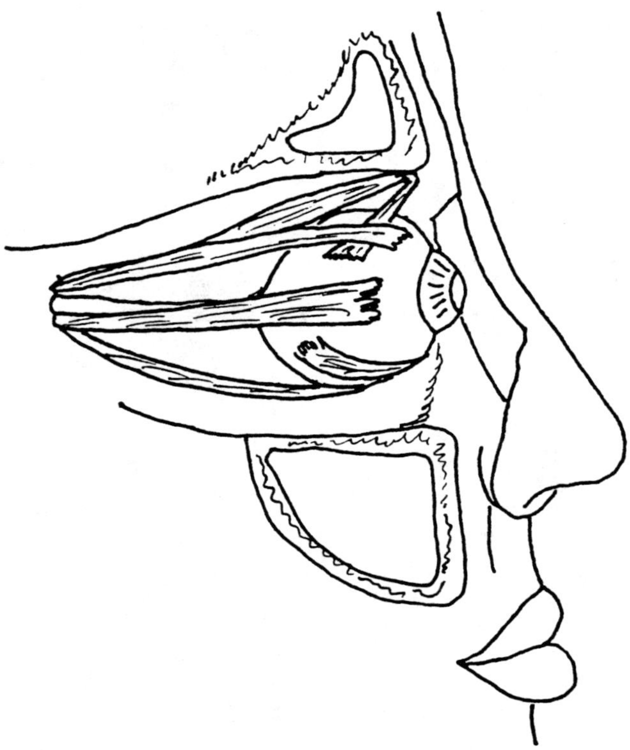

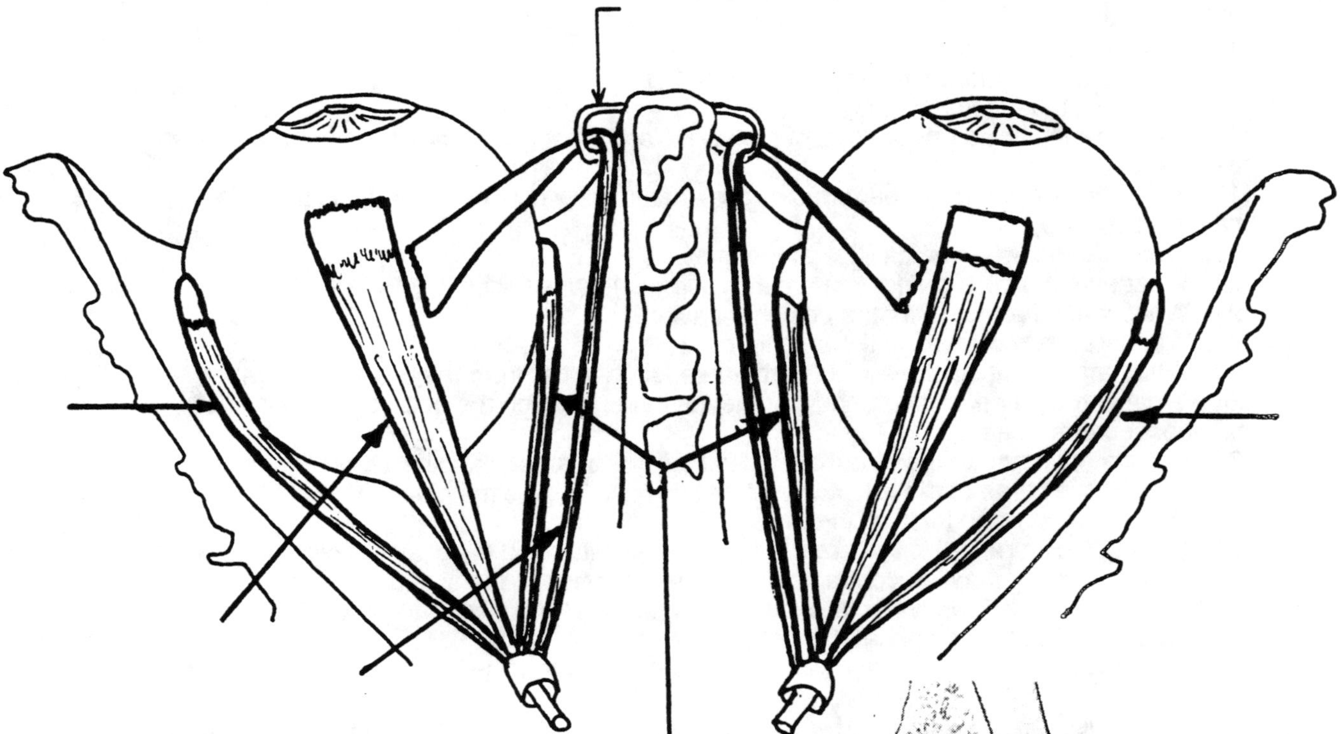

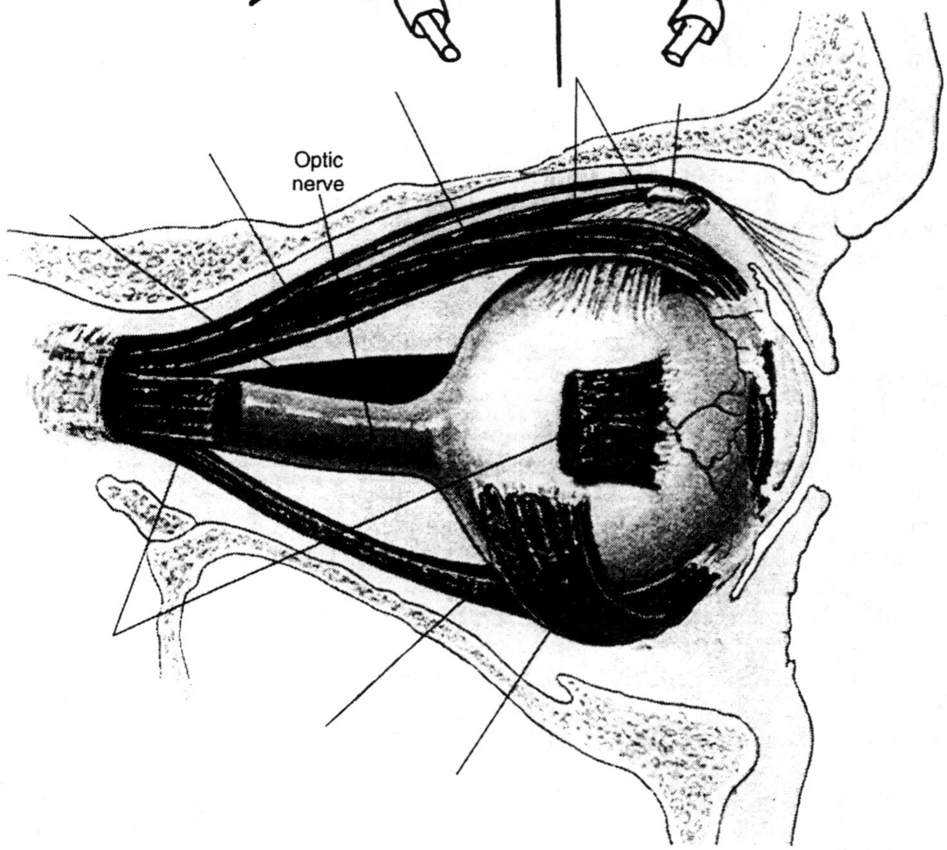
Optic
nerve

112. What are the structures of the eye that you can see ?
113. What are the associated structures of the eye that you can see ?
114. Where is the lacrimal gland located in the orbit ?
115. What is the function of the lacrimal gland ?
116. What are the functions of tears ?
117. What are the three tunics (layers) of the eye and what is their order from outside toward the inside of the eye ?
118. What thin, fragile, and clear structrure is anatomically continuous with the thick, tough, and opaque sclera ?
119. What are the two main cavities of the eye ?
120. What eye structure divides the anterior and posterior cavities ?
121. What is another name for the posterior cavity ?
122. What are the two areas of the anterior cavity ?
123. What anatomical structure divides the anterior chamber from the posterior chamber ?
124. What is the name of the opening in the iris through which the anterior and posterior chambers communicate ?
125. What is the name of the fluid that fills and inflates th anteior and posterior chambers ?
126. What is the name of the jelly-like fluid that fills and inflates the posterior cavity ?
127. What are the three intrinsic eye muscle ?
128. In what structure of the eye are the sphincter pupillae and the dilator pupillae located ?
129. If the sphincter pupillae contracts, then what happens to the size of the pupil ?
130. If the dilator pupilae contracts, then what happens to the size of the pupil ?
131. What is the overall purpose of the two iris muscles ?
132. Which iris muscle relaxes and which one contracts when you go from a light to a darker area ?
133. Which iris muscle relaxes and which one contracts when you go from a dark to a lighter area ?
134. Which intrinsic eye muscle controls the shape of the lens ?
135. What type of muscle are all three intrinsic eye muscles ?
136. Why are the intrinsic eye muscles called intrinsic ?
137. When you focus on a nearby object does your lens become more rounded, or flatter ?
137. When you focus on an object that is far away does your lens become rounded, or flatter ?
138. In what tunic of the eye is the fovea centralis located ?
138. Which tunic of the eye forms the optic disk ?
139. What is another name for the optic disk ?
140. As far as the presence of rods and cones is concerned, what is peculiar about the fovea centralis ? About the optic disk ?
141. Therefore, what is it that the fovea centralis and the optic disk have in common ?
142. Which tunic of the eye is the layer of photoreceptors ?
143. To what stimulus do these photoreceptors respond ?
144. Which of the three tunics is the tunic of blood vessels ?
145. Which tunic of the eye forms the optic nerve ?
146. What foramen in the skull does the optic nerve pass through in order to reach thebrain ?
147. What are the seven extrinsic eye muscles ?
148. Which one of these seven muscles is not attached to the eye ?
149. To what is this muscle attached ?
150. If you contract your two superior rectus muscles, then in which directions do your eyes move ?
151. If you contract your two inferior rectus muscles, then in which direction dor your eyes move ?
152. If you contract your two medial rectus muscles, then in which direction do your eyes move ?
153. If you contract your two lateal rectus muscles, then in which direction do your eyes move ?
154. You can contract both of you superior rectus muscles to simultaneously rotate both your eyes upward, or both of your inferior rectus muscles to simultaneously rotate both eyes downward, or contract both of your medial rectus muscles to simultaneously rotate your eyes medially. But, can you contract your lateral rectus muscles to simutaneously laterally rotate your eyes ?

155. What type of muscle are the extrinsic eye muscles ?
156. Why are the extrinsic eye muscles called extrinsic ?
157. To what tunic of the eye are six of the seven extrinsic eye muscle attached ?
158. What is the definition of the term abduction ?
159. Which of the movements of the eye would qualify as abduction of the eye ?
160. Therefore, which extrinsic eye muscle cause abduction ?.
161. Can you simtaneously abduct both of your eyes ?
162. If you contract your levator palpebra superioris, then what happens to your upper eyelid ?
163. What is the visual axis of the eye ?
164. If you follow the visual axis into the eye, then on what important area of the retina does it end ?
165. What area of the retina represents the place of most refined and acute vision ?
166. How is aqueous humor drained from the eye ?

III. The EAR

A. Functions of the ear

1.

2.

B. Gross structure of the ear

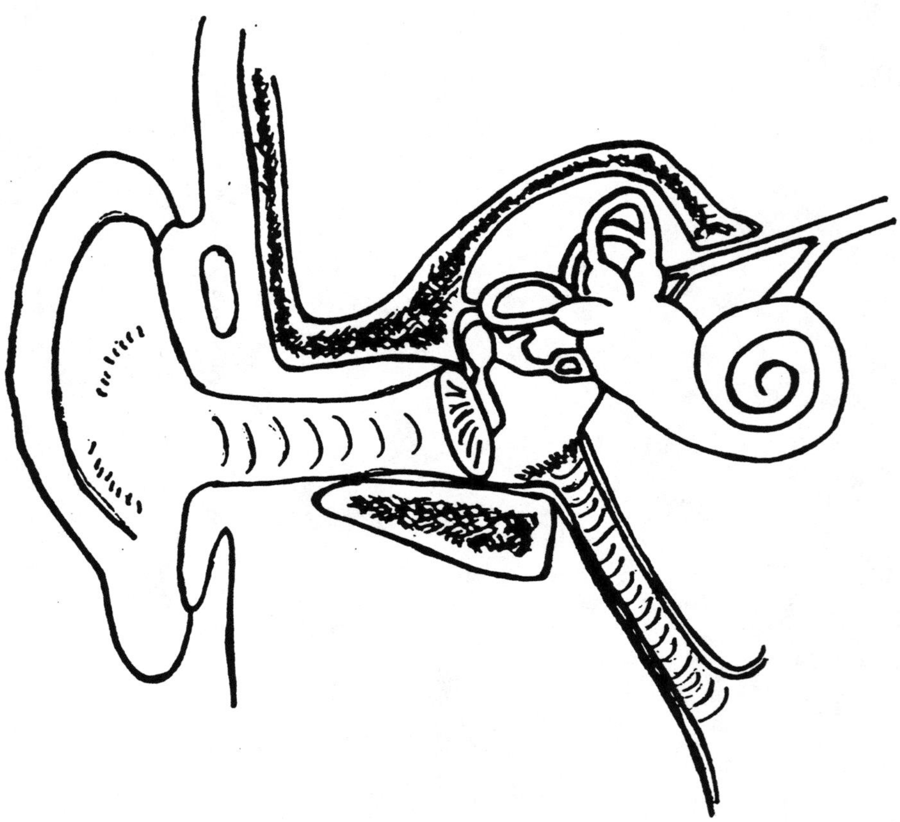

C. Anatomy of the **MIDDLE EAR**

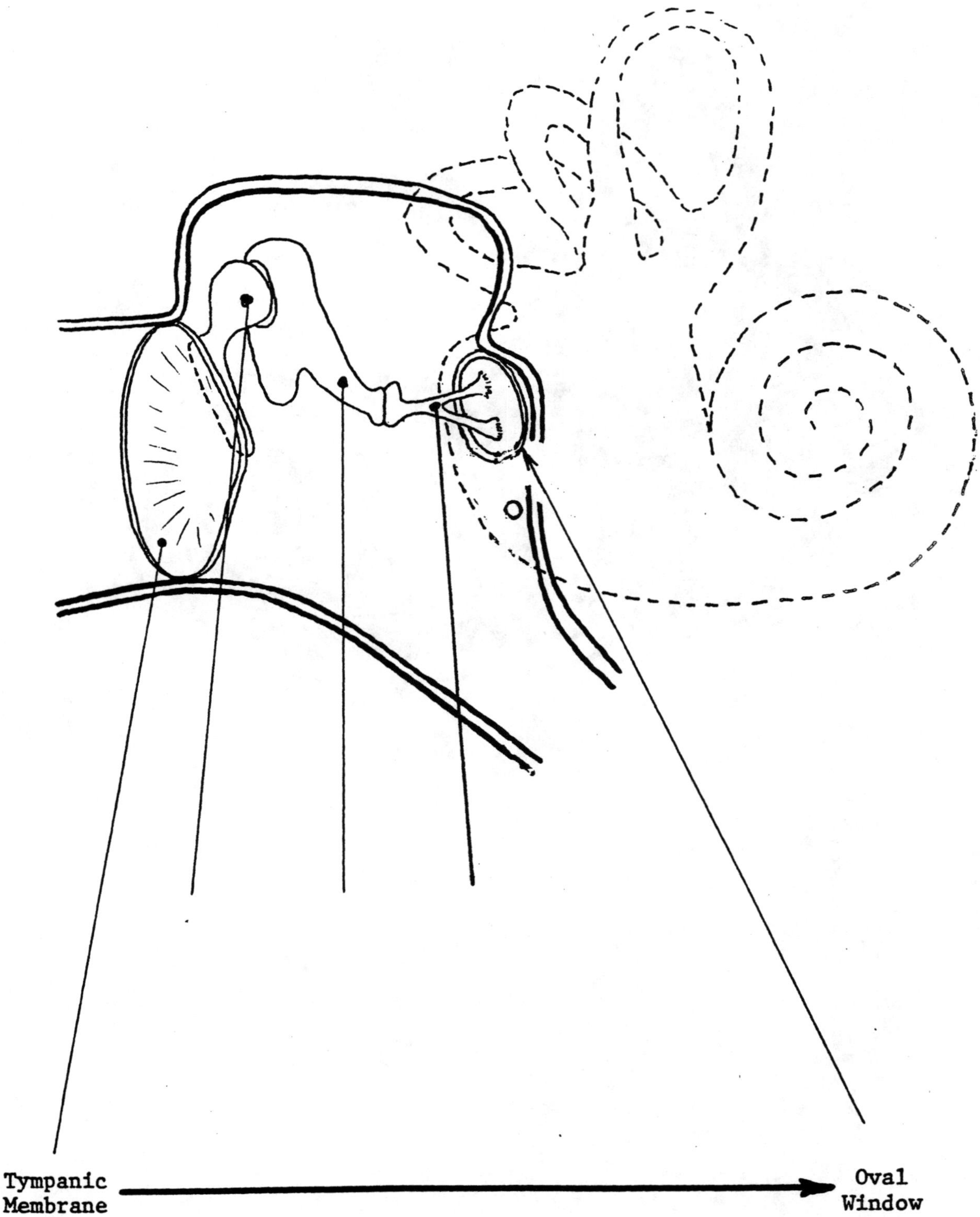

D. Anatomy of the **INNER EAR**

167. What are the two basic functions of the ear ?
168. What are the three major areas of the ear ?
169. What are the three structures of the external ear ?
170. Which of the external ear structures gathers in the compressional waves of the atmosphere ?
171. In what opening of the skull is the external auditory canal ?
172. What is at the end of the external auditory canal ?
173. Name the three ear ossicles.
174. Which of the three ear ossicles is attached to the tympanic membrane ?
175. Which of the ear ossicles is attached to and covers the oval window ?
176. Which ear ossicle forms joints with both the malleus and the stapes ?
177. What type of joint are these joints ?
178. In what area of what bone is the inner ear located ?
179. What type of bone is the temporal bone ?
180. What are the two major divisions of the inner ear ?
181. In actuality, what is the bony labyrinth ?
182. What is located in the spaces that we call the bony labyrinth ?
183. What are the three interconnected spaces that make up the bony labyrinth ?
184. What two structures of the membranous labyrinth are located in the vestibule ?
185. What tube emerges from the saccule ?
186. In what space of the bony labyrinth is the membranous cochlea located ?
187. What set of tubes emerge from the utricle ?
188. In what space of the bony labyrinth are the membranous semicircular canals located ?
189. What are the four structures that together make up the membranous labyrinth ?
190. What is the fluid that fills and inflates the structures of the membranous labyrinth ?
191. What is the fluid that is in the bony labyrinth spaces ?
192. What are two good and acceptable names for the tube between the middle ear and the nasopharynx ?
193. What is the function (purpose) of the auditory tube ?
194. What is one clinical problem that arises from having eustachian tubes ?
195. What is otitis media ?
196. Which inner ear structure functions in audition ?
197. Which inner structures function in vestibulation ?

198 - 213: For each of the following structures / areas of the ear, write an **A** if it is in the external ear, a **B** if it is in the middle ear, or a **C** if it is in the inner ear:

Vestibule______; Malleus_____; Pinna_____; Eustachian (auditory) tube_____; Perilymph_____; Tympanic membrane_______; Saccule_____; Bony cochlea______; Utricle_____; Incus_____; Membranous cochlea_____; Stapes_____; Endolymph_____; Bony semicircular canals_____; External auditory canal_____; membranous semicircular canals_____

214. What are the scala tympami and vestibuli and how are they created.
215. What is the purpose of the round window ?

IV. THE SENSE OF OLFACTION (SMELL)

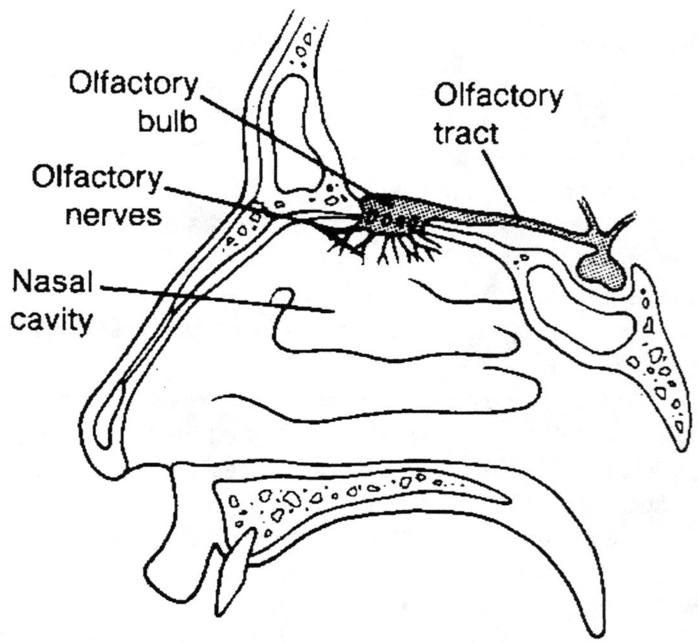

V. THE SENSE OF GUSTATION (TASTE)

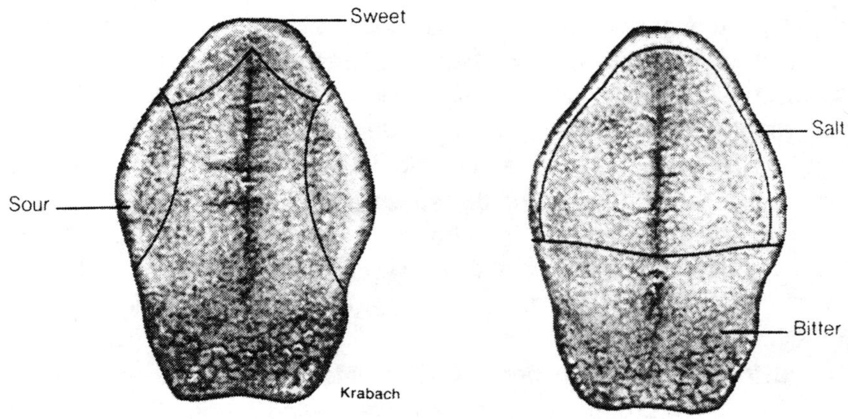

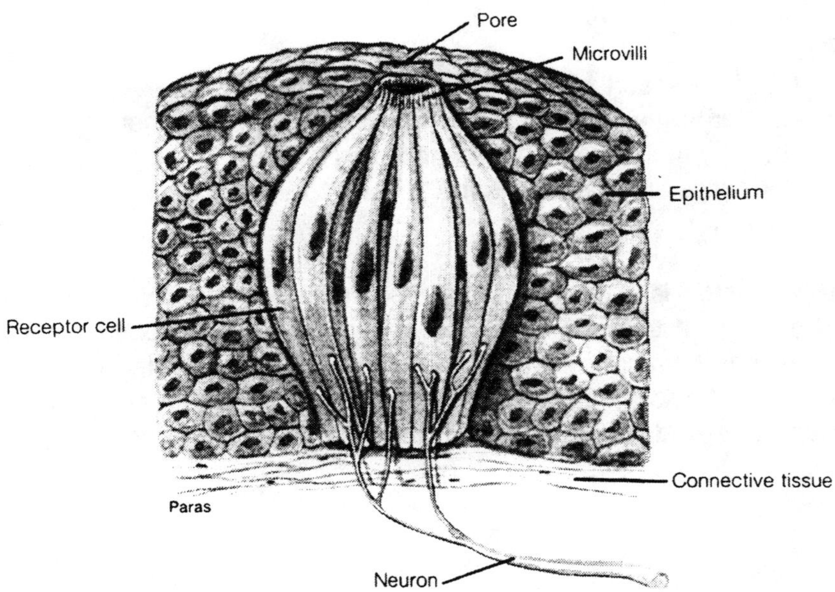

SYSTEMS THAT CONNECT AND / OR REGULATE SYSTEMS OF THE BODY

I. General Introduction

In order for your various systems to function together in an organized way, their activities (functions) must be regulated and integrated. As examples, you do not need to make and secrete HCl in your stomach constantly, only when food enters it. You do not need to contract your skeletal muscles constantly, but only a select few of the 650 that you need in order to accomplish the movement that you need or desire. Depending on the situation, you need at times to increase or decrease your heart rate and cardiac output, to shift blood to different parts of the body (to muscles during exercise, or to the digestive tract during a meal), to dilate or constrict the pupils, and for reproduction, males need to have penile erection and females need lubrication. These functions are just a few of those regulated and intergrated by the nervous system.

You also need to regulate such hormonal activities as the amount of the hormone thyroxine that your thyroid gland produces. The complicated and fascinating reproductive cycle consists of two cycles (ovarian cycle and uterine cycle) that must be regulated and properly integrated. So, not only does reproduction depend on the nervous system, but also, in females, the endocrine system. The many secretions from the cortex and medulla of the adrenal glands regulate several functions.

A hormone that is released from an endodrine gland does not have an effect on that gland but rather on another tissue or organ, called the target tissue or target organ, somewhere else in the body . The way that a hormone gets from its organ of origin to its target organ is through the blood. Hormones are released from the glands that make them into the circulating blood.

The circulatory system is a transport system through which not only hormones, but many other substances (O_2, CO_2, nutrients, urea, antibodies, etc), are transported from one part of the body to another. Although it has a few other functions (i.e., thermoregulation and immunity), your circulatory system is basically a transport system consisting of an enormous network of vessels (arteries, capillaries, and veins) through which blood is pumped by an incredible muscular pump, your heart.

II. The NERVOUS SYSTEM

A. General information and the purpose of the nervous system

Your nervous system is anatomically divided into a **CENTRAL NERVOUS SYSTEM** (**CNS**) and a **PERIPHERAL NERVOUS SYSTEM** (**PNS**). The central nervous system is easy to define for it consists of the **BRAIN and SPINAL CORD**, two large structures that are easily seen. You most likely recall, too, that the two anatomical components of the **CNS** are protected by your skeletal system.

Your **PNS** is, at this stage, more complicated to define. Practically everyone has heard of and seen pictures and diagrams of a brain and a spinal cord. But the components of the **PNS** are not well known to the general public because there are lots of them and they are spread throughout you body. Moreover, unlike your **CNS**, the components of your **PNS** are not protected by bones and therefore are more susceptible to injury. Additionally peripheral nerves can regenerate (though not 100 %), but those neurons in your **CNS** have a more difficult time and little or no regeneration occurs. However, the results of injury to your **PNS** are not nearly as debilitating and widespread as an equal injury to your **CNS**. Injury to small areas of your brain can result in all sorts of widespread deficits, such as loss of the ability to contract the skeletal muscles (paralysis) on the entire side of the body , or loss of sensations (sensory deprivation) from the entire side of the body.

The components of your **PNS** include such structures as the **1)** 31pairs of spinal nerves and spinal ganglia, **2)** the 12 pairs of cranial nerves, **3)** various nerve plexuses (brachial plexus, lumbosacral plexus), and **4)** peripheral nerves (radial nerve, ulnar nerve, axillary nerve, medial nerve, sciatic nerve, posterior tibial nerve, common peroneal nerve, and femoral nerve to name just a few). The best way to approach, study, and understand the **PNS** is to remember as you learn the nerves and plexuses that they are in the **PNS**.

Of all the systems of your body your nervous system is by far the most cpmplicated, mysterious, and intriguing, and therefore perhaps the most interesting. Certainly, it is one of the most challenging. Your entire nervous sustem (**CNS + PNS**) is made up of perhaps 150 billion **NEURONS** (**nerve cells**) that are connected to one another through very close (but not touching) encounters called **SYNAPSES**. Obviously, the enormous number of synapses between neurons, perhaps several trillion, makes for a complexity not seen in your other systems.

The **NEURON** is the anatomical unit of the nervous system. The reflex arc is the functional unit. Each neuron is composed of two types of components: **1)** The **CELL BODY** (**SOMA**), and the **2) PROCESSES** (projections that emerge from the cell body). The cell body contains all of the organelles normally found in all other types of cells and which function to keep the neuron alive. There are two types of process, and there are some very important differences between them. One type of process is called a **DENDRITE** (Dendrology is the study of trees, and dendrite means "little tree" which is what they look like under the microscope). The other type of process emerging from the cell body is the **AXON**.

YOU SHOULD BE AWARE THAT A SINGLE NEURON IS COMPOSED OF A SINGLE CELL BODY, THE SINGLE AXON EMERGING FROM THE CELL BODY AND ALL OF THE DENDRITES EMERGING FROM THE CELL BODY

So, what are the differences between an axon and a dendrite ? **1)** There is only one axon per neuron, but there can be **ONE OR MORE** dendrites per neuron; **2)** The axon has an enlargement at its end called the **SYNAPTIC KNOB** (**BOUTON; TERMINAL BUTTON; BOUTON TERMINEAUX**), in which are located many membranous vesicles called **SYNAPTIC VESICLES**. These vesicles store **NEUROTRANSMITTER SUBSTANCES** (**CHEMICALS**). The dendrite lacks a synaptic knob and therefore synaptic vesicles and neurotransmitter substances, a fact of great physiological importance.

The junction between two neurons is called a **SYNAPSE**. A synapse occurs when the axon of one neuron approaches extremely, extremely, extremely close to either the dentrite or cell body of another neuron. There is an extremely, extremely, extremely small space between the end of the synaptic knob, called the **PRESYNAPTIC MEMBRANE**, and the membrane of the cell body or dendrite, called the **POSTSYNAPTIC MEMBRANE**, called the **SYNAPTIC CLEFT**.

Although we will not dwell on the **nerve impulse** , you should nevertheless be aware that it is an electrochemical phenomonon that travels very swiftly (about 200, or so, mph) along the axon. When it reaches the synaptic knob it causes the release of neurotransmitter substances (chemicals) from their synaptic vesicles, where they are stored, into the synaptic cleft. The chemicals then diffuse very rapidly across the synaptic cleft and come into contact with the postsynaptic membrane of the next neuron. If enough of the neurochemical substance is released into the synaptic cleft and comes into contact with the next neuron, then a nerve impulse will be generated in that neuron. What happens if a nerve impulse travels out to the end of a dendrite ? Nothing. There is no neurotransmitter substances there to be released.

Although your nervous system is complicated, its complexity and mystery can be diminished and you can also get a real clue about the workings of your nervous system if you stop and consider the three things that you need to do in order to survive and get through the day safely. The three things that you must do, and the order in which you must do them, are: **1)** **DETECT**, **2)** **ANALYZE**, and **3)** **RESPOND**.

In your environment there are all sorts of physical and animate objects. Some of these objects are moving and may come into contact with your body (skin), and / or create compressional waves that enter your ears. Others perhaps reflect light waves that enter your eyes, or give off chemicals that you can taste or smell. You have **RECEPTORS** in your skin that detect when an object comes into **contact** with you. When contact does occur the receptors in your skin are stimulated and signals (nerve impulses) are sent from these stimulated receptors to your brain and, according to the type of contact (light, heavier, intense), the neurons in your brain consciously register the stimulus as touch, pressure, pain, or some combination of these. Your skin has receptors that detect the temperature of the environment.

If objects and / or events in your environment emit **light waves**, then these waves stimulate the rods and cones in your retinas (or retinae). The stimulated rods and cones initiate nerve impulses in neurons that lead to the brain for the conscious appreciation of the sensation of vision.

If some object in your environment moves, then that movement **disturbs (compresses) the molecules in the air** around it. This compression spreads away from the object and enters your ears to stimulate the hair cells in your membranous cochlea of your inner ear. This stimulation initiates nerve impulses that reach the brain where the neurons there will register this as the <u>sensation of sound</u>. **Chemicals** that enter the oral cavity (usually in food) stimulate the taste buds and your brain will register this stimulation as <u>taste</u>. Chemicals that enter your nose will stimulate the olfactory neuroepithelium which your brain will interpretate as <u>smell</u>.

It is obvious that we need to <u>draw distinctions between events, stimuli, and sensations</u>. Events are happenings that emit some disturbance that contains **energy.** It is the energy contained in what is emitted (energy in movement, energy in electromagnetic radiation, the energy in compressional waves, the energy contained in the chemical bonds of the molecules we taste and smell) that causes the <u>stimulation and activation of your receptors</u>. The stimulated receptor will initiate a nerve impulse in the sensory neuron to which it is attached. This nerve impulse travels to the brain to stimulate neurons there. It is these stimulated neurons in your brain that create the sensations that you are consciously aware of, the sensations of pain, temperature, touch, pressure, proprioception, vision, sound, taste, small, and balance. If some injury , such as spinal cord injury, interupts the movement of nerve impulses between your receptors and the neurons in your brain, then events in your environment still occur and your receptors are still stimulated, but the neurons in your brain are not stimulated and therefore you will experience no sensations.

There are three **<u>SENSORY NEURONS</u>** between a receptor and the neurons in your brain that are responsible for conscious appreciation that receptors have been stimulated. These are known simply as **<u>FIRST, SECOND, and THIRD SENSORY NEURONS</u>**.

It is not enough to only become consciously aware that your receptors have been stimulated. It is absolutely necessary that you make sense (**<u>ANALYZE</u>**) out of your general sensations (determine shapes, extent of pain, extent of temperature, textures, size, weight, etc.), and your special sensations (shades of colors, visual shapes, depths, velocities, pitch of sound, language, music, noxious tastes and smells, etc). So, once the neurons of initial conscious apprecition have been stimulated, they in turn send their nerve impulse to other neurons that perform analyses.

The neurons that perform the initial conscious sensory awareness in your brain are located in specific areas of your brain called **<u>PRIMARY SENSORY AREAS</u>**. The neurons that perform the analyses are located in other areas called **<u>ANALYTICAL (SECONDARY) AREAS</u>** .

Still, detection and analysis may not be enough to survive for there are many events occurring that will require you to **<u>RESPOND</u>** to them. The two types of structures, called **<u>EFFECTORS</u>**, that will allow you to respond events. Your two types of effectors are **1) <u>MUSCLES</u>**, and **2) <u>GLANDS</u>**. Therefore, muscle contractions and glandular secretions represent the only two ways that we can respond to events.

1. What are the two major divisions of the nervous system ?
2. What two major structures make up the central nervous system ?
3. What are the structures of the peripheral nervous system ?
4. What is the anatomical unit of the nervous system ?
5. What is the functional unit of the nervous system ?

6. How many pairs of spinal nerves are there ?
7. How many pairs of cranial nerves are there ?
8. Name some peripheral nerves.
9. Describe the anatomy of a single neuron.
10. How many axons can a neuron have ?
11. How many dendrites can a neuron have ?
12. What is the study of trees ?
13. What does the term dendrite mean ?
14. Why are dendrites called dendrites ?
15. What are some differences between an axon and a dendrite ?
16. What is a nerve impulse ?
17. What is a synapse ?
18. At the synapse do the two neurons involved touch one another ?
19. Where is the presynaptic membrane located ?
20. Where is the postsynaptic membrane located ?
21. What is the extremely small space located between the pre- and postsynaptic membranes called ?
22. Describe what a nerve impulse cause when it reaches the end of an axon.
23. Describe what happens when a nerve impulse reaches the end of a dendrite. Why ?
24. What are the three necessary things that you must do in order to survive the events in your environment ?
25. What structures does your body use to detect events in your environment ?
26. What is an event ?
27. What is a stimulus ?
28. What is a sensation ?
29. Can you receive and detect a stimulus, but not experience a sensation from it ?

30 - 45: Fill in the following table for correlating stimulus, receptor and location, and resulting sensation:

STIMULUS	RECEPTOR and LOCATION	RESULTING SENSATION
Light waves	30____________	31________
Compressional (sound) waves	32____________	33________
Chemicals in the air	34____________	35________
Chemicals in your food	36____________	37________
Light contact	38____________	39________
Heavier contact	40____________	41________
Intense contact	42____________	43________
Temperature changes	44____________	45________

46. How does a stimulus of a receptor get to your brain ?
47. How many sensory neurons are between a receptor and a primary sensory area of your brain?
48. What are the areas of the brain called that allow you to become conscious of a sensation ?
49. What are the brain areas called that do the analyses of your sensory information ?
50. With what two types of structures do you respond ?
51. What are muscles and glands collectively called ?

1. **The CENTRAL NERVOUS SYSTEM (CNS)**

    a. Components of the CNS

        1)

        2)

    b. Some anatomical characteristics of the CNS

        1)

        2)

        3)

        4)

2. **The PERIPHERAL NERVOUS SYSTEM (PNS)**

    a. Some components of the PNS

        1)

        2)

        3)

        4)

        5)

    b. Some anatomical characteristis of the PNS

        1)

        2)

        3)

        4)

C. THE NEURON

1. Also called a nerve cell
2. Basic anatomical unit of the nervous system
3. Anatomical characteristics

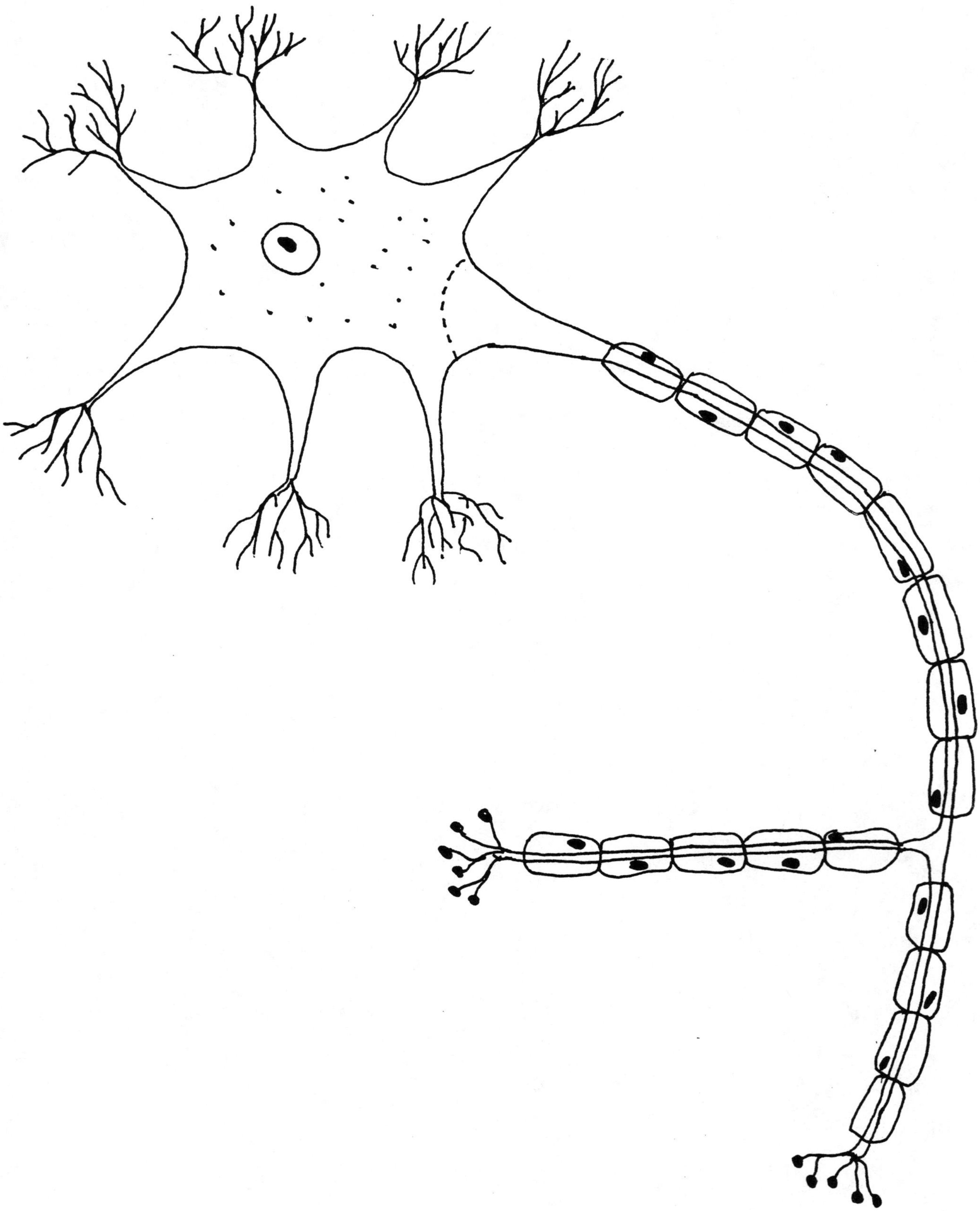

4. Not all neurons are myelinated as shown in the previous diagram. **Myelination** of neurons increases the rate of transmission of the nerve impulse. Generally, neurons with long dendrites or long axons will be myelinated, because conduction of the impulse would take too long without myelination. Each myelin coat is called a "neurolemma" and is produced by a single Schwann cell. The myelin is wrapped in layers around the axon as demonstrated in the following figure. The electrical (nerve) impulse does not travel through the thick myelin coat, but instead, "jumps" from node to node (saltatory conduction), which greatly increases the rate at which the impulse can be sent. You should notice that myelin is NOT part of the axon, but is wrapped around it.

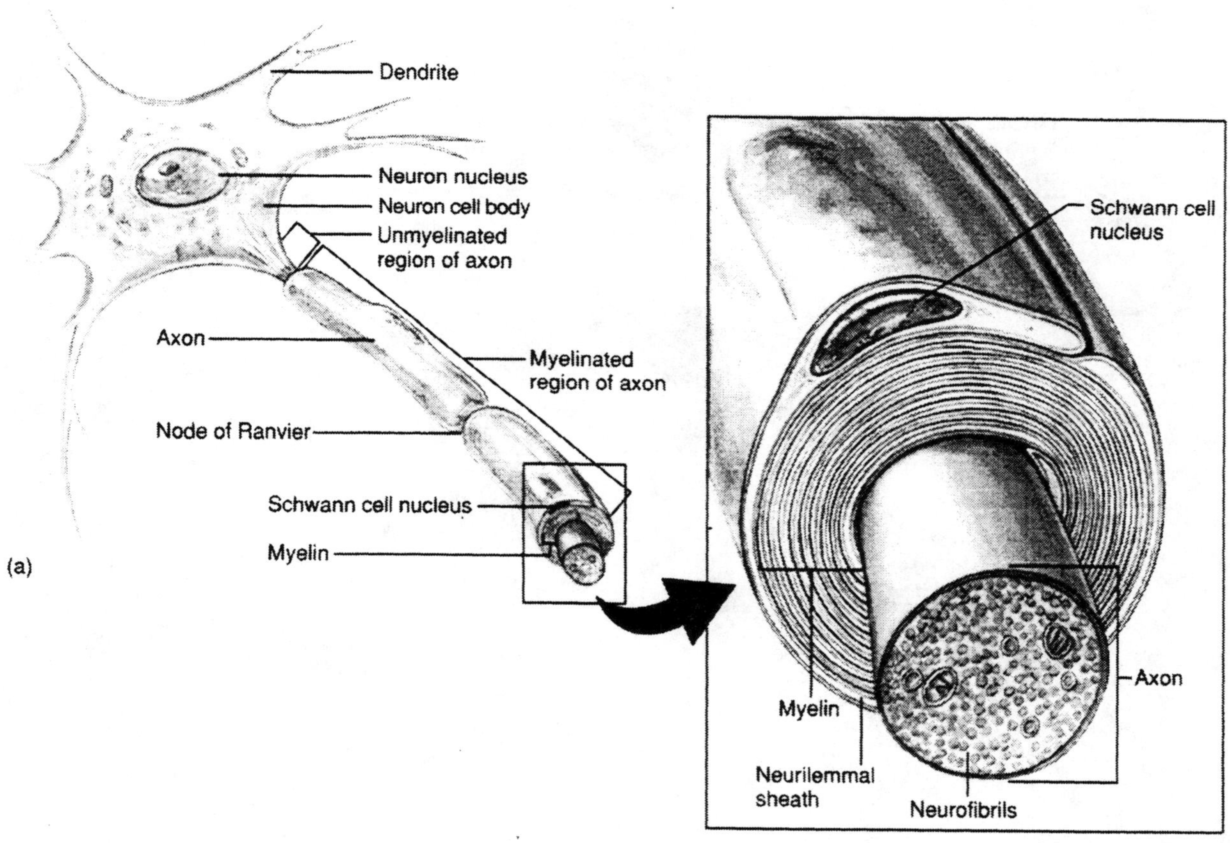

5. Classification of neurons on the number of processes emerging from the cell body.

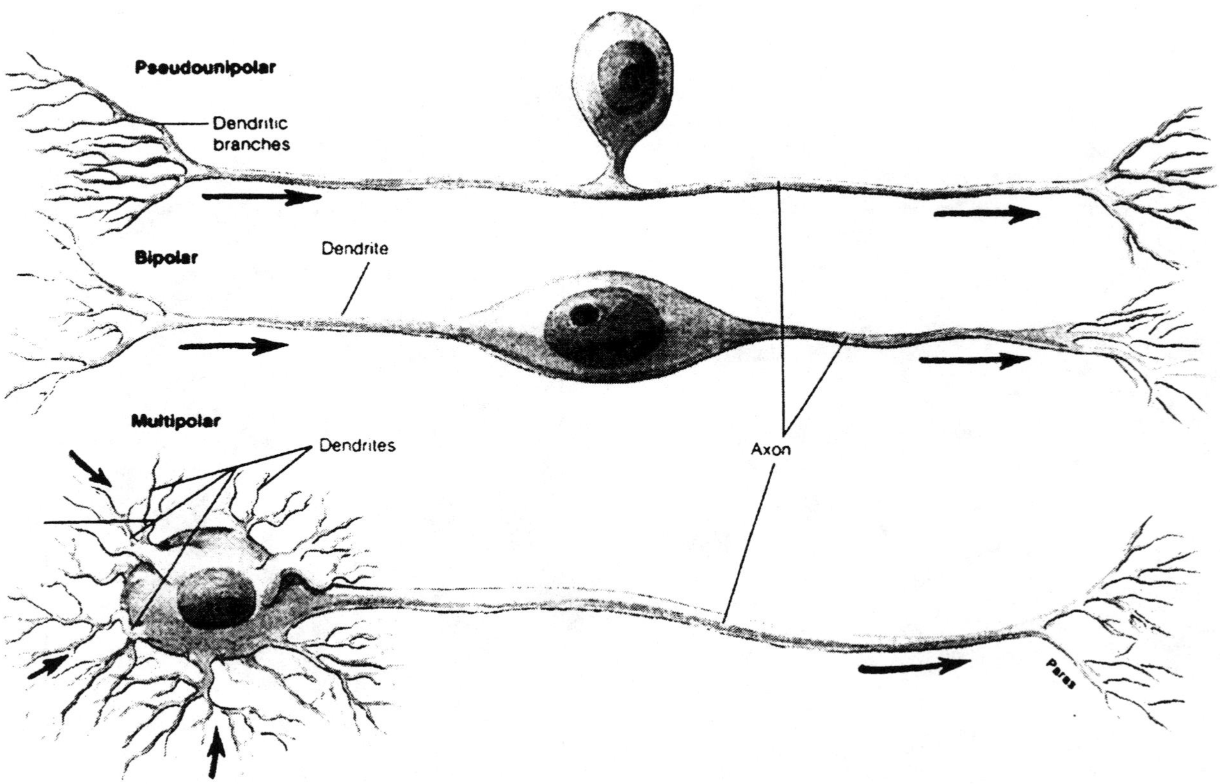

6. The **SYNAPSE** between two neurons

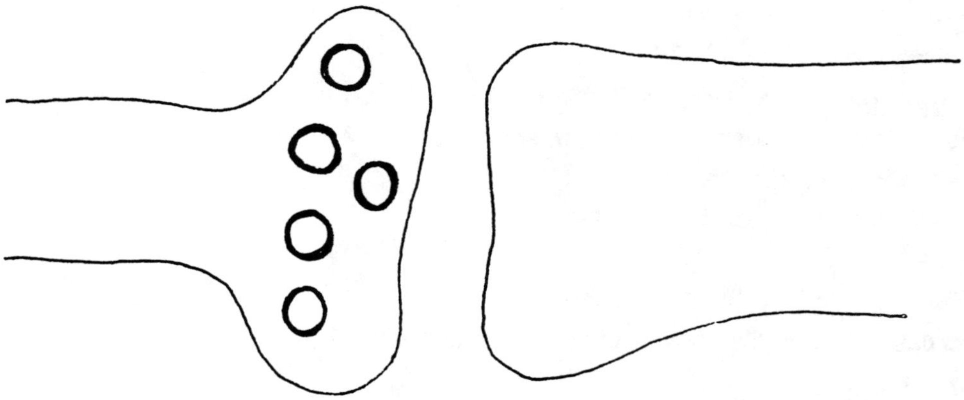

7. **Neuroglia**

Another type of non-nerve cell in the nervous system is the glial cell (or neuroglial cell). Not being neurons, these cells do not conduct nerve impulses but are essential to the neurons for their survival. The glial cells function to support, protect, and supply nourishment to the adjacent neurons that they surround and hold together like "glue". There are four (4) types of glial cells found in the nervous system

a.

b.

c.

d.

51. What are the two major divisions of the human nervous system ?
52. What are the two major components of the CNS ?
53. What are some anatomical characteristics of the CNS ?
54. What are some components of the PNS ?
55. What are some anatomical characteristics of the PNS ?
56. What is the name of the nerve cell ?
57. What is a Schwann cell ?
58. What is a node of Ranvier, and what is its relationship to the nerve impulse ?
59. What are some differences between the axon and a dendrite ?
60. How many axons can a neuron have ?
61. How many dendrites can a neuron have ?
62. What is myelin ?
63. What cell makes myelin ?
64. Classify neurons based on the number of neurons that the cell body has ?
65. What is a synapse ?
66. Where is the presynaptic neuron located ?
67. Where is the postsynaptic membrane located ?
68. Where are synaptic vesicles located ?
69. What is stored in the synaptic vesucles ?
70. The synaptic cleft is between what two membranes ?
71. On what basis are pre- and postsynaptic neurons termed as such ?

D. THE CENTRAL NERVOUS SYSTEM

1. the **BRAIN**

a. Gross anatomy of the brain - **LATERAL VIEW**

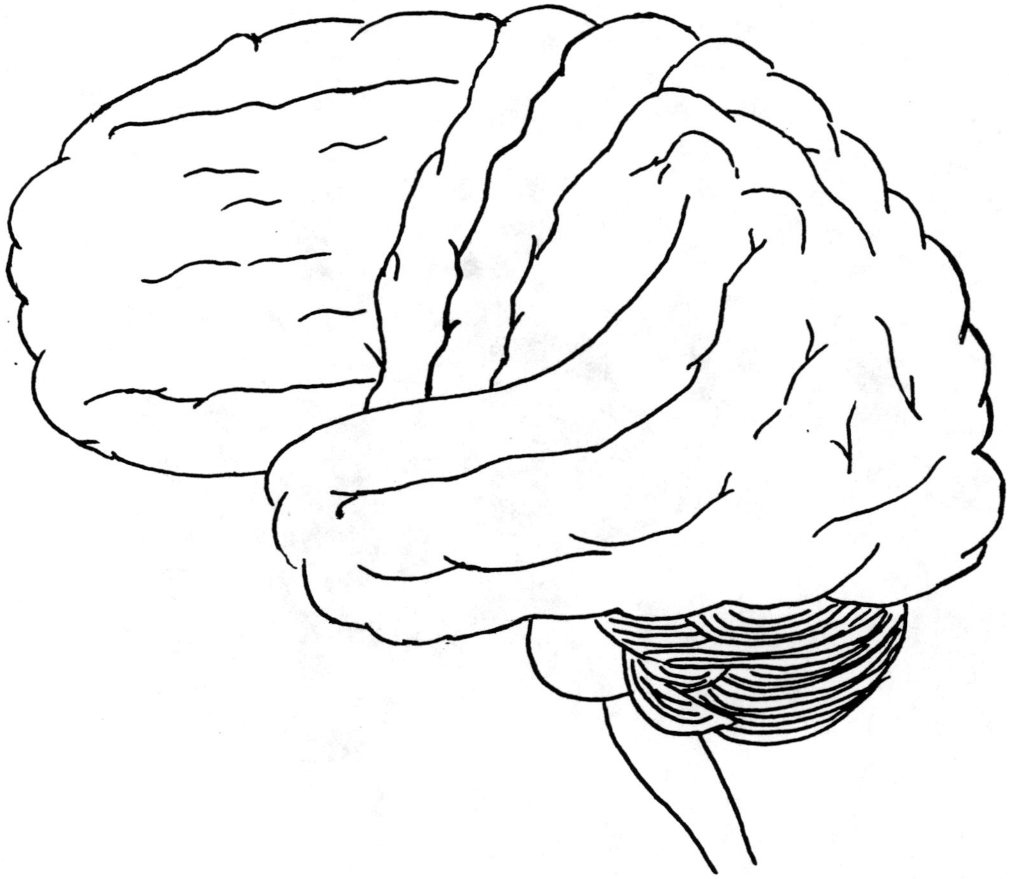

b. Gross anatomy of the brain - **MEDIAL VIEW**

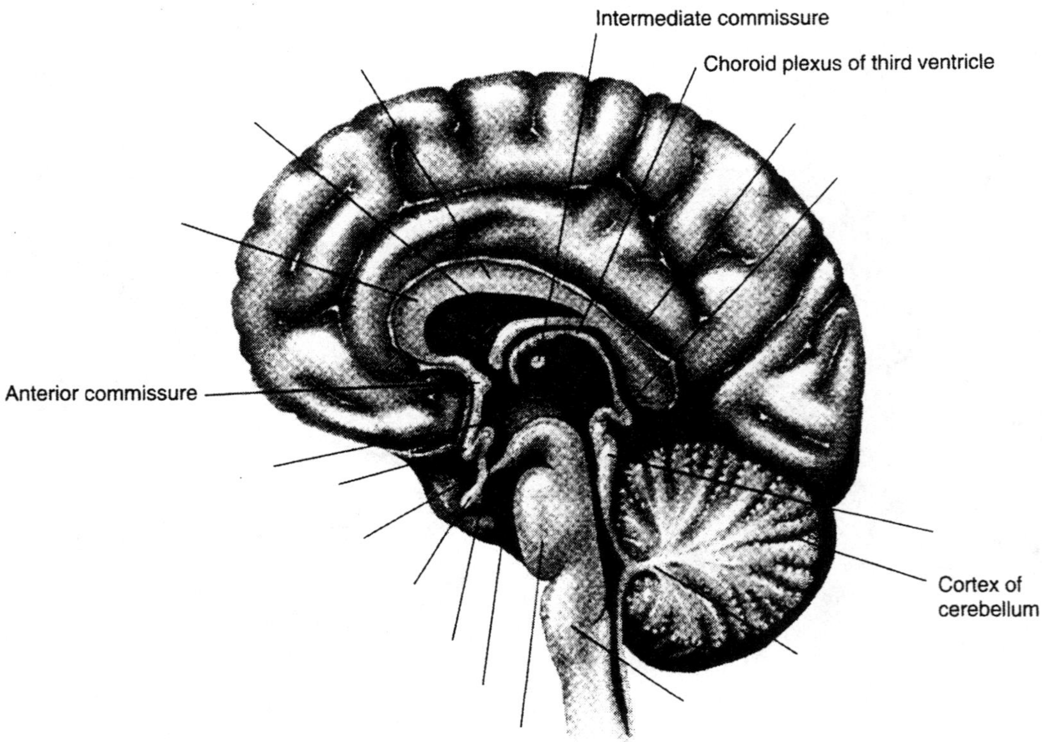

c. The five (5) major brain region

1) The **TELENCEPHALON** =

2) The **DIENCEPHALON** =

3) The **MESENCEPHALON** =

4) The **METENCEPHALON** =

5) The **MYELENCEPHALON** =

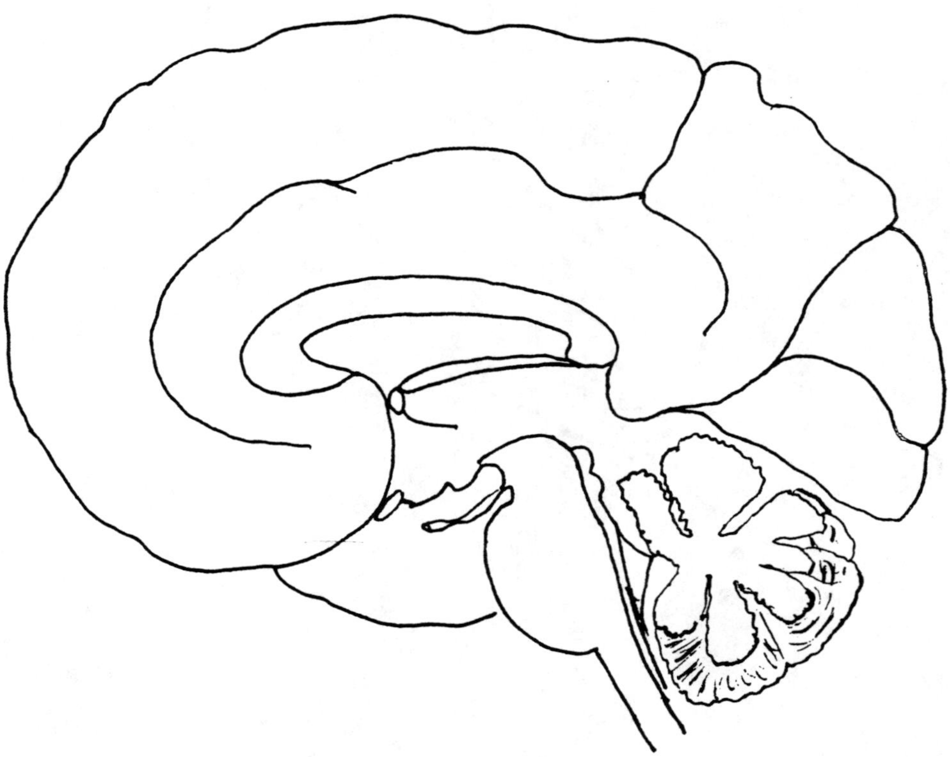

d. The **TELENCEPHALON**

1) The right and left **CEREBRAL HEMISPHERES**

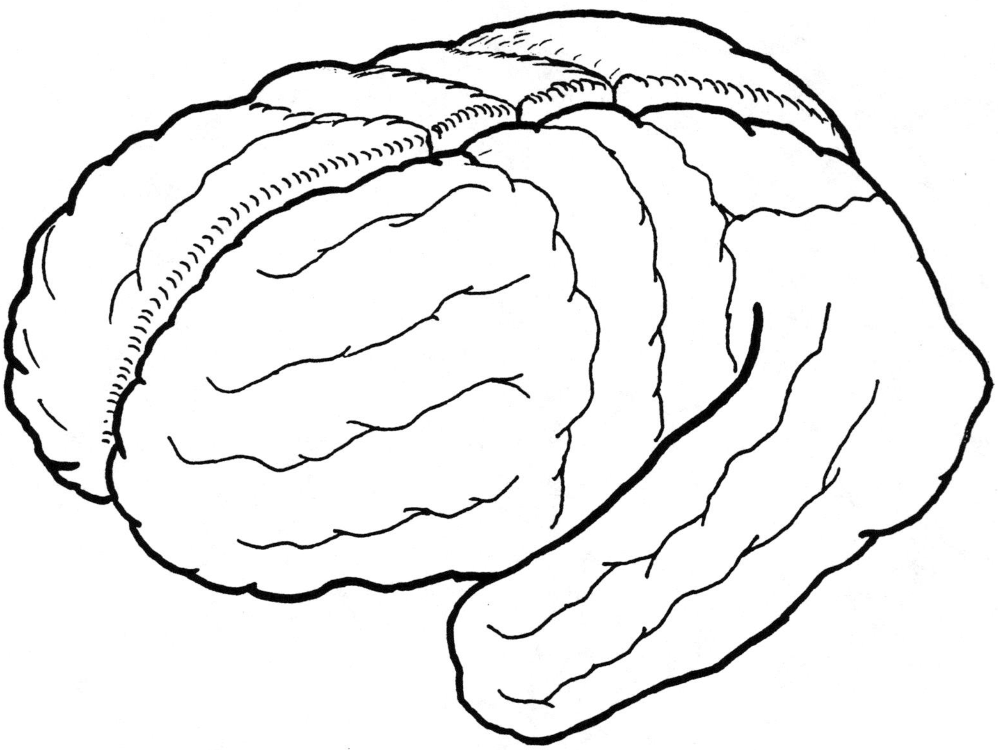

2) **LOBES** of the telencephalon (Cerebral hemispheres)

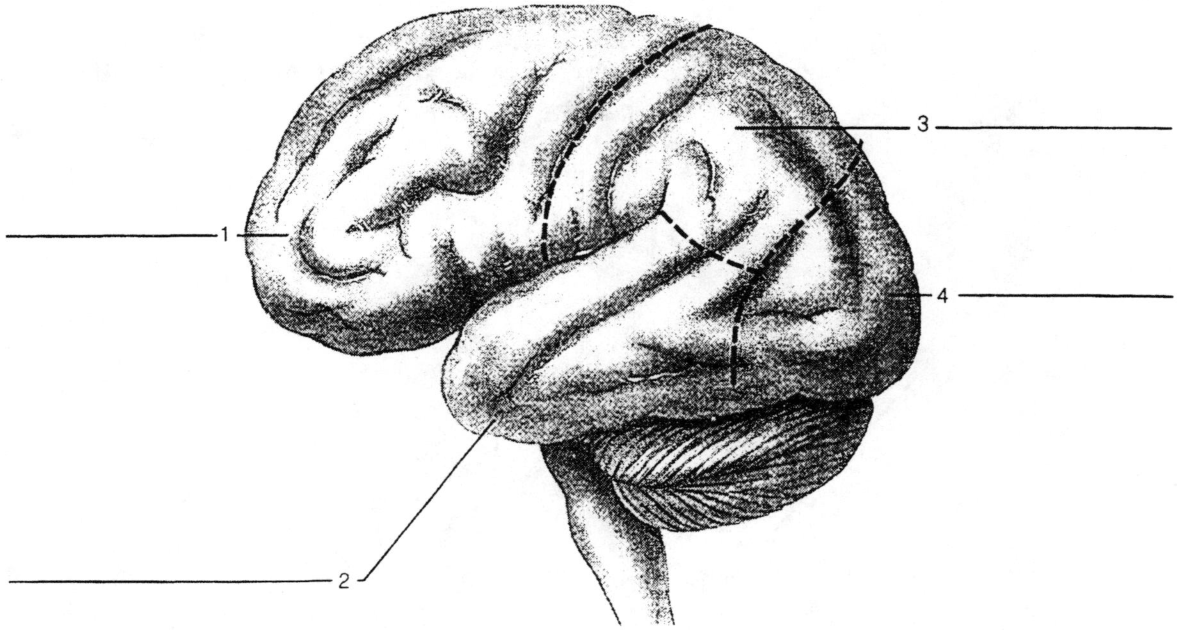

3) The cerebral cortex, white matter, and **basal ganglia**

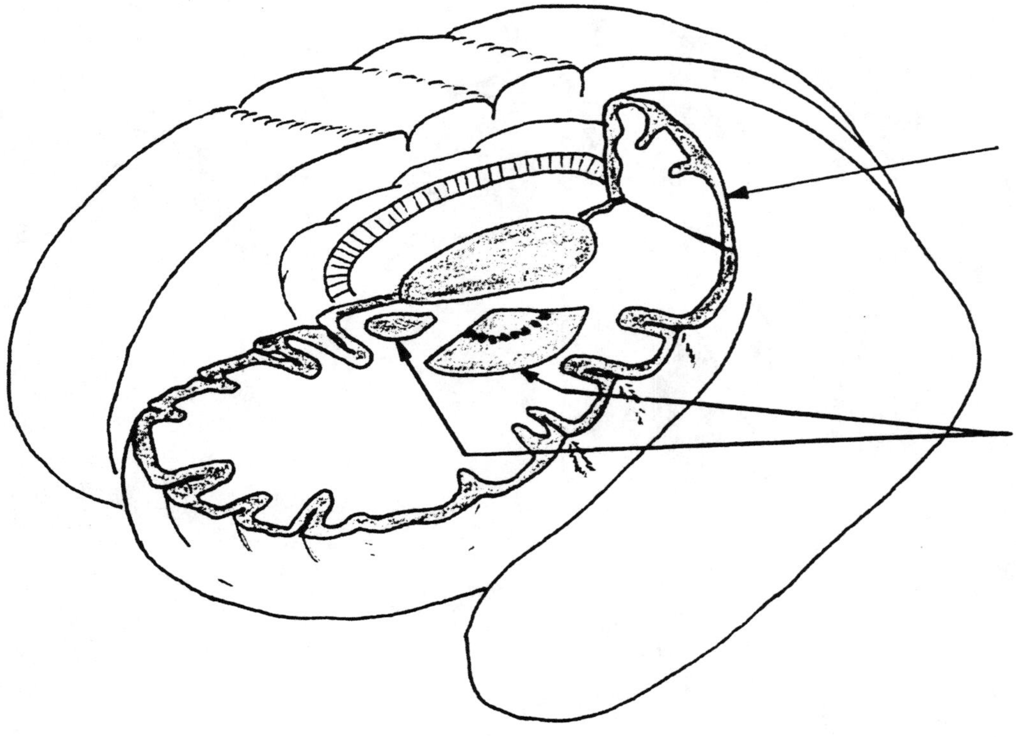

4) **SULCI, GYRI**, of the cerebral hemispheres and their functions
   a) **SULCUS** = A depression (crease, furrow) in the surface of the brain
   b) **GYRUS** = A raised area between two sulci
   c) Sulci and gyri on the **lateral surface** of the cerebral hemisphere

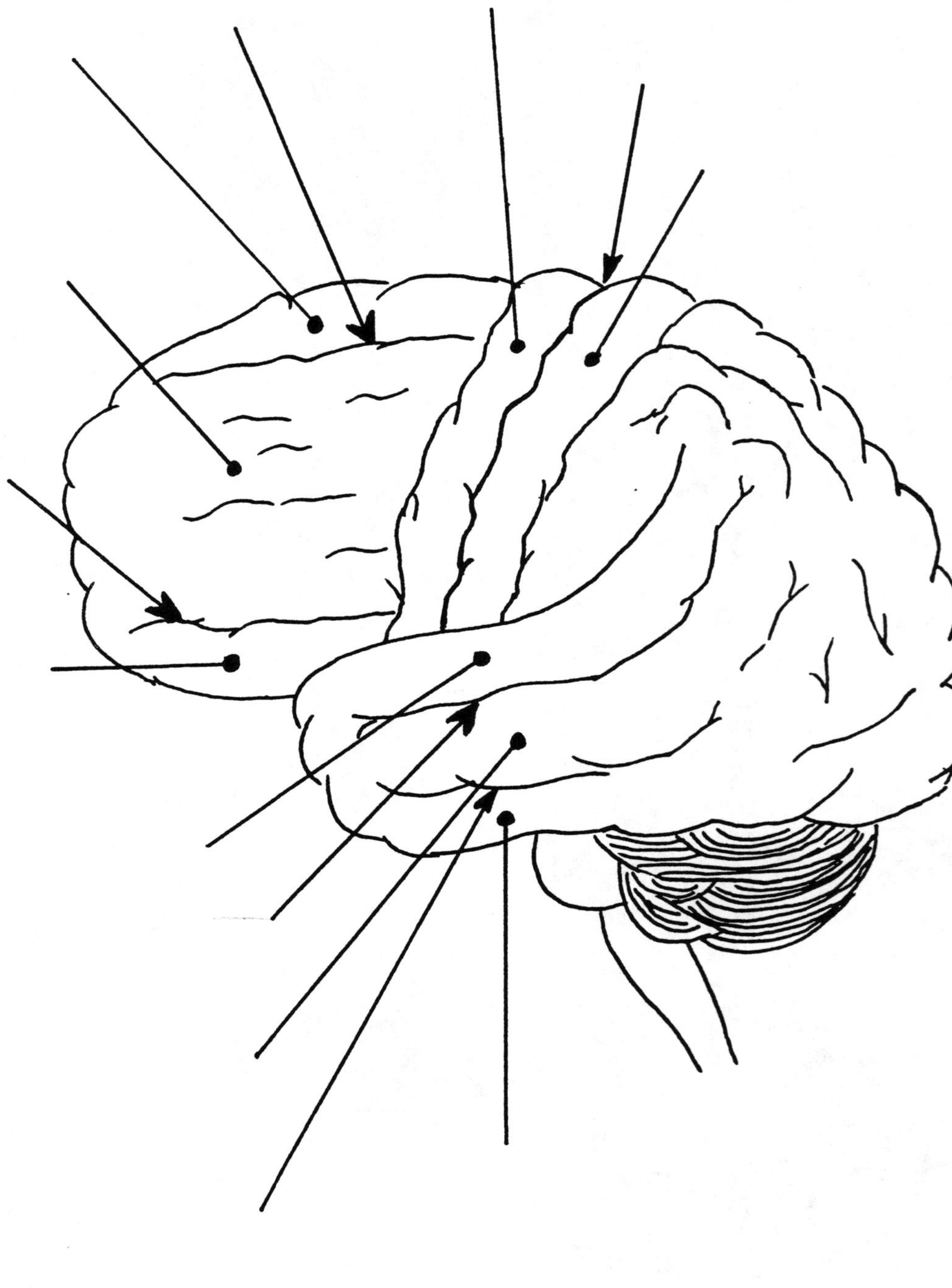

d) Sulci and gyri on the **medial surface** of the cerebral hemisphere

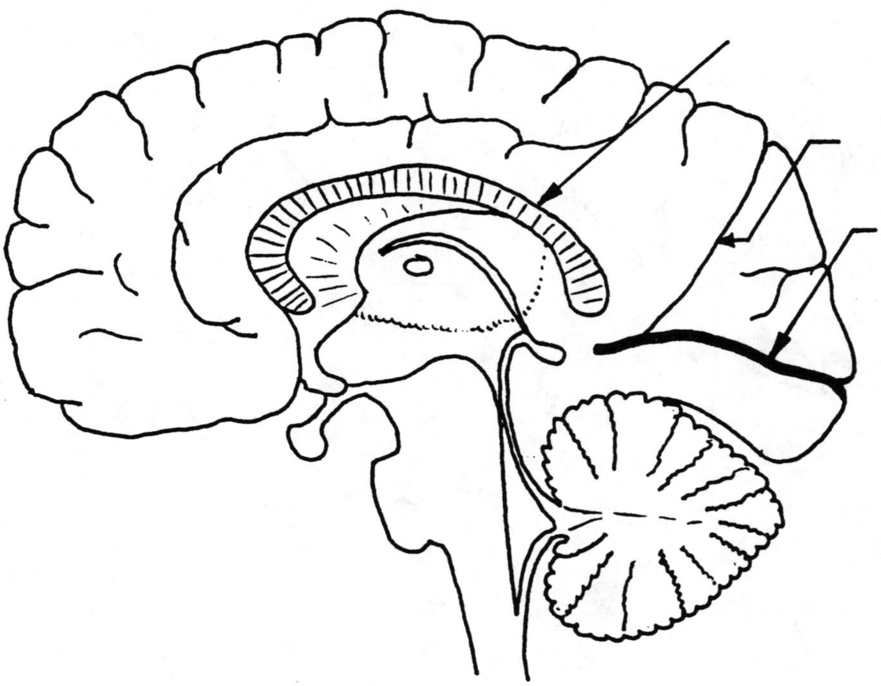

e. The **DIENCEPHALON**

1) Components

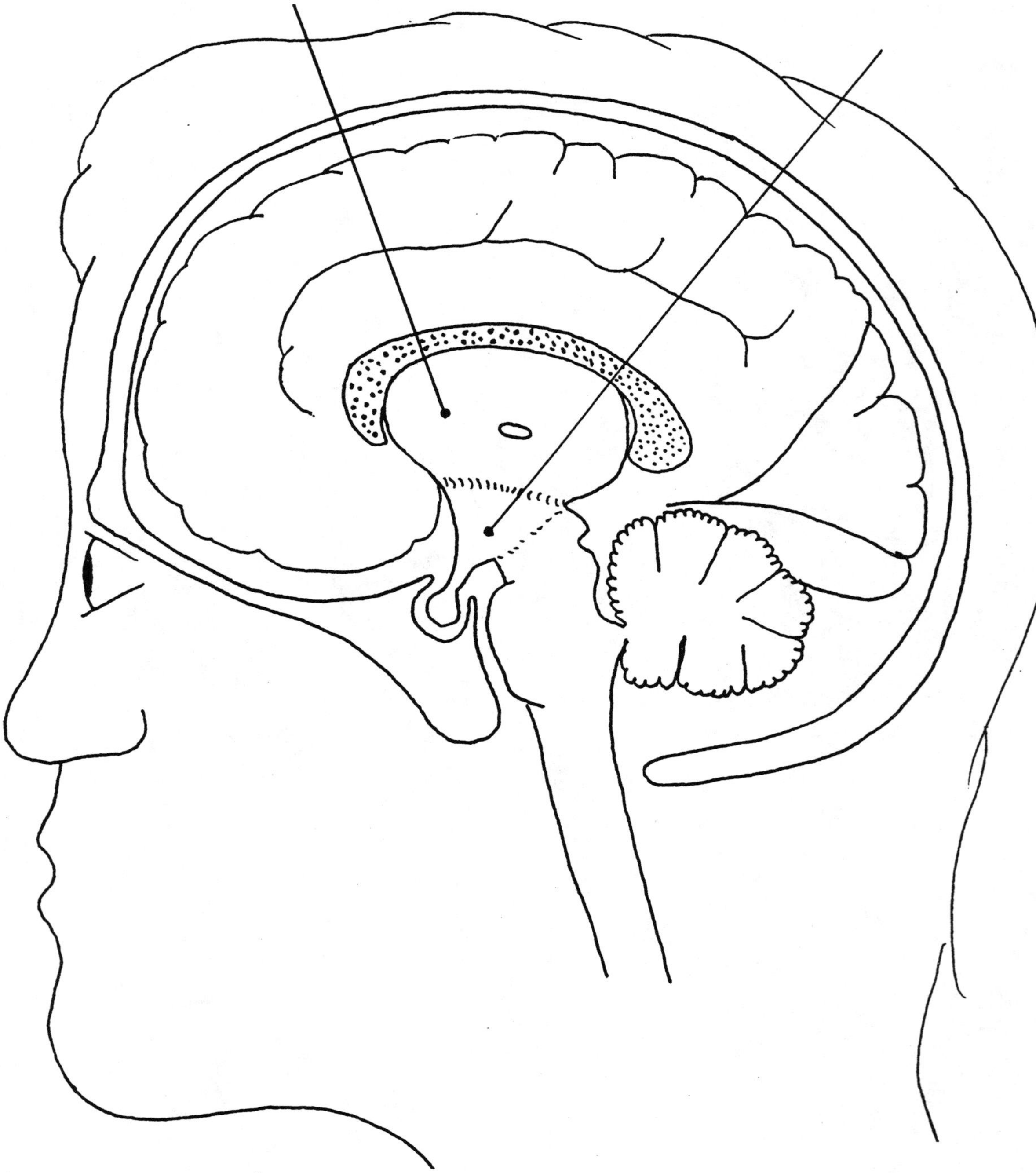

2) Functions of the components of the diencephalon

a) The **THALAMUS**

b) The **HYPOTHALAMUS**

3) What is the **CEREBRUM**

4) What is the **BRAINSTEM**

f. The **MESENCEPHALON**, **METENCEPHALON**, and **MYELENCEPHALON**

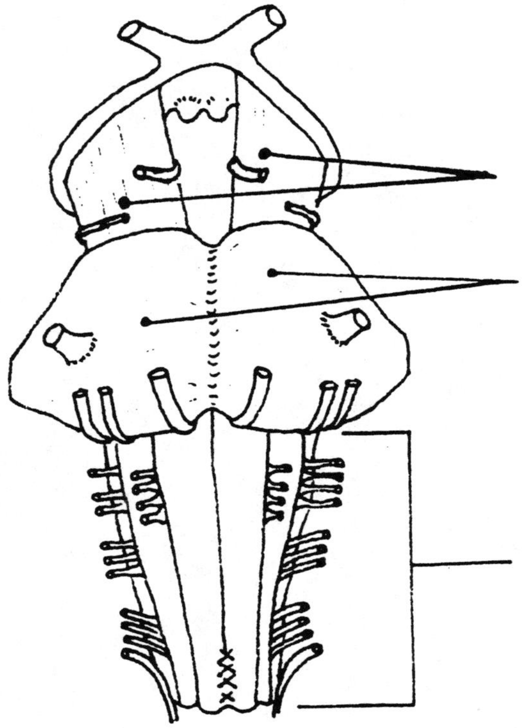

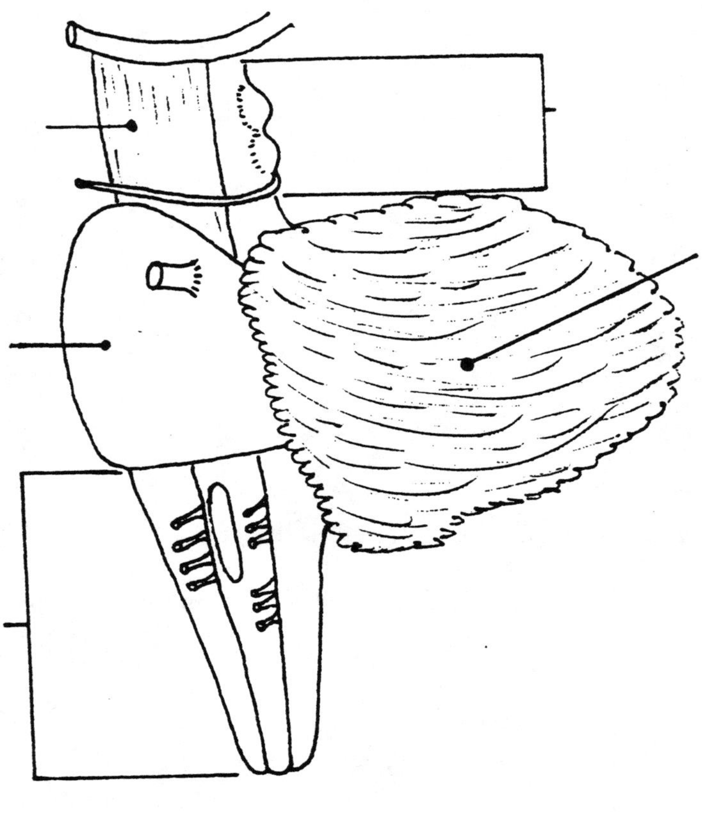

72. What structures are in the telencephalon ?
73. What structures are in the diencephalon ?
74. What is the function of the thalamus ?
75. What functions of the body does the hypothalamus regulate ?
76. What effectors of the body does the hypothalamus control ?
77. What structures are in the mesencephalon ?
78. What structures are in the metencephalon ?
79. What important body function does the cerebellum control ?
80. What is the major structure of the myelencephalon ?
81. Which of the five major brain regions makes uo the cerebrum
82. Which of the five major brain regions make up the brainstem ?
83. What seem to be some major differences between the right and left cerebral hemispheres ?
84. Name the four lobe of the telencephalon.
85. For what are these lobes named ?
86. What is the cerebral cortex ?
87. About how many neuron cell bodies are located in the cerebral cortex ?
88. What major function does the basal ganglia control ?
89. In what major brain region are the basal ganglia and the cerebral cortices located ?
90. What is the definition of a sulcus ?
91. What is the definition of a gyrus ?
92. Which lobe of the brain is responsible for initiating the contraction of your skeletal muscles ?
93. Exactly which gyrus of your frontal contains these upper motor neurons that initiate skeletal muscle contractions ? What is another name for this gyrus ?
94. Which lobe of your brain has the motor speech area ?
95. Exactly which gyrus of the frontal lobe is the motor speech area located on ?
96. The cerebral hemisphere that has the functional motor speech area is called the ----- hemisphere.
97. Which lobe is contains the primary sensory areas and the analytical area for the general sensations ?
98. Exactly which gyrus is it in your parietal lobe that is the primary sensory area for general sensations ?
99. In which lobule of the parietal lobe are your general sensations analyzed ?
100. In which lobe are the primary sensory area and the analytical area for vision located ?
101. At exactly which sulcus in the occipital lobe is the primary sensory area for vision ?
102. Where is the analytical area for vision ?
103. In which lobe are the primary sensory area and the analytical area for sound located ?
104. In exactly which gyrus of the temporal lobe is the primary sensory area for sound located ? The analytical area for sound ?
105. Which sulcus is the anatomical division between the frontal and parietal lobes ?
106. Which sulcus divides the parietal lobe into superior parietal lobule and inferior parietal lobule?
107. What are the components of the diencephalon ?
108. What is the very important function of the cerebellum ?
109. To which major brain region does the cerebellum belong ?
110. What is an important function of the medulla and the pons ?
111. To which major brain region does the pons belong ? The medulla ?
112. What is the cerebrum ?
113. What is the brainstem ?

2. The SPINAL CORD

a. Relationship of the spinal cord length to vertebral column length in the adult.

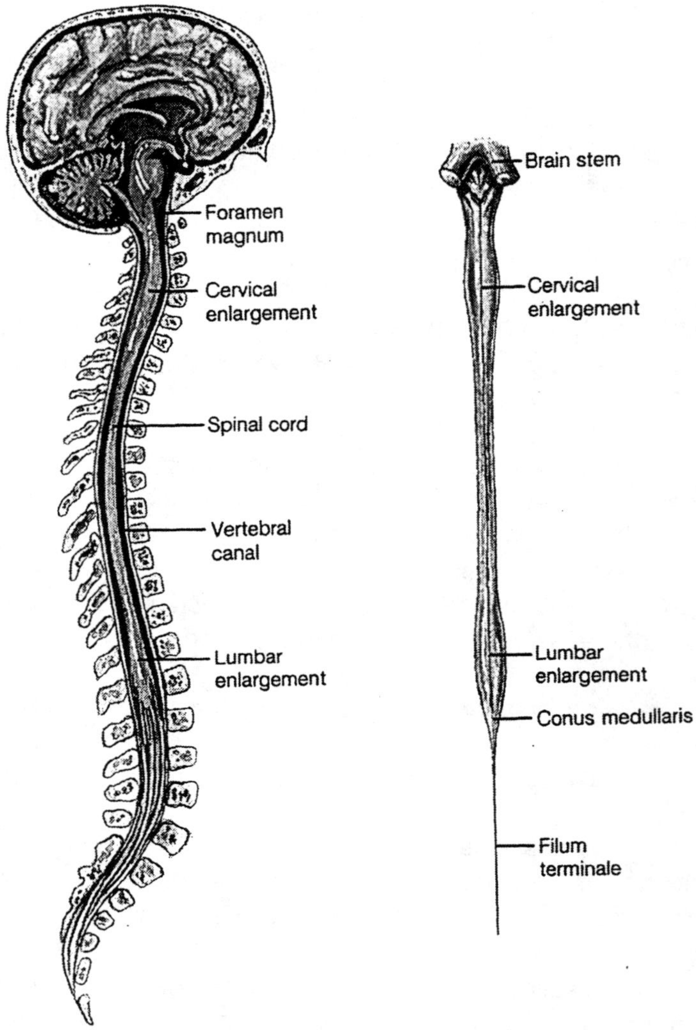

b. Posterior (dorsal) view of the spinal cord showing dorsal rootlets, dorsal root ganglia, the brachial plexus, and peripheral nerves in the upper limb

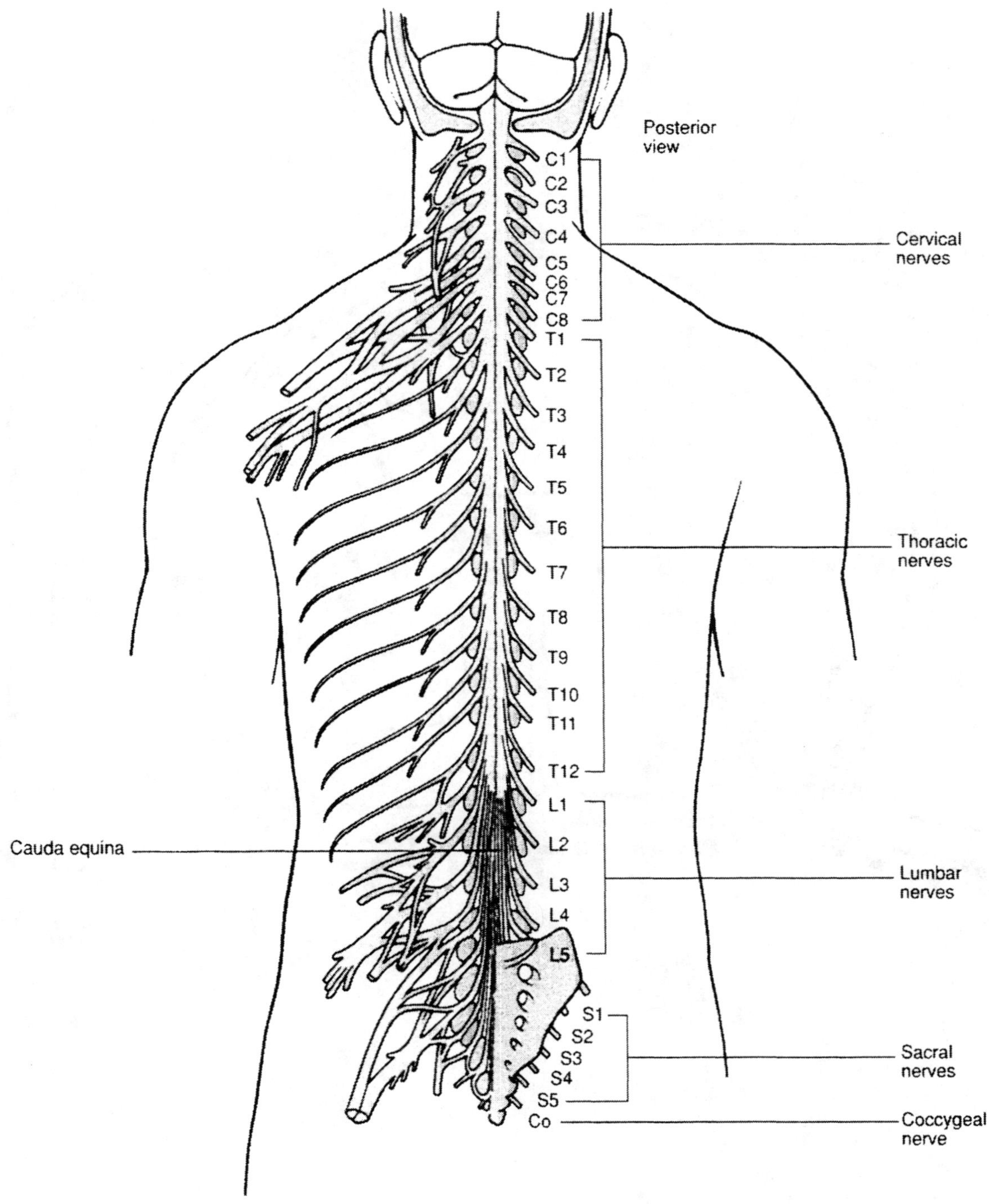

c. Cross section of the spinal cord

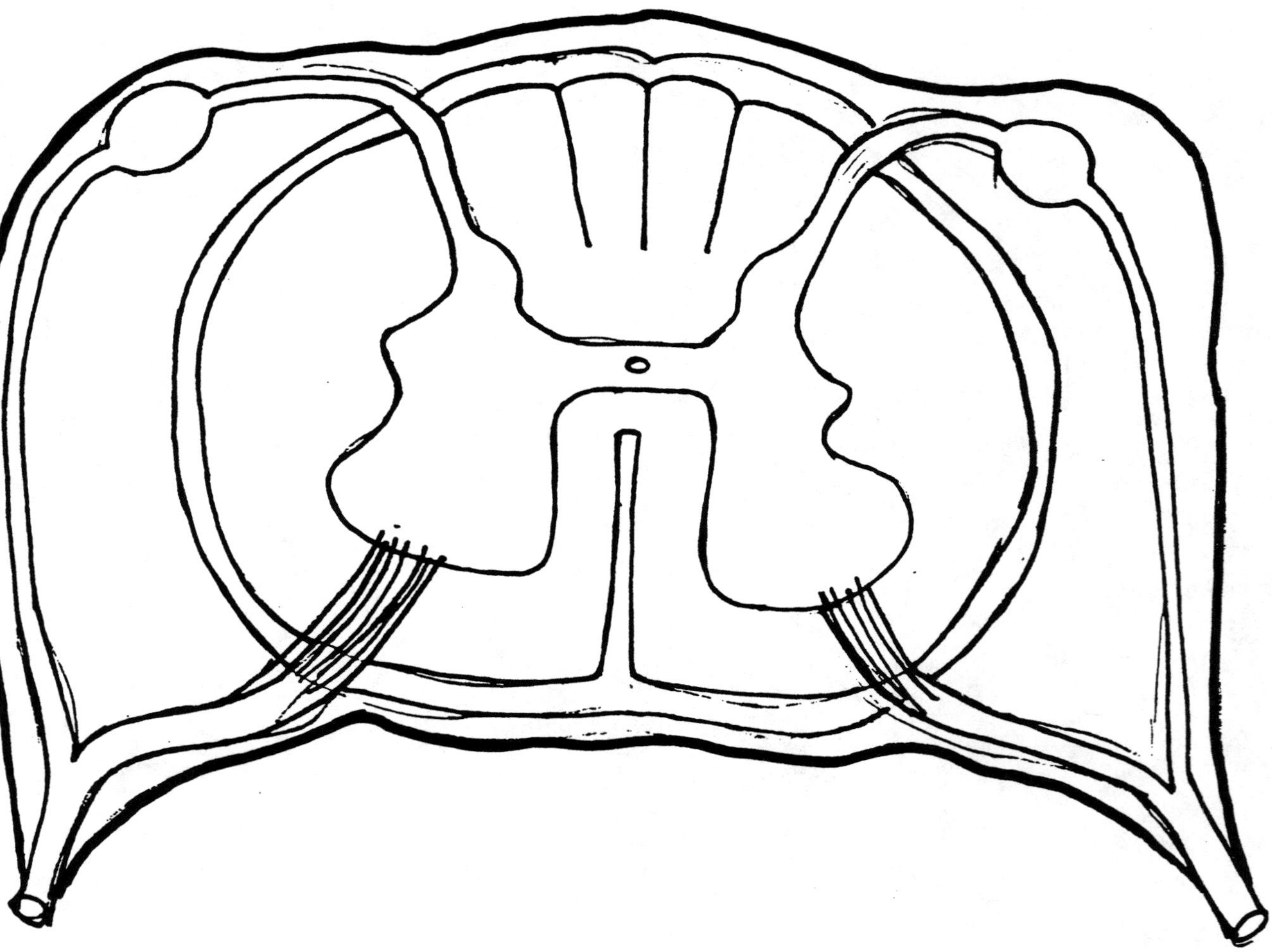

d. The reflex arc = The functional unit of the nervous system

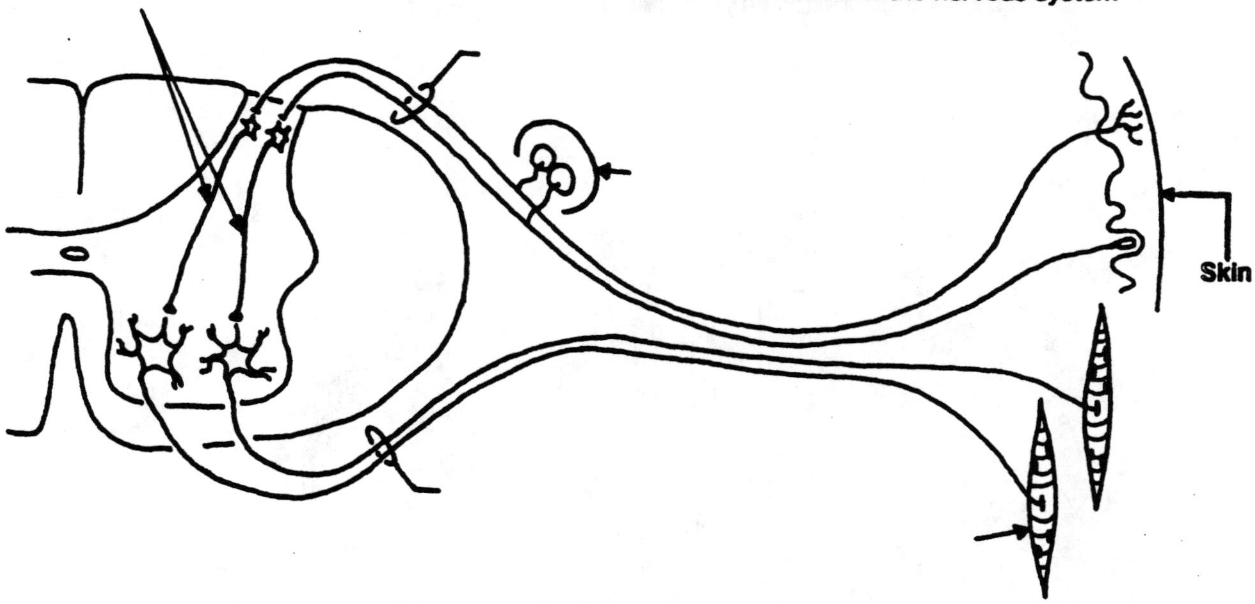

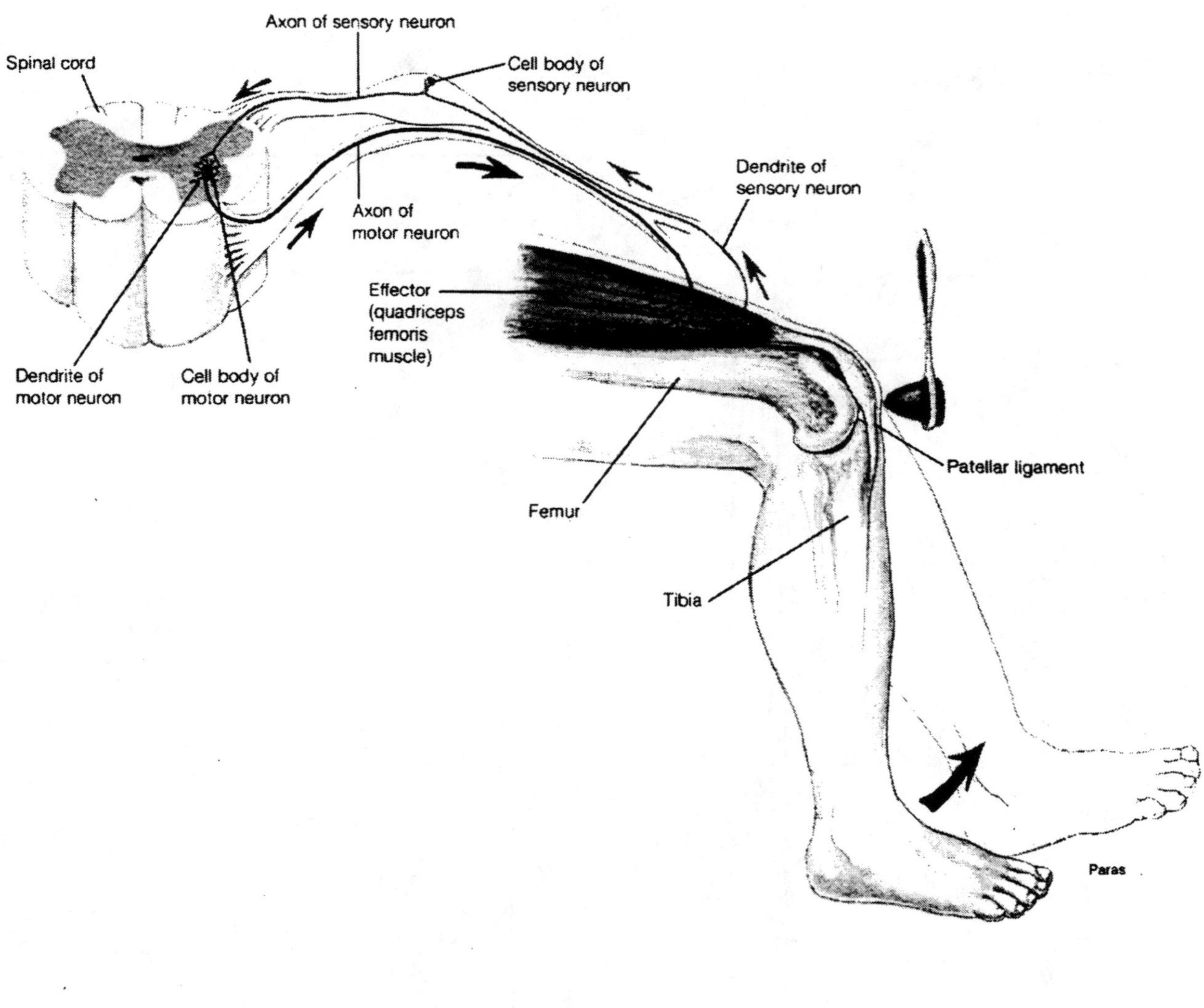

3. GENERAL SENSORY SYSTEMS and MOTOR SYSTEMS TO SKELETAL (VOLUNTARY) MUSCLES

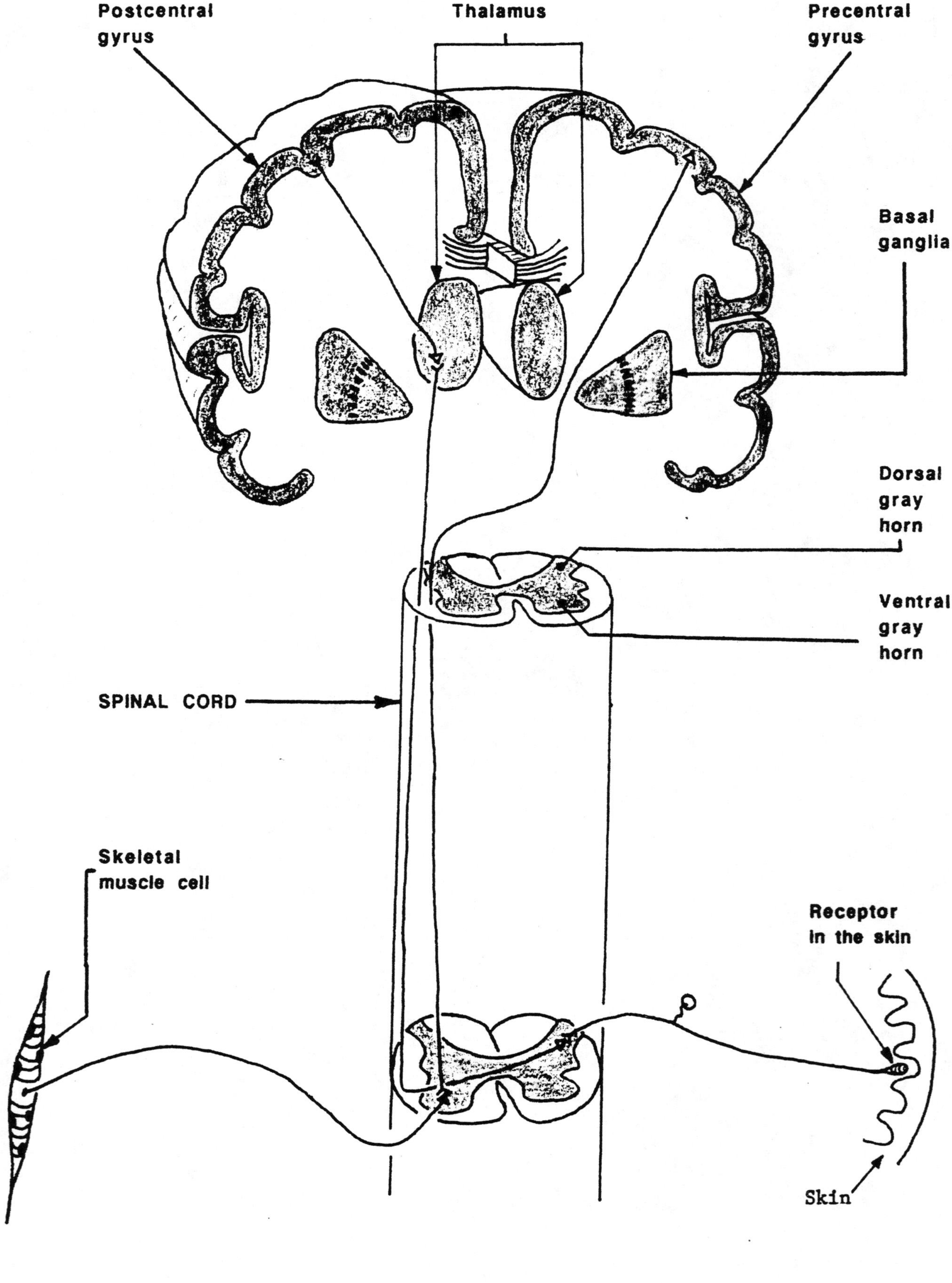

4. The **CRANIAL NERVES**

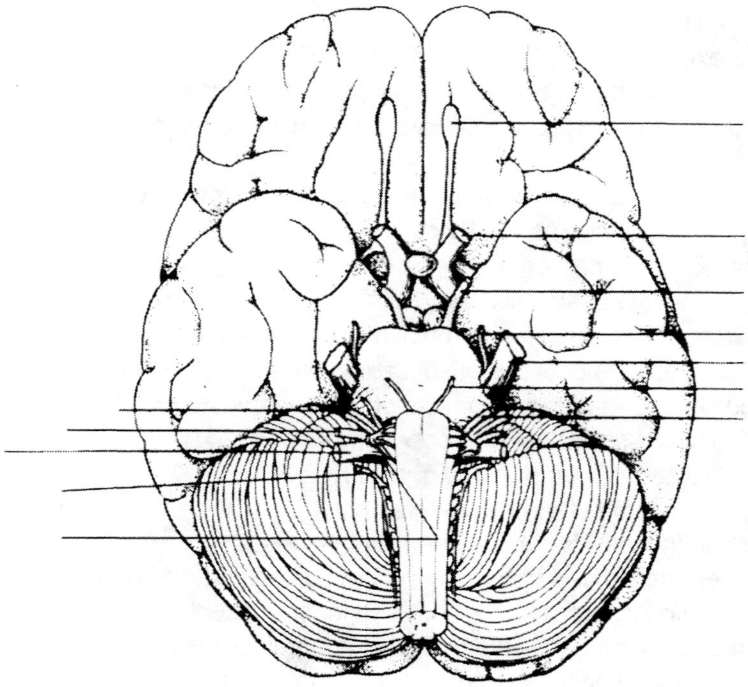

114. The spinal cord is anatomically continuous with which part of the brain ?
115. At what vertebral level in the adult does the spinal cord end ?
116. What is a spinal tap ?
117. Between what two vertebrae is a spinal tap usually performed ?
118. Why between L3 and L4 ?
119. What is the name of the fluid that surrounds the spinal cord and barain and is also located in the ventricles of the brain ?
120. How are spinal nerves attached to the spinal cord ?
121. What are the components of a typical spinal nerve ?
122. Are the spinal nerves and all of their components in the PNS or CNS ?
123. Is the brachial plexus in the PNS or CNS ?
124. What is the name of the peripheral nerve that passes behind the medial epicondyle ?
125. What is the name of the peripheral nerve that passes behind the humerus and into the posterior forearm ?
126. What is the name of the peripheral nerve that enters the anterior forearm ?
127. How many pairs of spinal nerves are there ?
128. Why arethey called spinal nerves ?
129. Are the peripheral nerves in the PNS ar CNS ?
130. Of what is the dorsal root ganglion composed (be specific)
131. Of what specifically are the dorsal rootlets of spinal nerves composed ?
132. Of what specifically are the ventral rootlets of spinal nerves composed ?
133. Of what specifically is a spinal nerve proper composed ?
134. What specifically is located in the dorsal gray horn ?
135. What specifically is located in the ventral gray horn ?
136. What specifically is in the white matter of the spinal cord ?
137. What are the three menninges that surround the spinal cord and brain ?
138. List the three menninges in order from outer to inner (closest to the cord and brain).
139. What is the name of the space between the dura mater and the arachnoid layer ?
140. What is the name of the space between the arachnoid and the pia mater ?
141. Which of the menninges is directly against the surface of the spinal cord and brain ?
142. Which of the three menninges is the thickest and toughest ?
143. Which of the three menninges is the thinnest and most fragile ?
144. Which of the three menninges is named after spiders ?
145. What is the anatomical unit of the nervous system ?
146. What is the functional unit of the nervous system ?
147. Describe a reflex arc.
148. How many sensory neurons are between a receptor and the postcentral gyrus ?
149. To exactly what are receptors connected ?
150. Where are the cell bodies of 1st sensory neurons located ?
151. Where are the axons of 1st sensory neurons located ?
152. Where are the cell bodies of 2nd sensory neurons located ?
153. Where are the axons of 2nd sensory neurons located ?
154. Where are the cell bodies of the 3rd sensory neurons located ?
155. To what specific gyrus do the axons of 3rd sensory neurons go ?
156. Where exactly does the synapse between the 1st and 2nd sensory neurons occur ?
157. Where exactly does the synapse between the 2nd and 3rd sensory neurons occur ?
158. Where exactly are the cell bodies of the upper motor neurons located ?
159. Where are the axons of upper motor neurons located ?
160. Where specifically are the cell bodies of lower motor neurons located ?

161. Where specifically are the axons of lower motor neurons located ?
162. To what specific effector is the end of the axon of a lower motor neuron attached ?
163. What are some names for this connection between the axon of the lower motor neuron and a skeletal muscle cell ?
164. Name and number the cranial nerves ?
165. How many pairs of cranial nerves are there ?
166. Which of the cranial nerves have no motor components (sensory only) ?
167. Which of the cranial nerves have no sensory components ?
168. Which of the cranial nerves have both motor and sensory components ?
169. Which cranial nerve conducts nerve impulses from the rods and cones ?
170. Which foramen does the optic nerve pass through ?
171. Which cranial nerve is for the sense of smell ?
172. There are two cranial nerves involved in taste. Which are they ?
173. Which conveys taste from the anterior 2/3 of the tongue ?
174. Which conveys taste from the posterior 1/3 of the tongue ?
175. Which cranial nerve is for the sense of sound ?
176. Which cranial nerve has the lateral rectus muscle as an effector ?
177. Which cranial nerve has the superior oblique as its effector ?
178. Which cranial nerve has the remaining extrinsic eye muscles as effectors ?
179. Which cranial nerve has the tongue muscles for effectors ?
180. Which cranial nerve has the muscles of facial expression as effectors ?
181. Which cranial nerve has the muscles of mastication (chewing) as effectors ?
182. Which cranial nerve has the trapezius and the sternocleidomastoid muscles as effectors ?
183. Which cranial nerve has the lacrimal glands as an effectors ?
184. Which cranial nerve has the submandibular and sublingual glands as effectors ?
185. Which cranial nerve has the parotid glands as effectors ?
186. Considering your answers to questions 172, 173, 174, and 184 and 185 what correlation do you see betwwen the cranial nerves that are sensory for taste and have salivary glands as receptors ?
187. Why do you think the same two cranial nerves responsible for taste are also responsible for salivation ?
188. Which cranial nerve is sensory for general sensations from the skin of the face ?

6. The **AUTONOMIC NERVOUS SYSTEM (ANS)**

a. Effectors of the **ANS** = smooth muscles, cardiac muscles and exocrine glands

b. The two divisions of the **ANS**

1) **SYMPATHETIC DIVISION**

2) **PARASYMPATHETIC DIVISION**

c. Interaction between these two divisions of the **ANS**

During normal activities, both divisions send nerve impulses to their effectors simutaneously. However, if certain events occur that cause stress, anger, fear, terror, or when a person eats a meal or goes to sleep, then either the sympathetic or parasympathetic division predominates and alters the activities of the various muscles and glands throughout the body.

EFFECTS OF THE ACTIVITY OF THE SYMPATHETIC AND PARASYMPATHETIC DIVISIONS OF THE ANS ON THE ACTIVITIES OF VARIOUS ORGANS OF THE BODY

SYMPATHETIC DIVISION	PARASYMPATHETIC DIVISION

d. General anatomy of the **ANS**

1) Ganglia of the **ANS**

a) Definition of a ganglion

b) Autonomic ganglion =

c) Classification of ANS ganglia

1)) Prevertebral ganglia =

2)) Paravertebral ganglia =

3)) Terminal ganglia =

2) The locations of the **PREGANGLIONIC NEURON** and the **POSTGANGLIONIC NEURON** in the **ANS**

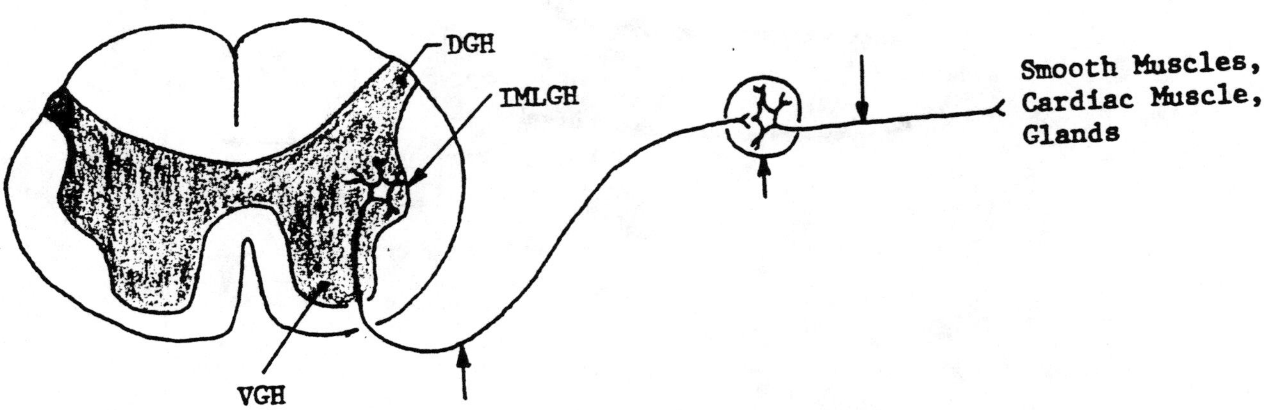

e. Spinal nerves showing **ANS** sympathetic and parasympathetic preganglionic and postganglionic neurons

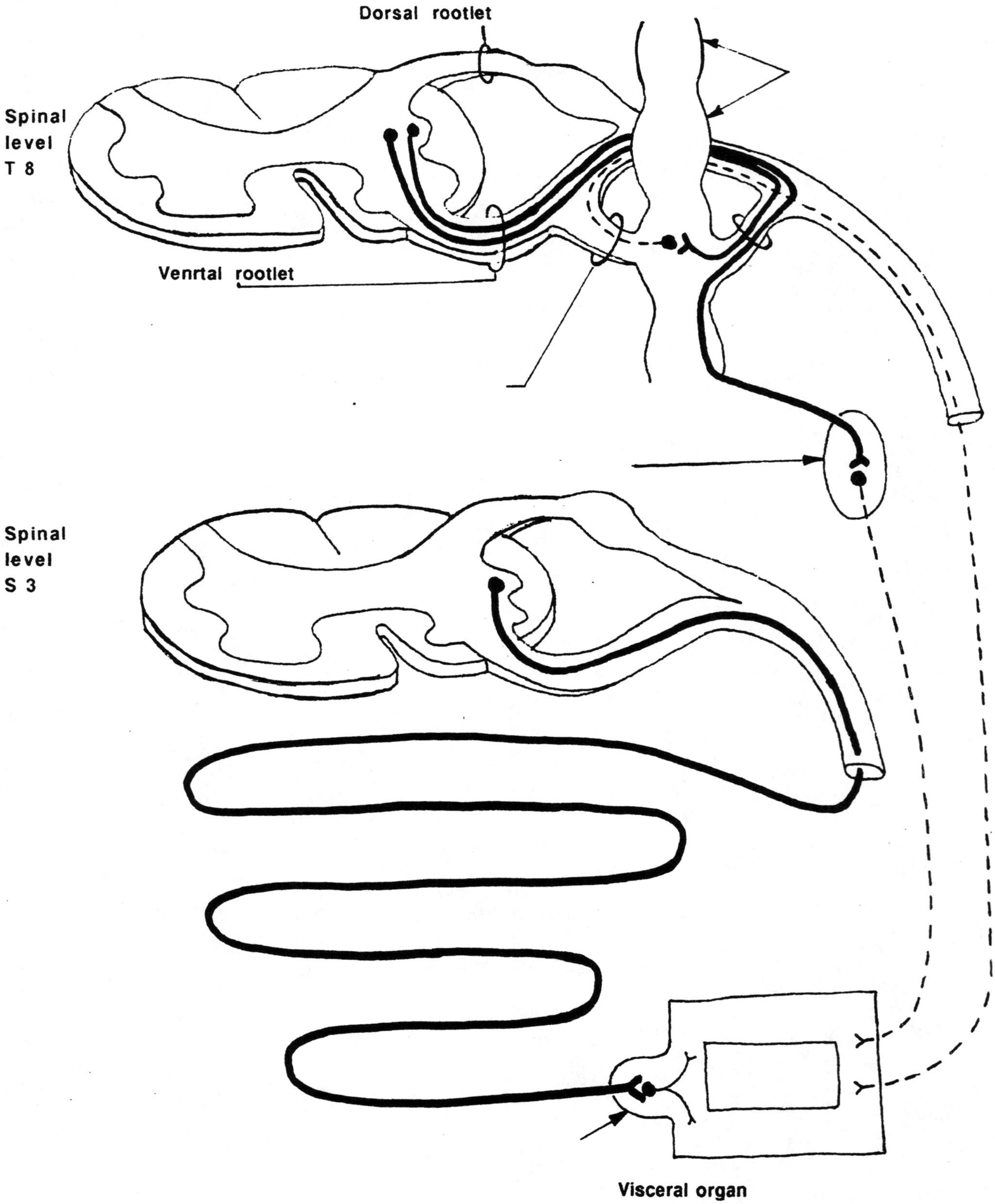

f. **OVERVIEW OF THE ANS**

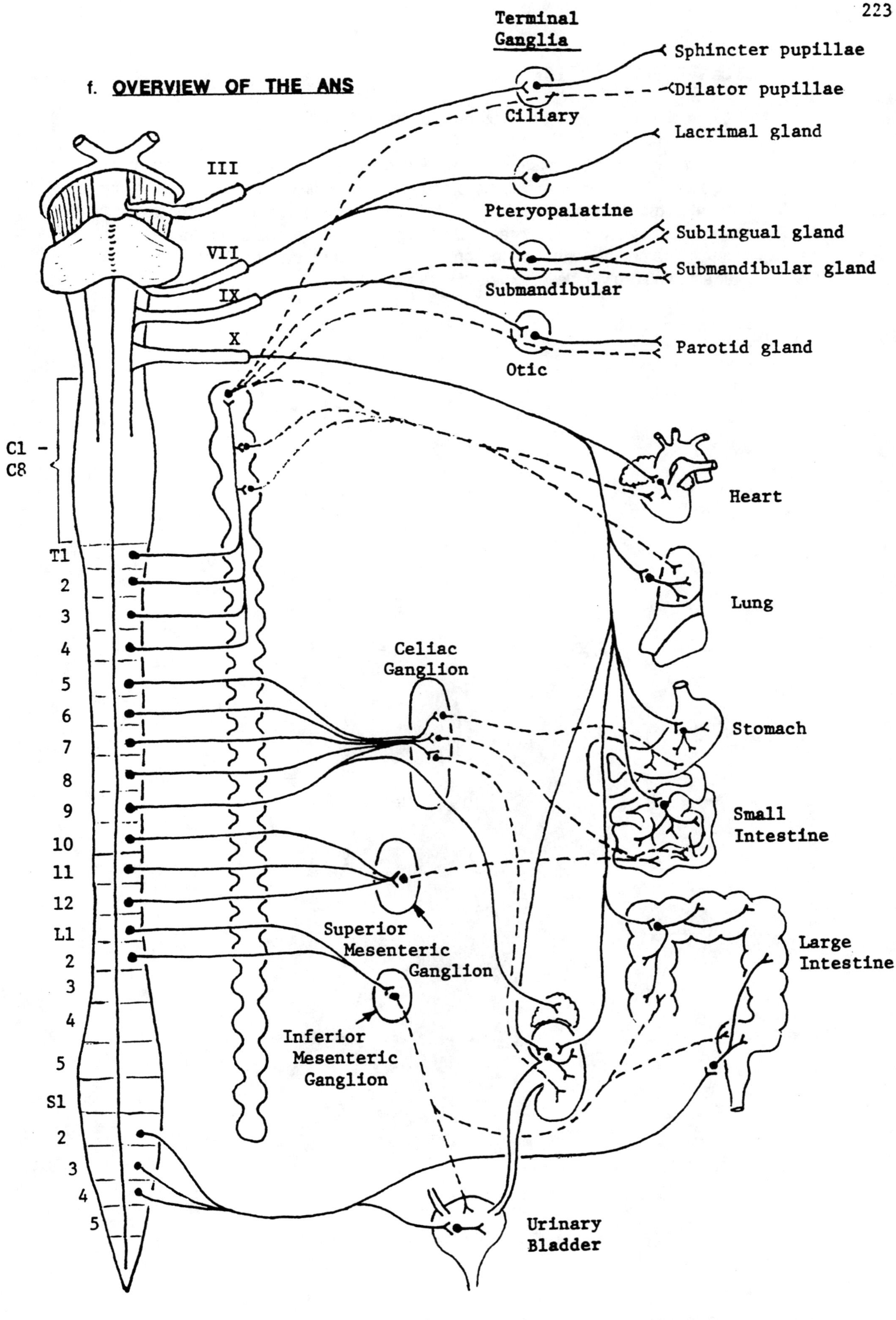

III. The ENDOCRINE SYSTEM

A. General information and the purpose of the endocrine system

Your endocrine system is a collection of glands throughout your body that together regulate a number of metabolic processes. Glands of the endocrine system are ductless, meaning that they secrete their products (hormones) directly into the blood of the circulatory system. The circulatory system then transport the hormones to the "targe tissues", or "target organs". The primary primary functions of the endocrine system are metabolic control, reaction to stress, and growth, development and reproduction.

B. Location of the endocrine glands

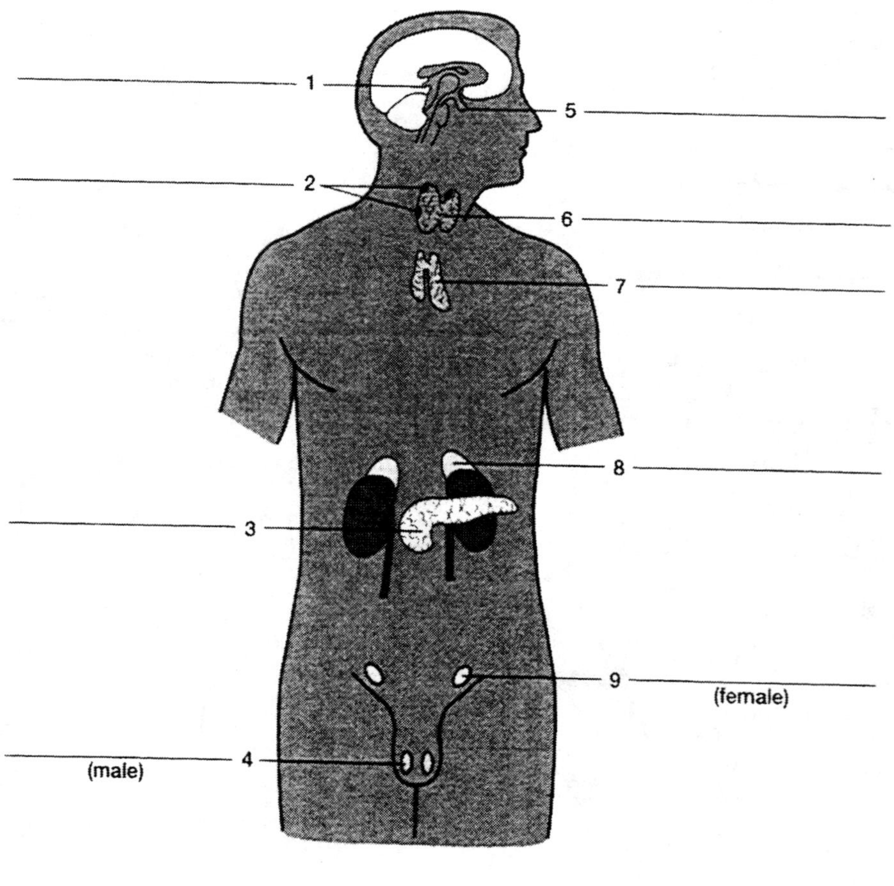

C. The major endocrine glands

1. The **<u>PITUITARY GLAND (HYPOPHYSIS)</u>**

a. Location =

b. Hormones produced and released

1) **<u>Oxytocin</u>**

2) **<u>Antiduretic Hormone (ADH)</u>**

3) **<u>Thyroid Stimulating Hormone (TSH)</u>**

4) **<u>Adrenocorticotropic Hormone (ACTH)</u>**

5) **<u>Luteinizing Hormone (LH)</u>**

6) **<u>Follicle Stimulating Hormone (FSH)</u>**

7) **<u>Prolactin</u>**

8) **<u>Growth Hormone (GH)</u>**

2. **The THYROID GLAND**

a. Location =

b. Hormones made and released

1) **Thyroxine**

2) **Triiodothyronine**

3) **Calcitonin**

3. **The PARATHYROID GLANDS**

a. Location =

b. Hormone made and secreted

1) **Parathyroid hormone**

4. **<u>The ADRENAL GLANDS</u>**

a. Location =

b. Areas of the adrenal gland

1) Adrenal cortex

2) Adrenal medulla

c. Hormones of the adrenal cortex

1) <u>Corticoids</u>

2) <u>Aldosterone</u>

3) <u>Sex hormones</u>

d. Hormones of the adrenal medulla

1) <u>Epinephrine</u>

2) <u>Norepinephrine</u>

5. **<u>The PANCREAS</u>**

a. Location =

b. Hormones made and secreted by the pancreas

1) **<u>Insulin</u>**

2) **<u>Glucagon</u>**

3) **<u>Somatostatin</u>**

7. **<u>The GONADS</u>**

a. Locations

1) **Ovaries**

2) **Testes**

b. Hormones made and released by the ovaries

c. Hormone made and released by the testes

IV. The CARDIOVASCULAR SYSTEM

A. General information and the purpose of the cardiovascular system

Your **CIRCULATORY SYSTEM** is composed of blood vessels (**ARTERIES**, **CAPILLARIES**, and **VEINS**) through which **BLOOD** is being pumped by a truly fantastic muscular pump, your **HEART**.

Only a few of the trillions of cells in your body are without a blood supply. That is good because each of your cells is a tremendous chemical factory that requires nutrients and oxygen (O_2) to make ATP in order to support the necessary chemical reactions which are critical for staying alive and for growth and cell division. Also, each of your cells, thanks to the Krebs citric acid cycle, is producing carbon dioxide (CO_2) which is a toxin (poison).

Nutrients and O_2 must be delivered to your cells, and CO_2 must be removed from them and eventually eliminated from the body. As blood flows by a cell in a capillary, the nutrients and O_2 leave the blood and enter the cell, and the CO_2 in the cell enters the blood. The cell solves its problems by these exchanges with the blood, but the newly acquired CO_2 creates a problem for the blood because its presence tends to make the blood more acid. If the CO_2 in the blood is not removed and accumulates, then it leads to a serious condition known as **acidosis**.

A major portion of the CO_2 in your blood is removed, and acidosis is prevented, when blood flows through the capillaries surrounding the alveoli of your lungs. **It is here at the capillary / alveoli junction that O_2 enters your blood and CO_2 leaves it.**

Your **HEART** is a muscular pump made up of four chambers whose walls are composed of **cardiac muscle**. The chambers consist of **RIGHT ATRIUM, LEFT ATRIUM, RIGHT VENTRICLE,** and **LEFT VENTRICLE**. The muscle of the heart is often referred to as the **myocardium**.

There are several ways to categorize the four chambers. One way is to call them the **upper chambers** (the right and left atria), and the **lower chambers** (the right and left ventricles). Another useful way is to divide the heart into a right side (**RIGHT HEART**), and a left side (**LEFT HEART**). The **right heart** is composed of the right atrium and right ventricle, and the **left heart** is composed of the left atrium and left ventricle. Viewing the chambers of the heart as upper and lower chambers will allow you to understand the **cardiac cycle**, and considering the heart as being composed of a right heart and a left heart will allow you to understand the interrelationship between **systemic, coronary, and pulmonary circulations**.

If we take a look at the path that a red blood cell (RBC) can take from either the right or left ventricle of the heart, then we discover that from the left ventricle an RBC can go out to either **1)** the cells in the organs of the systems of your body (**SYSTEMIC CIRCULATION**), or **2)** the cardiac muscle cells of the myocardium of your heart (**CORONARY CIRCULATION**). From the right ventricle, an RBC enters the pulmonary trunk to go to your lungs (**PULMONARY CIRCULATION**).

Systemic circulation is for delivering O2 to the cells of your systems. This circulation begins with oxygeneated blood in the left heart (left ventricle), goes to the capillaries around cells where O_2 is given to the cells, and CO_2 is taken from the cells by the blood. The unoxygenated blood then enters the veins of systemic circulation and returns to the right heart (right atrium). From the right atrium the unoxygenated blood goes into the right ventricle, which begins **PULMONARY CIRCULATION**. Blood pumped from the right ventricle enters the pulmonary trunk, then into the pulmonary arteries and circulates to the lungs and the capillaries around the alveoli where O_2 enters the blood and CO_2 leaves the blood and enters the alveoli. Oxygenated blood then retuns to the left heart (left artium) through the pulmonary veins. The left atrium is the end of pulmonary circulation. The freshly oxygenated blood goes from the left atrium to the left ventricle.

Not all of the oxygenated blood in your left ventricle enters your systemic circulation. Some of the blood that enters the aorta from the left ventricle goes into the two coronary arteries and their branches and finally into the capillaries in the **myocardium** to supply O_2 to these muscle cells and to pick up their CO_2. Like the unoxygenated blood in your systemic circulation, the unoxygenated blood is returned to your right heart (right atrium), and from here this blood passes to the right ventricle to begin pulmonary circulation.

Both systemic and coronary circulations begin in the left ventricle and end in the right atrium. Pulmonary circulation begins in the right ventricle and ends in the left atrium. You should notice that, in the adult, in order for blood to go from the right heart to the left heart it must go to the lungs first.

In systemic circulation and in coronary circulation O_2 is depleted from the blood and CO_2 is added. Just the opposite happens in pulmonary circulation because O_2 is added to and CO_2 is removed from the blood.

The cardiac cycle is basically a cycle that requires about 0.75 seconds to complete. Actually, when you are taking your pulse you are measuring how many cardiac cycles your heart is going through per minute. There are 60 seconds in a minute, 3,600 seconds in an hour, 86,400 in a day, 31,536,000 in a year, and, if you live 70 years then you will have lived 2,207,520,000 seconds.

How many cardiac cycles will your heart go through in these 70 years ?

During a cardiac cycle, the **two upper chambers** (right and left atria) contract at essentially the same time. The unoxygenated blood in the adult right atrium is pumped through the right atrioventricular valve (THE tricuspid valve) and fills up the relaxed right ventricle. The oxygenated blood in the left atrium is pumped through the left atrioventricular valve (mitral, or bicuspid valve) and fills up the relaxed left ventricle. The atria relax and then the ventricles contract at about the same time in order to pump blood out of them. Obviously, you do not want blood to go right back up into the atria from where it has just come. To prevent this from happening, the two atrioventricular valves, which are open when the atria contract, are held closed by the papillary muscles and chorda tendineae when the ventricles contract. The blood in the right ventricle has no recourse but to be pumped past the pulmonary valve and into the pulmonary trunk, and the blood in the left ventricle has no recourse but to be pumped past the aortic valve and into the aorta. Therefore, when the ventricles are contracting there are several things going on: **1)** the two atrioventricular valves are closed, **2)** the two semilunar valves (pulmonary and aortic valves) are open, and **3)** the relaxed atria are refilling with blood (the right atrium with unoxygenated blood from the body and the heart, and the left atrium with oxygenated blood from the lungs).

There are <u>four cardiovascular valves</u>. Two of them are anatomically situated between the atria and the ventricles and for that reason are called **the atrioventricular valves (AV valves).** The **right AV valve** is also frequently called **<u>THE</u> tricuspid valve**, and the **left AV valve** is frequently called the **mitral or bicuspid valve.**

The other two cardiovascular valves are in the base of the great arteries attached to the ventricles. These are the **semilunar valves**, of which <u>there are two</u>, the **aortic valve** and the **pulmonary valve.** It is the function of both of the semilunar valves to prevent blood from re-entering the ventricles due to gravity when the ventricles relax.

Therefore, the overall purpose of the four valves (atrioventricular + semilunar) is to keep blood flowing in one direction, not only through the heart, but throughout the body.

1. What are the five major components of your cardiovascular system ?
2. What two things do your cells require ?
3. What is one of the consequences of having too much CO_2 in your blood ?
4. What are the four chambers of the heart ?
5. What type of muscle is the muscle of your heart ?
6. What is the muscle of your heart often called ?
7. What are the two upper chambers of the heart ?
8. What are the two lower chambers of the heart ?
9. What are the two chambers of the right heart ?
10. What are two good and acceptable names for the valve that is between the two chambers of the right heart ?
11. What are the two chambers of the left heart ?
12. What are three good and acceptable names for the valve between the two chambers of the left heart ?
13. What is systemic circulation ?
14. In which chamber of the heart does systemic circulation begin ?
15. In which chamber of the heart does systemic circulation end ?
16. What is coronary circulation ?
17. In which chamber of the heart does coronary circulation begin ?
18. In which chamber of the heart does coronary circulation end ?
19. As far as O_2 content is concerned, what is the difference between the blood that leaves the left ventricle and the blood that returns to the right atrium ? What is the reason for this difference ?
20. As far as CO_2 content is concerned, what is the difference between the blood that leaves the left ventricle and the blood that returns to the right atrium ? What is the reason for this difference?
21. The blood in the right atrium of the adult heart that is unoxygenated and high in CO_2 goes where next ?
22. Which valve of the heart does this blood have to pass through in order to get to the right ventricle ?
23. The right atrium is the beginning of which circulation ?
24. What is the difference between the O_2 and CO_2 content of the blood that leave the right ventricle and the blood that enters the left atrium ?
25. Where does the blood go between its trip from the right ventricle to the left atrium ?

26. Where do both systemic circulation and coronary circulation begin ?
27. Where do they both end ?
28. The two AV valves are open when which two chambers contract ?
29. The two AV valves are closed when which two chambers contract ?
30. Why must the AV valves be held closed when the ventricles contract ?
31. How are the two AV valves held closed ?
32. What are the names of the two semilunar valves ?
33. The semilunar valves are open when which two chambers contract ?
34. The two semilunar valves are closed when which two chambers relax ?
35. What is the purpose of closing the semilunar valves when the ventricles relax ?
36. What is the overall purpose of the four heart valves ?
37. About how many cardiac cycles will an average heart go through in 70 years ? Its an amazing number , almost too many to believe isn't it ? But believe it !

B. **BLOOD**

1. Portions of the blood

   a. Fluid portion = **PLASMA**

   b. Cellular portion = **FORMED ELEMENTS**
      1) Erythrocytes = Red blood cells (RBCs)

      2) Leucocytes = White blood cells (WBCs)

      3) Thrombocytes = Platelets

2. Functions of the blood (and therefore of the cardiovascular system)

   a. Transport of O_2 to cells and tissues ____________________

   b. Transport of nutrients to cells and tissues____________________

   c. Transport of CO_2 from the cells to the lungs____________________

   d. Transport of urea from the liver to the kidneys____________________

   e. Transport of hormones from their glands of origin to their target tissues
   ____

   f. Thermoregulation____________________________

   g. Provision of immunity____________________________

3. Formed elements of the blood

a. **The ERYTHROCYTE (Red Blood Cell ; RBC)**

1) Most abundant of the formed elements
2) A mature human erythrocyte has no nucleus
3) Has a lifespan of about 120 days
4) Contains millions of the a protein molecule, hemoglobin, which binds oxygen

b. **LEUCOCYTES**

1) **GRANULAR LEUCOCYTES**

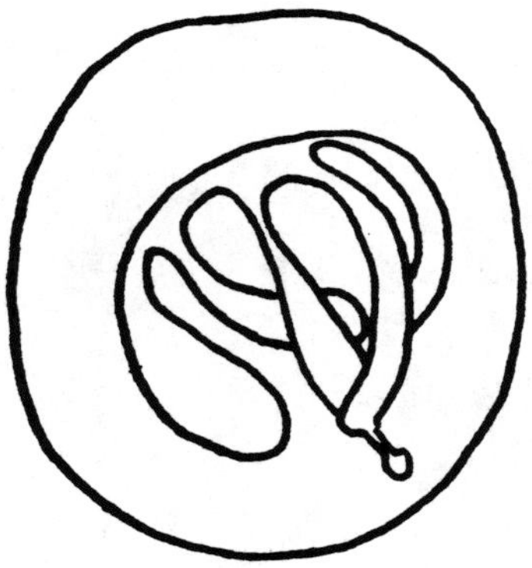

NEUTROPHIL

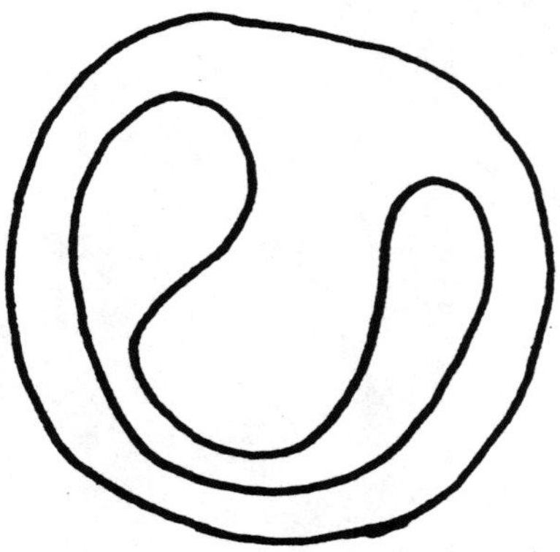

EOSINOPHIL

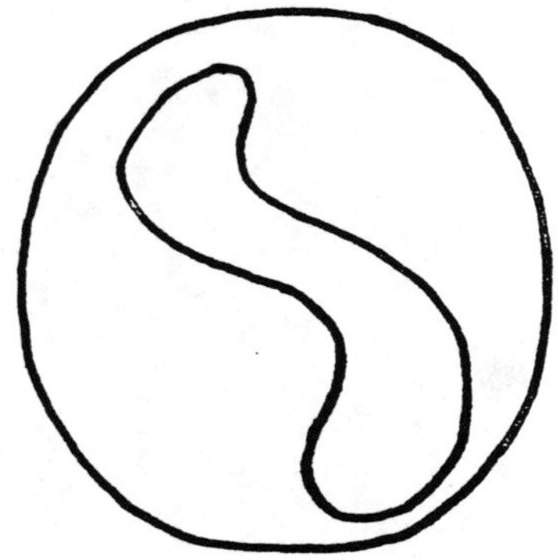

BASOPHIL

2) AGRANULAR LEUCOCYTES

LYMPHOCYTE

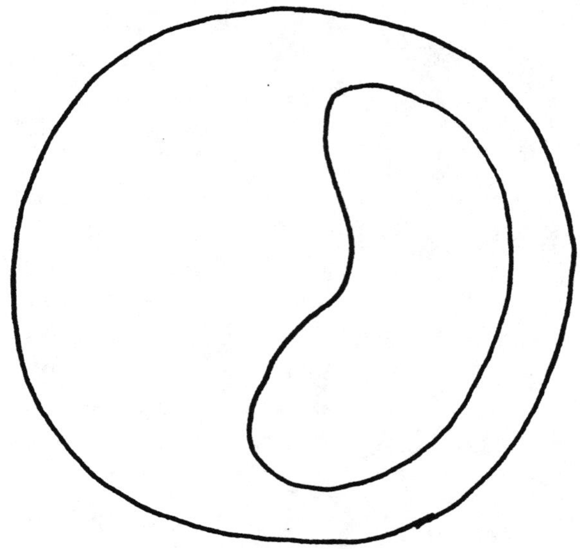

MONOCYTE

c. THROMBOCYTES

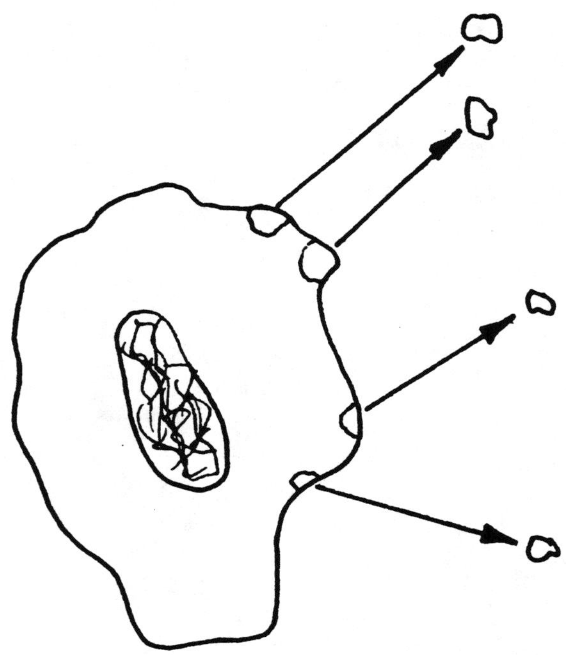

4. **PLASMA**

a. The components of plasma

1)

2)

3)

4)

5)

6)

b. The components of serum

c. Therefore, the difference between plasma and serum is:

C. Arteries, veins, capillaries and the regulation of blood flow through a capillary

1. Definitions

a. ARTERY =

b. ARTERIOLE =

c. VEIN =

d. VENULE =

e. CAPILLARY =

2. The walls of arteries and veins

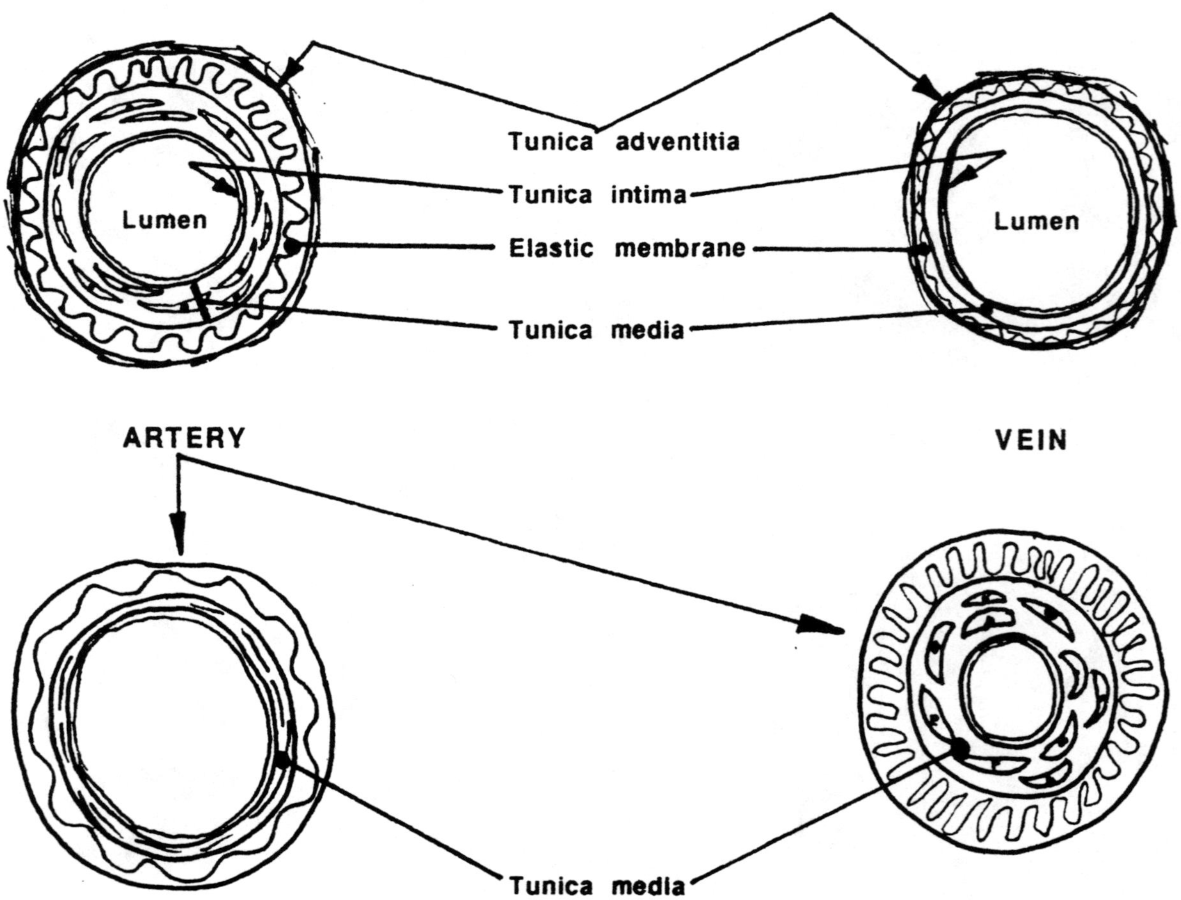

3. Regulation of blood flow through the capillary

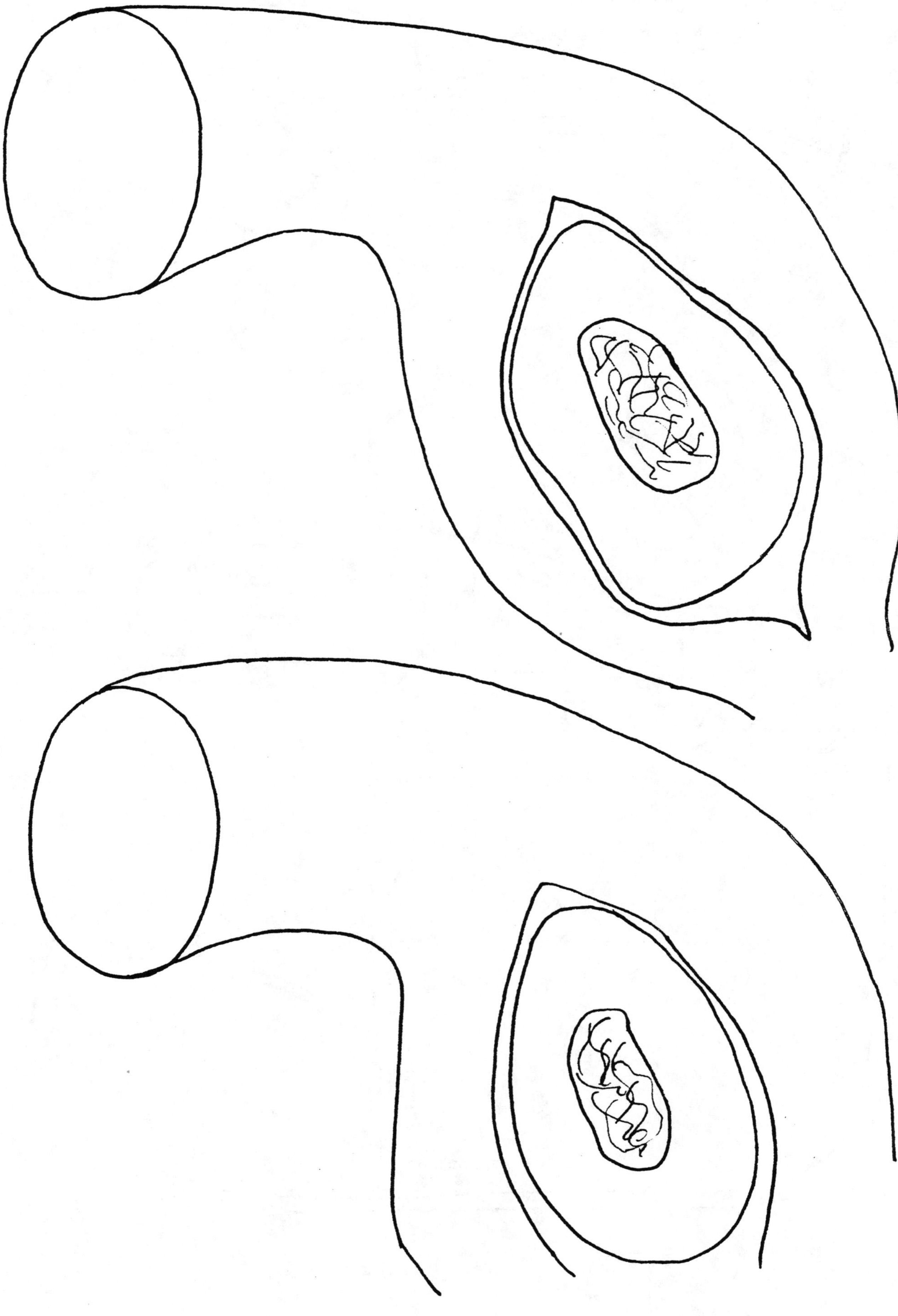

D. The general circulatory scheme

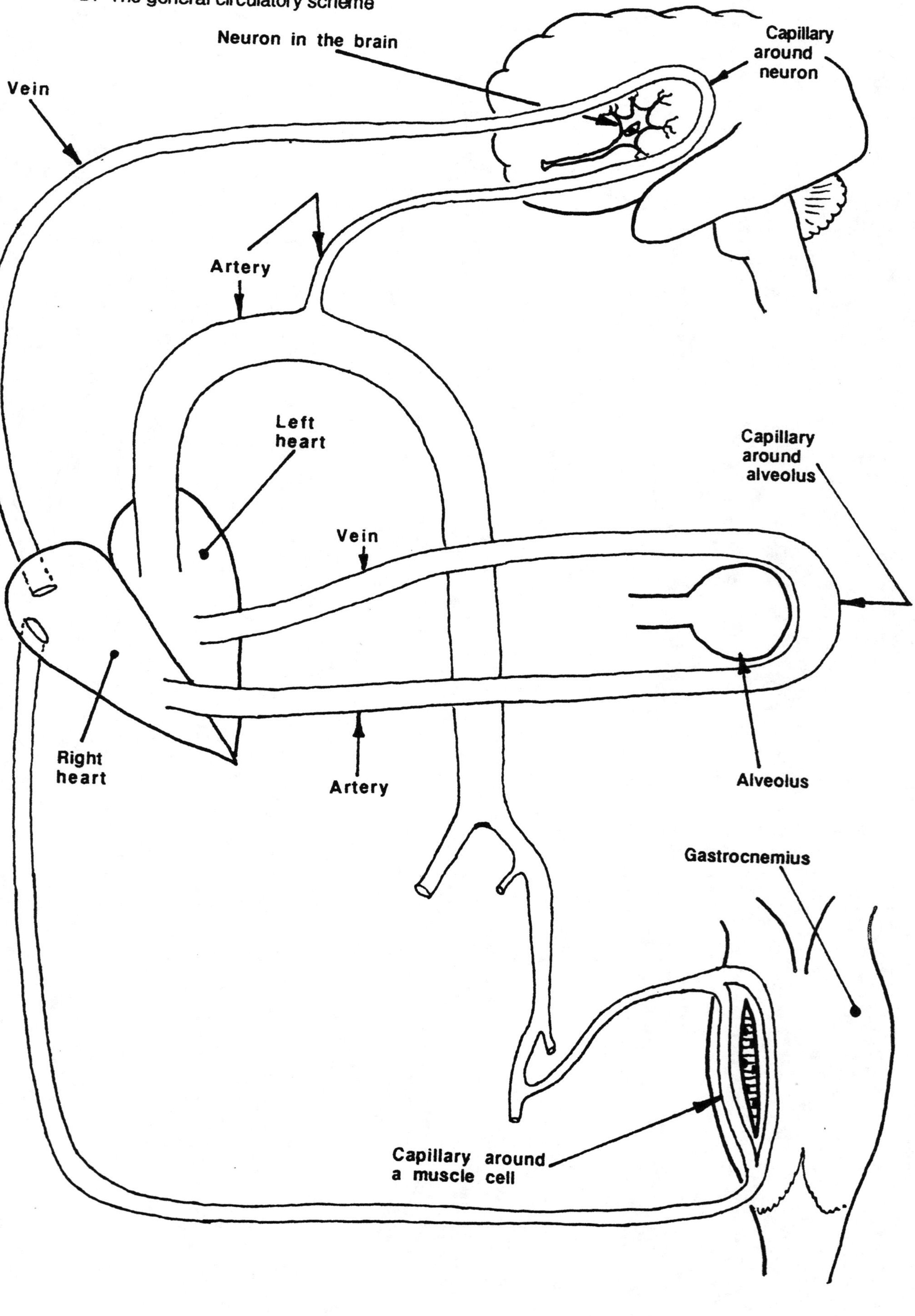

E. Circulatory schemes in the **ADULT**

1. **SYSTEMIC AND PULMONARY CIRCULATIONS**

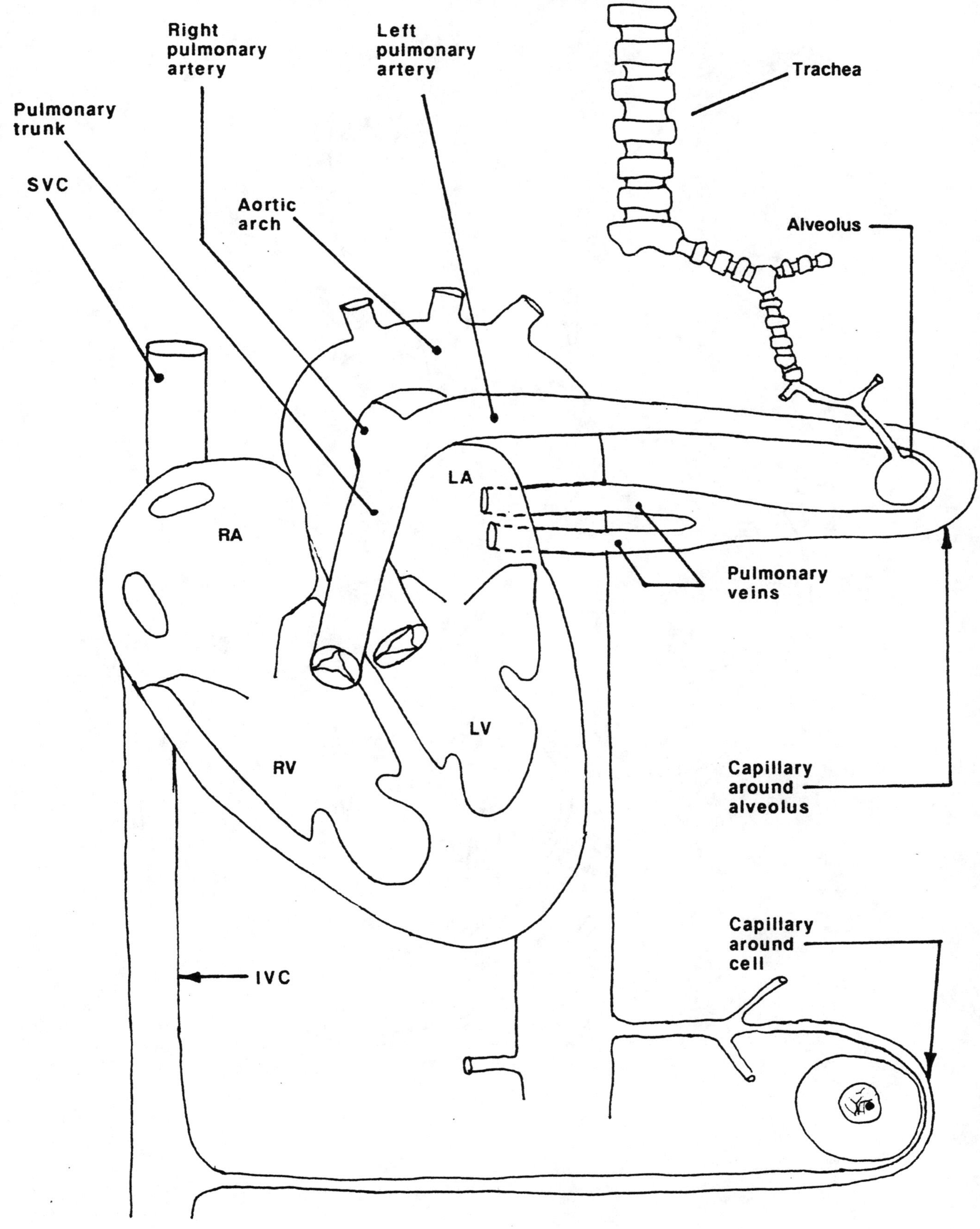

2. **CORONARY CIRCULATION**

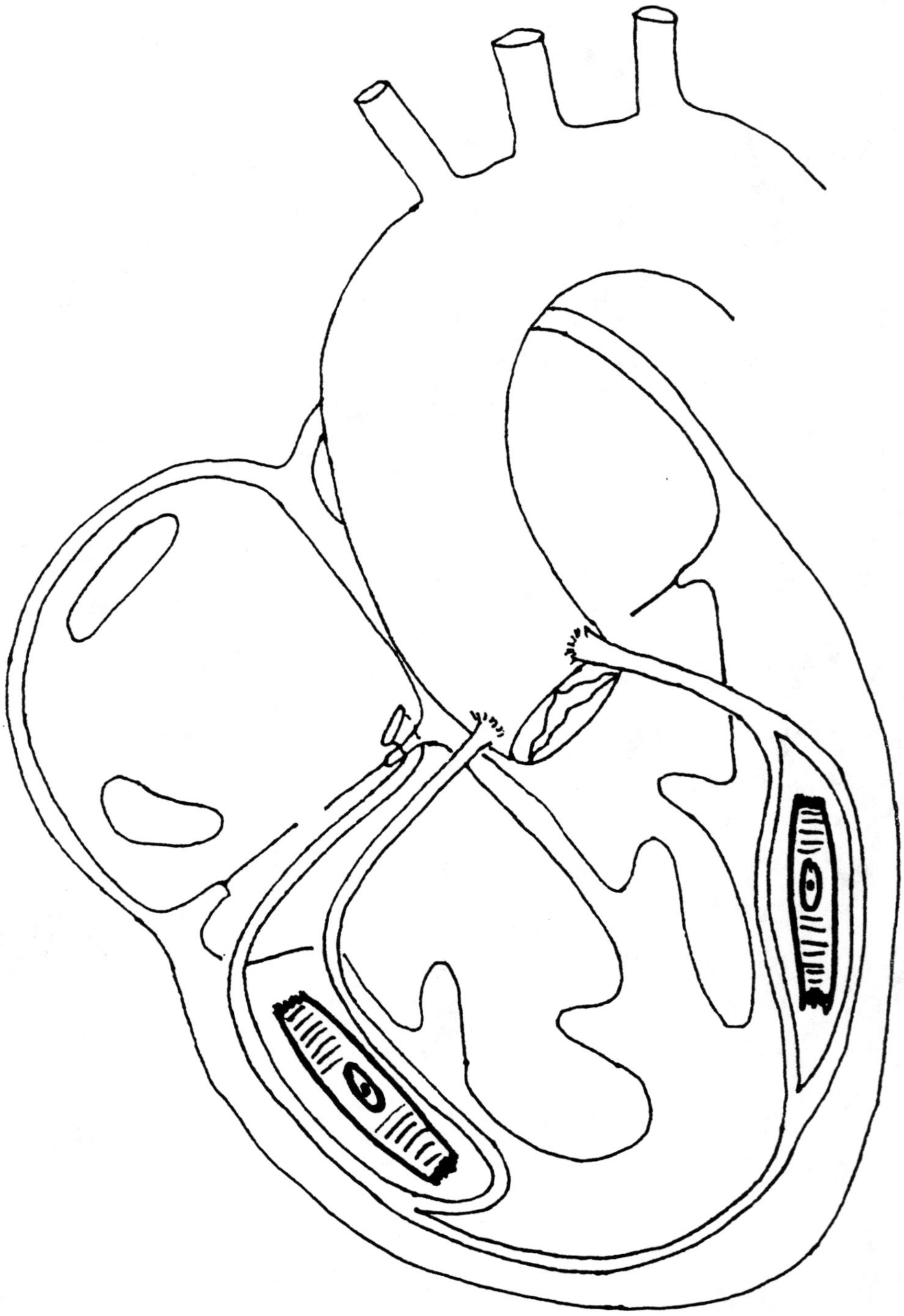

3. Some facts and generalities that can be derived from the previous diagrams

   a. FIRST THE **FACTS**

      1) The superior vena cava, inferior vena cava, and the coronary sinus are veins that contain unoxygenated blood flowing from the body and heart into the right atrium

      2) The pulmonary veins contain oxygenated blood flowing from the lungs into the left atrium

      3) The pulmonary trunk contains unoxygenated blood flowing out of the right ventricle toward the lungs

      4) The aorta contains oxygenated blood flowing out of the left ventricle and toward the cells of the body and the heart

   b. NOW THE **GENERALITIES**

      1) The **veins** that are attached to the heart **are attached to the atria**, not the ventricles

      2) The **arteries** that are attached to the heart **are attached to the ventricles**, not the atria

      3) The blood that is flowing in the **veins** of the body eventually enters **either the right or left atrium**

      4) The blood that **leaves** the heart does so through **arteries**

      5) The **right side of the adult heart** contains **unoxygenated blood** flowing through its two chambers

      6) The **left side of the adult heart** contains **oxygenated blood** flowing through its two chambers

F. Valves of the heart, the cardiac cycle, and the reasons for a unidirectional blood flow in the body

1. Valves of the heart

a. General anatomy of a valve

1) Definition of a **CUSP** =

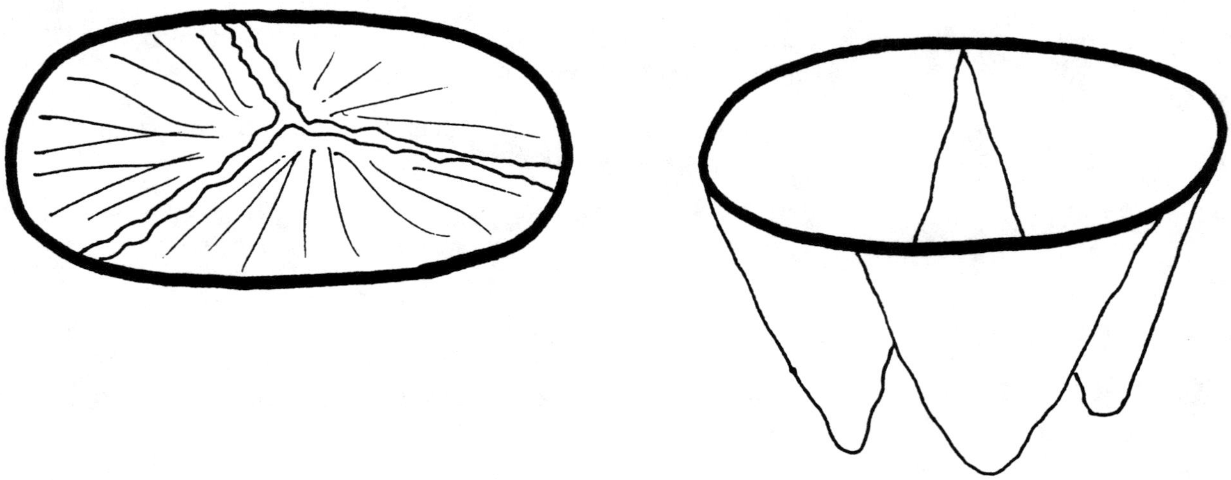

b. <u>**SEMILUNAR AND ATRIOVENTRICULAR VALVES**</u>

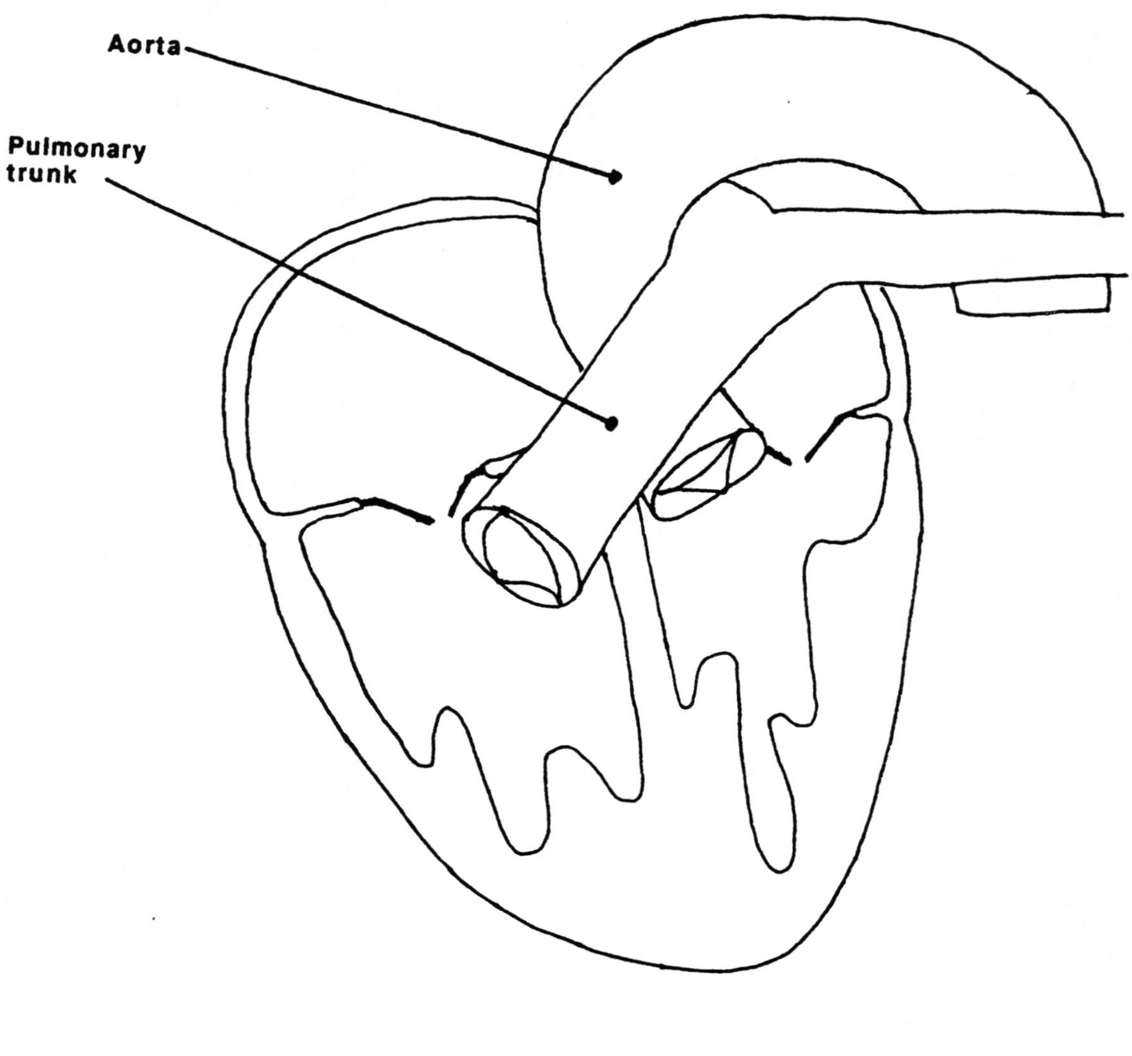

2. The cardiac cycle and the functions of the valves

a. The cardiac cycle

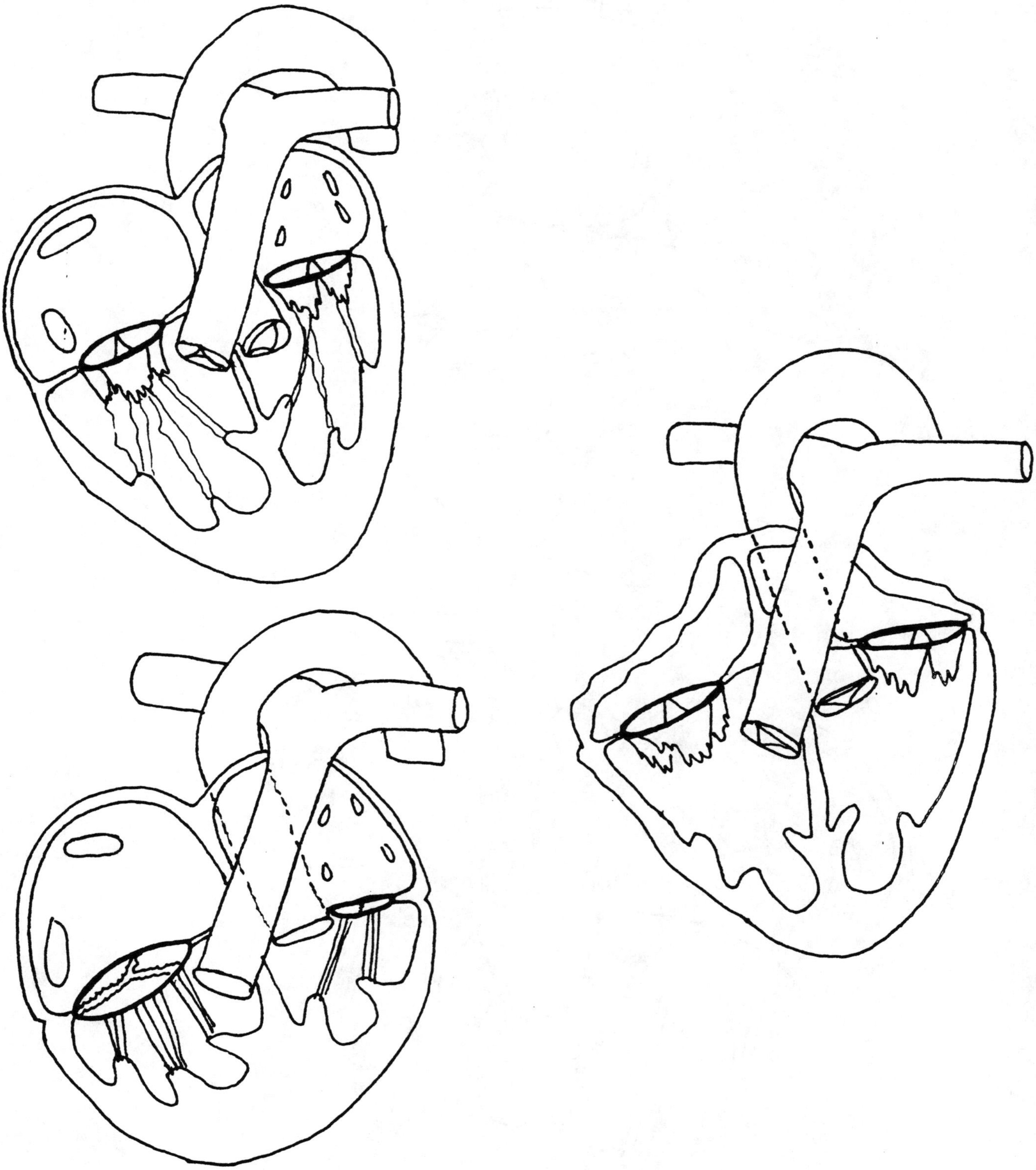

b. Functions of the valves

1) Of the **ATRIOVENTRICULAR VALVES**

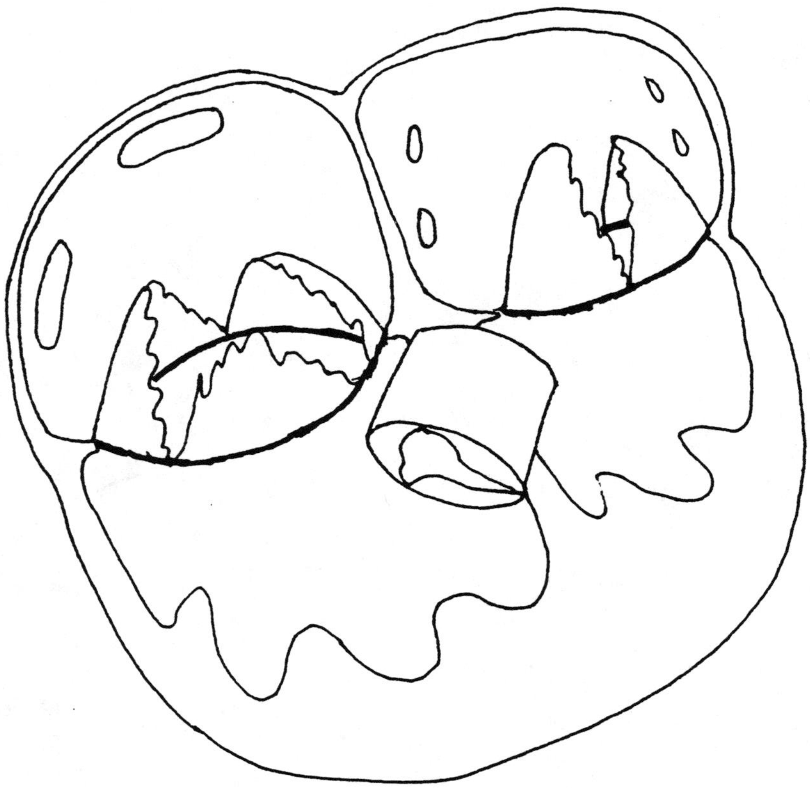

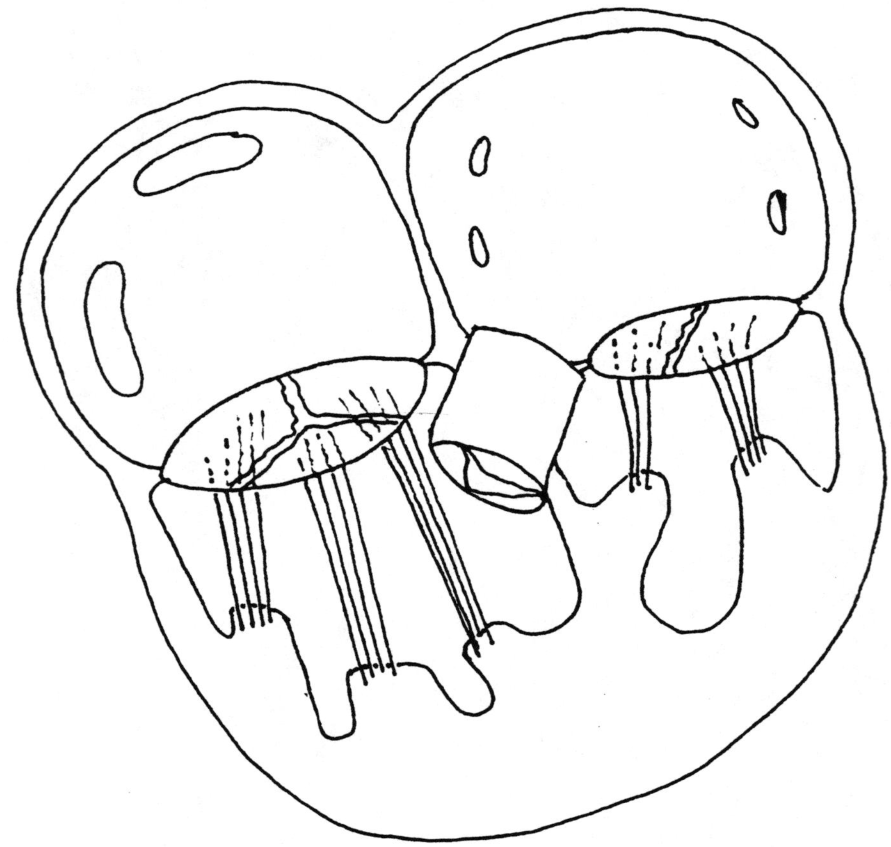

2) Function of the **SEMILUNAR VALVES**

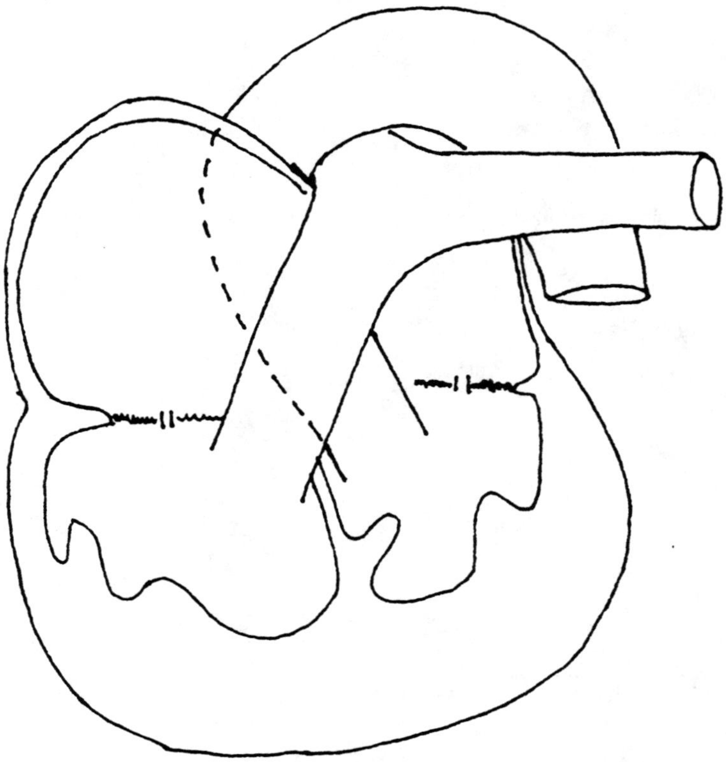

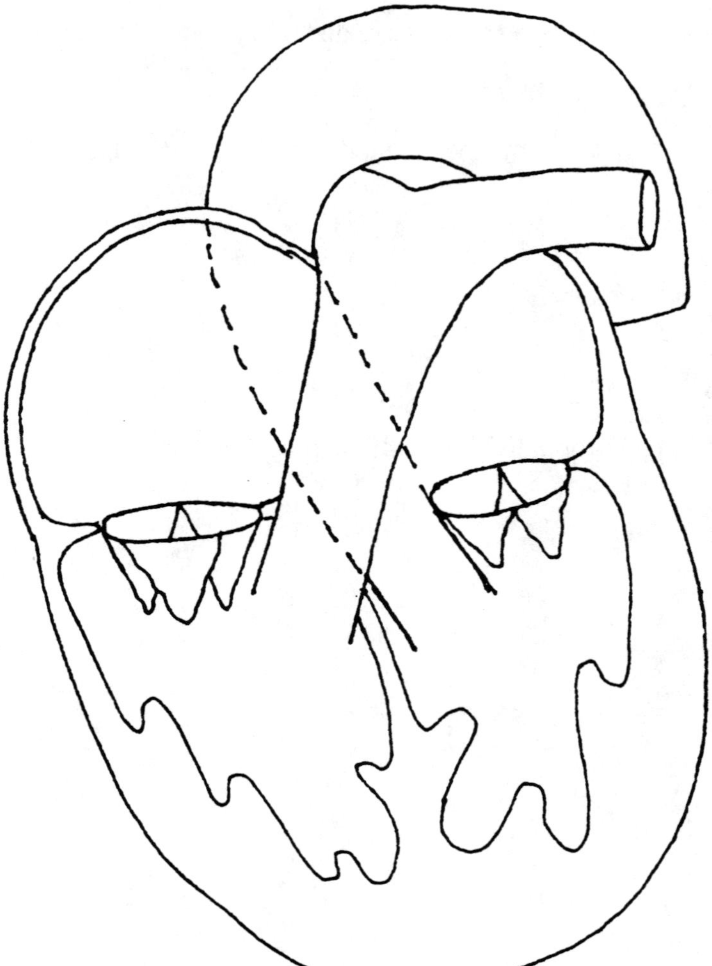

G. Some common congenital heart malfunctions and their consequences

1. Atrial septal defect

2. Patent ductus arteriosus

3. Ventricular septal defect

4. Tetraology of Fallot

a.

b.

c.

d.

RESULT OF ANY OF THESE DEFECTS =

I. The innervation of the heart and the conductive system of the heart

1. Innervation of the heart

a. Parasympathetic innervation by the **VAGUS NERVE (CN X)**

b. Sympathetic innervation through the spinal cord

2. The conductive system of the heart.

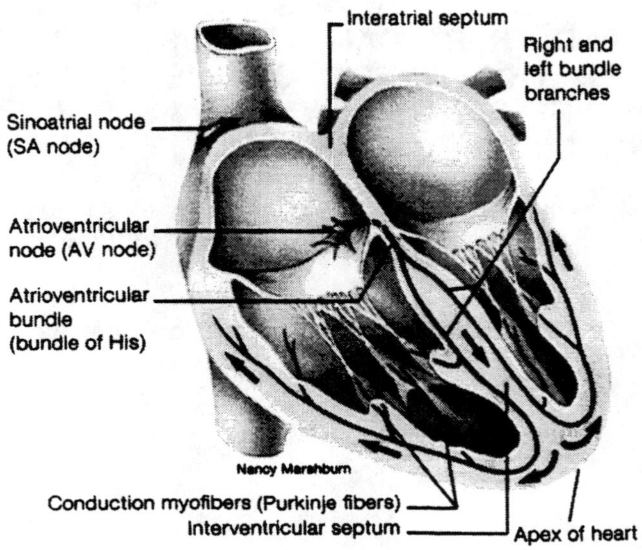

H. FETAL CIRCULATION

1. In the fetus the systemic and coronary circulations are the same as in the adult. In addition, the cardiac cycle of the fetus is the same as in the adult. However, **pulmonary circulation in the fetus is greatly diminished**.

2. The reason for having a diminished pulmonary circulation in the fetus is

BECAUSE THE FETUS IS DEVELOPING UNDER A FLUID (AMNIOTIC FLUID) AND THERE CAN OBVIOUSLY BE NO O_2 / CO_2 EXCHANGE BETWEEN THE ALVEOLI OF THE RESPIRATORY SYSTEM AND THE CIRCULATORY SYSTEM AS OCCURS IN THE LUNGS FOLLOWING BIRTH

3. Because there is no reason to send large quantities to the fetal lungs there **are two anatomical devices to divert blood in the fetal right heart away from the lungs.**

a. **The anatomical device inside the fetal heart =**

b. **The anatomical device outside the heart =**

4. **The cells in the rapidly growing fetal body have a great need for nutrients and O_2. Since the lungs do not function for O_2 / CO_2 exchange, and the fetus cannot eat, or eliminate the urea that its liver makes, then there must be an additional anatomical device that takes over the functions of the respiratory, digestive, and urinary systems. This anatomical device is the:**

5. Diagram of fetal circulation

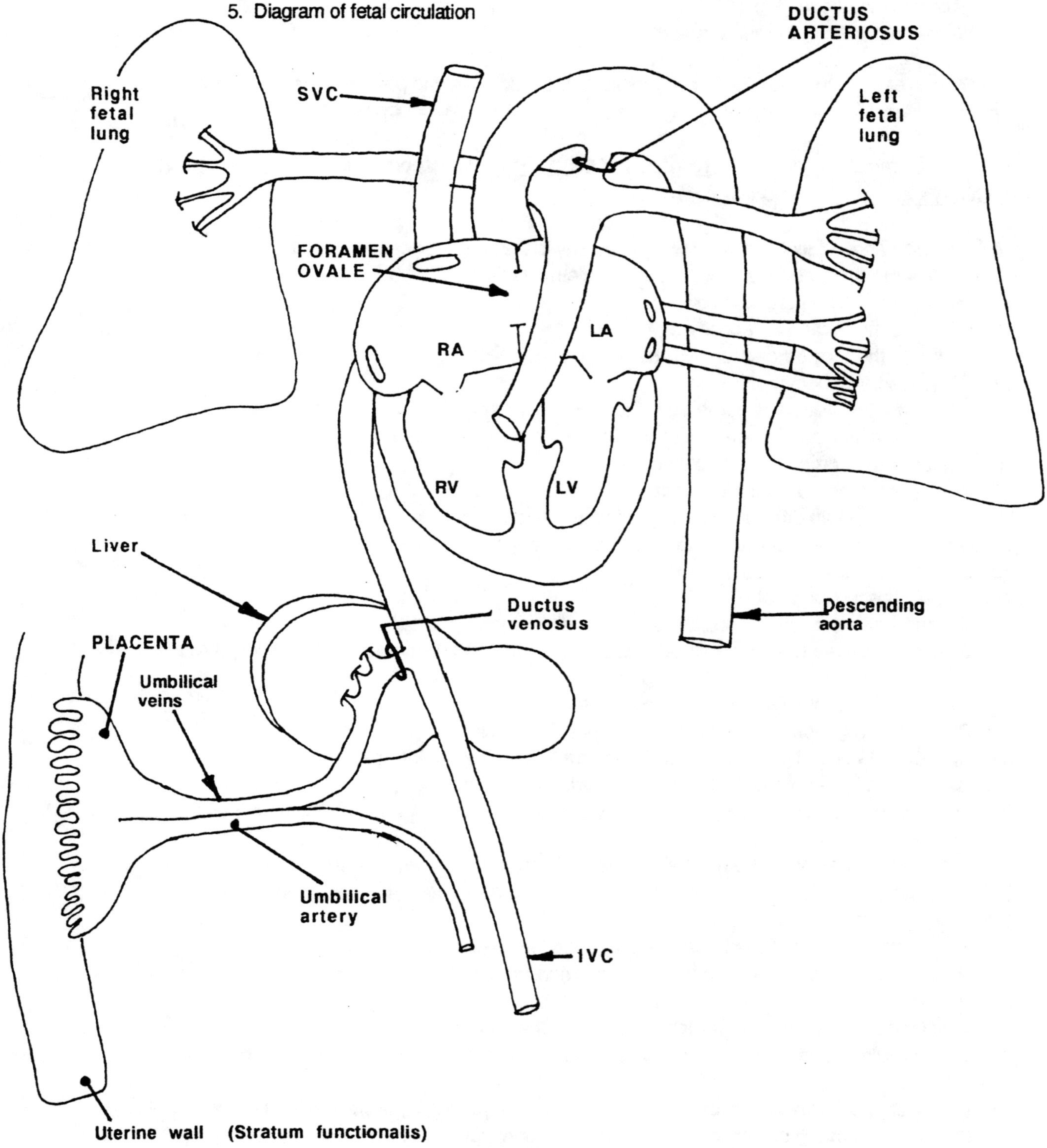

1. What are the two main portions of blood ?
2. What are the three formed elements ?

3 - 7: For each of the following blood functions mark an **A** if it is performed by the plasma, a **B** if performed by the erythrocytes, or a **C** if performed by the leucocytes:

Provision of immunity_______; Transport of O_2_______; Transport of urea_______; Transport of CO2_______; Transport of hormones_________.

8. From the above functions what would you say is the major function of blood ?
9. What is the most abundant of the formed elements ?
10. What are the three granular leucocytes ?
11. What are the agranular leucocytes ?
12. Which of the formed elements binds oxygen ?
13. What are thrombocytes ?
14. Where do thrombocytes come from (how are they formed) ?
15. What is another good name for a thrombocyte ?
16. What are the components of plasma ?
17. What are the components of serum ?
18. What is the single difference between plasma and serum ?
19. Define the following terms: Artery, arteriole, vein, venule, capillary
20. Describe the wall of an artery and vein.
21. What is the tunica media ?
22. Describe how the tunica media is similar to the sphinter pupillae.
23. Describe how the tunica media operates to increase and decrease the blood flow through a capillary.
24. What are vasoconstriction and vasodilation ?
25. Describe in a general way systemic, coronary, and pulmonary circulations.
26. Describe in detail these three circulations and how they are interrelated.
27. What type of blood vessels attach to the atria of the heart ?
28. The superior vena cava contains what type of blood and from where in the body has this blood come ?
29. Into which chamber of the heart does the blood from the superior vena cava flow ?
30. The inferior vena cava contains what type of blood and from where in the body does this blood come ?
31. Into which chamber of the heart does the blood in the inferior vena cava flow ?
32. What type of blood is in the coronary sinus and from where in the body does this blood come from ?
33. Which chamber of the heart does the blood in the coronary sinus flow into ?
34. What type of blood is in the pulmonary veins and from where in the body does this blood come ?
35. In the adult, what type of blood, oxygenated or unoxygenated, flows into the right atrium ?
36. In the adult, what type of blood, oxygenated or unoxygenated, flows into the left atrium ?
37. To which circulation do the superior vena cava and the inferior vena belong ?
38. To which circulation does the coronary sinus belong ?
39. To which circulation do the pulmonary veins belong ?
40. What types of vessels are those that are attached to the ventricles ?
41. Which artery receives blood from the right ventricle ?
42. Is this blood in the adult pulmonary trunk oxygenated or unoxygenated ?

43. Which artery receives blood from the left ventricle ?
44. Is the blood in the aorta oxygenated, or unoxygenated ?
45. Where is the unoxygenated blood in the pulmonary trunk going ?
46. Where is the blood in the aorta going ?
47. To which circulation does the pulmonary trunk belong ?
48. To which two circulations does the aorta belong ?
49. Considering your answers to questions **28 - 33, 35,** and **41, 42,** what can you say about the type of blood in the right heart of the adult - oxygenated, or unoxygenated ?

50. Considering your answers to questions **34, 36, 43, and 44**, what can you say about the type of blood in the left heart of the adult - oxygenated or unoxygenated ?
51. What is a cusp ?
52. What are the two types of valves ?
53. What are the names of the two semilunar valves and where are they located ?
54. How many cusps does each semilunar valves have ?
55. Why are semilunar valves called semilunar ?
56. What are all the names of the atrioventricular valves ?
57. How many cusps does the right AV valve have ?
58. How many cusps does the left AV (mitral) valve have ?
59. Describe briefly the sequence of atria and ventricle contraction during the cardiac cycle.
60. During the simultaneous contraction of the two ventricles during the cardiac cycle are the two AV valves open or closed ?
61. During the simultaneous contraction of the ventricles during the cardiac cycle are the two semilunar valves open or closed ?
62. During the contraction of the ventricles how are the two AV valves held closed ?
63. What are the chorda tendineae and papillary muscles and what are their functions ?
64. Why is there a need to keep your AV valves closed when your ventricles contract ?
65. When the ventricles contract are the semilunar valves open or closed ?
66. Are chorda tendineae attached to the cusps of the AV valves ?
67. Are they attached to the cusps of the semilunar valves ?
68. What do you think would be the consequences of having the chorda tendineae attached to the cusps of the semilunar valves ?
69. When do the two atria fill up with blood ?
70. What prevents blood from flowing back into the ventricles due to gravity when the two ventricles relax ?
71. When the ventricles contract what prevents blood from being pumped back into the atria ?
72. Which of the three circulations is greatly diminished in the fetus ?
73. Why ?
74. What are the two anatomical devises that divert (shunt) large quantities of blood away from the heart in fetal circulation ?
75. Which of these devices is inside the heart ?
76. What does the foramen ovale become following birth ?
77. Which of the devices is outside the heart ?
78. What does the ductus arteriosus become following birth ?
79. Which three systems does the placenta substitute for in the fetus ?
80. With whose circulation does the fetus exchange O_2, CO_2, nutrients and urea ?
81. What is an atrial septal defect ?
82. What is its result ?
83. What is a patent ductus arteriosus ?

84. What is the result of a patent ductus arteriosus ?
85. What is a ventricular septal defect ?
86. What is its result ?

87 - 90: What are the four abnormalities in the tetraology of Fallot ?

91. What is the result of the tetraology of Fallot ?
92. Which cranial nerve innervates the heart ?
93. What is the sinoatrial nerve ?
94. What is another name for the SA node ?
95. Why is it also called the pacemaker ?
96. How many signals per minute does the SA node (pacemaker) send out ?
97. What are the events that a signal from the SA node causes to happen ?
98. Where is the SA node located ? How many signals / minute does it emit ?
99. What is the function of the AV node ?
100. What does the AV node cause to happen when it sends out a signal ?
101. Trace a signal from the SA node over the conductive system to the musculature (myocardium) of the ventricles.

J. CIRCULATORY ROUTES

1. CORONARY CIRCULATION

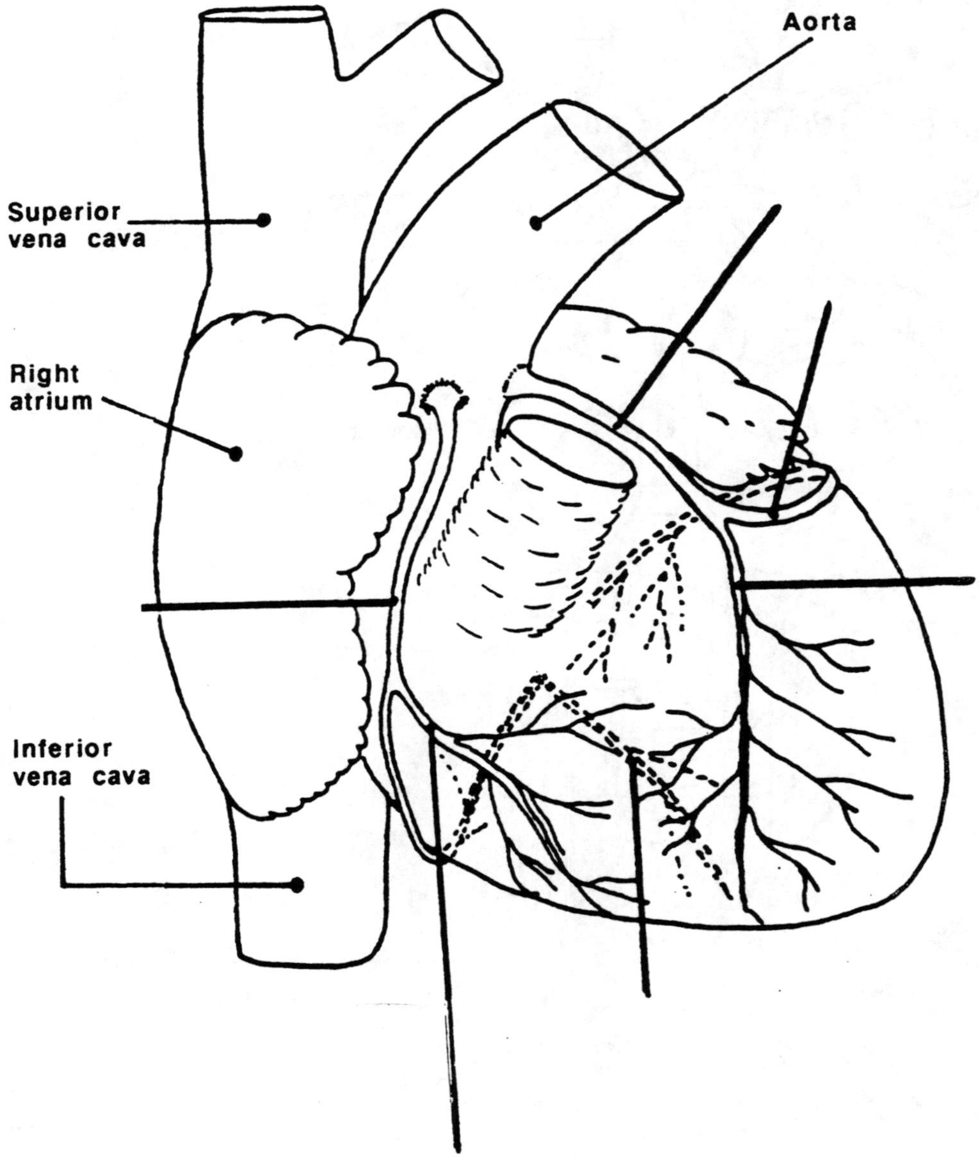

2. CIRCULATION TO THE BRAIN

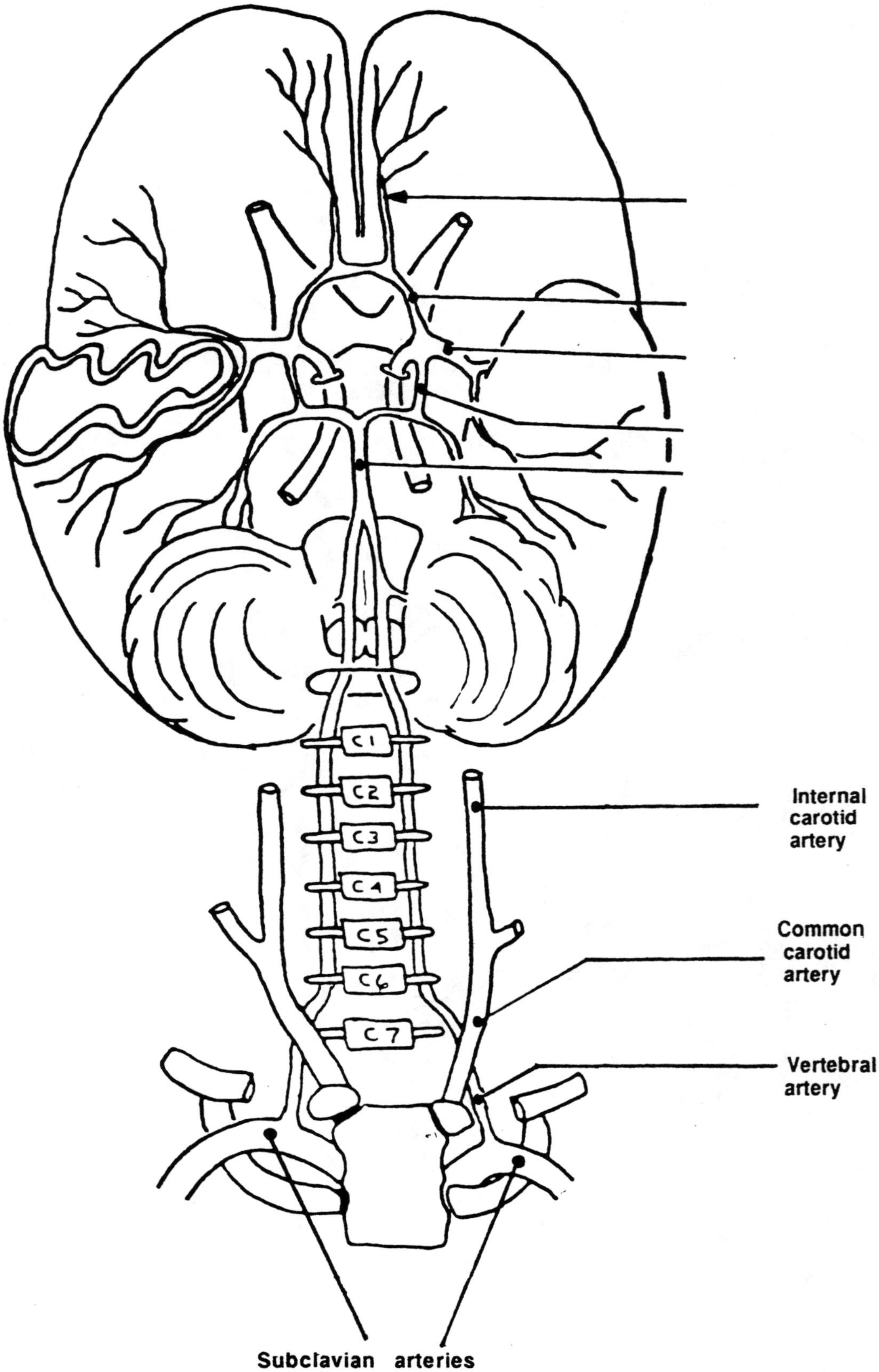

3. CIRCULATION TO THE UPPER LIMB

a. ARTERIAL SUPPLY

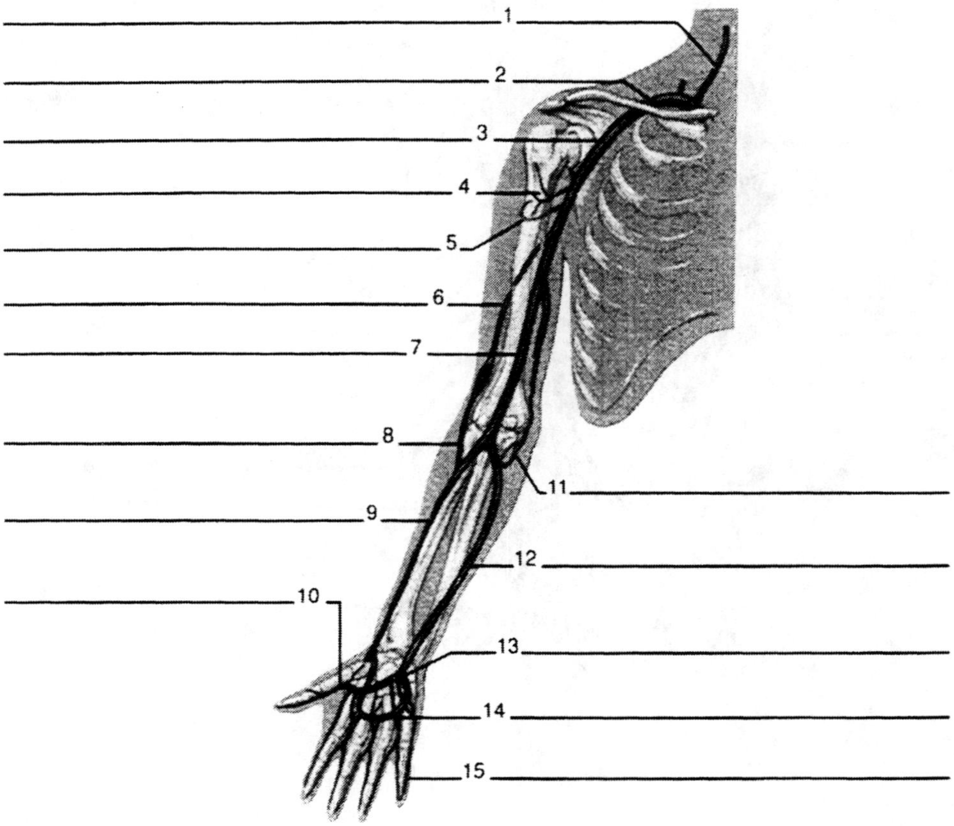

b. ARTERIES AND VEINS AT THE CUBITAL FOSSA

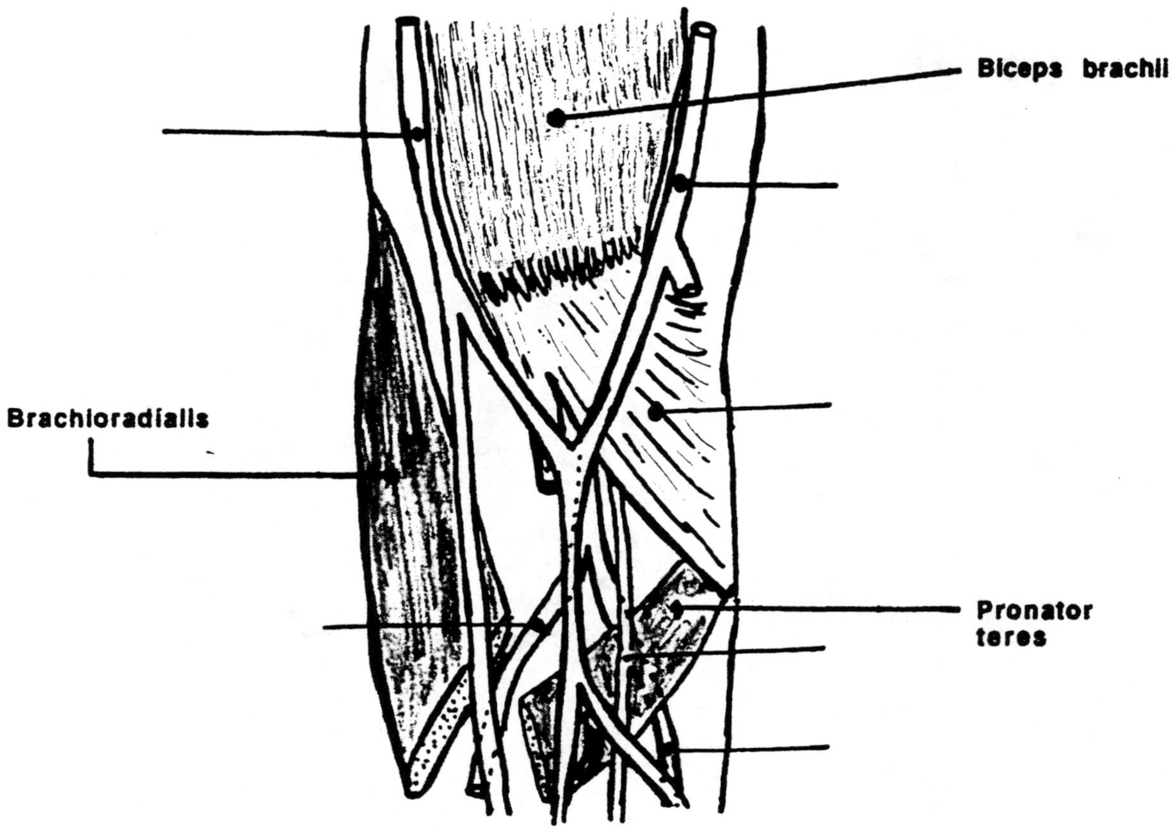

3. CIRCULATION TO THE STOMACH, SPLEEN, and LIVER

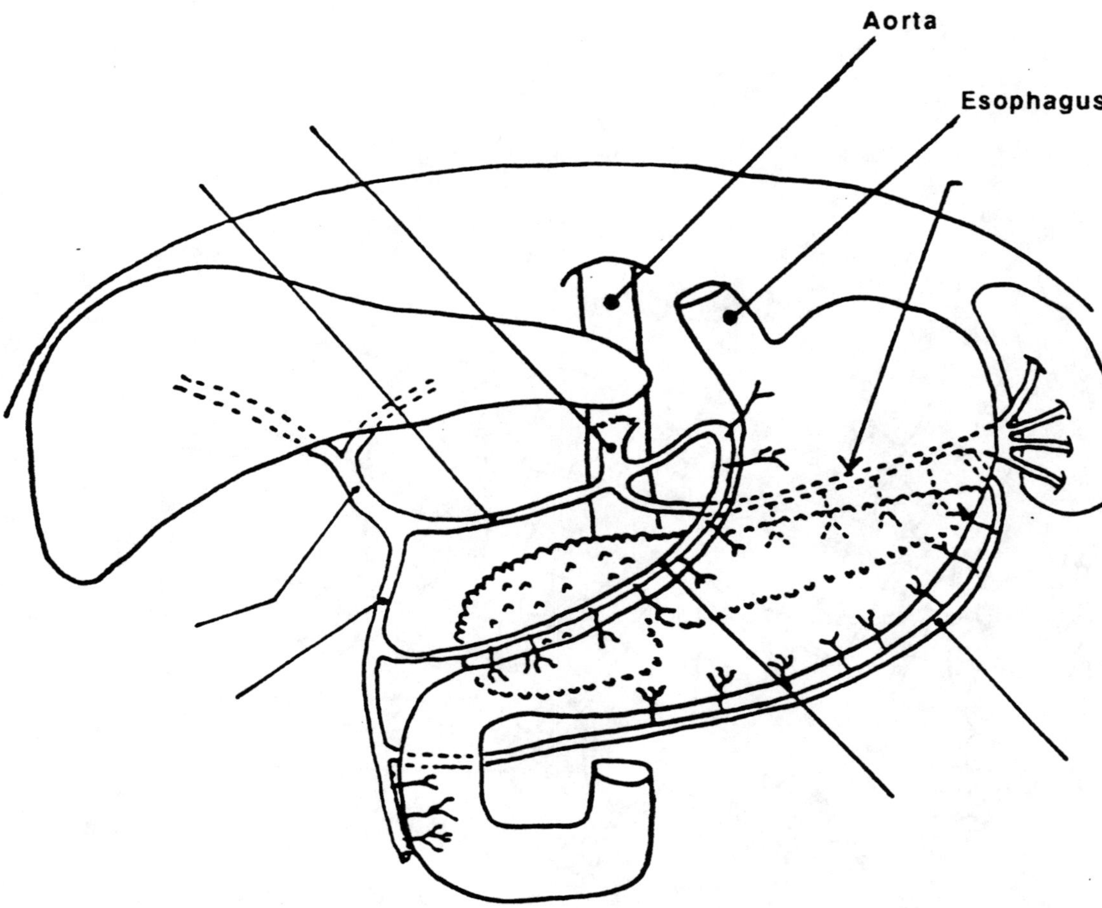

4. CIRCULATION TO THE INTESTINAL TRACT

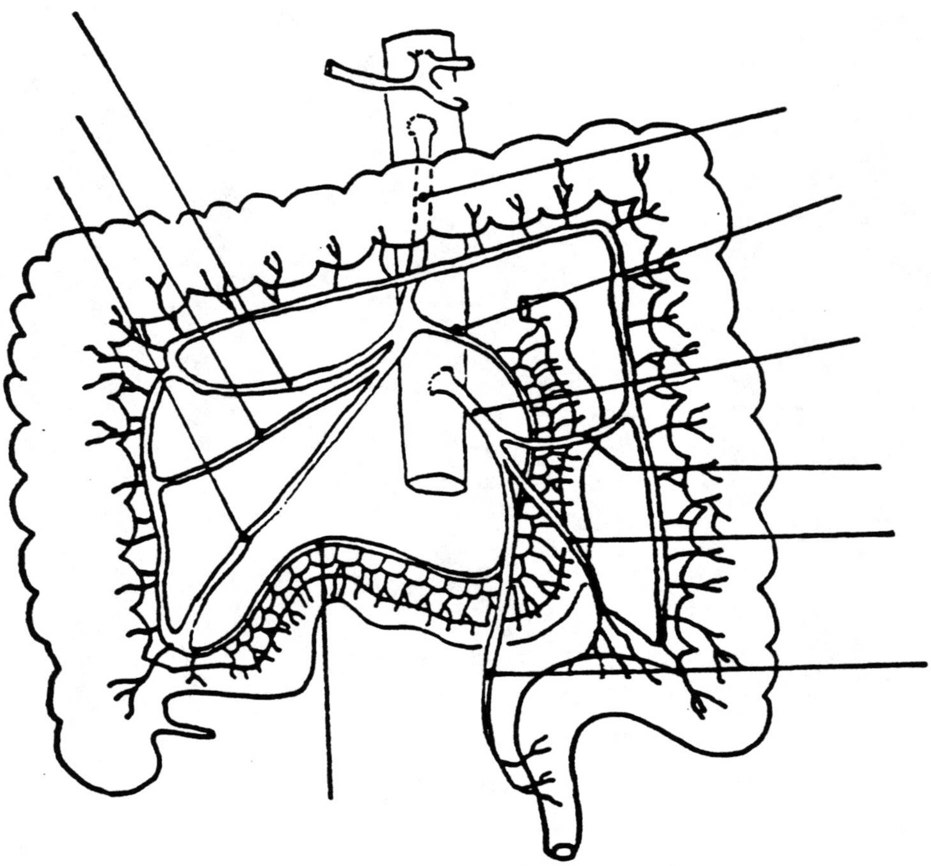

5. HEPATIC PORTAL CIRCULATION

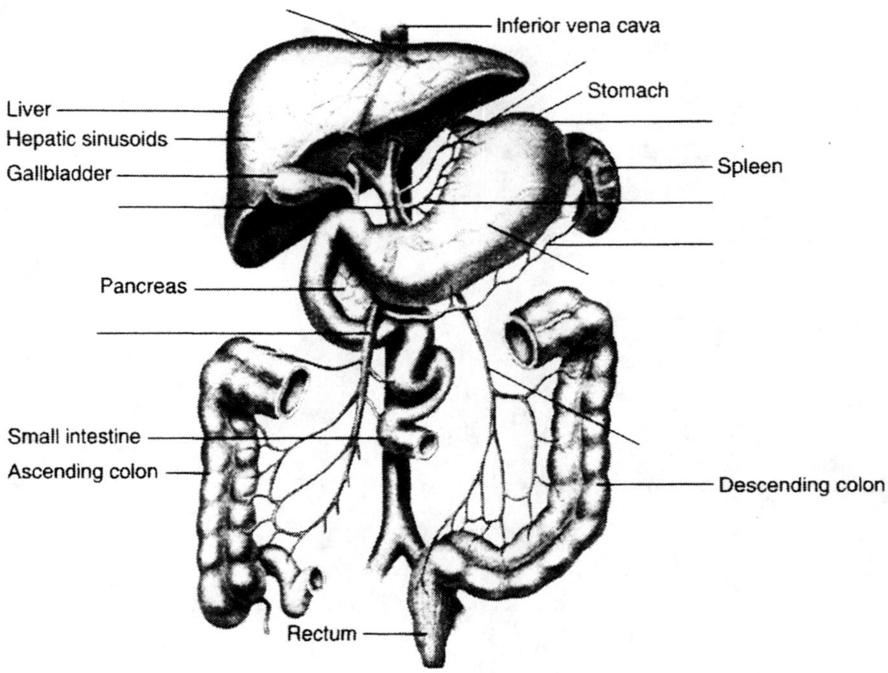

6. CIRCULATION OF THE LOWER LIMB

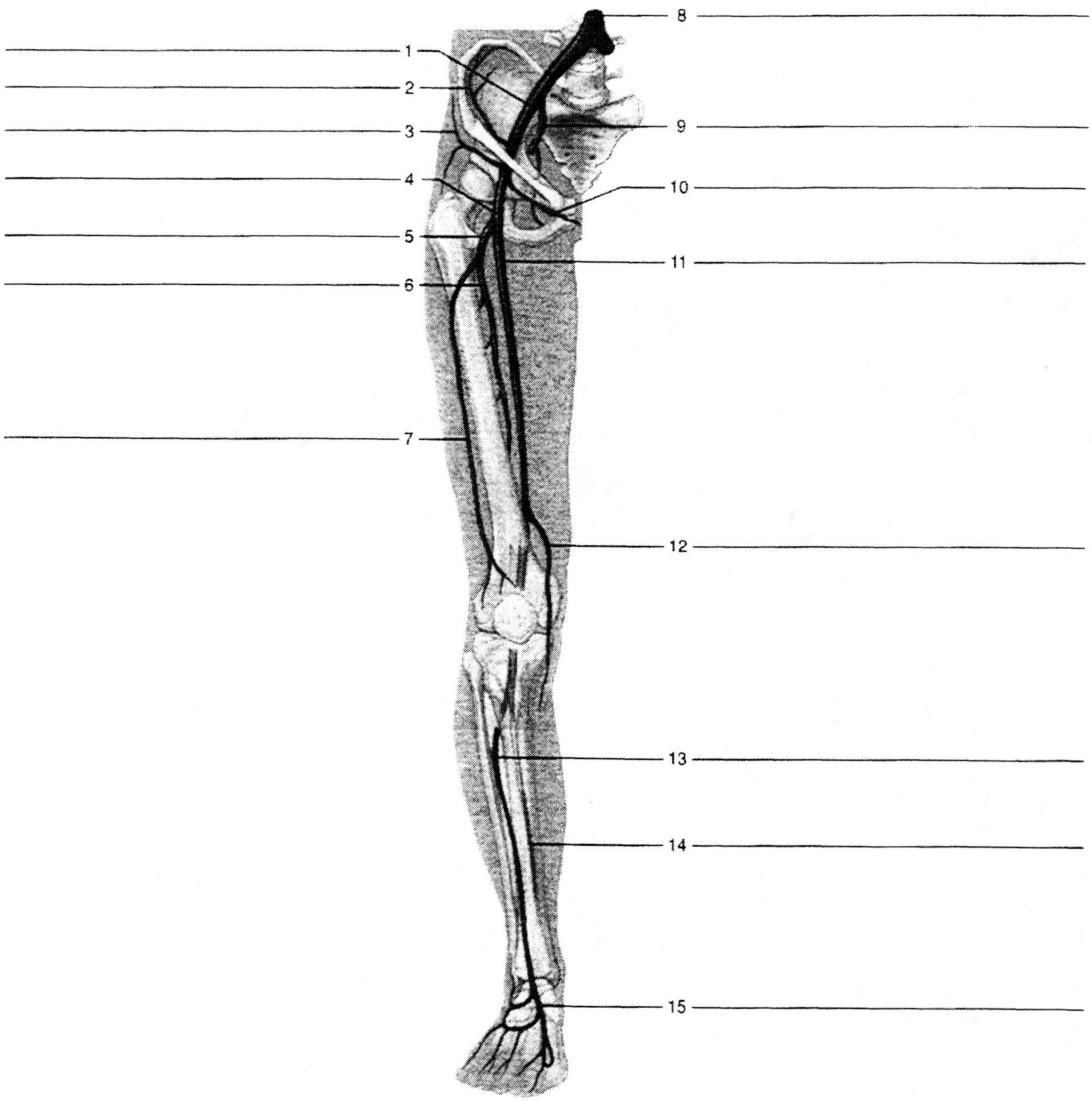

Anterior view

V. The LYMPHATIC SYSTEM

A. General information and the purpose of the lymphatic system

Your **lymphatic system** is a network of lymphatic vessels and organs that are responsible for filtering foreign substances from the blood and protecting your body from diseases.

Lymph is the fluid that flows through the capillaries and other vessels of the lymphatic system. Basically, lymph is composed of excess water and proteins that have entered it from body tissues. Fluids are collected from the interstitial (extracellular) spaces into **lymphatic capillaries,** which flows into larger **lymphatic vessels**, and eventually, the collected fluid is emptied back into the circulatory system through veins in the neck. **Lymph nodes** are situated throughout the body at specific sites along the lymphatic vessels to filter foreign substances and dead bacteria from the lymph.

B. Other important parts of the lymphatic system

1. Tonsils

2. Thymus gland

3. Spleen

4. Peyer' patches

C. The four (4) important functions of the lymphatic system

1.

2.

3.

4.

LABORATORY 1 - CYTOLOGY

LABORATORY ON THE CELL AND CELL CYCLE

INTRODUCTION : Cells are too small to be seen without the aid of some sort of magnifying device. The device that we use is the microscope. Basically, there are two types of microscopes, the **light microscope** and the **electron microscope**. The light microscope uses light waves to pass through the cells, while the electron microscope uses electrons to pass through the cell. The light microscope is fairly uncomplicated and uses glass lenses to focus the light waves, but the electron microscope uses magnets to focus the electrons. The main difference between the light microscope and electron microscope is the power of magnification. A very good light microscope can magnify an object perhaps 1000 - fold, but an electron microscope can magnify an object some 200,000 - to 400,000 - fold.

With the light microscope you can see the major, and larger, cellular structures, such as the **plasma membrane**, **nucleus**, and the **nucleolus**. With the electron microscope, however, you can see in great detail the same structures that you can see in the light microscope, and in addition you can see the small structures inside the cell that are termed the **organelles**. The organelles are structural elements inside the cell that perform specific functions that keep the cell alive and allow the cell to perform a specific function. Some organelles common to all cells include the **nucleus** and **nucleolus** (which can also be seen, though not in great detail, in the light microscope) the **mitochondria**, the **endoplasmic reticulum**, the **Golgi apparatus**, and the **ribosomes**. A very powerful electron microscope can reveal the outlines of protein molecules and the DNA molecule.

OBJECTIVES : The objectives of this laboratory are to:

1. Identify cellular structures and organelles on kodachrome slides

2. Identify the cellular structures that can be seen through a light microscope

3. Identify the various stages of the cell cycle of plant and animal cells as presented under the microscope, or on kodachrome slides

EXERCISES :

1. Your Laboratory Instructor will show you kodachrome slides that will demonstrate the cellular structures that can be seen with the light microscope and with the electron microscope.

a. Cellular structures that can be seen through the light microscope:

1) **Plasma membrane**	4) **Nucleoplasm**
2) **Cytoplasm**	5) **Nucleolus**
3) **Nuclear envelope (membrane)**	6) **Chromatin**

b. Cellular structures that can be seen through the electron microscope:

1) Those same structures that can be seen with the light microscope

2) Organelles located in the <u>cytoplasm</u>

a) **Mitochondria**
b) **Endoplasmic reticulum**
c) **Golgi apparatus**
d) **Ribosomes**

3) Organelles located in the <u>nucleus</u>

a) **Nucleolus**
b) **Chromatin**

2. Your Laboratory Instructor will show you a series of kodachrome slides that will demonstrate the stages of the cell cycle that will include interphase, mitosis, and cytokinesis. From these slides you will be responsible for identifying :

a. Stages of the cell cycle

1) **Interphase**

2) <u>Mitotic stages</u> of the cell cycle

a) **Prophase**
b) **Metaphase**
c) **Anaphase**
d) **Telophase**

3) **Cytokinesis**

b. Cellular structures

1) **Mitotic spindle apparatus**
2) **Astral rays**
3) **Chromosomes**
4) **Equitorial plate**

3. Your Laboratory Instructor will demonstrate how to use the light microscope and how to go about bringing an object into focus under low, medium, and high powers.

4. Following the kodachrome slide demonstration and the light microscope use demonstration, obtain a light microscope and look at the slides that have been provided for you at your table.

a. Choose one of the slides that is on your table and mount it on the microscope stage. Secure the slide with the stage clips.

b. Rotate the objective lenses to low power.

c. Try to focus the slide so that it is clear to you. **IF YOU ARE UNABLE TO FOCUS THE SLIDE, THEN NOTIFY YOUR LAB INSTRUCTOR**. Your Lab Instructor will give you guidance through the focusing process.

d. After you have focused under low objective power, rotate the objective lenses to the next highest power. Using the fine adjustment knob - not the coarse adjustment knob - refocus the slide.

e. Repeat step **d** using the highest power objective lens.

f. Repeat steps **a - e** using another slide.

g. On one of the slides select some cells to look at and find the:

1) **Plasma membrane**

2) **Cytoplasm**

3) **Nucleus**

4) **Nucleolus** (cannot be seen in all cells)

5. A box of slides containing slides of various tissue types is at your table. These slides are numbered and you are responsible for keeping them in order. Please do report to your Lab Instructor if a slide gets broken, or if one is missing, so that it can be replaced.

Slides 1 and 2 are slides of the cell cycle

SLIDE 1: **The cell cycle of a plant cell (Allium Root Tip).** There are several cells to see on this slide of plant cells, many of which are in various stages of the cell cycle. By moving the slide around find as many stages of the cell cycle as possible. In the diagram below, fill in the blanks for each of the indicated cell cycle stages :

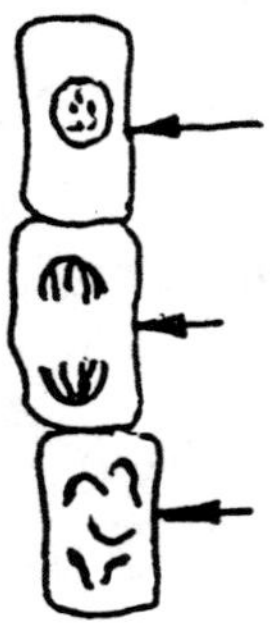

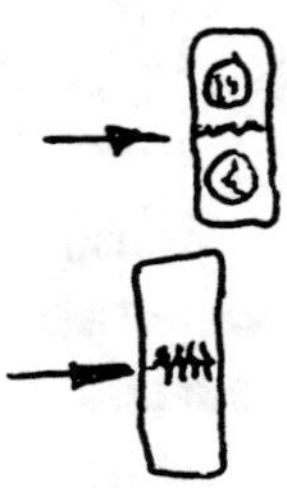

SLIDE 2 : <u>The cell cycle of an animal cell (Cells in the Whitefish Blastula)</u>. The blastula is an early stage of development that occurs in vertebrates. Under the microscope the whitefish blastula will look like the following:

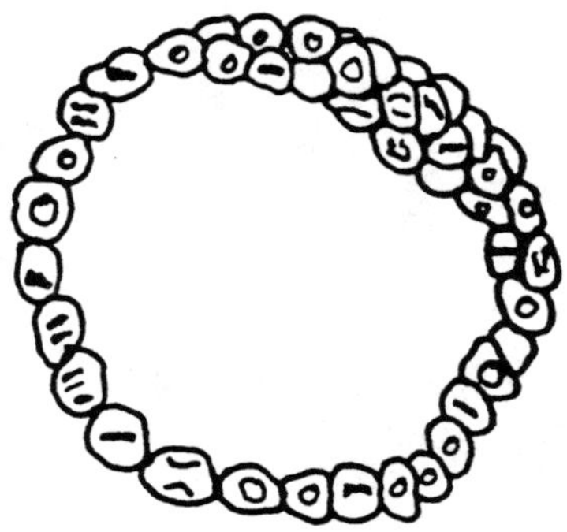

If you look at the blastula through low power, then you will be able to see several stages of the cell cycle. By switching the objective lenses to high power and refocusing you will be able to see these stages in much more detail. Try to find the stages under high power that are diagrammed below.

In addition to observing the cell cycle stages, you should look for and identify : **Chromosomes, the mitotic spindle apparatus, the equitorial plate, and astral rays.**

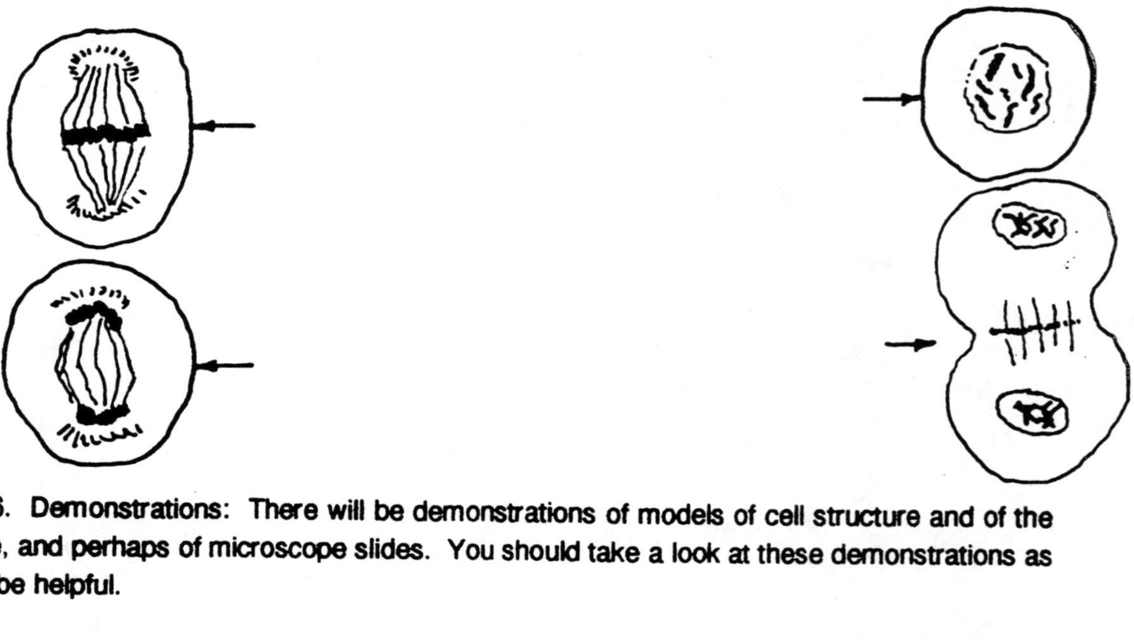

6. Demonstrations: There will be demonstrations of models of cell structure and of the cell cycle, and perhaps of microscope slides. You should take a look at these demonstrations as they will be helpful.

WHAT YOU MUST KNOW AND BE ABLE TO DO AFTER THESE LABORATORY EXERCISES

1. Be able to identify on kodachrome slides the following structures that can be seen through the light microscope : **Plasma membrane, cytoplasm, nuclear membrane, nucleus, chromatin, nucleolus.**

2. Be able to identify through a light microscope the same structures listed in 1.

3. Be able to identify on kodachrome slides the structures that can be seen through the electron microscope : **Plasma membrane, cytoplasm, mitochondria, Golgi apparatus, ribosomes, nuclear membrane, nucleoplasm, chromatin, nucleolus.**

4. Be able to identify cell structures on the demonstration models.

5. Be able to identify the various stages of the cell cycle on kodachrome slides and under the microscope. You should be able to do this for <u>both plant and animal cells.</u>

6. Be able to identify the following structures both on kodachrome slides and under the microscope: **Chromosomes, equitorial plate, mitotic spindle apparatus, astral rays**

7. Be able to identify the cell cycle stages on the demonstration models.

8. **Be able to answer questions about the cell, cellular structures (organelles), the cell cycle and its stages from information that you received in lecture and your reading assignments.**

LABORATORY 2 - HISTOLOGY

HISTOLOGY OF THE EPITHELIAL TISSUES AND CARTILAGE

INTRODUCTION: A tissue is an aggregation of similar cells. What this means, of course, is that the cells in a tissue all have the same function and therefore a given tissue will perform a specific and basic function, such as **secretion or absorption** (epithelial tissues), **connection and support** (connective tissues, such as connective tissue proper, blood, cartilage, and bone), **contraction** to create movements (the muscle tissues), and **propagation of an electrochemical impulse** along its plasma membrane (nervous tissue).

There are no specific laboratories on the organs because these are incorporated into discussions and laboratories on the systems. Nevertheless, an organ can be defined as an aggregate of several tissues which together perform a specific function which none of the individual tissues alone could perform. Let's take the stomach as a good example. The stomach has three tissues that make up it structure. There is an **epithelial tissue** that lines the inner surface of the stomach. The function of this tissue is to **1)** make and secrete the very strong acid called Hydrochloric Acid (HCl), and **2)** to make and secrete an enzyme that partially digests the proteins in your diet. Then there are three layers of **smooth muscle tissue** in the wall of the stomach that are responsible for contracting and thereby causing churning, mixing, and mechanical action when food is in the stomach. These two types of tissues (epithelial and muscle) are held together and fixed in place by **connective tissue** that permeates throughout the wall of the stomach. Together, these three tissues allow the stomach to perform a specific function (partial digestion of protein and the production of a liquid form of food called chyme) that neither of these tissues alone could accomplish. Furthermore, although there are epithelial tissues found in many places in the body, the epithelial tissue lining the inside of the stomach is the only one that makes and secretes HCl and the enzyme that digests proteins.

OBJECTIVES : The objectives of this laboratory are to :

1. Identify the various epithelial tissues on kodachrome slides and microscope slides

2. Identify cartilage as a tissue, and to recognize the matrix , lacunae, and chondrocytes. These must be identified on both kodachrome and microscope slides

EXERCISES:

1. Your Laboratory Instructor will show you kodachrome slides of the following tissues and discuss their structures and locations :

a. **Epithelial Tissues**

1) Simple squamous
2) Simple cuboidal
3) Simple columnar
4) Pseudostratified columnar
5) Stratified squamous
6) Stratified cuboidal
7) Stratified columnar
8) Transitional

With each epithelial tissue your Instructor will point out the general features of the epithelial tissues, namely, the free surface, the basement membrane, and the fact that there is little interstitial space.

b. **Cartilage**

1) Matrix 2) Lacunae 3) Chondrocytes

2. From the box of microscope slides choose the slides that demonstrate the various epithelial tissues and, using these slides, focus and study these tissues. Be sure that you can identify each type and also identify the free surface and the basement membrane.

3. From the slide box choose slides that show cartilage and look closely at these under the microscope. You should be able to recognize cartilage, and also to recognize the matrix, lacunae, and chondrocytes.

4. There may be demonstrations of these tissues set up for you to look at.

WHAT YOU MUST KNOW AND BE ABLE TO DO AFTER THESE LABORATORY EXERCISES

1. Be able to identify the various epithelial tissues when you see them on either kodachrome or microscope slides.

2. Be able to identify the free surface and the basement membrane when you see them on kodachrome or microscope slides.

3. Be able to identify cartilage tissue when you see it on either kodachrome or microscope slides.

4. Be able to identify the matrix, lacunae, and chondrocytes of cartilage when you see them on either kodachrome or microscope slide.

5. **Be able to answer questions from lecture and reading assignments concerning the tissues and structures seen in these laboratory exercises**

LABORATORY 3 - HISTOLOGY

HISTOLOGY OF CONNECTIVE TISSUE PROPER, BONE TISSUE, MUSCLE TISSUE, AND NERVOUS TISSUE

INTRODUCTION : In Laboratory 2, you studied one type of connective tissue, cartilage. In this laboratory you will study the rest of the connective tissues (Connective tissue proper, and bone) as well as muscle and nerve tissues.

Connective tissue proper includes 1) **loose (areolar) connective tissue,** 2) **dense connective tissue,** 3) **elastic tissue,** 4) **reticular tissue,** and 5) **adipose tissue.** It is a rather pervasive tissue, found almost everywhere in the body. The protein fibers of **loose connective tissue** form networks that support blood vessels and nerves: those of **dense connective tissues** are parallel, densely packed and can be found in ligaments, which bind bone to bone, and tendons, which bind muscles to bones. **Elastic tissue** is found, as you might well guess, in structures that undergo stretching and contracting, such as artery walls, walls of hollow organs (stomach, intestine), vocal cords, trachea , and bronchi. **Reticular tissue** is found mainly inside many soft organs as a support system. **Adipose tissue** is located beneath the skin, around the kidneys (where it is termed perirenal fat), around the eyeball (where it is termed periorbital fat), and in the mesentery in the abdominal cavity.

Bone is the rigid form of connective tissue. You will later be studying the overall structure of bones, but here we want to look at the microscopic structure of bone and discover that actually bone is, despite its static look, quite dynamic. The chemical composition of bone is one in which calcium and phosphorus have precipitated around a collagenous fiber system. It is the calcium phosphate that makes the bone rigid. As you will see, compact bone is not completely solid, but is made up of numerous small tunnels through which blood vessels and nerves penetrate throughout compact bone. An anatomical unit of these tunnels is termed an **OSTEON**, or **HAVERSIAN SYSTEM**, and in the compact bone of any of the 206 bones in the body there are many osteons.

There are three types of **muscle tissue** , designated as **smooth, cardiac, and skeletal.** Additionally, smooth and cardiac muscle types come under the category of **involuntary muscle**, and skeletal muscle is classified as **voluntary muscle**. As far as general locations are concerned, smooth muscle is located interiorly, in the walls of blood vessels and the walls of hollow organs (esophagus, stomach, intestinal tract, uterus, urinary bladder, gall bladder), and in the eye. Skeletal muscles are located more exteriorly in the body, many just below the skin. Why do you think skeletal muscles are called skeletal muscles ? Unlike either smooth or skeletal muscles, cardiac muscles is confined to a specific location, the heart, where it is often called the **myocardium**. Muscle cells are also called muscle fibers.

The basic cell of **nervous tissue** is the **neuron**. Nervous tissue is very complicated, mainly because of the structure of the neuron and the trillions of connections, called synapses, between them. Your brain is made up of about 100 billion neurons. As an example of the complexity of nervous tissue, consider the thin layer of neurons called the cerebral cortex, located on the outer surface of your brain. In your own cerebral cortex there are 10 - 14 billion neurons all connected to one another through perhaps almost a trillion synapses. It is through these synapses among these neurons that you think, generate your emotions, have your fantasies, store your memories, contract your skeletal muscles, form your creativity , produce your dreams.

OBJECTIVES: The objectives of this laboratory are to:

1. Identify the various types of connective tissue proper on both kodachrome and microscope slides

2. Identify the components of the Haversian system on kodachrome and microscope slides, and on models

3. Identify the three types of muscle tissue on both kodachrome and microscopic slides

4. Identify the neuron on both kodachrome and microscopic slides

EXERCISES:

1. Your Laboratory Instructor will show you kodachrome slides along with a discussion of the following:

a) The histology of the various forms of connective tissue proper

b) The histology of compact bone tissue

c) The three forms of muscle tissues

d) The structure of the neuron

2. From the box of microscope slides choose the slides that show connective tissues proper and, using these slides, focus and study these tissues to the extent that you will be able to recognize them.

3. From the box of slides select the slides that shoe compact bone tissue. With these slides make sure that you find and identify the following components of the Haversian system (Osteon):

Haversian canal **Canaliculi**

Matrix

Lacunae **Lamellae**

4. From the box of slides choose the slides that show the three types of muscle tissue. Look at these slides until you can confidently recognize each type.

5. From the box of slides select the slides that show neurons and look at these.

6. Demonstrations of these tissues may be set up for you to look at.

WHAT YOU MUST KNOW AND BE ABLE TO DO AFTER THESE LABORATORY EXERCISES

1. Be able to identify the various connective tissues proper when you see them on either kodachrome or microscope slides.

2. Be able to identify bone tissue and the components of the Haversian system (Haversian canal, canaliculi, lacunae, matrix, and lamellae) when you see them on either kodachrome slides, microscope slides, or models.

3. Be able to identify the three types of muscle tissue when you see them on either kodachrome of microscope slides.

4. Be able to recognize a neuron on either kodachrome or microscope slides.

5. **Be able to answer questions from lecture and reading assignments concerning the tissues and structures that you learn in these laboratory exercises.**

LABORATORY 4 - SYSTEMS

THE REPRODUCTIVE SYSTEMS

INTRODUCTION: Although they start out looking anatomically quite similar, within a few weeks of intrauterine development it is possible to begin to see some anatomical differences between male and female reproductive systems. Keep this in mind, that we all start out looking the same, because you will then be able to understand that there is still a great deal of anatomical similarity between women and men as adults. As an example, it is indeed true that grown, adult men actually have a vagina and uterus. However, these two structures have become so diminished in size during the development of the male reproductive system that not only could the uterus not even remotely sustain development, but the name is changed. And it is equally true that women have a "scrotum", but the ovaries do not descend as do the testes in the male, so the name gets changed in females. We can try to keep up with the anatomical equivalent structures as we go along. Some will be obvious (glans clitoris and glans penis, for example), but some will require a bit of thinking (the labia minora and the penile urethra are equivalent structures, but that will take some thinking to understand why).

You may rightfully think it very odd that there are two reproductive systems. It requires only one respiratory system to properly breathe, only one digestive system to get properly nourished, only one urinary system to properly rid the body of urea and maintain your blood pressure and fluid balance. "I think, therefore I am", so thinking requires only a single nervous system. You can move your own body around in space, so actually only your own muscular and skeletal systems are needed. But, try as you might, you cannot properly reproduce by yourself. As odd and peculiar as it may seem when you really stop to consider it, proper reproduction for us humans requires two individuals.

Reproduction begins with the **fertilization** of the **ovum** (egg) by a single **spermatazoa** (sperm) to produce a fertilized egg called the **zygote**. The best place for fertilization to happen in humans is in the **ampulla region of the fallopian (uterine; ovarian) tube**. For seven to nine days following fertilization the individual drifts down the fallopian tube and finally emerges out into the uterine cavity. By this time, cell division has created an individual of many cells that is termed a **blastocyst**. It is the blastocyst that is first capable of implantation and it will implant regardless of its location. A tubal pregnancy occurs if movement down the fallopian tube is impeded and the blastocyst stage of development occurs while in the fallopian tube.

Proper reproduction requires a lot of things to occur: **1)** Spermatzoa must be made in the **testes**, **2)** ova must be made in the **ovaries**, **3)** one sperm and an egg must somehow be united for fertilization to occur (in the vast majority of human cultures this can only be done by **sexual intercourse**. However, in some technologically advanced cultures, getting a sperm and egg together for fertilization **can** be accomplished outside the body (and then the blastocyst implanted into the endometrium) , **4) implantation** after fertilization must occur (hopefully into the endometrium of the upper half of the **uterus**), **5)** growth and development should occur in the **uterus** to term, birth (**parturition**) through the **vagina**, and then **6)** nourishment after birth (**mammary glands**).

If you stop to look at these six things that are required for proper reproduction, then you must be overwhelmed by the fact that men are needed for precious few of them, just two, in fact. However, women are needed for all. Truly, the burden of reproduction falls with women, and truly the meaning of reproduction is very complicated for humans.

In this laboratory on the reproductive systems you will be studying the structures that allow men and women to accomplish the above six steps. As you take a look at these structures you might want to consider which step each structure participates in (with the exception of step 6, which is, of course, performed by the mammary glands).

PRELABORATORY PREPARATION: You should complete this worksheet before you come to lab. It will allow you to get somewhat organized about what you will be seeing, and it will be your list of things that you have to know for this laboratory exercise on the reproductive systems. By using the diagrams given under the heading of **EXERCISES** (below), fill in the blanks below :

FEMALE REPRODUCTIVE SYSTEM

Structures of the External Reproductive System

1.____________________
2.____________________
3.____________________
4.____________________

Structures of the Internal Reproductive System

1.____________________
2.____________________
3.____________________
4.____________________
5.____________________
6.____________________
7.____________________

Urinary and Digestive System Structures on the model

1.____________________
2.____________________

Pouches

1.____________________

MALE REPRODUCTIVE SYSTEM

Structures Found Outside the Abdominopelvic Cavity

1.____________________
2.____________________
3.____________________
4.____________________
5.____________________
6.____________________
7.____________________
8.____________________

Tubes

1.____________________
2.____________________
3.____________________
4.____________________
5.____________________
6.____________________
7.____________________

Glands

1.____________________
2.____________________
3.____________________

Urinary and Digestive Structures on the model

1.____________________
2.____________________
3.____________________
4.____________________

OBJECTIVES: The objectives of the laboratory are to:

1. Identify the structures on the female reproductive model
2. Identify the structures on the male reproductive model

EXERCISES:

1.Obtain one of the models of the female reproductive system and, using this diagram of the model given below, find the labeled structures

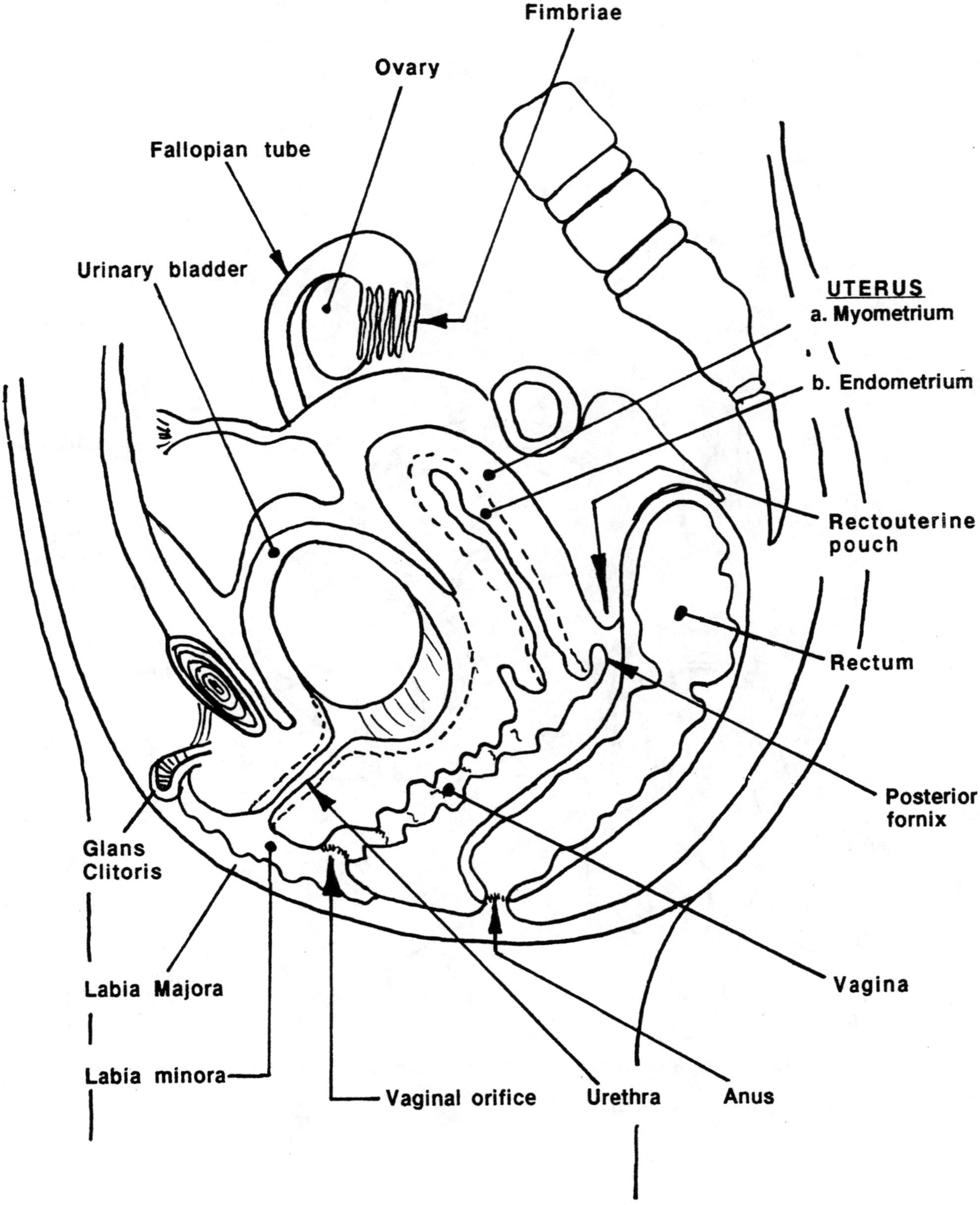

2. Obtain one of the models of the male reproductive system and, using the diagram of the model given below, find the labeled structures

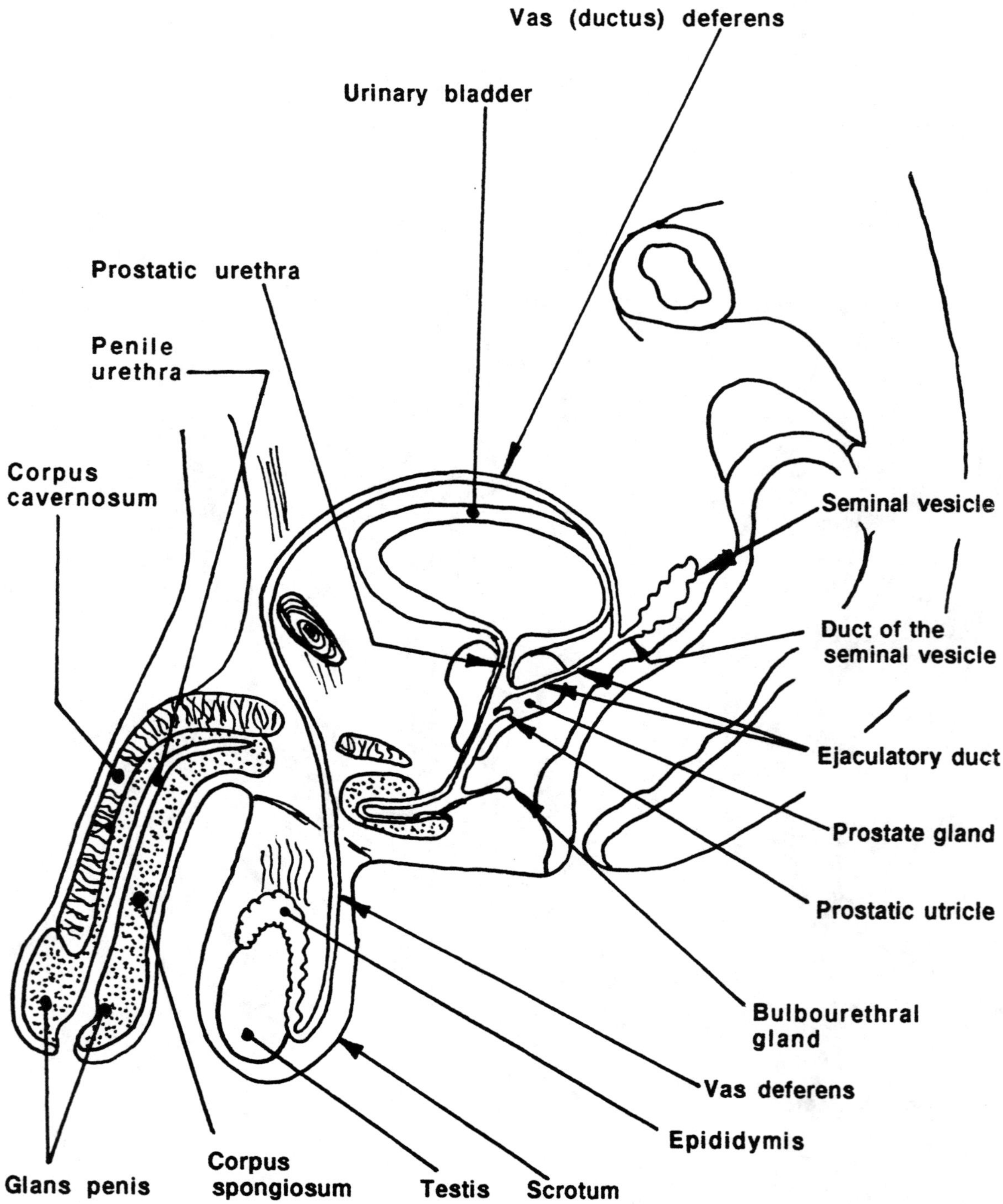

QUESTIONS TO ANSWER AND THINK ABOUT (Some of these questions can be answered directly from the models, but some will require you to look closely at your lecture notes and reading assignments) :

1. What are the six events required for successful reproduction among humans ?

2. What reproductive systems structures are required for each event ?

3. What are the four external structures of the female reproductive system ?

4. What are the seven internal structures found in the female reproductive system , and what are their functions ?

5. What is the clinical importance of the close proximity of the posterior fornix and the rectouterine pouch?

6. For the following female reproductive system structures give the anatomical equivalent in the male reproductive system: a) Penis: b) glans penis; c) labia majora; d) labia minora e) vagina and uterus; e) ovaries

7. What are the functions of the bulbourethral gland, the prostate gland, and the seminal vesicles?

8. In women do the reproductive and urinary systems share any structures ? If so, then what are they ?

9. What are eight structures that can be found outside the abdominopelvic cavity in the male reproductive system ?

10. Where are the testes located ?

11. Why are the testes located outside the abdominopelvic cavity ?

12. Why is the scotum so wrinkled ?

13. What are two good functions of the testes ?

14. In men do the reproductive and urinary systems share any structures ? If so, then what are they?

15. Trace the path that a sperm cell takes from the epididymis to the fallopian tube.

16. With the simple surgical procedure of vasectomy, what male reproductive structures are cut and tied (ligated) ?

17. With the somewhat more complicated surgical procedure of tubal ligation, what female reproductive structures are cut and ligated (tied) ?

18. Who has the major reproductive burden among us humans ?

19. Who produces the most gametes, males or females ?

20. Look closely at the model of the female reproductive system. Can you see why the frequency of voiding of urine increases during the last trimester of pregnancy ?

21. The penis on the model of the male reproductive system is in a "flaccid" state. What are the anatomical reasons and the mechanism for erection ? Which of the structures on this model are involved in penile erection ?

22. What are a few common factors that can temporarily interfere with normal penile erection ?

23. Compare the urethras of the female and male. Is the female urethra in the clitoris ? Is the male urethra in the penis ? Who has the longer urethra ? What do you think this difference in urethral length means as far as urinary tract infections are concerned ?

24. Where is the best place for fertilization to occur ?

25. Where is the best place for implantation of the blastocyst to occur ?

26. Do you think it ever happens that fertilization occurs, but implantation does not ?

27. Therefore, does fertilization always lead to pregnancy ?

28. What is a tubal pregnancy ?

29. What is the myometrium ?

30. What are the two layers of the endometrium ?

WHAT YOU MUST KNOW AND BE ABLE TO DO AFTER THESE LABORATORY EXERCISES

1. Be able to identify the labeled structures on the female reproductive system models.

2. Be able to identify the labeled structures on the male reproductive system models.

3. **Be able to answer questions from lecture and reading assignments concerning these reproductive systems structures that you see on the models in laboratory.**

LABORATORY 5 - SYSTEMS

THE AXIAL DIVISION OF THE SKELETAL SYSTEM

INTRODUCTION: A general definition of a system is that it is made up of organs. This is clear with a system such as your digestive system, which has such organs as salivary glands, liver, stomach, pancreas, esophagus, small and large intestines, and gall bladder, or your nervous system, which has the brain and a spinal cord, or your reproductive system which, depending on your gender, has such structures and organs as the clitoris, penis, vagina, uterus, vas deferens, ovaries, testes, and so on.

But we encounter a slight difficulty when we want to consider what the organs of the skeletal system are. It turns out to be quite arbitrary. There is no anatomical law that says that a system has to be made up of multiple organs, in which case we could simply state that all the bones together make up an organ. Or, if there was a law enacted that dictated that a system had to have more than one organ, then we just simply say that each bone is an organ making the skeletal system a 206 organ system ! Some might then argue that all bones have equal functions thereby making them identical and turning the skeletal system into **the skeletal organ.** Some might even take this arguement to silliness and argue that bones do not even qualify as organs, that each one is no more than a tissue. Now we have to call the skeletal system **the skeletal tissue** ! Can all this be resolved and get us back to a skeletal system ?

There are 206 bones in the body, give or take a few. Some individuals will develop one or two extra coccygeal bones, or an extra vertebra, or an extra pair of ribs (usually cervical ribs), or some Wormian bones. Sesamoid bones in our tendons come and go according to the stress tendons are subjected to. So, 206 is an average number, not a cold and hard fact. Nevertheless, each of your bones contains two types of bone called **COMPACT (DENSE) BONE**, and **SPONGY (CANCELLOUS) BONE**. Compact bone surrounds spongy bone.

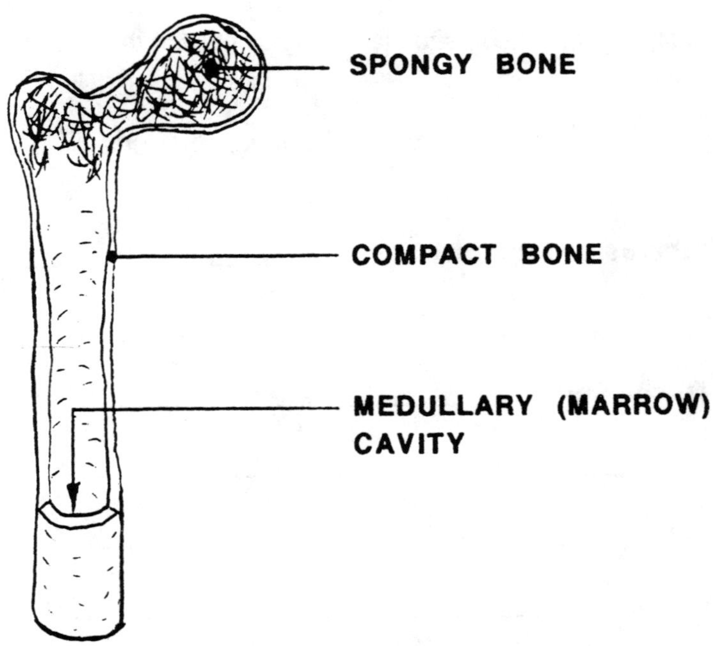

Proximal end of the Femur

Bones can be easily classified on their overall shape as **LONG**, **SHORT**, **FLAT**, or **IRREGULAR**.

Some of the functions of bones are obvious, but some are not. Bones obviously give your body **SUPPORT and FORM**. Bones obviously provide places for the **ATTACHMENT OF SKELETAL MUSCLES** for movement, and bones obviously are for the **PROTECTION OF SOFT ORGANS** (Brain, spinal cord, lungs, heart, liver, stomach, spleen, uterus, urinary bladder. Oddly, your kidneys are not so well protected by your bones). Not so obvious is that your bones are a **STORAGE PLACE FOR CALCIUM AND PHOSPHORUS**, two extremely important minerals. Another not so obvious function of your bones is that some of the long bones provide for a place (Medullary, or marrow cavity) for some **BLOOD CELL FORMATION**, at least during childhood.

Perhaps this last function gets us out of a dilemma, for indeed not all bones in the body participate in blood cell formation during childhood and adolescence, which means that not all bones are functionally the same. In this way you have, for at least the early part of your life, blood - forming bones, and non blood - forming bones. That is, we now have two organs - bones that make blood, and bones that do not, making a true skeletal **system**.

There are two major divisions to your skeletal system, the **AXIAL DIVISION**, and the **APPENDICULAR DIVISION**. The axial division (often simply referred to as the **axial skeleton**) is made up of about 80 bones that include the **bones of the skull and face, the vertebrae, ribs, sternum, and the hyoid**. The appendicular division (often referred to as the appendicular skeleton) contains about 126 bones and is made up of the **pectoral girdle and the bones of the upper limb, and the pelvic girdle and the bones of the lower limb**.

OBJECTIVES: The objectives of this laboratory are to:

1. Identify the bones of the axial skeleton

2. Identify the specified bone markings on bones in the axial skeleton

DEFINITIONS: Using your text and notes write out a brief definition of the following terms:

1. **Tuberosity / tubercle** ______________________________

2. **Condyle** ______________________________

3. **Trochanter** ______________________________

4. **Spine** ______________________________

5. **Fossa** ______________________________

6. **Foramen** ______________________________

7. **Meatus** ______________________________

PRELABORATORY PREPARATION : You should complete this worksheet before you come to lab. It will not only allow you to get a bit organized about what is coming up, but it will be your list of things that you have to know for this laboratory exercise on the axial skeleton. By using the diagrams on the next eight (8) pages (those under **EXERCISES**), fill in the blanks below:

SKULL AND FACE

Bones	Projections (Processes) and Condyles	Sutures
1__________	1__________	1__________
2__________	2__________	2__________
3__________	3__________	3__________
4__________	4__________	4__________
5__________	5__________	5__________
6__________		
7__________		
8__________		
9__________		
10__________		

Foramina / Fissures / Canals

Anterior View	Inferior View	Interior View
1__________	1__________	1__________
2__________	2__________	2__________
3__________	3__________	3__________
4__________	4__________	4__________
5__________	5__________	5__________
6__________	6__________	6__________
7__________		7__________
8__________		8__________

VERTEBRAE AND VERTEBRAL COLUMN

"Typical" Vertebra	Foramina	Types and number of each
1__________	1__________	1__________
2__________	2__________	2__________
3__________		3__________
4__________		4__________
5__________	Curvatures	5__________
6__________	1__________	
7__________	2__________	
8__________	3__________	

1. What is found only in cervical vertebrae ____________________ 2. What is found only in thoracic vertebrae ____________________ 3. What is C1 also called ________ 4. What is C2 also called ________ 5. C7 is also called what ____________________

EXERCISES:

1. Bones, sutures, and projections of the skull and face

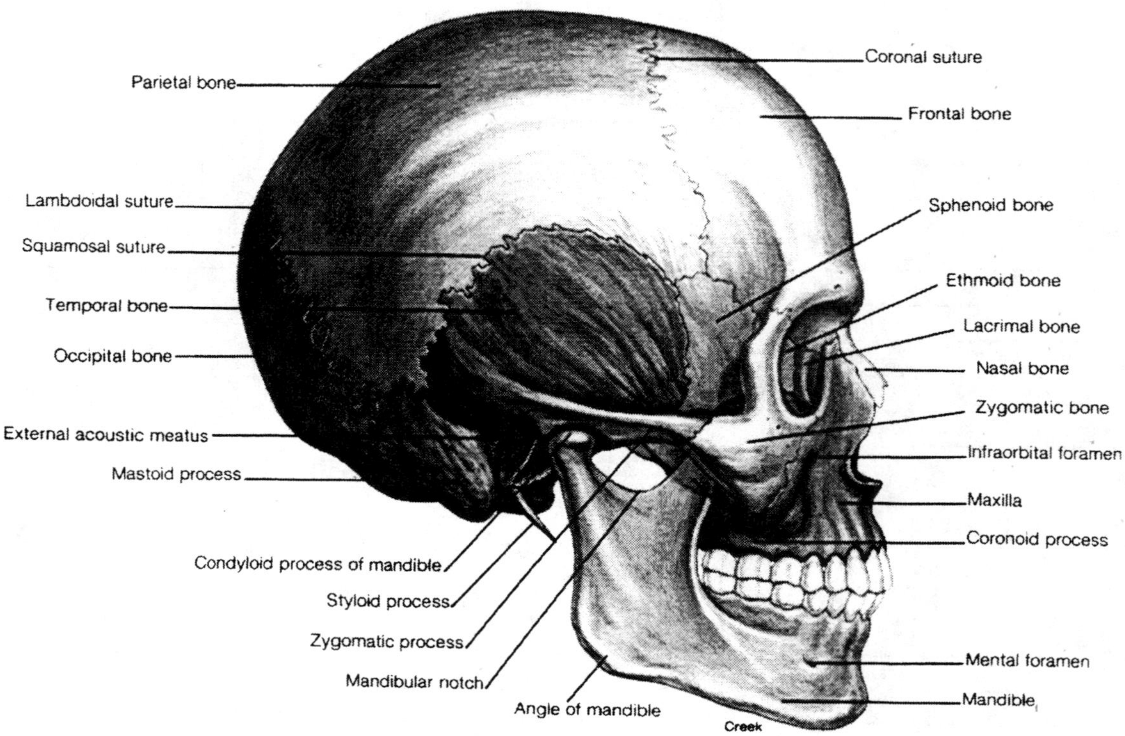

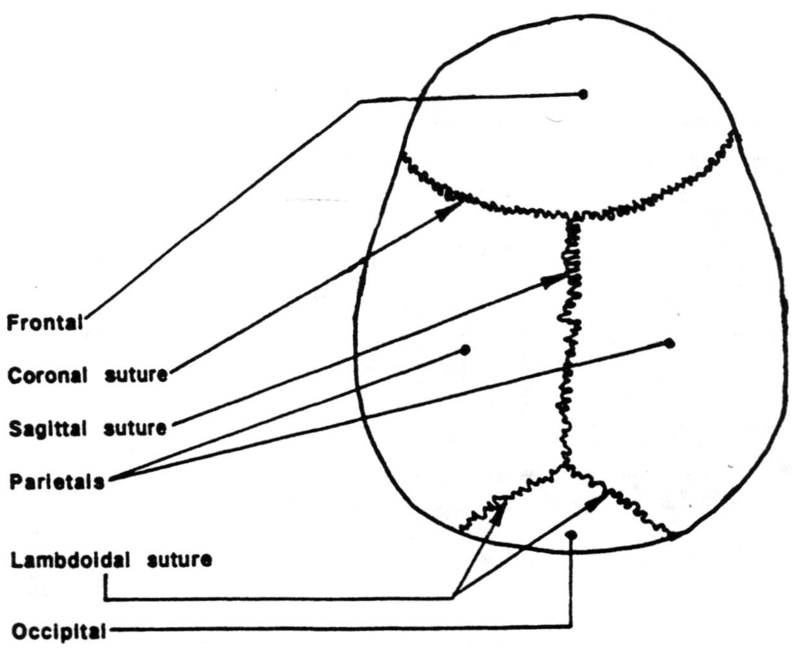

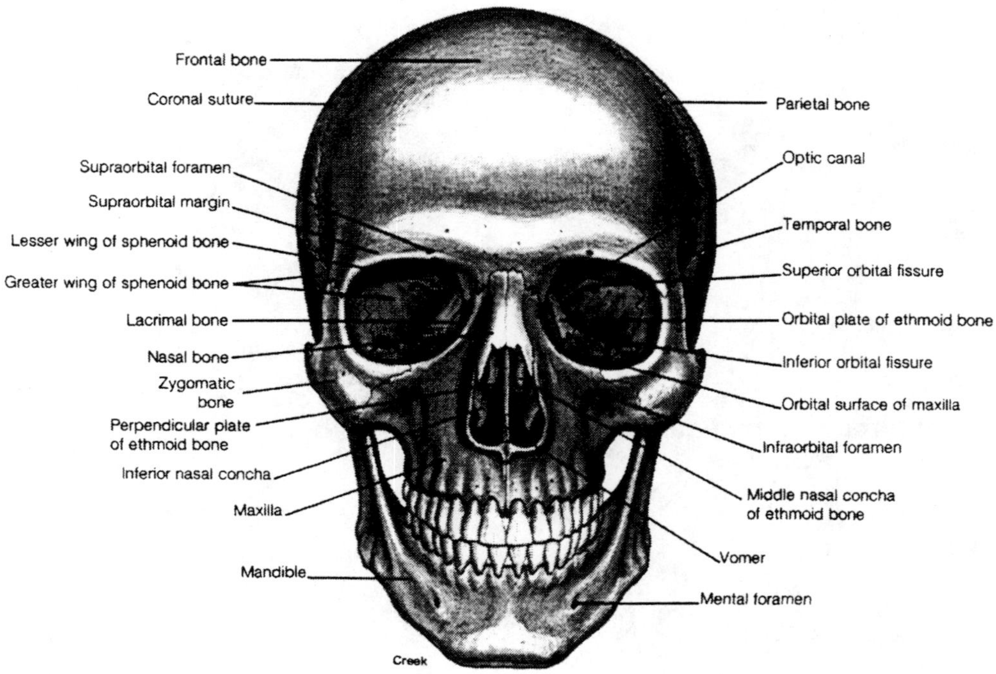
Frontal bone
Coronal suture
Supraorbital foramen
Supraorbital margin
Lesser wing of sphenoid bone
Greater wing of sphenoid bone
Lacrimal bone
Nasal bone
Zygomatic bone
Perpendicular plate of ethmoid bone
Inferior nasal concha
Maxilla
Mandible
Parietal bone
Optic canal
Temporal bone
Superior orbital fissure
Orbital plate of ethmoid bone
Inferior orbital fissure
Orbital surface of maxilla
Infraorbital foramen
Middle nasal concha of ethmoid bone
Vomer
Mental foramen
Creek

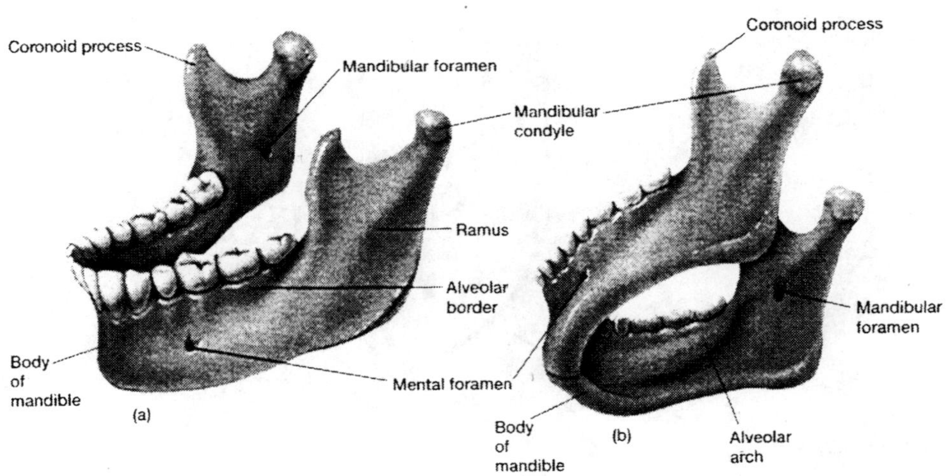
Coronoid process
Mandibular foramen
Mandibular condyle
Ramus
Alveolar border
Body of mandible
Mental foramen
(a)
Coronoid process
Mandibular foramen
Body of mandible
(b)
Alveolar arch

2. Foramina, canals, and fissures

a. Those that can be seen on an anterior view of the skull

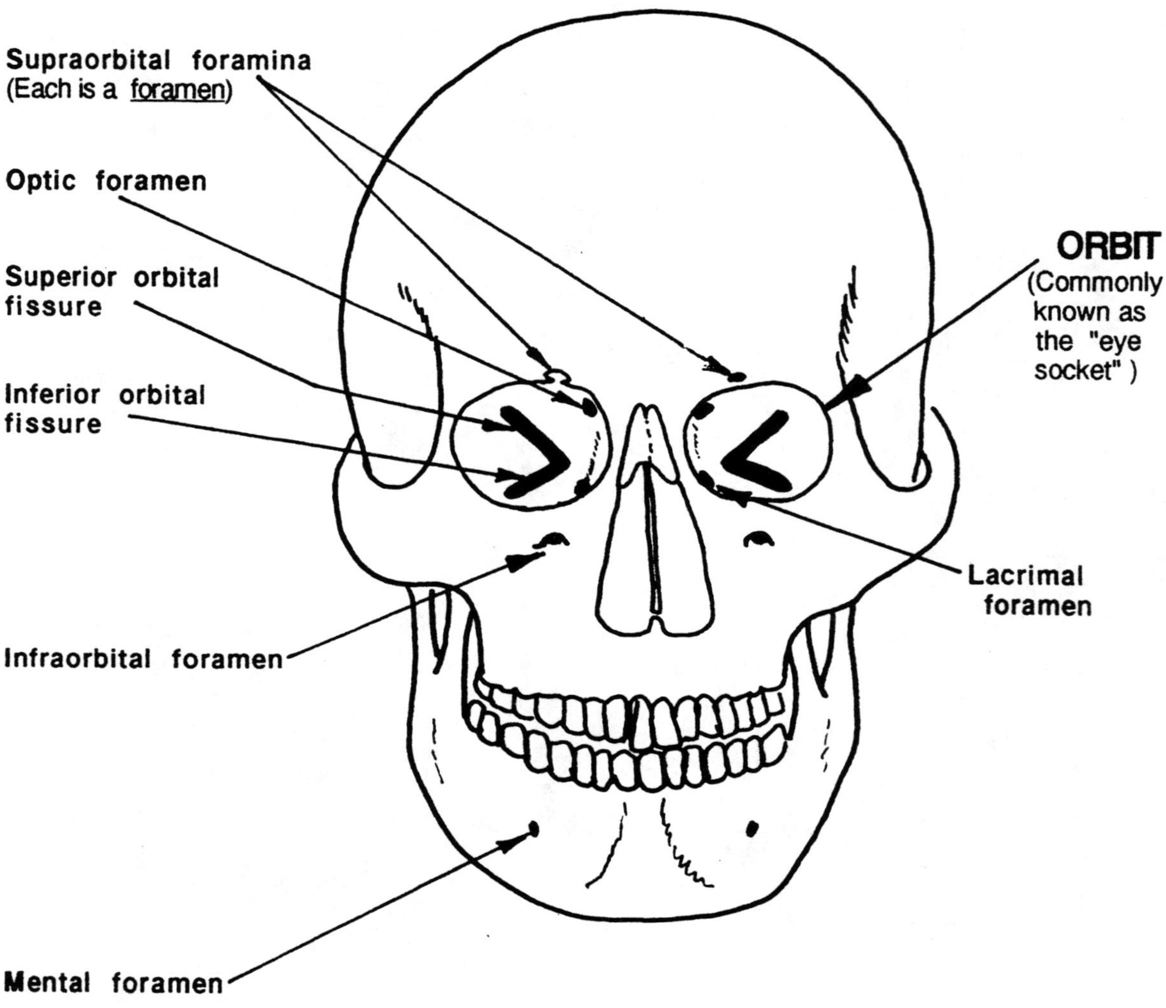

b. Those that can be seen on an inferior view of the skull

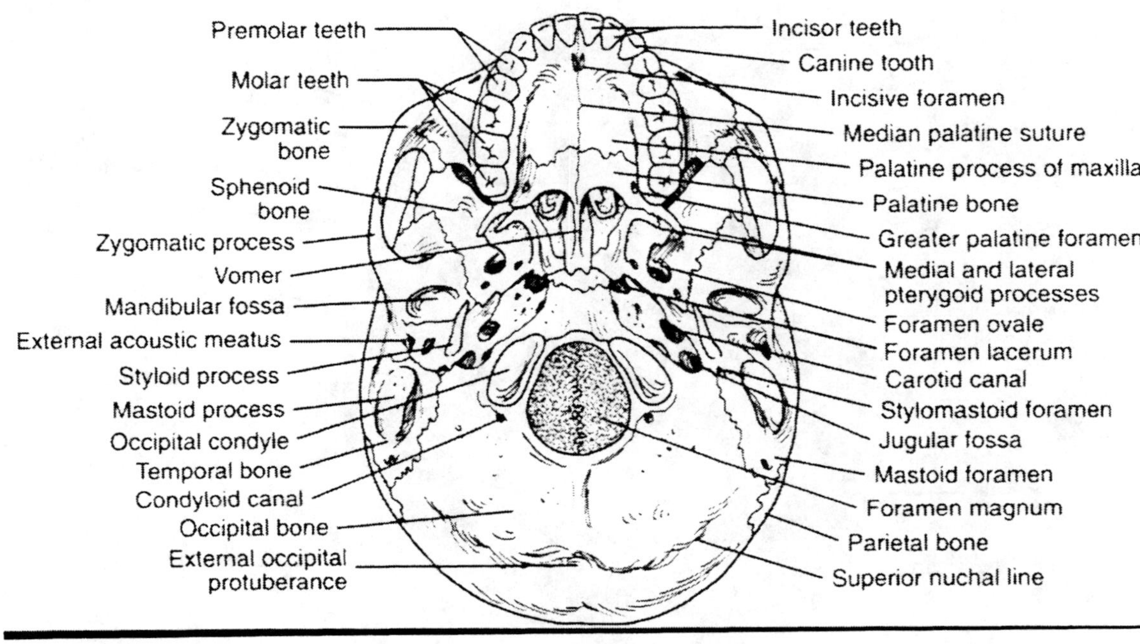

c. Those that can be seen on an interior view of the skull

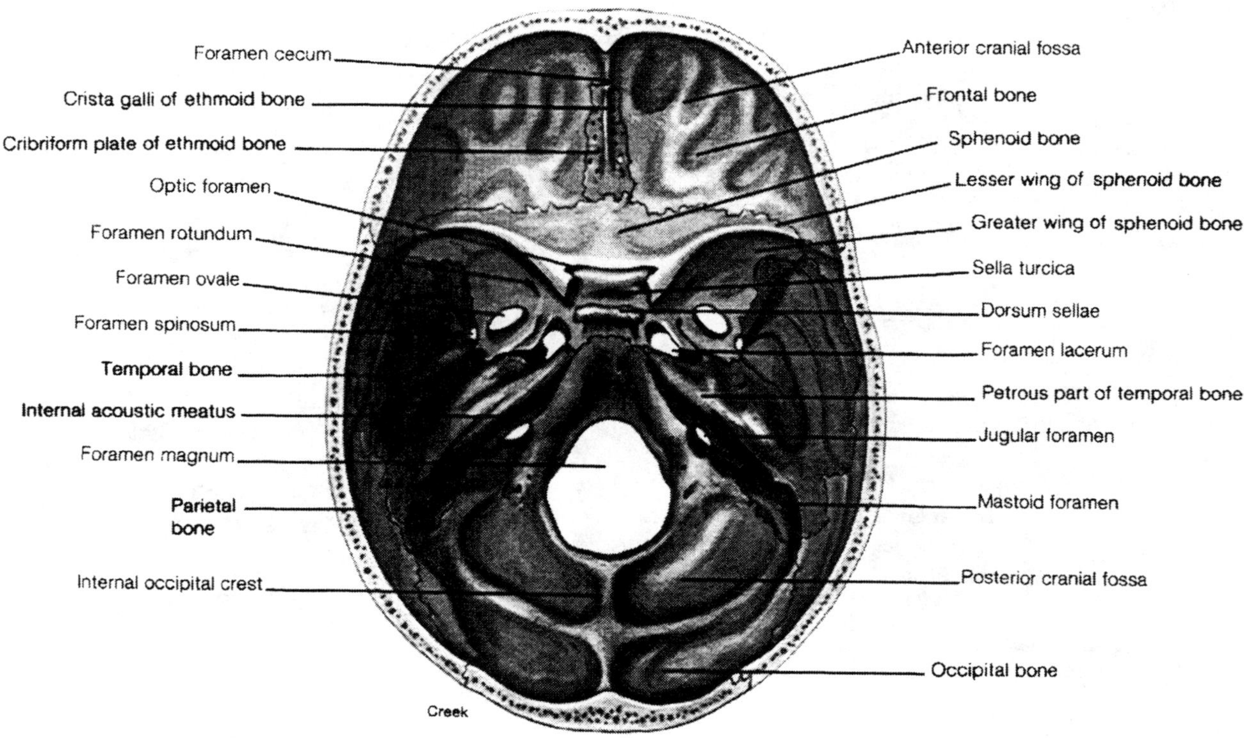

3. The vertebrae

a. Structure of a "typical" vertebra

All of these structures can be found on all vertebrae except the atlas (C1), which lacks both a spinous process and a body. You will be responsible for identifying these structures on the three types of vertebrae.

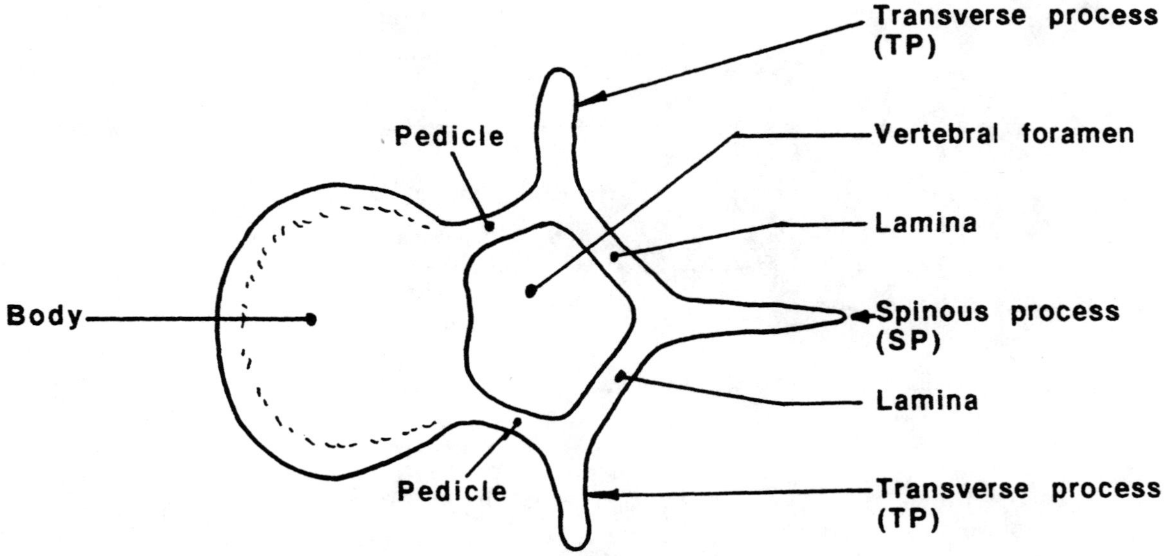

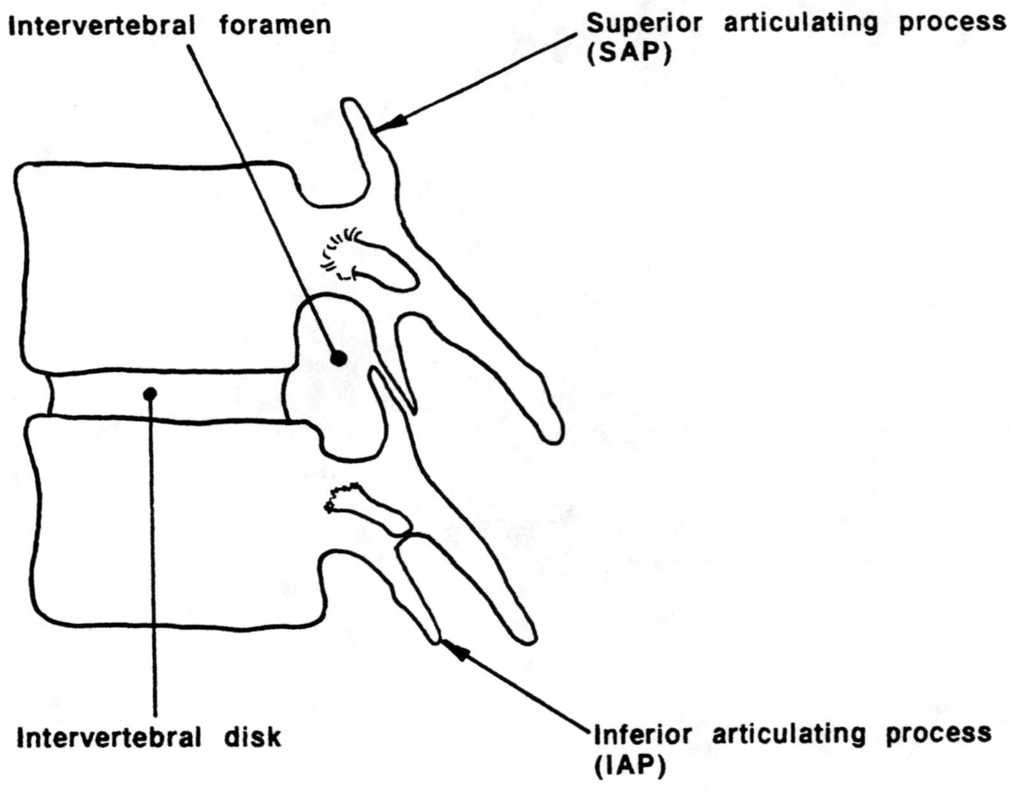

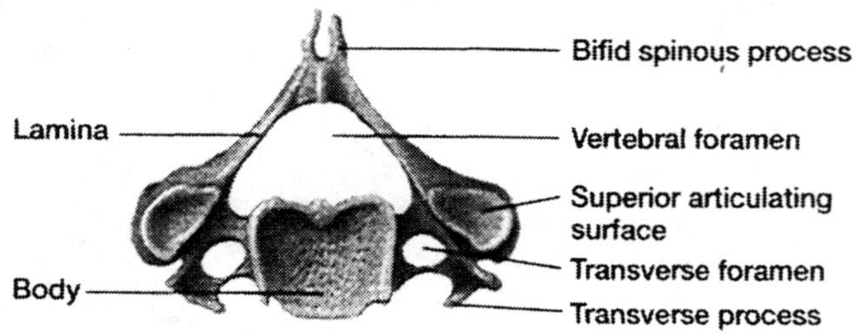

(a)
Cervical vertebra

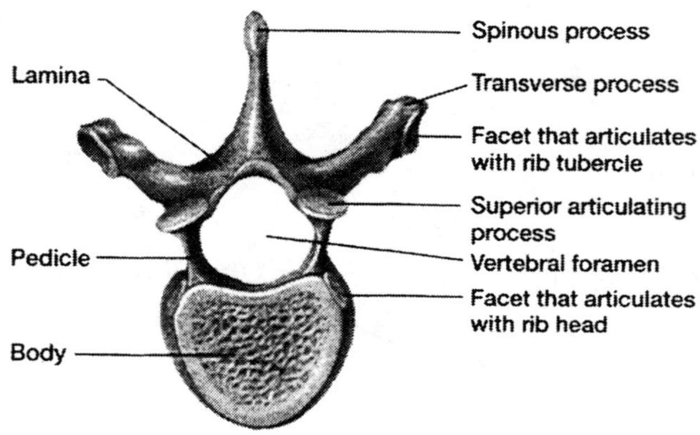

(b)
Thoracic vertebra

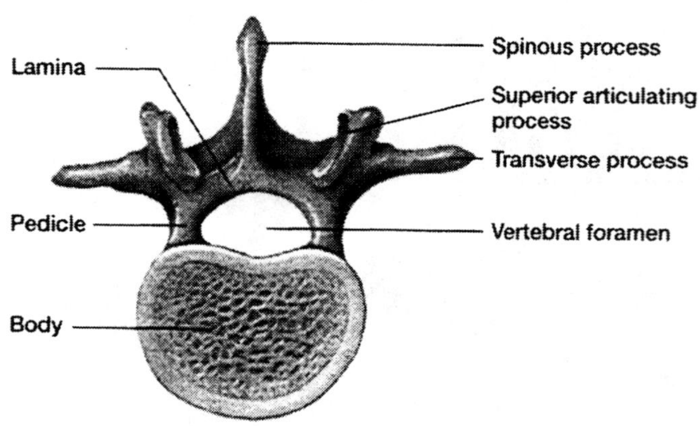

(c)
Lumbar vertebra

4. The vertebral column and its curvatures

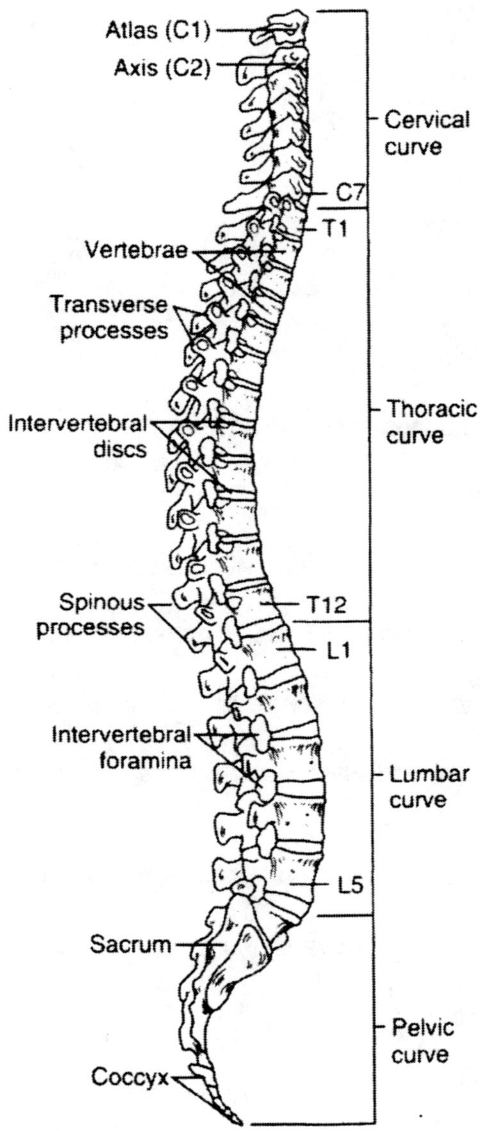

QUESTIONS TO ANSWER AND THINK ABOUT (Some of these questions can be answered directly from looking at the bones, but some will require you to look closely at your lecture notes and reading assignments):

1. What is the definition of a tissue ?

2. What is the definition of an organ ?

3. What is the definition of a system ?

4. How many organs are there in the skeletal system ?

5. How many bones are in the human skeleton ?

6. What two types of bone are found in all 206 bones ?

7. Where are sesamoid bones found ?

8. What are the classifications of bones based on shape ?

9. What are the functions of bone ?

10. What are the two major divisions of the human skeletal system ?

11. How many bones are in the axial skeleton, and, in general, what are these bones ?

12. How many bones are in the appendicular skeleton, and, in general, what are these bones ?

13. The bones and bone markings in this laboratory exercise (Laboratory 5) belong to which division ?

14. Name the flat bones of the skull. What important soft organ are these bones protecting ?

15. Name some irregular bones that you have encountered in this laboratory exercise.

16. Completely classify the suture as a joint.

17. Why is it good that sutures are synarthrotic fibrous joints ?

18. Name the sutures that you had to learn in this laboratory exercise.

19. What bone separates your nose into two parts ?

20. The occipital condyles are right next to what foramen ?

21. In what bone are your upper teeth located ? Your lower teeth ?

22. You might have to look this up in the dictionary, but nevertheless, what does the term mental refer to ?

23. Think about your own skull and then determine what is the only bone of the skull that moves.

24. What is the orbit ?

25. What do the two prefixes "supra" and "infra" mean ?

26. What is the purpose of having foramina, canals and fissures in your skull ?

27. What is the singular form for the term foramina ?

28. What is the plural form of the term foramen ?

29. What are the bones of the hard palate ?

30. Classify vertebrae based on their shape.

31. What important structure is found in the vertebral foramen ?

32. An intervertebral disk is located between what parts of adjacent vertebrae ?

33. Completely classify the intervertebral disk as a joint.

34. What part of a vertebra can you easily feel in yourself or someone else ?

35. What is the one distinguishing anatomical feature of the cervical vertebrae ?

36. What is the distinguishing anatomical feature of thoracic vertebrae ?

37. How many cervical vertebrae are there ? Thoracic vertebrae ? Lumbar vertebrae ? Sacrals ? Coccygeals ?

38. Rank the vertebrae in size from smallest to largest.

39. What is C1 also called ? What is C2 also called ? What is C7 also called ?

40. Reach around and at the base of your neck feel the prominent spinous process. To which vertebra do you think this process belongs ?

41. S1 - S5 are as a group called what ?

42. What inportant structures pass through the intervertebral foramina ?

43. Which vertebra has neither a body nor a spinous process ?

44. With what does the atlas articulate ?

45. What are the three curvatures of the vertebral column ?

46. What flat bones articulate with the thoracic vertebrae ?

47. If a pair of ribs attach to each thoracic vertebra, then how many pairs of ribs are in your body ?

48. Consult your text and determine what true, false and floating ribs are.

49. Name five soft organs that the ribs and sternum protect.

50. What important soft structure is protected by the vertebrae ?

WHAT YOU MUST KNOW AN BE ABLE TO DO AFTER THESE LABORATORY EXERCISES

1. Be able to recognize and correctly spell the labeled bones and bony landmarks for the skull, face and vertebrae.

2. Be able to answer the questions above that deal with the axial skeleton and what you have learned in laboratory .

3. Be able to answer questions from your lecture notes and reading assignments that relate to the structures that you encounter in this laboratory exercise.

LABORATORY - 6 - SYSTEMS

THE APPENDICULAR DIVISION OF THE SKELETAL SYSTEM

INTRODUCTION: The **APPENDICULAR DIVISION** of the skeletal system (sometimes referred to simply as the appendicular skeleton), contains 126 bones, and includes the bones of the pectoral girdle plus the bones of the upper limbs, and the bones of the pelvic girdle plus the bones of the lower limbs. Are there any long bones in the axial skeleton ? Well, there are plenty of them in the appendicular skeleton. Another question is whether there are any irregular and flat bones in the appendicular skeleton. Perhaps you can answer these questions as you go along in the exercises in Laboratory 6.

When you stop to consider movements you notice that it is in your limbs that you get the most types of movements and the greatest range of motion of movements. There are of course movements at joints in your axial skeleton, but these cannot begin to compare to those available to you in your appendicular skeleton. You should notice, too, that the more distally you go in your limbs, the more refined you can make your movements. That is, the movements that you can make with your fingers are much more refined and precise than those movements that you can make at your shoulder joint. The same is true in your lower limbs; movements of the foot are more refined than those involving the hip joint. Why is that ?

OBJECTIVES: The objectives of this laboratory are to:

1. Identify the areas of the upper and lower limbs
2. Identify the bones of the appendicular skeleton
3. Identify the specified bone markings on bones in the appendicular skeleton

DEFINITIONS Using your text and notes write out a brief definition of the following terms:

1. **Tuberosity/tubercle**____
2. **Trochanter**____
3. **Condyle**____
4. **Neck**____
5. **Trochlea**____
6. **Fossa**____
7. **Malleolus**____
8. **Spine**____
9. **Notch**____

PRELABORATORY PREPARATION: Laboratory 6 on the appendicular skeleton will require you to learn about **63 bones** and , in addition, many bone markings on each. Obviously, there is a lot to do in this laboratory, so it really would be helpful to you and make your time much more efficient if you fill out these blanks before you come to lab. By using the diagrams under the heading of **EXERCISES**, fill in the blanks below :

Pectoral Girdle Bones
1_______________
2_______________

Pelvic Girdle Bones
1_______________
2_______________
3_______________

Bone of the arm
1_______________

Bone of the thigh
1_______________

Bones of the forearm
1_______________
2_______________

Bones of the leg
1_______________
2_______________

Bones of the hand
1_______________
2_______________
3_______________

Bones of the foot
1_______________
2_______________
3_______________

The Scapula

a. **Angles**
1_______________
2_______________

b. **Fossae**
1_______________
2_______________
3_______________
4_______________

c. **Projections**
1_______________
2_______________
3_______________

d. **Borders**
1_______________
2_______________

The Humerus

a. **Tuberosities**
1_______________
2_______________
3_______________

b. **Epicondyles**
1_______________
2_______________

c. **Fossae**
1_______________
2_______________

d. **Necks**
1_______________
2_______________

e. **Grooves**
1_______________

f. **Rounded or Smooth Areas**
1_______________
2_______________
3_______________

The Radius

1_______________
2_______________
3_______________
4_______________

The Ulna

a. **Projections**
1____________
2____________
3____________

b. **Notches**
1____________
2____________

c. **Smooth areas**
1____________

The Hand (Manus)

a. **Carpals**
1____________________
2____________________
3____________________
4____________________
5____________________
6____________________
7____________________
8____________________

b. **Metacarpals I - V**
c. **Phalanges (14 bones)**

The Pelvic Girdle

a. **Bones**

1____________________
2____________________
3____________________

b. **Bony Landmarks**

1____________________
2____________________
3____________________
4____________________
5____________________
6____________________
7____________________

The Femur

1__________________
2__________________
3__________________
4__________________
5__________________
6__________________
7__________________

The Fibula

1__________________
2__________________

The Tibia

1__________________
2__________________
3__________________

The Foot (Pes)

a. **Tarsals**

1__________________
2__________________
3__________________
4__________________
5__________________
6__________________
7__________________

b. **Metatarsals I - V**

c. **Phalanges (14 bones)**

EXERCISES:

1. Appendicular skeleton bones in the full skeleton

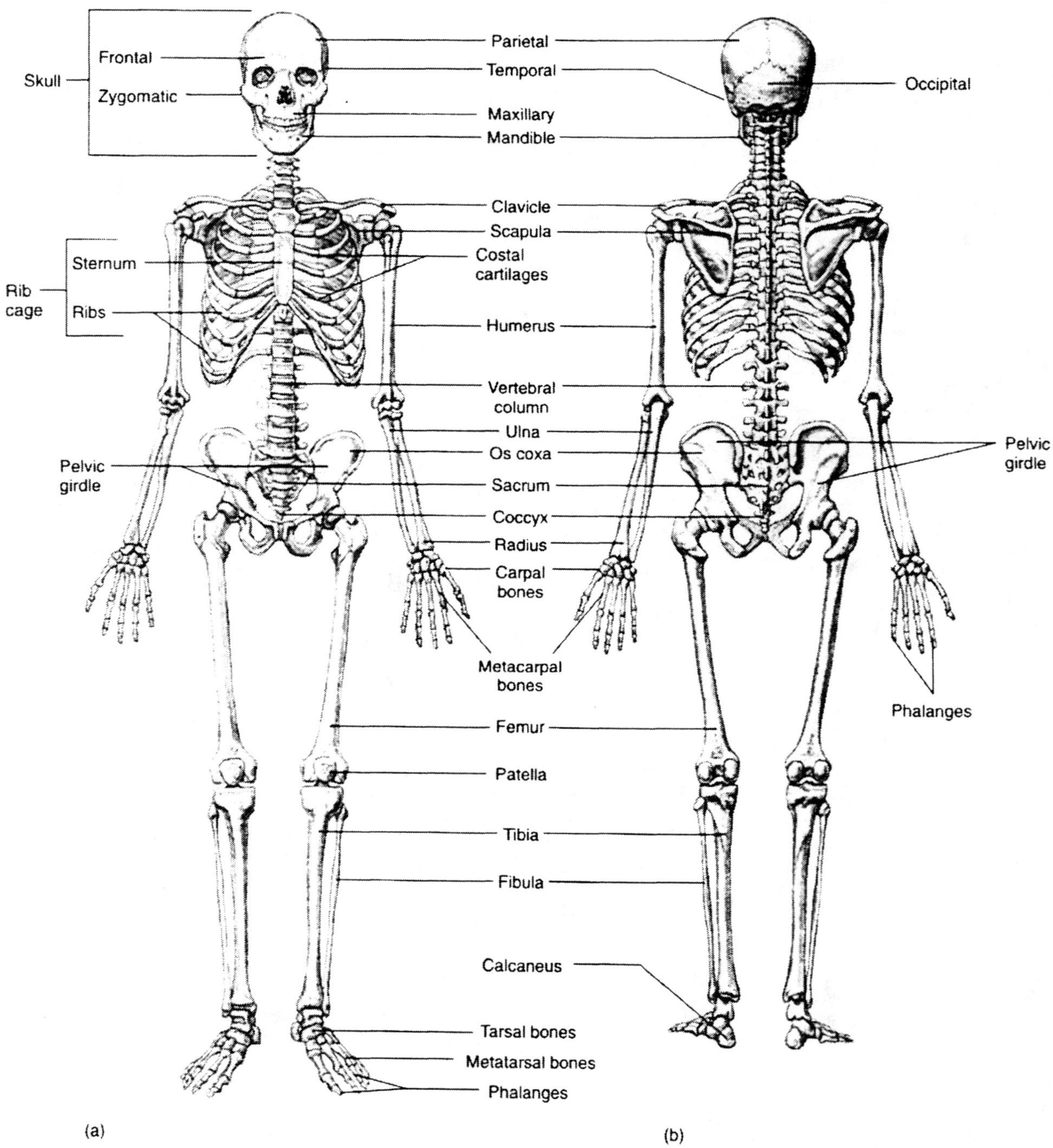

2. Bones and bone markings of the scapula

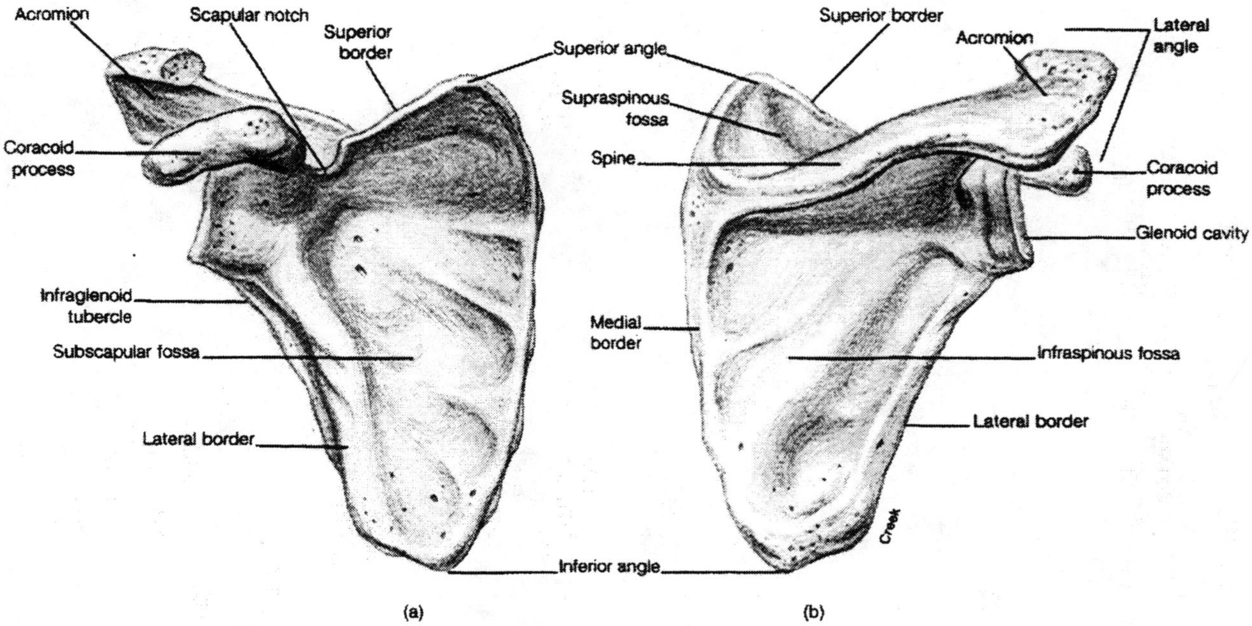

3. The bone of the arm (**THE HUMERUS**)

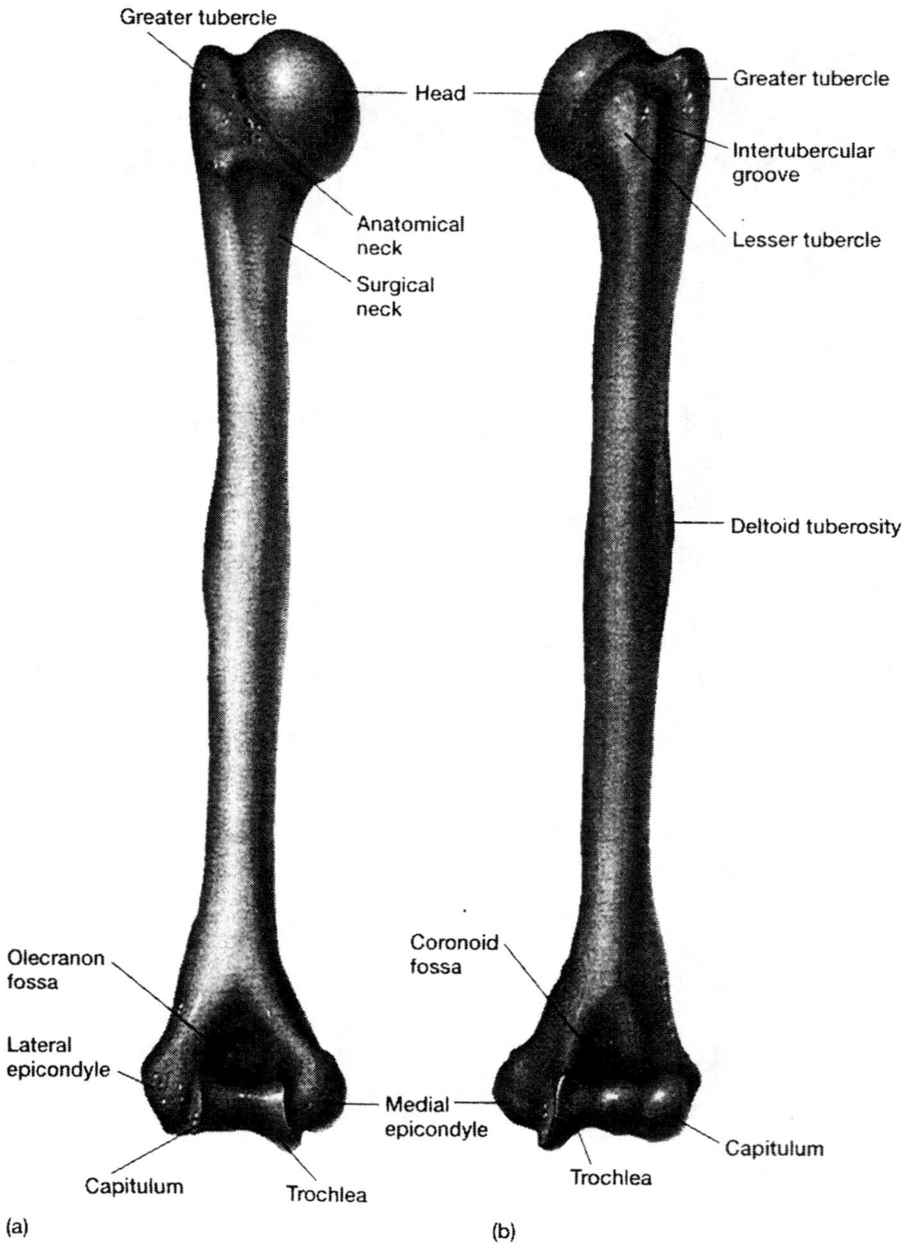

4. The bones of the forearm (**THE RADIUS AND ULNA**)

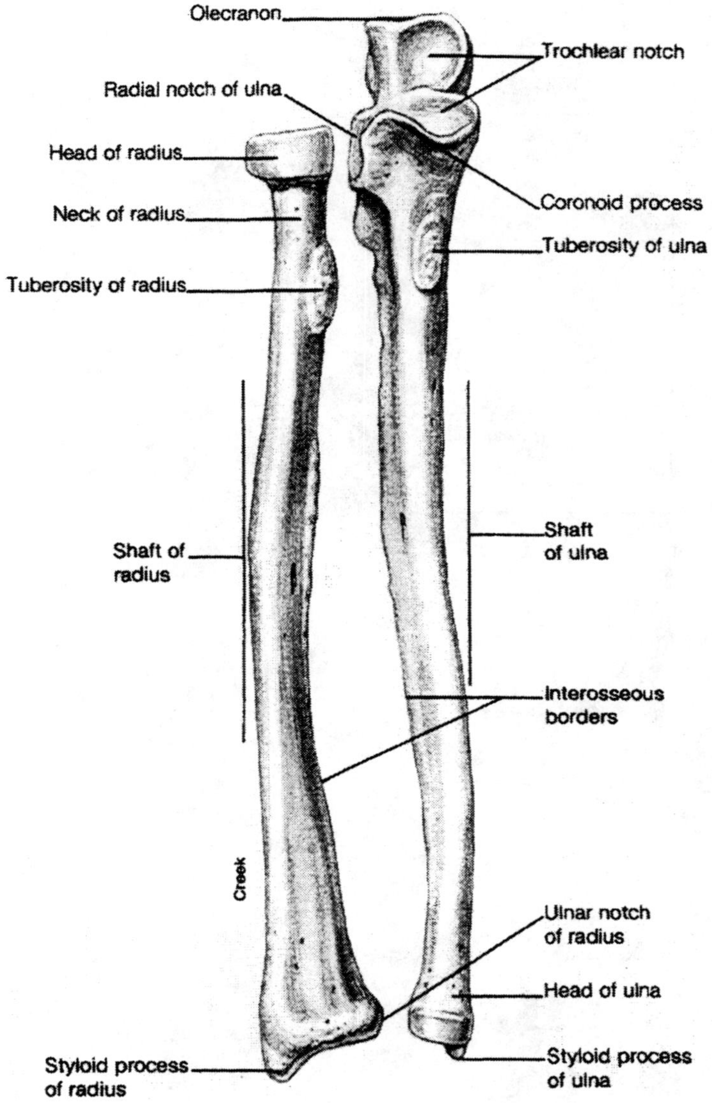

5. The bones of the hand (**CARPALS, METACARPALS, and PHALANGES**)

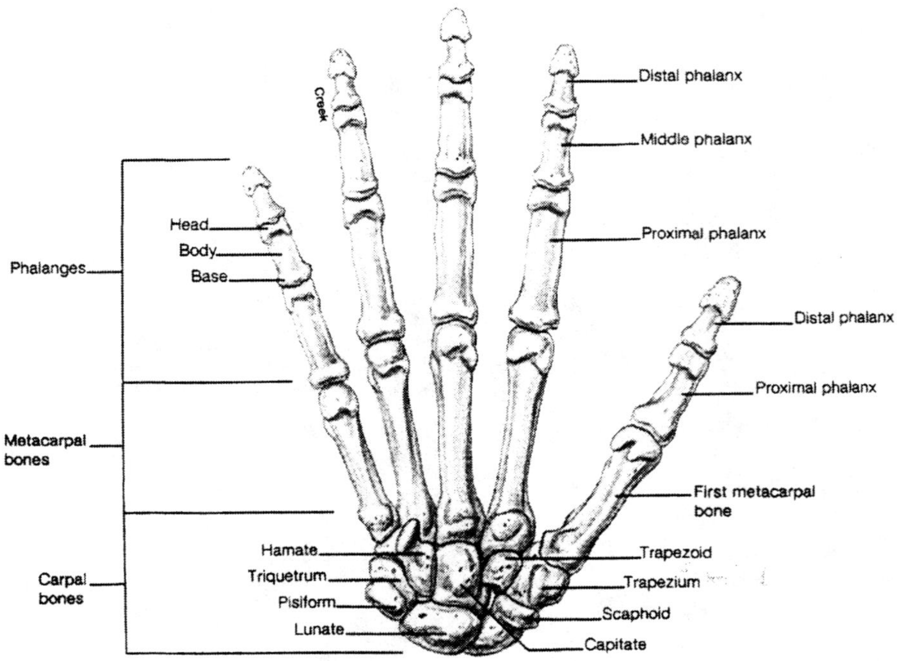

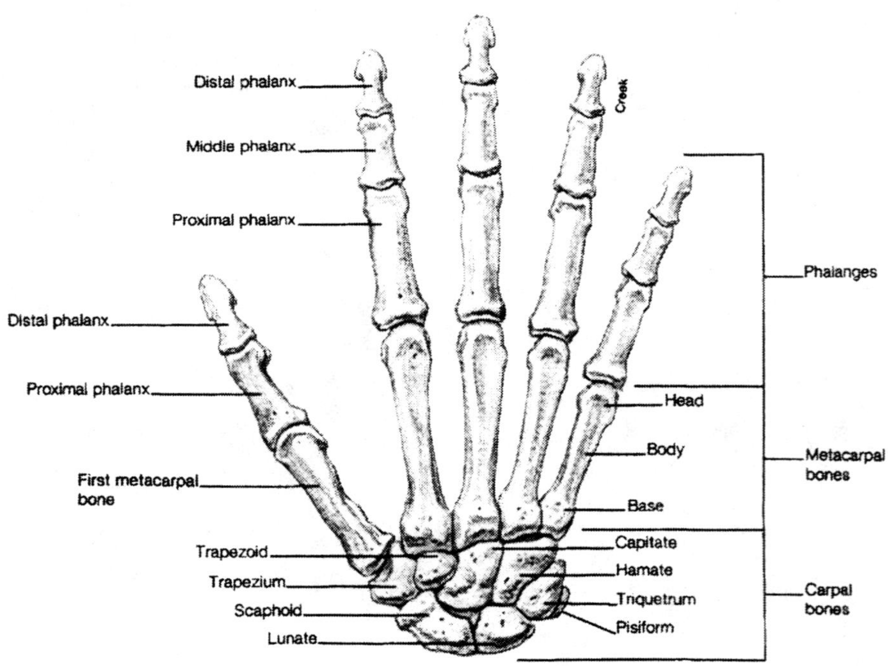

6. Bones of the pelvic girdle (**ILIUM, ISCHIUM, and PUBIS**)

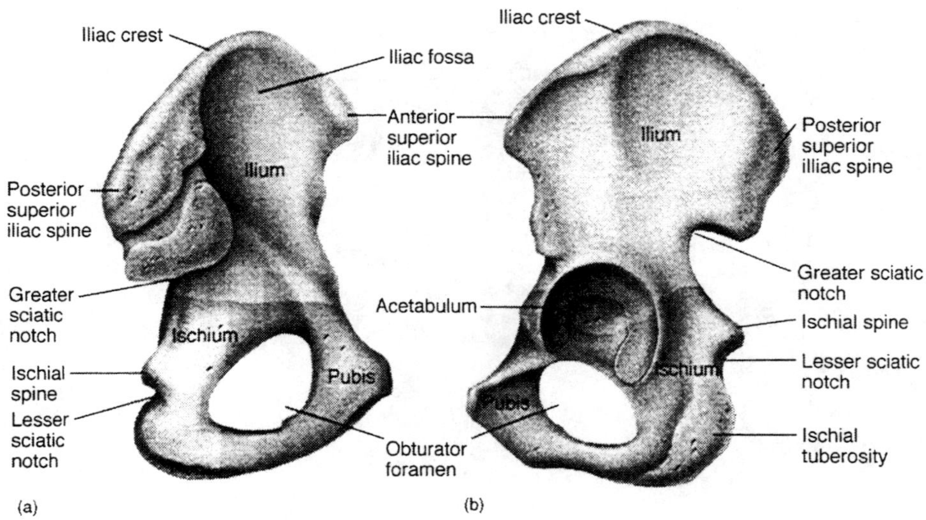

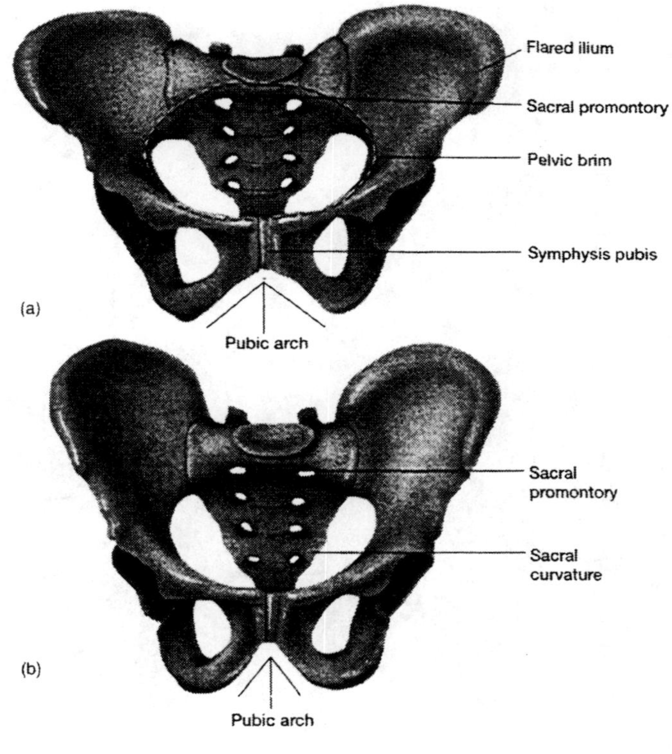

7. The bone of the thigh (**THE FEMUR**)

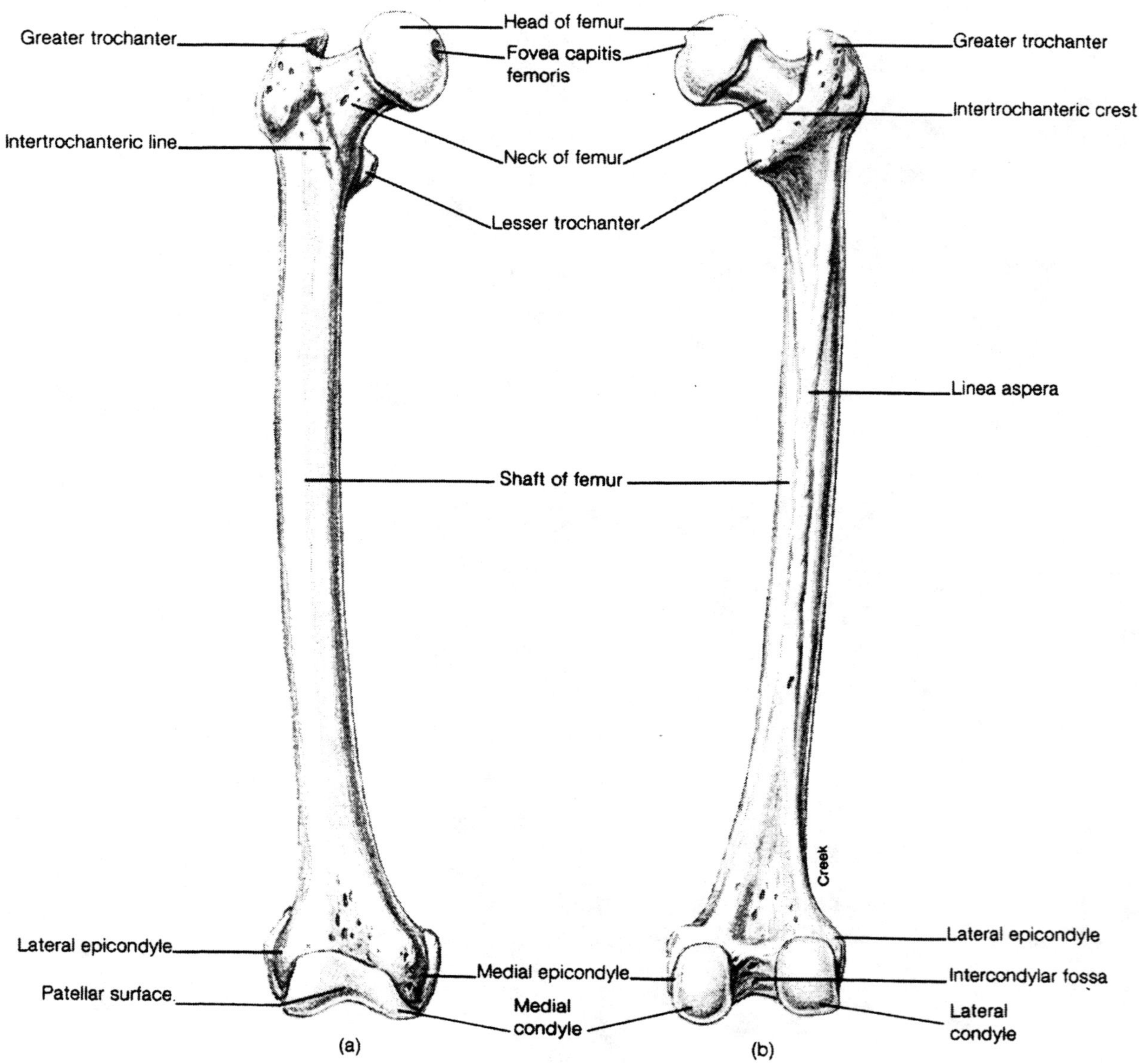

8. **Bones of the leg (The <u>TIBIA and FIBULA</u>)**

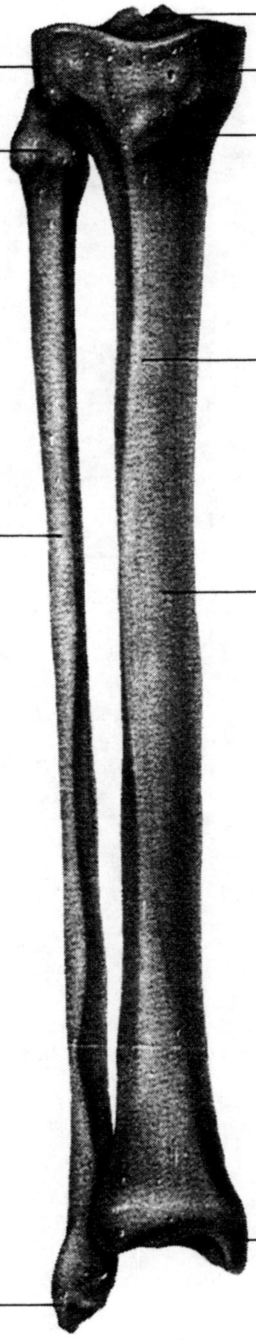

9. Bones of the foot (**TARSALS, METATARSALS, PHALANGES**)

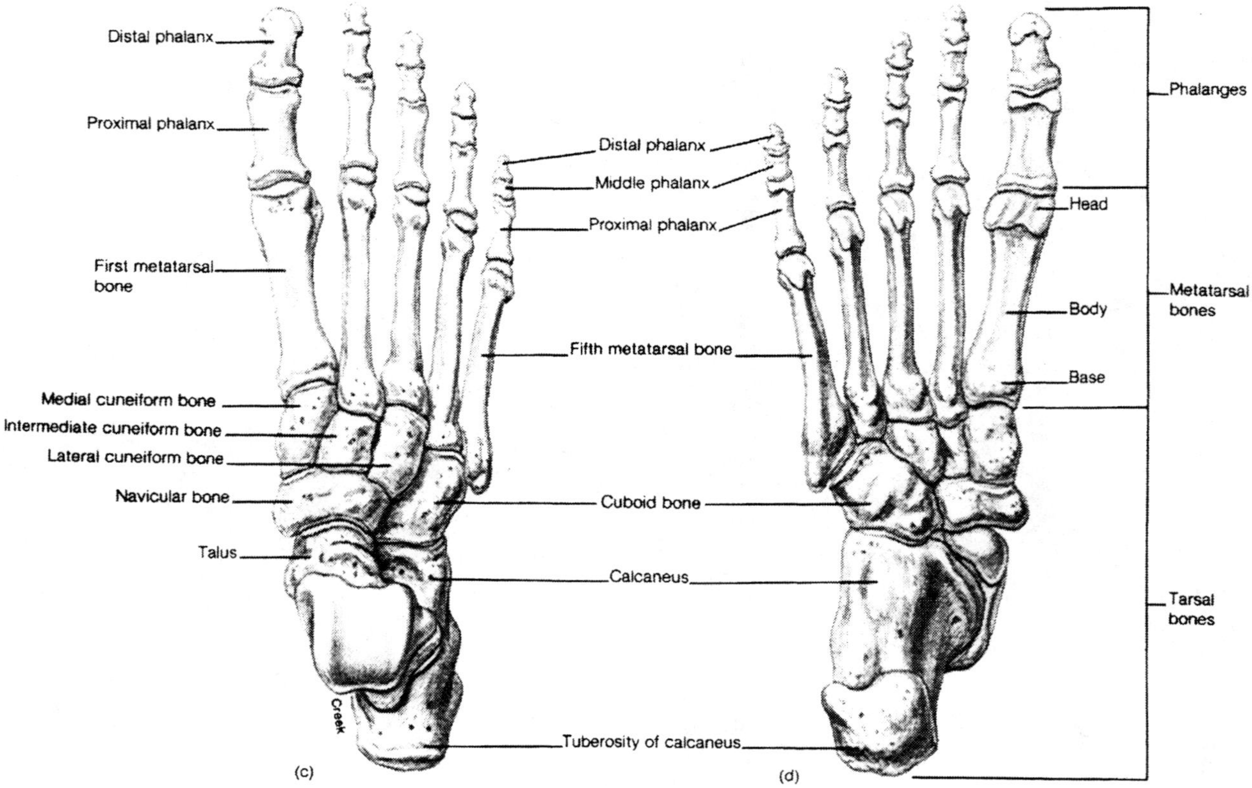

QUESTIONS TO ANSWER AND THINK ABOUT (Some of these questions can be answered directly from looking at the skeleton and the bones, but some will require you to look closely at your lecture notes and reading assignments):

1. Can you notice in your limbs that as you go from proximal to distal areas that the bones become smaller and more numerous ?

2. How many bones are in each upper limb ? Therefore, how many are in both upper limbs?

3. How many bones are in each lower limb ? Therefore, how many are in both lower limbs ?

4. How many bones are in the pectoral girdle ? Therefore, how many are in both girdles ?

5. Here is a trick question. First of all, what **<u>are</u>** the bones of the pelvic girdle ? Now, how many bones do you think are in the adult pelvic girdle - **six,** or **two** ?

6. Sum up the bones in both pectoral girdles, the pelvic girdle, and all four limbs. Did you come up with 130 ? It should be 126, so if you got 130, then where do you think you went wrong?

7. List the long bones in the appendicular skeleton.

8. List the short bones in the appendicular skeleton.

9. What is an obviously flat bone in the appendicular skeleton ?

10. There are actually some irregular bones in the appendicular skeleton. What are these bones ?

11. What type of joints are the hip and shoulder joints ?

12. Try to feel the following bony landmarks on yourself, or someone else: Acromion, spine of the scapula, superior and inferior angles of the scapula, medial border of the scapula, clavicle, medial and lateral epicondyles of the humerus, head of the radius, olecranon process, styloid processe of the radius and ulna, pisiform, iliac crest, anterior superior iliac spine, ischial tuberosities, greater trochanter, epicondyles of the femur, patella, condyles of the tibia, tibial tuberosity, medial malleolus, head of fibula, lateral malleolus.

13. The patella and pisiform are what type of bones ?

14. What specifically fits into the glenoid fossa ? Into the acetabulum ?

15. Why is the medial border of the scapula also called the vertebral border ?

16. Why is the ventral surface of the scapula also called the costal surface ? What does the term costal refer to ?

17. Are you going to get the cora**C**oid process and the coro**N**oid processes confused ? What is the difference betweem them ?

18. The head of humerus fits into what depression on the scapula ?

19. What projection fits into the coronoid fossa of the humerus when you bend your upper limb at the elbow (flexion at the elbow) ?

20. Is the coronoid process on the radius or ulna ?

21. What projection fits into the olecranon fossa of the humerus when you straighten out your upper limb at the elbow (extension at the elbow) ?

22. Is the olecranon process on the radius or ulna ?

23. What does the term trochlea mean ?

24. What depression on the ulna does the trochlea of the humerus fit into ?

25. What rounded structure on the humerus does the head of the radius fit onto ?

26. Which of the carpals is a sesamoid bone ?

27. How many middle phalanges are there ? Which digit (finger) is missing a middle phalanx ?

28. The metacarpophalangeal joints are between what two set of bones ?

29. The proximal interphalangeal joints are between what two sets of bones ?

30. The distal interphalangeal joints are between what two sets of bones ?

31. The two largest and strongest bones in the entire skeleton are located in the appendicular skeleton. What is the name of these two bones ?

32. The condyles of the femur articulate with what ?

33. In the leg, which bone is most lateral, the tibia or the fibula ?

34. In the forearm, in the anatomical position, which bone is most lateral, the ulna or the radius ?

35. Which of the digits (toes) lacks a middle phalanx ?

WHAT YOU MUST KNOW AND BE ABLE TO DO AFTER THESE LABORATORY EXERCISES

1. Be able to recognize and correctly spell the labeled bones and bony landmarks of the pectoral and pelvic girdles, and the upper and lower limbs.

2. Be able to answer the questions above that deal with the appendicular skeleton and what you have learned in laboratory.

3. Be able to answer questions from your lecture notes and reading assignments that relate to the structures that you encountered in this laboratory exercise.

LABORATORIES 7, 8 - SYSTEMS

THE MUSCULAR SYSTEM

INTRODUCTION: There are three types of muscle in the body - skeletal, smooth, and cardiac. The muscular system that we will investigate is the one of skeletal muscles.

There are approximately 650 muscles in the human body, but, like bones, most occur in pairs, so there are only about 325 names to contend with. You will encounter about 60 -70 muscles in these two laboratory exercises.

With the exception of just a few , skeletal muscles are attached to bones, hence their designation as skeletal muscles. The basic characteristic of all three types of muscles is that they shorten their lengths, a process termed contraction. In most cases, contraction of muscles causes movement of some sort. As examples, contraction of the smooth muscle in the esophagus and the small and large intestines cause movement of food/drink and feces along their way, contraction of smooth muscles in the stomach wall causes a great deal of mechanical churning of the food in the stomach, and contraction of the cardiac muscle in the wall of the heart results in moving blood around the body. Contraction of skeletal muscles, since they are attached through tendons to your bones, cause the various movements of the skeleton that not only are necessary to get you through your daily routines, but amaze you at times with their refinement, dexterity, and precision.

You will study mainly the muscles in the upper and lower limbs and a few in the trunk. This is not to indicate that the trunk muscles are unimportant, for without them you would discover it very difficult to stand erect and to have correct posture. This realization brings us to a general consideration of muscle actions in the various parts of the body. Those in the trunk allow you to stand erect, allow you to have proper posture, and allow you to balance your body when you are using one limb or another. These are obviously very important functions. The muscles in yourr lower limbs are mainly for locomotion along the surface of this planet. You might want to consider here that using the muscles of your trunk and lower limbs for their activities really don't require a lot of mental thought. How many times have you walked up a flight of stairs, or manuvered yourself through a crowded shopping mall without really overtly thinking about it. Yet, your skeletal muscles in your trunk and lower limbs balanced and moved you along quite well. But your upper limbs are different. When you stop to think of it, you use your upper limbs to perform the purposeful movements, and many of these movements you have to think about and adjust while you are doing them. It has probably not escaped your notice that you use your lower limb to almost unconsciously press the gas pedal of your car , but it takes a good deal more conscious attention to steer your car with your upper limbs. How difficult would it be to drive if the steering wheel was located on the floor board and you had to steer with your lower limbs and press the gas pedal with you hand.

In considering movements, you can notice in you own limbs that in going from proximal to distal that your movements become more refined, and precise. You will notice this especially with your upper limbs.

OBJECTIVES: The objectives of these laboratories are to learn the various muscles in the trunk, upper limbs, and lower limbs

DEFINITIONS: Below are some definitions for the varios movements. Please fill in the blanks as these movements may not be discussed in detain in lecture:

Flexion____

Extension____

Flexion at the shoulder____

Extension at the shoulder____

Flexion at the hip____

Extension at the hip____

Abduction____

Adduction____

Pronation____

Supination____

Dorsiflexion____

Plantar flexion____

PRELABORATORY PREPARATION: You should complete this worksheet before you come to lab. It will allow you to work more efficiently and to complete your work earlier and with less confusion. By using the diagrams on the following pages (those under **EXERCISES**), fill in the blanks below:

MUSCLE GROUP LOCATIONS

Flexors at the elbow________________________

Extensors at the elbow________________________

Flexors at wrist________________________

Extensors at wrist________________________

Flexors at the metacarpophalangeal joints________________________

Extensors at the metacarpophalangeal joints________________________

Flexors at the interphalangeal joints________________________

Extensors at the interphalangeal joints________________________

Flexors at hip________________________

Extensors at hip________________________

Abductors at hip________________________

Adductors at hip________________________

Flexors at the knee________________________

Extensors at the knee________________________

Dorsiflexors________________________

Plantar flexors________________________

AGONISTS

Flexor at the shoulder
1________________________________

Abductors at the shoulder
1________________________________
2________________________________

Extensor at the shoulder
1________________________________
2________________________________

Flexors at the elbow
1________________________________
2________________________________
3________________________________

Extensor at the elbow
1________________________________

Flexors at the wrist
1________________________________
2________________________________
3________________________________
4________________________________
5________________________________

Extensors at the wrist
1________________________________
2________________________________
3________________________________
4________________________________

Flexors at the metacarpophalangeal and interphalangeal joints
1________________________________
2________________________________

Extensors at the metacarpophalangeal interphalangeal joints
1________________________________
2________________________________
3________________________________
4________________________________

Flexors at the hip
1________________________________
2________________________________

Extensors at the hip
1________________________________
2________________________________
3________________________________
4________________________________

Abductors at the hip
1________________________________
2________________________________

Adductors at the hip
1________________________________
2________________________________
3________________________________
4________________________________
5________________________________

Flexors at the knee
1________________________________
2________________________________
3________________________________

Extensors at the knee
1________________________________
2________________________________
3________________________________
4________________________________

Dorsiflexors
1________________________________
2________________________________

Plantar flexors
1________________________________
2________________________________

EXERCISES:

1. General areas of the limbs

You should be able to identify the following general areas on the upper and lower limb models:

a. Upper Limb Model

1) Anterior arm
2) Posterior arm
3) Anterior (flexor) forearm
4) Posterior (extensor) forearm

b. Lower Limb Model

1) Anterior pelvis
2) Posterior pelvis
3) Anterior thigh
4) Posterior thigh
5) Medial thigh
6) Anterior leg
7) Posterior leg

2. You should also be able to identify the following non-muscular structures on the limb models:

a. Upper Limb Model

1) Clavicle
2) Coracoid process
3) Medial border of scapula
4) Acromion
5) Spine of scapula
6) Humerus
7) Medial epicondyle
8) Olecranon process

b. Lower Limb Model

1) Iliac crest
2) Ischial tuberosity
3) Anterior superior iliac spine
4) Greater trochanter
5) Patella
6) Tibia
7) Lateral malleolus
8) Medial malleolus

3. Muscles of the Anterior Aspect of the Upper Limb

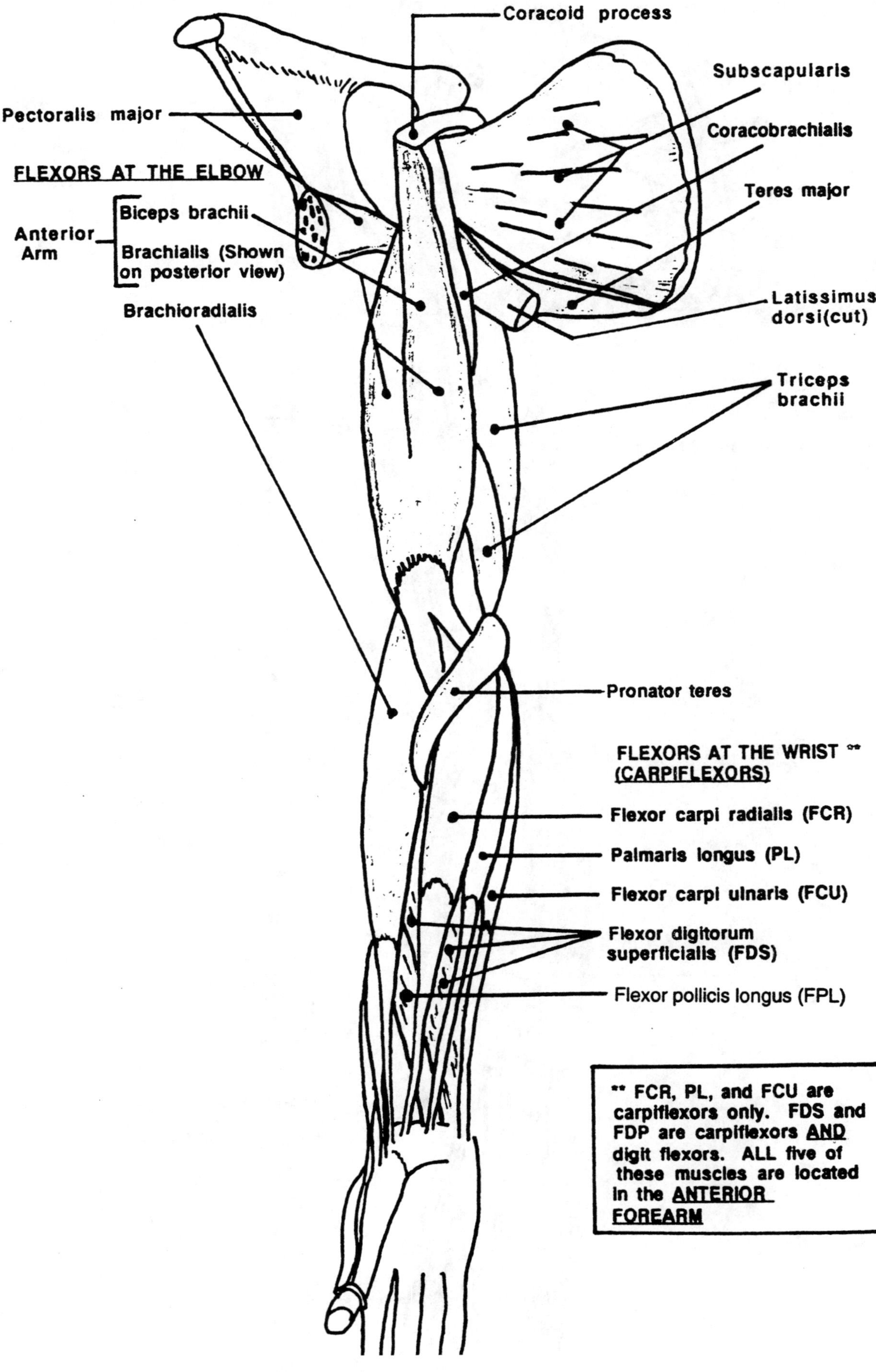

** FCR, PL, and FCU are carpiflexors only. FDS and FDP are carpiflexors AND digit flexors. ALL five of these muscles are located in the ANTERIOR FOREARM

4. Muscles on the Posterior Aspect of the Upper Limb

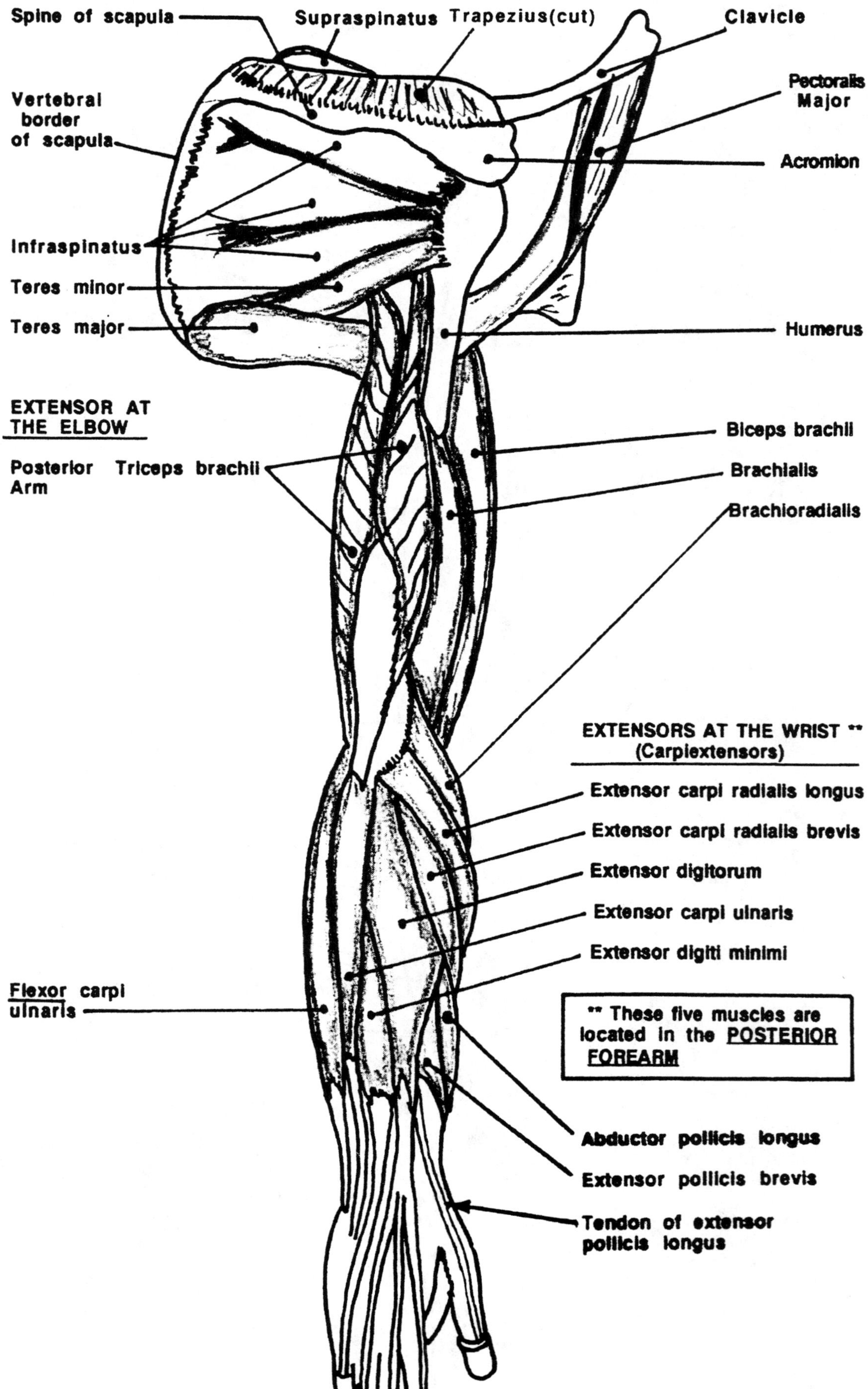

5. Muscles of the Anterior Pelvis, Anterior and Medial Aspects of the Thigh, and the Anterior Leg

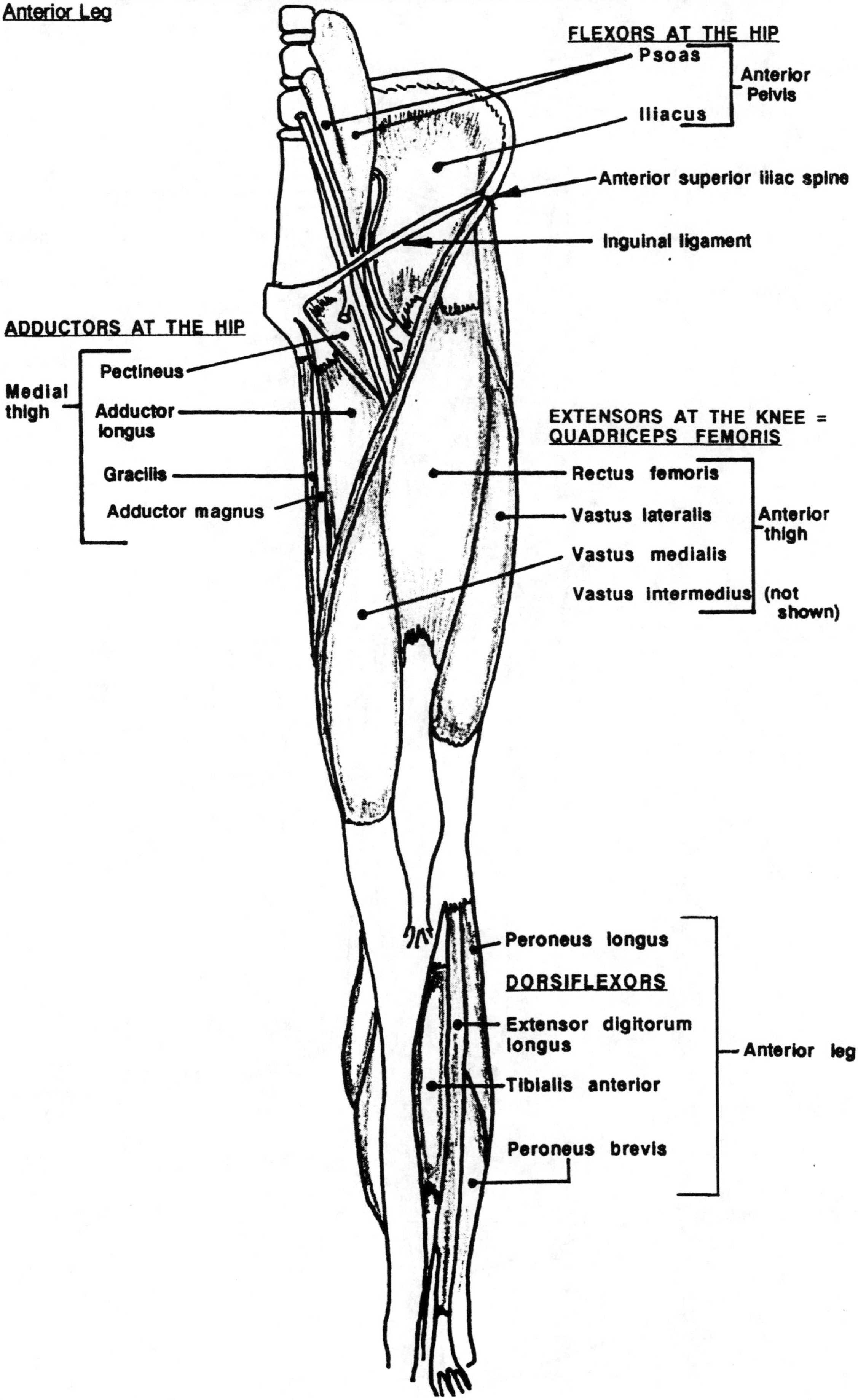

6. Muscles of the Posterior Pelvis, Posterior Thigh, and Posterior leg

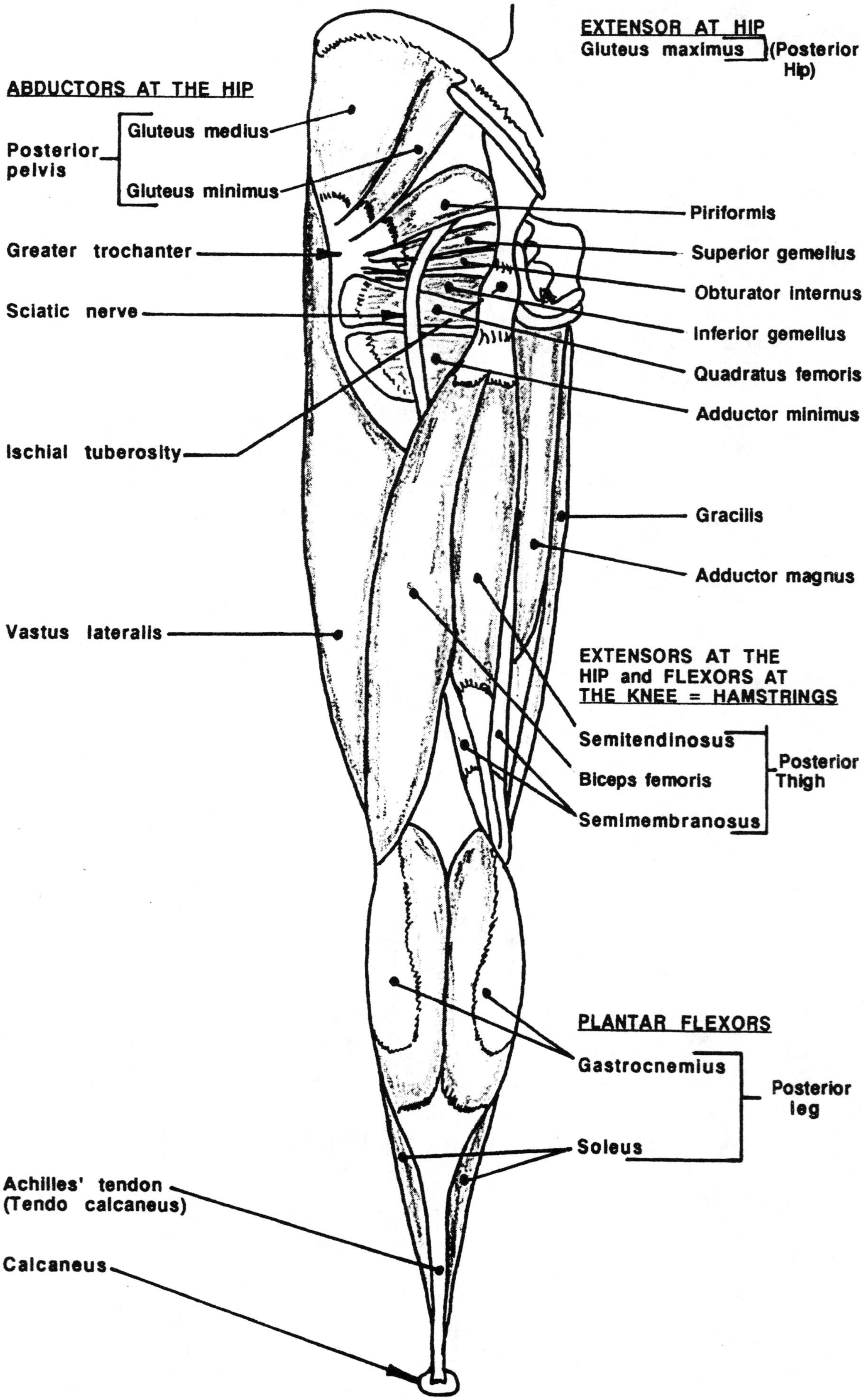

7. Muscles on the torso models

Find as many of the listed muscles as you can on the available torso models

a. Masseter
b. Trapezius
c. Latissimus dorsi
d. Gluteus maximus
e. Deltoid
f. Pectoralis major
g. Teres major
h. Teres minor
i. Infraspinatus

QUESTIONS TO ANSWER AND THINK ABOUT (Some of these questions can be answered directly from looking at the muscles, but some will require that you consult your lecture notes and / or reading assignments):

1. Can you find the following areas on the upper limb model: Anterior arm, posterior arm, anterior forearm, posterior forearm ?

2. What is another name for the anterior forearm ?

3. What is another name for the posterior forearm ?

4. Can you find the following areas on the lower limb model : Anterior pelvis, posterior pelvis, anterior thigh, posterior thigh, medial thigh, anterior leg, posterior leg ?

5. Can you find the following non-muscle structures on the upper limb model: Clavicle, coracoid process, medial border of the scapula, acromion, spine of the scapula, humerus, medial epicondyle, and olecranon process ?

6. Can you find the following non-muscle structures on the lower limb model: Iliac crest, ischial tuberosity, anterior superior iliac spine, greater trochanter, patells, tibia, lateral malleolus, and medial malleolus ?

7. Can you notice that the muscles in the upper limb that cause opposing (antagonistic) movements at a given joint are on opposite sides of the limb ?

WHAT YOU SHOULD KNOW AND BE ABLE TO DO AFTER THESE LABORATORY EXERCISES

1. Be able to recognize and correctly spell the non-muscle structures on both limb models.

2. Be able to recognize and correctly spell all of the muscles given in the four diagrams.

3. **Be able to answer questions from lecture about the locations and actions of these muscles.**

LABORATORY 9 - SYSTEMS

THE RESPIRATORY SYSTEM

INTRODUCTION: The purpose of the respiratory system is to provide a way to get the **O_2** from the atmosphere to the **alveoli (air sacs)** of the lungs so that the O_2 can be taken into the circulatory system, and to pick up the **CO_2** from the blood and ventilate it out to the atmosphere. What this means is that when you inhale (inspiration) you breathe in O_2, and when you exhale (expiration) you breathe out CO_2. That is because when air is in the alveoli of your lungs there is a **O_2 / CO_2 exchange** between your respiratory system and your circulatory system.

The two major divisions of your respiratory system are the **UPPER RESPIRATORY TRACT** and the **LOWER RESPIRATORY TRACT**. The upper respiratory tract includes 1) the **nasal cavity**, 2) the **oral cavity**, 3) the three areas of **the pharynx**, and 4) **the larynx**. Any structure that is in these structures is of course in the upper respiratory tract. The lower respiratory tract includes 1) the **trachea**, 2) the **bronchi**, 3) the **bronchial tree**, 4) the **alveoli**, and 5) the **lungs**.

The Larynx consists of nine cartilages located in the anterior aspect of the neck. In fact, you can easily see the protrusion of the thyroid cartilage, commonly referred to as the "Adam's apple". For hormonal reasons, the thyroid protrusion is more prominent in males than females. Of these nine laryngeal cartilages, three of them (**arytenoids, corniculates, and cuneiforms**) are paired, and three (**thyroid, cricoid, and epiglottis**) occur as single cartilages.

Within the larynx are the **vocal cords**. There are two types of these : 1) True vocal cords, and 2) False vocal cords. You use your true vocal cords for speech and your false vocal cords to seal off the respiratory system ("hold your breath") when you pick up something heavy.

Obviously, you want only air to enter your larynx. However, when you start to swallow, food and drink are suddenly and with great force, pushed into the laryngopharynx, and it is from here that food and drink have the opportunity to either enter your larynx, where they are **not** supposed to go, or into the esophagus (part of the digestive system) where they **are** supposed to go. Having something solid (popcorn, peanut, coins, chunks of food) enter your larynx can be fatal. Nevertheless, every time you swallow, you put youself at risk of death. When we get to the digestive system you will discover how the **epiglottis** is involved in sealing off the larynx during swallowing.

The **hyoid**, which is a bone, is connected to the thyroid cartilage by the thyrohyoid membrane. The hyoid is on the larynx models that you will look at in lab, but it is not considered part of the larynx.

The **trachea** begins the lower respiratory tract. It is a fairly short tube (5 - 6 inches) that is located just behind and about half way down your sternum. The top part of the trachea protrudes just above the jugular notch of the manubrium (superior bone of the sternum) where it is fairly superficial, and therefore makes an excellent place to inset a tube (tracheotomy) to aid those in respiratory trouble. The trachea is held open by the **C- rings**, which are also called **cartilaginous arches**. The last cartilage of the trachea, at the branch point, is called the **carina** .

There are two short **primary bronchi**, right and left. There are some anatomical differences between them that will be discussed in lecture. Each primary bronchus gives rise to several smaller **secondary bronchi**, which give rise to several smaller **tertiary bronchi**, which then give rise to several smaller **bronchioles**. The bronchioles are not the end of the **bronchial tree**, but they are the last structures of the tree that you can see on the models in lab.

The bronchi contain C - rings, but beginning with the bronchioles the tubes in the rest of the bronchial tree are made of smooth muscle and have no C - rings.

The **lungs** are located in the pleural cavities. Although they are of about equal size and shape, they do have some anatomical differences. The right lung has a horizontal fissure and a middle lobe, neither of which the left lung has, but the left lung has a cardiac notch that is not present in the right lung. They both have an apex, base, superior lobe, inferior lobe, oblique fissure, and the components of the bronchial tree. In each lung there are approximately 300 - 350 million air sacs called **alveoli.**

OBJECTIVES: The objectives of this laboratory are to:

1. Be able to identify and correctly spell the structures on the larynx / trachea model

2. Be able to identify and correctly spell the structures on the larynx / trachea / bronchi model

3. Be able to identify and correctly spell the structures on the lung models

4. Be able to identify and correctly spell the respiratory structures on the torso models

PRELABORATORY PREPARATION: You should complete this worksheet before you come to lab. It will allow you to work more efficiently and to complete your work earlier and with less confusion. By using the diagrams on the following pages (those under **EXCERCISES**), fill in the blanks below.

Name the four models that you will look at in lab

____
____
____
____

Lung Models

Right Lung
1________________
2________________
3________________
4________________
5________________
6________________
7________________

Left Lung
1________________
2________________
3________________
4________________
5________________
6________________

Larynx / Trachea model

a. Cartilages of larynx
1____________________
2____________________
3____________________
4____________________
5____________________

b. Non cartilage larynx structures
1____________________
2____________________

c. Bone
1____________________

d. Trachea structures
1____________________

Torso models
1____________________
2____________________
3____________________
4____________________
5____________________
6____________________
7____________________
8____________________
9____________________

Larynx / trachea / bronchi model

a. Larynx structures
1________________________
2________________________
3________________________
4________________________
5________________________
6________________________
7________________________
8________________________

b. Trachea structures
1________________________
2________________________

c. Bronchi
1________________________
2________________________
3________________________

d. Bone
1________________________

e. Gland
1________________________

f. Membrane
1________________________

EXERCISES:

1. Structures of the larynx

a. Anterior view of the larynx / trachea model

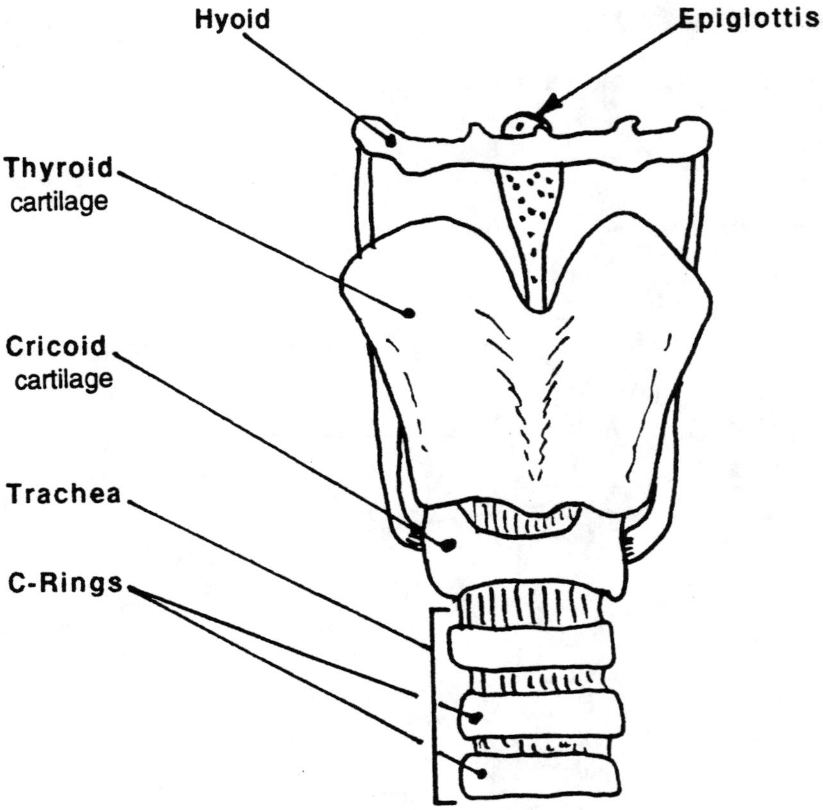

b. Posterior view of the larynx model

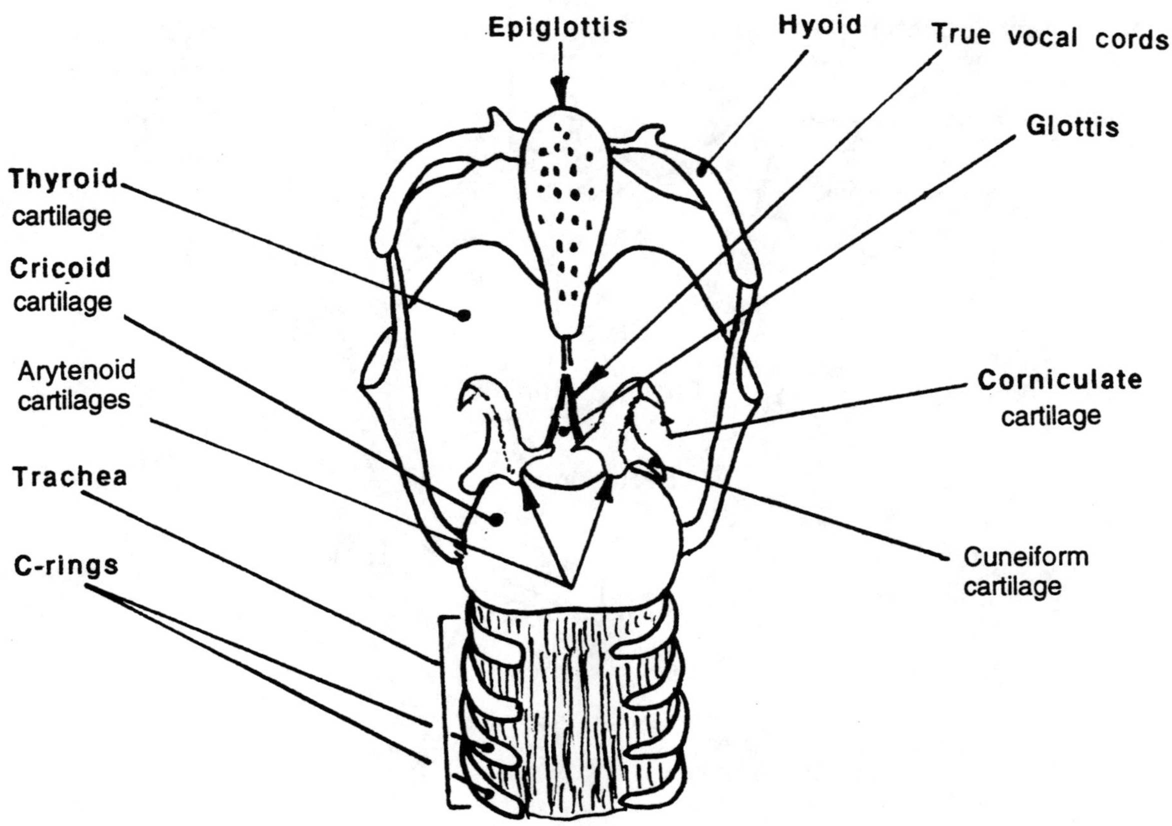

2. Structures of the larynx, trachea, and bronchi

a. Anterior view of the larynx / trachea / bronchi model

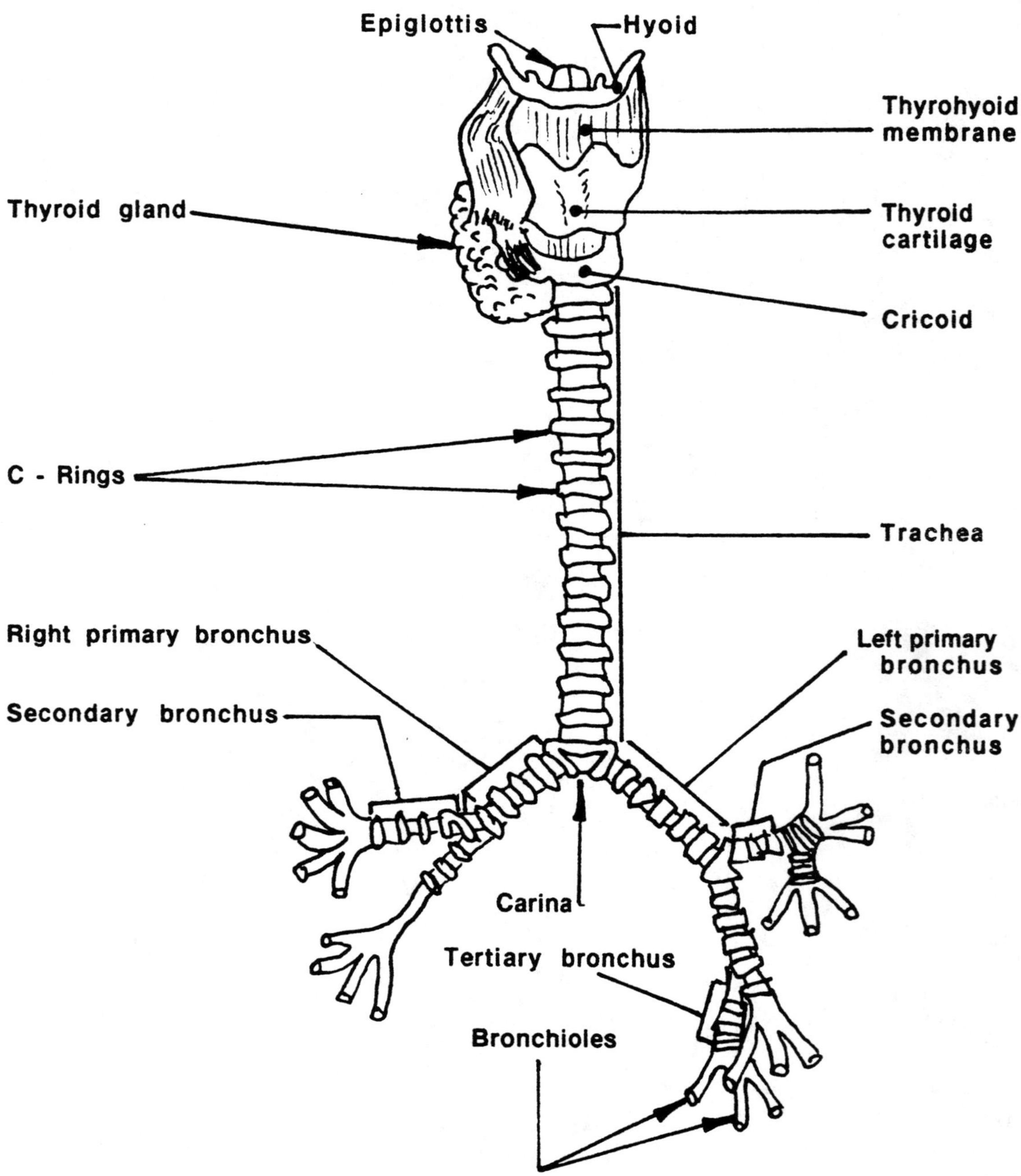

b. Internal and lateral view of the larynx / trachea / bronchi model

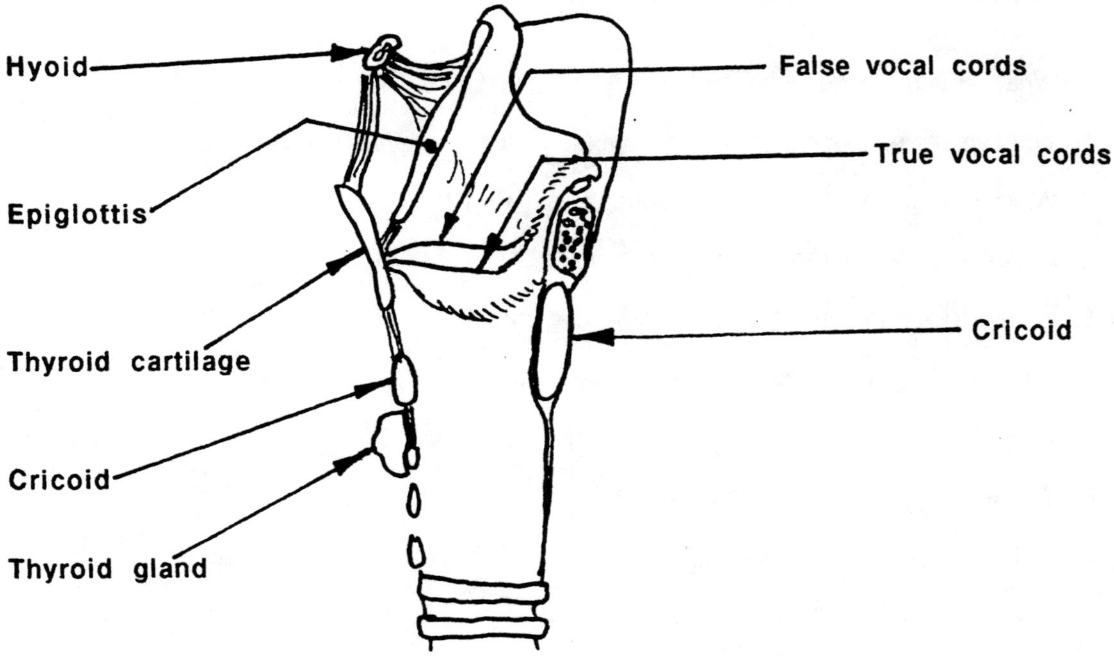

3. Respiratory structures on the torso models

On the torso model find the following respiratory structures:

a. Nasal cavity
b. Conchae
c. Oral cavity
d. Nasopharynx
e. Larynx
f. Trachea and C - rings
g. Carina
i. Primary bronchi
j. Lungs

4. Structures of the lungs

a. Right Lung
   1) Superior lobe
   2) Middle lobe
   3) Inferior lobe
   4) Horizontal fissure
   5) Oblique fissure
   6) Apex
   7) Base

b. Left lung
   1) Superior lobe
   2) Inferior lobe
   3) Oblique fissure
   4) Cardiac notch
   5) Apex
   6) Base

QUESTIONS TO ANSWER AND THINK ABOUT (Some of these questions can be answered directly from looking at the respiratory models, but some will require you to consult your lecture notes and reading assignments):

1. What cavities are the lungs in ?
2. What are the nine cartilages of the larynx ? Which are paired, and which are unpaired ?
3. Which laryngeal cartilage is used during swallowing ?
4. Is the hyoid considered a part of the larynx ?
5. Is the larynx in the upper, or lower respiratory tract ?
6. Is the cricoid in the upper, or lower respiratory tract ?
7. How long is the trachea ?
8. Which bone is the trachea behind ?
9. What cartilage is also called the "Adam's apple" ?
10. What is the branch point of the trachea called ?
11. What are the names of the two thin membranes of the pleural cavities and lungs ?
12. Can you trace a molecule of O_2 from the atmosphere to the alveoli and list all of the respiratory structures that it will pass or go through on the way ?
13. Name the structures that have C - rings.
14. Name the structures in the bronchial tree that do not have C - rings.
15. What does the right lung have that the left lung does not ?
16. What does the left lung have that the right lung does not ?
17. List all of the structures that the right and left lungs have in common ?
18. To what two laryngeal cartilages are the true vocal cords attached ?
19. What separates the oral and nasal cavities ?
20. If food and drink are in the laryngopharynx, then where is it possible for them to go next during the swallowing process ?
21. How do you pronounce pharynx ?
22. How do you pronounce larynx ?

WHAT YOU SHOULD KNOW AND BE ABLE TO DO AFTER THIS LABORATORY:

1. Be able to identify and correctly spell structures on the following models:

    a. Larynx model

    b. Larynx / trachea / bronchi model

    c. Torso model

2. Be able to answer questions about these structures from your lecture notes and reading assignments.

LABORATORY 10 - SYSTEMS

THE DIGESTIVE SYSTEM

INTRODUCTION: The digestive system is composed of a long tube (**ALIMENTARY CANAL**) through which food and drink pass while being digested by **ENZYMES** that are made and secreted by some of the **ACCESSORY DIGESTIVE ORGANS**.

The digestive system begins with the **oral cavity** (mouth) and ends at the **anus**. In between, in order, are the **oropharynx**, **laryngopharynx**, **esophagus**, **stomach**, **small intestine** (consisting of the **duodenum**, **jejunum**, and **ileum**), **large intestine** (consisting of the **cecum**, **ascending colon**, **transverse colon**, **descending colon**, and **sigmoid colon**), the **rectum**, and, just above the anus, the **anal canal**. Getting through the human alimentary canal is about a 25 foot trip.

The purpose of the digestive system is to digest the complex molecules in food into their fundamental components so that they can be absorbed into the blood and lymphatic circulations for distribution to the rest of the body. The actual digestion of food molecules is done by the action of enzymes.

We can now go back through the alimentary canal and say a little about what occurs in each part of the digestive system. When food is introduced into the **oral cavity** you chew it (**mastication**), which is done by your **teeth.** Actually, what you are doing with the chewing mechanism is grinding your food to increase the surface area of the food for more efficient digestion. The digestive event that occurs in the mouth is partial digestion of two complex carbohydrates, starch and glycogen. This digestive event is made possible by the secretion of **saliva**, which not only moistens your food, but contains an enzyme calle amylase that attacks starch and glycogen. Saliva, and therefore amylase, is made by the **salivary glands** (parotid, sublingual, and submandibular glands). Obviously, your teeth and salivary glands are accessory digestive organs for the oral cavity.

Once you have masticated your food enough you then do something that is very common, but actually quite dangerous, and that is **swallowing**. You might want to notice that when you swallow that you move the collected food (called a bolus) to the back of your mouth, and then with great rapidity and force push the bolus of food into your **oro- and laryngopharynx.** What a dangerous place for food to be under such pressure and at such speed. Where can this food and drink go ? If you do not elevate your soft palate up against your nasopharyngeal wall, then liquids can easily enter your nasal cavity. That can be a little embarrassing, but normally you won't die from it. If you do not elevate your larynx at precisely the correct time, then the base of your tongue, as it pushes the food / drink into your pharynx cannot also push your epiglottis down over the entrance into your larynx. This is now a potentially fatal situation for instead of food / drink entering your esophagus, where it's supposed to go, it can go instead into your larynx. Liquids may not be too bad, making you simply sputter, but solid food can obstruct the glottis, a potentially fatal situation. So, with swallowing, as with the forward pass in football, three things can happen, and two of them are bad.

If everything goes as it should, as it usually does, then food / drink enter the **esophagus** because that is the only place left to go. The esophagus is a muscular tube connecting the laryngopharynx to the stomach. In order to get to the stomach, the esophagus passes posterior to the trachea and heart and then through an opening in the thoracic diaphragm called the esophageal hiatus, the location of the common hiatal hernia . There is a thickening of the smooth musculature at the distal end of the esophagus called the **cardiac,** or **lower esophageal sphincter** . The purpose of this sphincter is to relax and allow the bolus of food / drink to enter the stomach, and then to contract in order to prevent contents in the stomach from re-entering the esophagus.

Once food / drink are in the **stomach** they remain there for several hours, and it is during this time that hydrochloric acid (HCl) and a digestive enzyme are released into the stomach, and that there is a great deal of mechanical action by the three layers of smooth muscle which mixes and churns stomach contents. The digestive event that occurs in the stomach is partial digestion of the proteins in your food.

After several hours of digestion and churning and mixing you have a liquid form of food called **chyme** that can now move past your **pyloric sphincter** (located at the distal end of the stomach) into the **duodenum of the small intestine**. In the duodenum digestion is completed by digestive enzymes that are made and secreted into the duodenum by the **pancreas**. Not all nutrients produced by the digestive process are the same, and we can categorize them into water - soluble and water - insoluble nutrients. The water - soluble nutrients are absorbed across the villi and into blood vessels and then are transported directly to the **liver**. The water - insoluble nutrients are another matter. In order for them to be absorbed out of the small intestine, they must be solubilized by **bile**, which is continuously made by the liver and stored in between meals by the **gall bladder**. When chyme enters the duodenum the gall bladder ejects its bile down the common bile duct into the duodenum where it solubilizes the water - insoluble nutrients. The solubilized water - insoluble nutrients are absorbed across the villi, but instead of entering the blood, they enter a central lacteal and are transported to the left subclavian vein by the thoracic duct.

By the time contents get to the ileum and enter your **cecum of the large intestine**, the majority of the nutrients have been absorbed, leaving behind the ingredients in you diet for which you have no digestive enzymes . Cellulose is a good example. You take in lots of cellulose in your diet, especially if you eat plenty of plants (vegetables, grains, etc.) However, unlike termites, you lack the enzymes to digest cellulose, so it passes right on through your digestive system. Cellulose turns out to be great food for termites and it is precisely the cellulose in the wood in the foundation of your house that they are after.

The functions of your large intestine are at least two: 1) To make feces out of the indigestible nutrients in your diet that come into it from the ileum, and 2) to reabsorb water back into your body . Most likely you have noticed already that if you are **inside** your alimentary canal, then you are **outside** your body . As food / drink pass through your alimentary canal, various amounts of fluids are taken from your body and put into your alimentary canal. You do not want to permanently lose this fluid, and so it is the function of you large intestine to get that fluid back. There are all sorts of things that interfere with the proper reabsorption of fluids, such as bacterial and viral infections, and various drugs. If not enough water is reabsorbed, then you get diarrhea, an extremely dangerous condition in babies. If too much water is reabsorbed, then there is the condition of constipation, uncomfortable, but at least not life-threatening.

OBJECTIVES: The objectives of this laboratory are to:

1. Be able to identify and correctly spell the structures on the tooth model
2. Be able to identify and correctly spell the structures on the stomach model
3. Be able to identify and correctly spell the structures on the duodenal / pancreas model
4. Be able to identify and correctly spell digestive structures on the torso models

PRELABORATORY PREPARATION: You should complete this worksheet before you come to lab. It will allow you to work more efficiently and to complete your work earlier and with less confusion. By using the diagrams on the following pages (those under **EXERCISES**), fill in the blanks below:

Name the five models that you will look at in lab
1__________
2__________
3__________
4__________
5__________

Tooth model
a. Areas
1__________
2__________
3__________

b. Structures
1__________
2__________
3__________
4__________
5__________

Liver / gall bladder model
a. Lobes of the liver
1__________
2__________
3__________
4__________

b. Biliary duct system
1__________
2__________
3__________
4__________
5__________

Torso models
a. Structures above the stomach
1__________
2__________
3__________

b. Stomach and stomach areas
1__________
2__________
3__________
4__________

Stomach model
a. Areas
1__________
2__________
3__________
4__________

b. Curvatures
1__________
2__________

c. Folds
1__________

d. Sphincters
1__________

e. Duodenum

Duodenum and structures
a. Areas
1__________
2__________
3__________
4__________

b. Other structures
1__________
2__________

The ileum __________

The pancreas
a. Areas
1__________
2__________
3__________

b. Ducts
1__________
2__________

Duodenal / pancreas model
a. Organs
1__________
2__________

b. Duodenal areas
1__________
2__________
3__________
4__________

c. Duodenal folds
1__________

d. Openings
1__________

e. Pancreas areas
1__________
2__________
3__________

f. Pancreas ducts
1__________
2__________

Salivary glands
1__________
2__________
3__________

Large intestine
1__________
2__________
3__________
4__________
5__________
6__________
7__________
8__________
9____

EXERCISES:

1. Tooth model

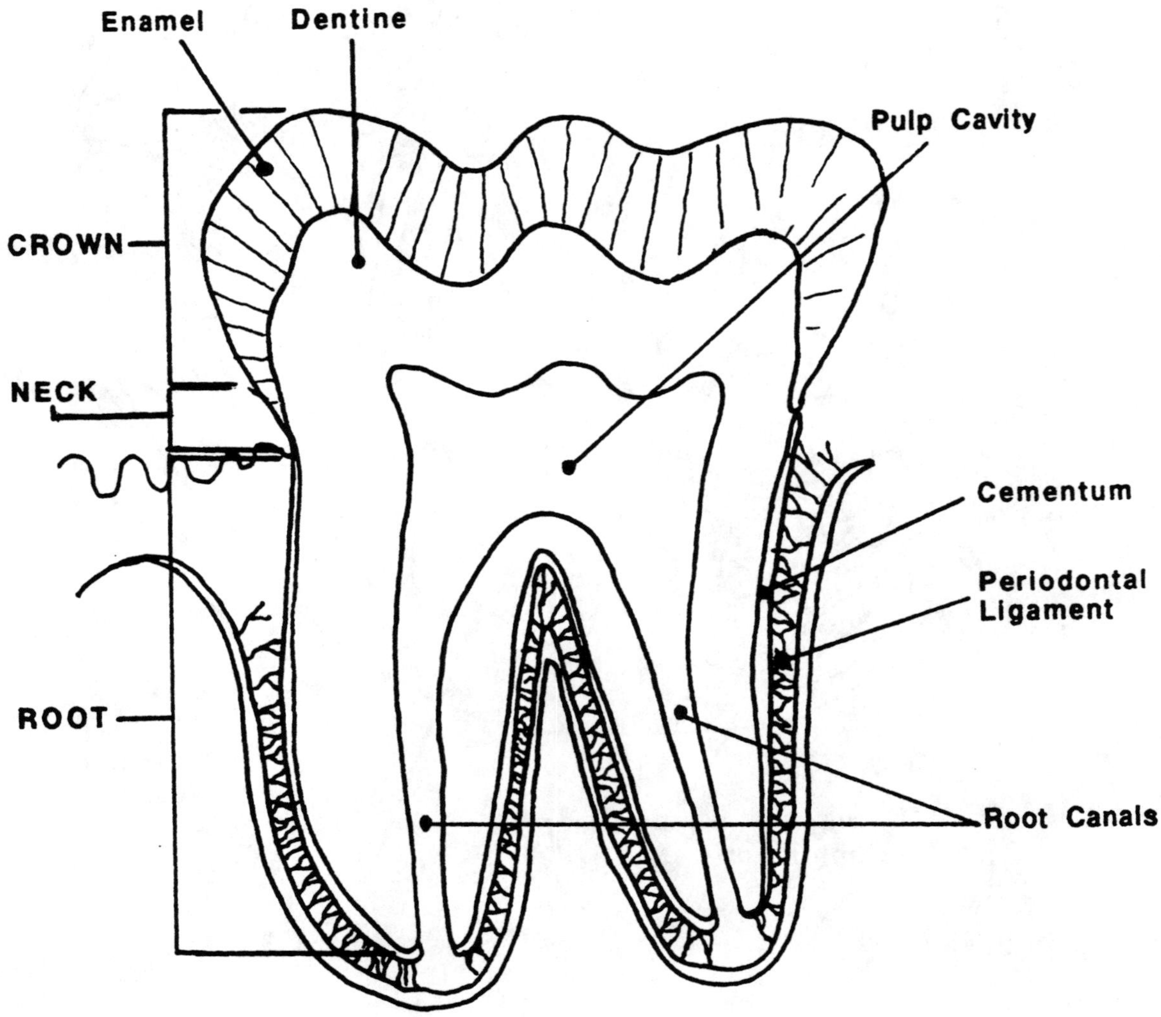

2. Stomach Model

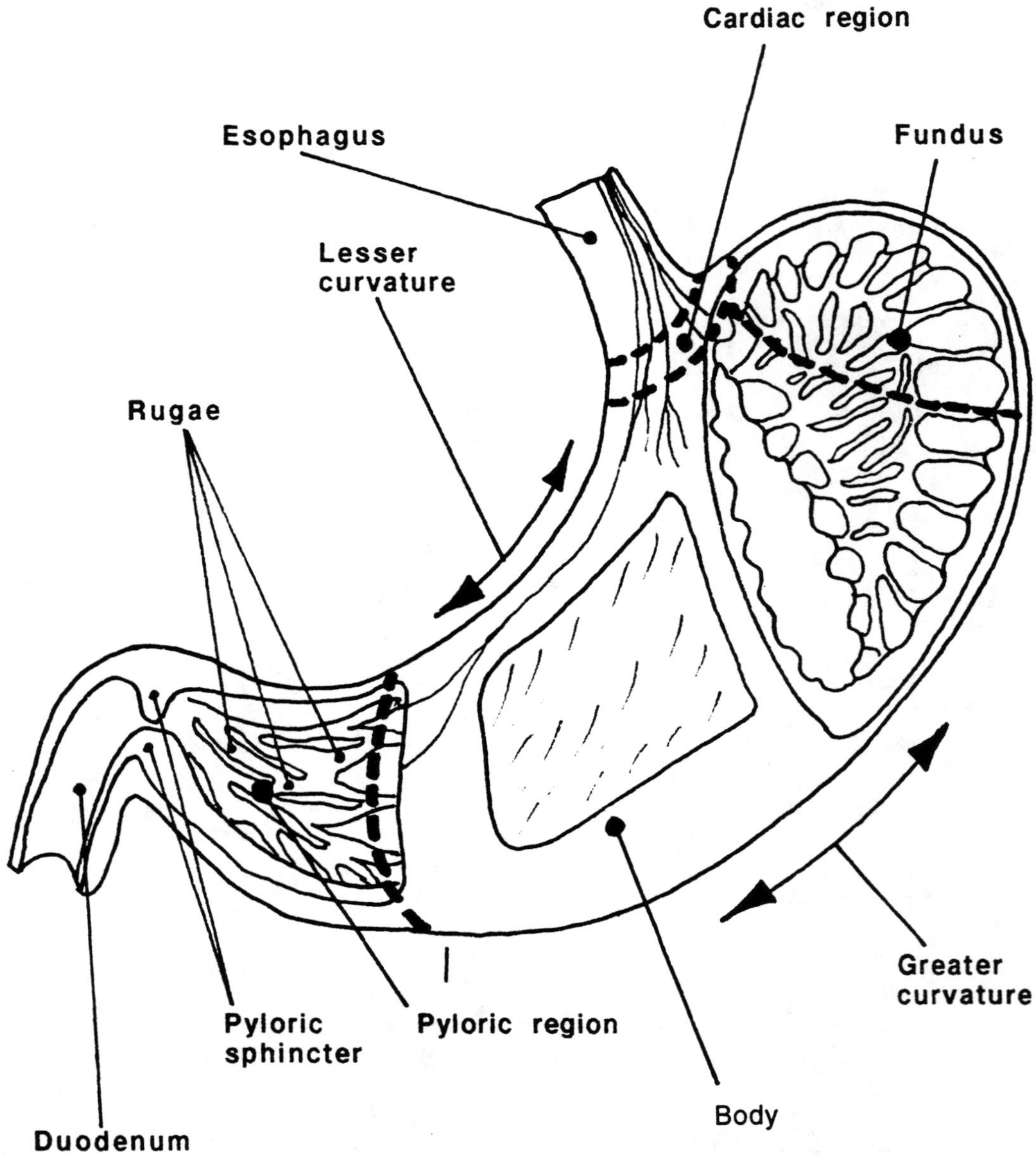

3. Structures of the duodenal - pancreas model

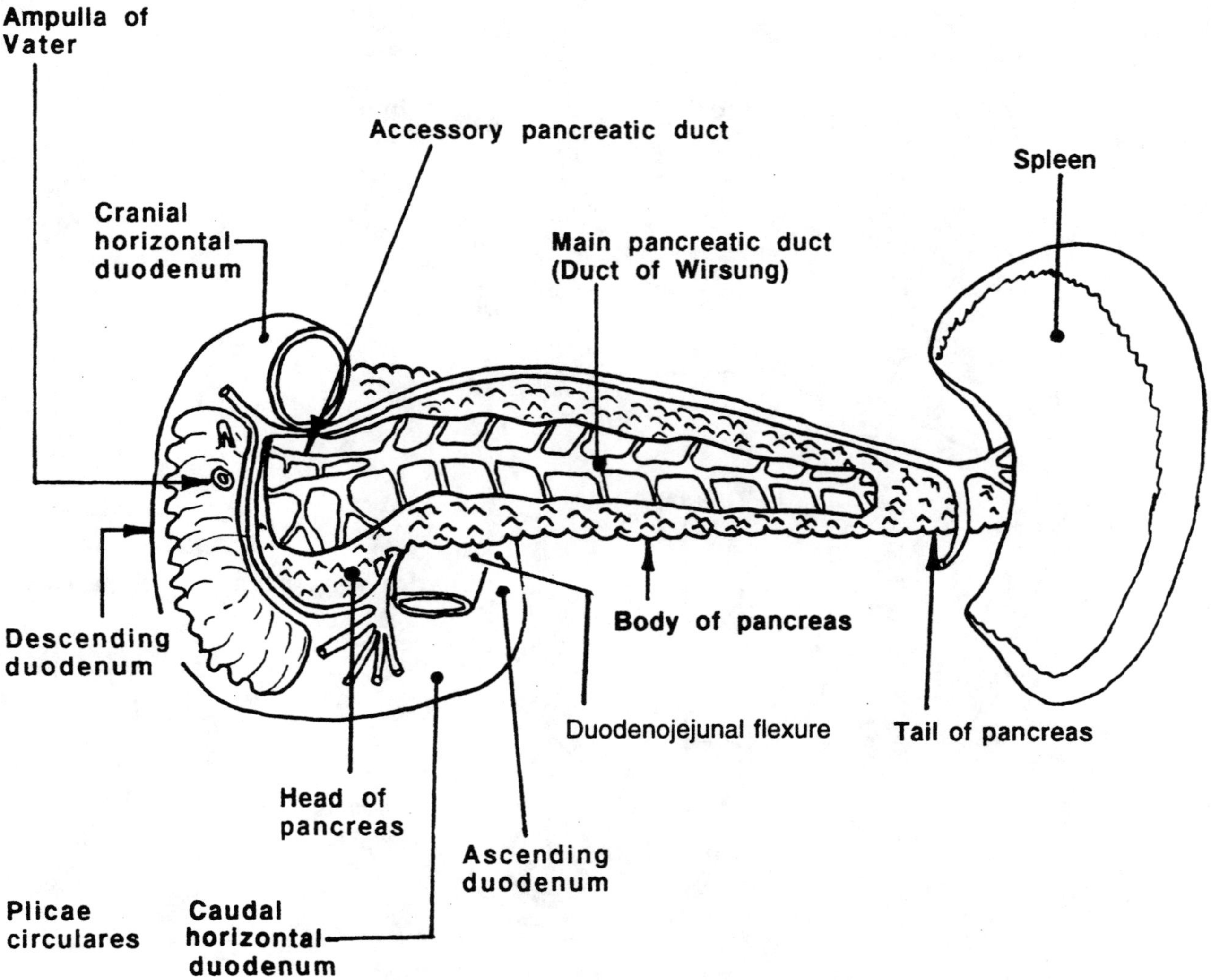

4. Liver models

There are several liver models in the lab. Some of these are incomplete as to the structures that you can see. However, be sure that you look at them all and try to understand the structures on each.

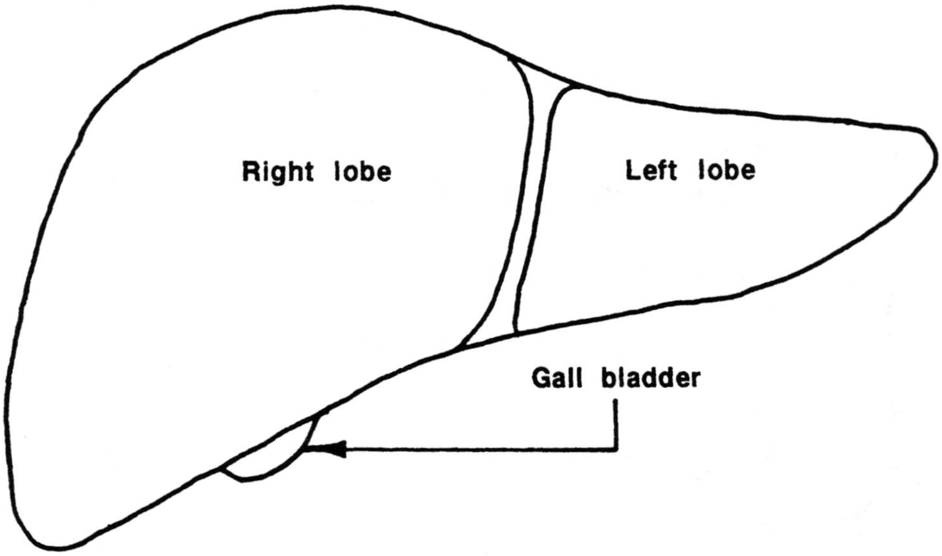

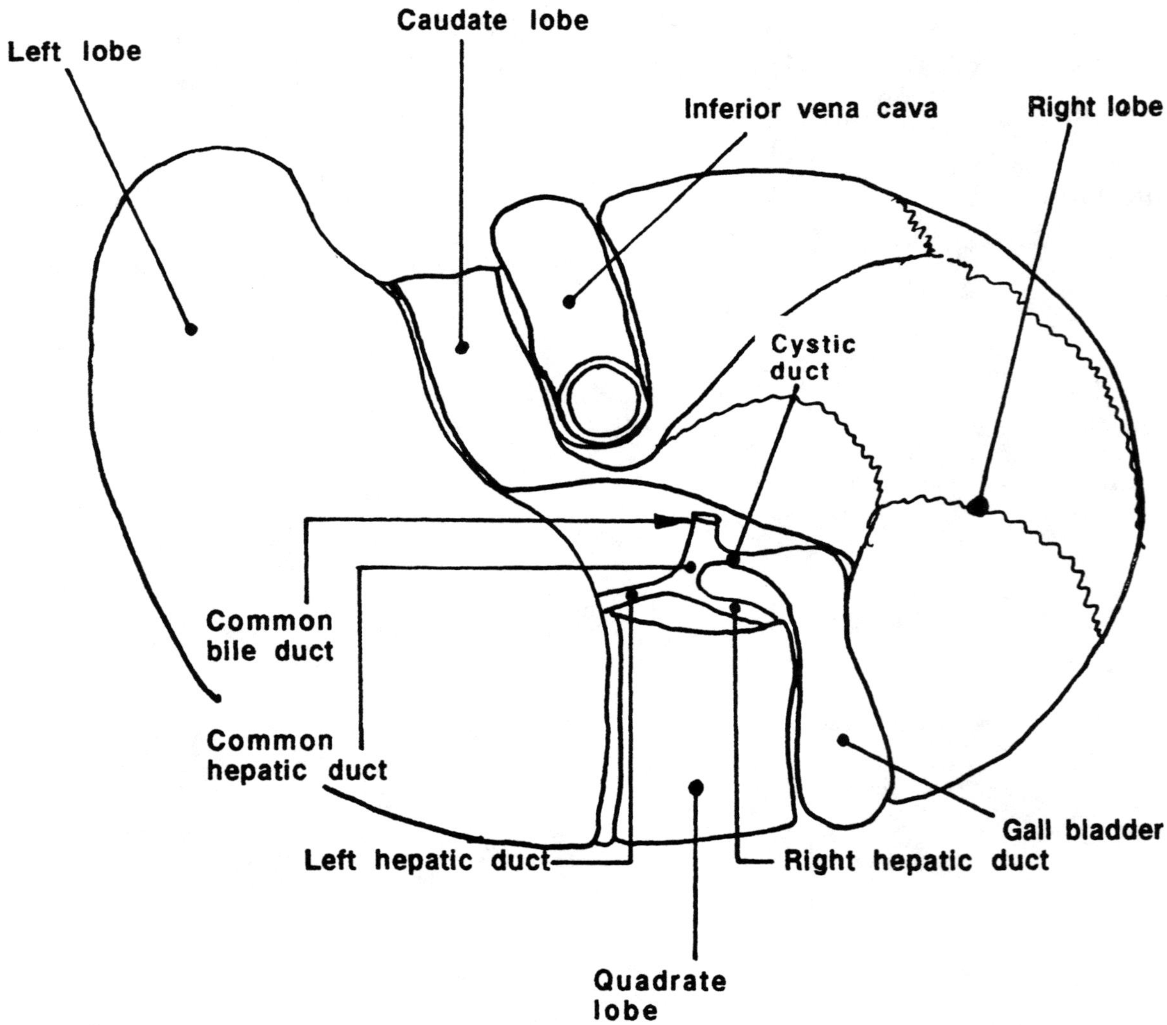

5. Digestive structures on the torso models

There are several torso models in the lab, and, like people, each is a little different in how it looks. Nevertheless, you should make every effort to look at all of the models available to you and learn the following structures listed below. This would be an excellent exercise in which to use your textbook diagrams since the torso models are not drawn in this manual:

a. Oral cavity and structures associated with it
  1) Oral cavity
  2) Salivary glands
    a) Parotid gland (Note also the Duct of the parotid gland)
    b) Sublingual gland
    c) Submandibular gland

b. Oropharynx

c. Esophagus

d. Esophogeal hiatus

e. Stomach
  1) Areas
    a) Cardiac
    b) Fundus
    c) Body
    d) Pyloric
  2) Rugae
  3) Pyloric sphincter

f. Small intestine
  1) Duodenum
    a) Areas
      1)) Cranial horizontal
      2)) Descending
      3)) Caudal horizontal
      4)) Ascending
  2) Duodenojejunal flexure
  3) Beginning of jejunum
  4) End of ileum

g. Large intestine
  1) Cecum
  2) Appendix
  3) Ascending colon
  4) Transverse colon
  5) Descending colon
  6) Sigmoid colon
  7) Rectum
  8) Haustrae
  9) Taeniae coli
  10) Right colic (hepatic) flexure
  11) Left colic (splenic) flexure

h. Pancreas
  1) Head
  2) Body
  3) Tail

i. Liver and gall bladder
  - See liver model structures

j. Spleen

QUESTIONS TO ANSWER AND THINK ABOUT (Some of the questions can be answered directly from looking at the digestive system models, but some will require that you consult your lecture notes and / or reading assignments):

1. When you smile at someone, then what parts of your teeth does (s)he see ?
2. What part of a tooth is imbedded in an alveolus ?
3. What structure anchors your teeth to the bone that they are in ?
4. What two bones are your teeth in ?
5. What is the purpose of your teeth ?
6. What are the functions of the salivary glands ?
7. For what are the three salivary glands named ?
8. How is liquid kept out of the nasal cavity during swallowing ?
9. How is food / drink kept out of the larynx during swallowing ?
10. What anatomical areas do the respiratory and digestive structures have in common ?
11. What kind of muscle is in the wall of the esophagus ?
12. What kind of muscle is in the wall of the stomach ?
13. What are the purposes of the cardiac and pyloric sphincters ?
14. What type of muscle makes up these two sphincters ?
15. Which bones protect the stomach, liver, and spleen ?
16. Rugae can be found in the stomach and in what other organs of the body ?
17. What is the function of rugae in all of these organs ?
18. If the cardiac sphincter malfunctions and goes into a spasm (cardiospasm) and allows HCl reflux into the esophagus, then great pain results behind the sternum. What is this pain commonly termed ?
19. What is the name of the opening in the thoracic diaphragm that the esophagus passes through?
20. What area of the abdominal cavity is the liver located in ?
21. What area of the abdominal cavity are the stomach and spleen located in ?
22. As far as the digestive system is concerned, what is the function of the liver ?
23. Where is the bile made by the liver stored in between meals ?
24. Where is the gall bladder located ?
25. How do the digestive enzymes made in the pancreas get into the duodenum ?
26. How does the bile made in the liver get into the gall bladder ?

27. How does the bile in the gall bladder get into the duodenum ?

28. What is the purpose of bile in the small intestine ?

29. What are the three areas of the small intestine in going from the pyloric sphincter to the cecum of the large intestine ?

30. Where do the water-soluble nutrients immediately go after their absorption from the small intestine ?

31. Where do water-insoluble nutrients go after they have been absorbed out of the small intestine ?

32. In what abdominal area are the cecum and the appendix located ?

33. What digestive event occurs in the oral cavity ?

34. what digestive event occurs in the stomach ?

35. Bile is made continuously by the liver. What happens to this bile if the gall bladder has been surgically removed ?

36. What digestive function is impaired if the gall bladder is removed ?

37. Where does a hiatal hernia occur ?

38. What large blood vessel passes through the liver ?

39. What part of the pancreas is next to the duodenum ? Next to the spleen ?

40. The spleen is on some of the lab models, but does it have any digestive functions ?

41. On which side of the abdominal cavity is the ascending colon ? The descending colon ?

42. Why is the right colic flexure also called the hepatic flexure ?

43. Why is the left colic flexure also called the splenic flexure ?

44. Why is the sigmoid colon called sigmoid ? That is, what does sigmoid mean ?

45. What are the two very important functions of the large intestine ?

WHAT YOU MUST KNOW AND BE ABLE TO DO AFTER THIS LABORATORY:

1. Be able to identify and correctly spell the structures on the following models:

   a. Tooth model

   b. Stomach model

   c. Duodenal / pancreas model

   d. Liver model

   e. Torso models

2. **Be able to answer questions about these structures from information in your lecture notes and reading assignments.**

LABORATORY 11 - SYSTEMS

THE URINARY SYSTEM

INTRODUCTION: **Urea** is a small molecule made by normal metabolism in your liver. Regardless of the fact that urea is made by the body, it is nevertheless a toxin (poison) and for that reason it must be eliminated not only from the liver cells, but from the body. As blood is flowing through your liver, the liver cells pass the urea across their plasma membranes and into the blood, and with that passage one problem is solved, that of cellular toxicity. But still a problem remains and that is to remove urea from your blood.

Removal of urea from your blood is done by your kidneys. **The kidneys** are the main functional structures of your urinary system. Not only do your kidneys **1)** constantly remove the urea in your blood by a filtration process and create a fluid (**urine**), but they also **2)** regulate the fluid balance in your body, **3)** regulate your long-term, daily blood pressure, **4)** regulate the balance of calcium and phosphorus in your blood, and **5)** participate in the acid-base balance of your blood.

The remainder of the urinary system structures are for accommodating the urine that is made by your kidneys. The **ureter** is a tube that allows for transport of the urine from the kidney to the bag where it is stored, **the urinary bladder.** The urinary bladder fortunately has rugae which allow it to expand. You can store several hundred milliliters (cubic centimeters) of urine in your urinary bladder without becoming too uncomfortable. However, there will come a time when you will want to void this urine to the outside of your body . The single tube that leads from the urinary bladder to the outside of the body is your **urethra**. However, due to the development of two reproductive systems (female and male), the urethra becomes incorporated into the penis in the male. Therefore, the urethra is longer in males than females and for this reason urinary tract infections can spread from the urethra to other urinary system structures (urinary bladder and kidneys) much faster and easier in females than in males.

There are essentially four major areas to a kidney: 1) **Renal cortex**; 2) **Renal medulla**: 3) the spaces called **calyces**, and 4) the space called the **renal pelvis**. The filtration of your blood occurs in the renal cortex. The renal medulla is made up of two structures, the renal pyramids , and the renal columns.

If your blood is going to be filtered, then it must have a way to flow out to the renal cortex . Getting your blood out to the renal cortex for filtration is the job of the arterial blood flow system of the kidney. Each of your kidneys has a single **renal artery** bringing blood from the abdominal aorta into the kidney. The renal artery branches into several smaller **interlobar arteries** that pass through the renal medulla toward the renal cortex. However, at the medulla - cortex junction the interlobar arteries branch into many smaller **arcuate arteries.** Each of the arcuate arteries give off numerous smaller **interlobular arteries** that penetrate into the renal cortex. The interlobular arteries then give off numerous and very small **afferent arterioles.** At the end of each afferent arteriole is a single small rounded plexus of blood vessels called a **glomerulus.** There are in each of your kidneys about one million afferent arterioles, so how many glomeruli do you have in each kidney ?

Surrounding each of your glomeruli is a hollow bulb called the **Bowman's capsule** (BC). Bowman's capsule belongs to the urinary system and the glomerulus belongs to the circulatory system. It is precisely here, at the **glomerulus - Bowman's capsule junction**, that the circulatory and urinary systems meet one another and it is precisely at this junction that almost all of the molecular contents (urea, nutrients, small proteins, ions, vitamins, etc.) get filtered out of blood flowing through the glomerulus and into the Bowman's capsule. The removal of molecular contents from your glomerular blood into Bowman's capsule is called **FILTRATION** .

There is a single small tube, the **proximal convoluted tubule** (PCT), that emerges from the back of Bowman's capsule. The filtrate in Bowman's capsule flows out of Bowman's capsule into this tubule. The PCT straightens out to form a tubule, the **loop of Henle** (LOH), that loops down into the renal medulla region and then back up into the cortex. That portion of the LOH that goes into the medulla is called the **descending limb of the loop of Henle,** and that portion of the LOH that comes back up into the cortex region is termed the **ascending limb of the loop of Henle.** At the distal end of the ascending limb, the tubule becomes twisted again and this is called the **distal convoluted tubule** (DCT). A DCT will connect with a **collecting tubule (duct)**. In fact a single collecting tubule (duct) will receive many DCTs.

When we look at these structures in a numerical way we see that there is a one to one correspondence between the glomerulus, BC, PCT, LOH, and the DCT. In fact, in each of your kidneys there are approximately one million of each. The BC, PCT, LOH, and DCT represent a filtration unit, and so each one of your kidneys has about one million of these filtration units. Each of these filtration (functional) units is called a **NEPHRON**, and therefore each of your kidneys has about one million nephrons. Each nephron is composed of **one BC**, **one PCT**, **one LOH**, and **one DCT**. Some authors and texts will include the glomerulus as part of the nephron. **The collecting tubule (duct) is not considered part of the nephron.**

Urine is formed by your nephrons, flows through the collecting tubule into the minor calyces, into the major calyces, into the renal pelvis, then on out the ureter and into the urinary bladder.

OBJECTIVES: The objectives of this laboratory are to:

1. Be able to identify and correctly spell the urinary system structures on the torso model

2. Be able to identify and correctly spell the structures on the kidney models

3. Be able to identify and correctly spell the structures of the nephron / collecting tubule model

EXERCISES:

1. Kidney models

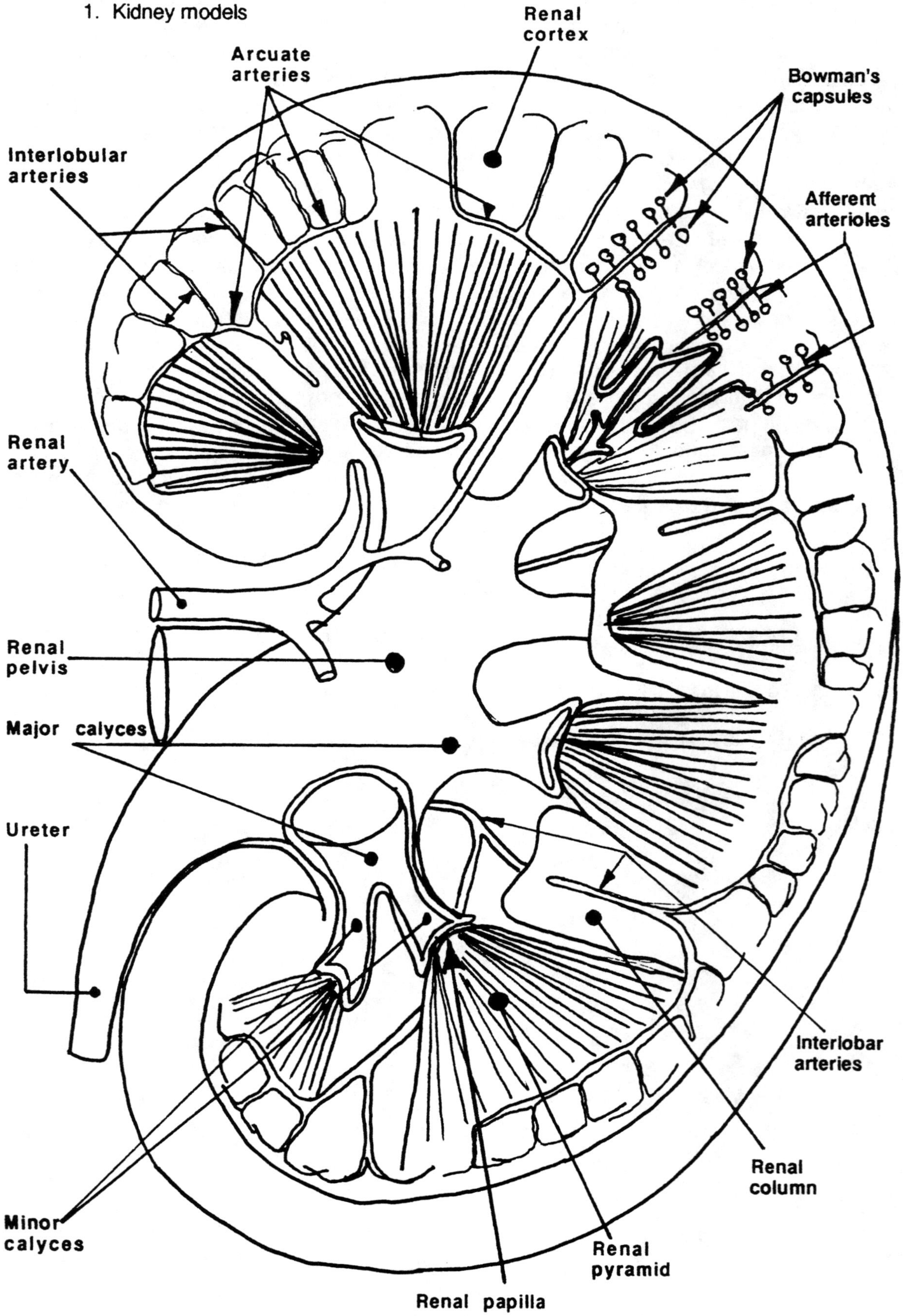

2. Nephron models

a. Nephron / pyramid model

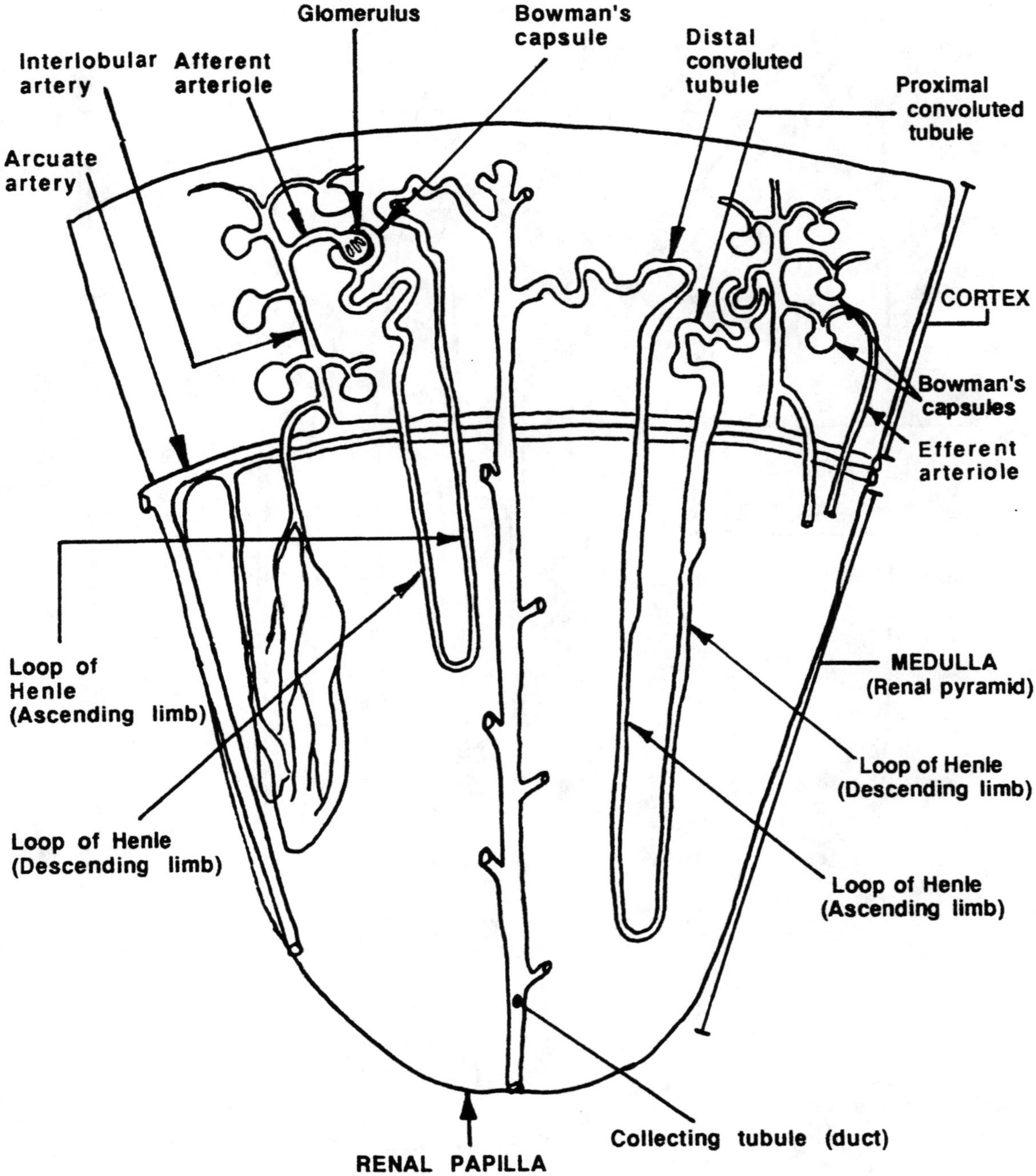

b. Nephron model

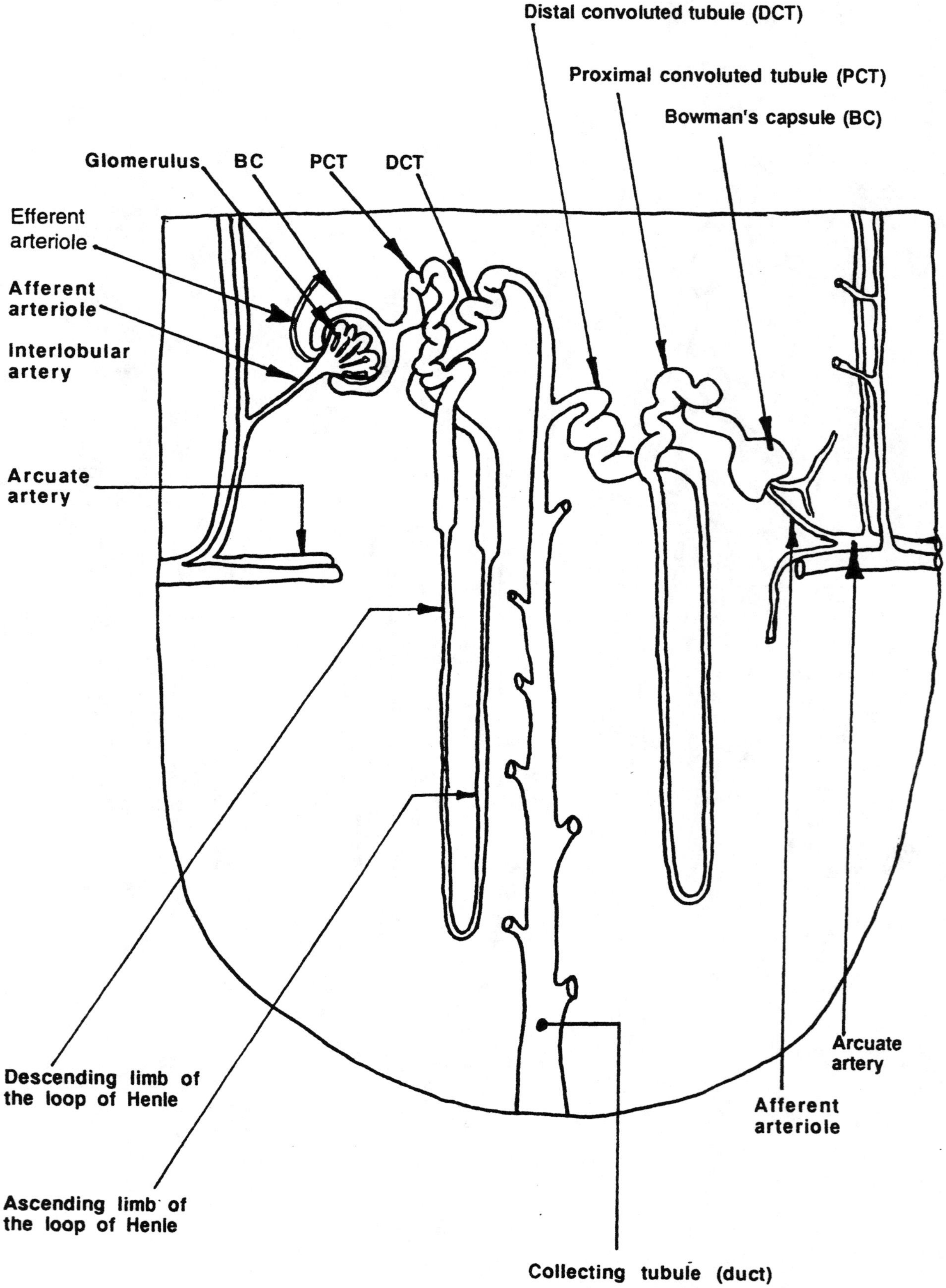

PRELABORATORY PREPARATION: You should complete this worksheet before you come to lab. It will allow you to work more efficiently and to complete your work earlier and with less confusion. By using the diagrams on the following pages (those under **EXERCISES**), fill in the blanks below:

Torso models
1____________________
2____________________
3____________________

Kidney models
a. Areas
1____________________
2____________________

b. Spaces in the kidney
1____________________
2____________________
3____________________

c. Tubules in renal pyramid
1____________________

d. Tube leading out of the kidney
1____________________

e. Arteries / arterioles
1____________________
2____________________
3____________________
4____________________
5____________________

f. Thin outer covering of the kidney
1____________________

g. Rounded tip of renal pyramid
1____________________

h. Areas of the renal medulla
1____________________
2____________________

Nephron models
a. Circulatory system components
1____________________
2____________________
3____________________

b. Nephron components
1____________________
2____________________
3____________________
4____________________
5____________________

c. Tubules in renal pyramid
1____________________

3. Torso models

On the torso models in the laboratory find the following three urinary structures

a. Kidneys b. Ureters c. Urinary bladder

QUESTIONS TO ANSWER AND THINK ABOUT (Some of the questions can be answered directly from looking at the urinary system and torso models, but some will require that you consult your lecture notes and / or reading assignments):

1. In what organ is urea made ?
2. What are the areas of the kidney ?
3. What is the name of the thin outer covering on the kidneys ?
4. What are the functions of the kidneys ?
5. What exactly is filtration ?
6. In what region of the kidney does filtration occur ?
7. Trace a molecule of urea in the renal artery out to the glomerulus.
8. What structure surrounds a glomerulus ?
9. The glomerulus belongs to what system of the body ?
10. The Bowman's capsule belongs to what system of the body ?
11. Exactly where does filtration occur in the kidney ?
12. What are the components of a nephron ?
13. You can clearly see that a collecting tubule (duct) is on the nephron models, but is the collecting tubule a component of a nephron ?
14. Approximately how many nephrons are there per kidney ?
15. Look at the kidney model. What are some general statements that you could make about vessel size and number in going from the renal artery to the glomerulus ?
16. What two structures make up the renal medulla ?

17. Trace a molecule from the aorta to the outside of the body , and in doing so list all ot the structures that it passes or goes through.

18. Which urinary system structure is involved in "passing" a kidney stone ?

19. Who has the longer urethra, females, or males ?

20. Why ?

21. What is one important result of males having the longer urethra ?

22. Can you see the differences in size, number, and location of the interlobar and interlobular arteries ? What are these differences ?

WHAT YOU SHOULD KNOW AND BE ABLE TO DO AFTER THIS LABORATORY:

1. Be able to identify and correctly spell the structures on the kidney models.

2. Be able to identify and correctly spell the structures on the nephron / renal pyramid model.

3. Be able to identify and correctly spell the structures on the nephron model.

4. **Be able to answer questions about these structures from information in your lecture notes and reading assignments.**

LABORATORY 12 - SYSTEMS

THE INTEGUMENTARY SYSTEM

INTRODUCTION: Exactly how important is your skin ? Well, it depends on how old you are. But, consider this: You can lose 50 % of your lung capacity (one lung), 50% of your urinary system (one kidney), probably 50 % of your intestinal tract, and half of your brain, and still survive and function quite well. But if you lose 50 % of your skin surface, then the chances are very good that you will not survive.

There are now burn treatment centers in most large and moderate sized cities that do outstanding work with burn victims. However, if someone is unable to obtain treatment at a burn center, then the chances of survival are approximated by the following equation:

% chances of survival = 100 % - (Age + % of total skin lost)

Obviously, if a 10 year old person and a 50 year old person both suffer burns over 40% of their bodies, then the 10 year old has a much better chance of survival (50 %), than the 50 year old (10 %).

The functions of the skin are 1) **protection**, 2) **thermoregulation** (regulation of body temperature), 3) **excretion**, 4) formation of **Vitamin D** and 5) **reception** of the general sensations.

Skin has two major areas. There is the **dermis**, and above that the **epidermis**. The **epidermis** has five layers (strata), and from the outermost to the innermost they are 1) the **stratum corneum**, 2) the **stratum lucidum** (present only in the skin of palms of the hands and soles of the feet), 3) the **stratum granulosum**, 4) the **stratum spinosum**, and 5) the **stratum basale**. The stratum basale and the stratum spinosum are combined and called the **stratum germinitivum**.

When you stop to think of it, your skin is constantly being subjected to various types of frictions and abrasions (mostly very mild). The wind, normal movement of your clothes over the surface of your skin, showers and bathing are just some of the daily activities that rub away our skin. Having this happen creates an obvious problem for you. How do you replace the skin that gets rubbed away ?

It is the stratum corneum that is worn away by friction and which must be replaced. The replacement of the worn away cells of the stratum corneum is done by migration of cells from the lucidum or granulosum into the corneum. The cells of the lucidum that migrate up into the corneum must be replaced and these replacements come out of the granulosum, which of course leaves the granulosum deficient in cells. The cells of the granulosum are replaced by cells migrating up from the spinosum and basale. The cells of these two strata that migrate out into the granulosum are replaced by cell divisions among the cells of the spinosum and basale. So, ultimately, the cells in the stratum corneum have been generated by the cells of the stratum germinitivum, hence its name. As the cells migrate up through the epidermis to assume their position on the surface they are in the process of dying, and the outermost cells of the stratum corneum are composed of dead cells.

Below the epidermis is the thicker **dermis**, which is composed of loose connective tissue, blood vessels, lymphatic vessels, hair follicles, and sweat glands. You might want to consider how closely associated you skin is to your brain inasmuch as you obtain an enormous amount of information about your environment through your skin. In your dermis are located the receptors for the general sensation (free nerve endings for **pain** and **temperature**, Pacinian corpuscles for **pressure**, and Meissner's corpuscles, hair follicle plexuses, and Merkel's disks for **touch**).

Below the dermis is the **hypodermis**, or **subcutaneous layer**, a layer that is thicker and better supplied with blood vessels and nerves than is the dermis. Also this layer contains the subcutaneous fat of the body. However, the hypodermis (subcutaneous layer) is not considered part of the skin.

OBJECTIVES: The objectives of this laboratory are to:

1. Be able to identify and correctly spell the structures on the skin model

PRELABORATORY PREPARATION: You should complete this worksheet before you come to lab. It will allow you to work more efficiently and to complete your work earlier and with less confusion. By using the diagram of the skin given under the **EXERCISE** section, fill in the following blanks:

Major areas of the skin
1____________________________
2____________________________

Structures in the dermis
a. Receptors for general sensations
1____________________________
2____________________________

b. Glands
1____________________________
2____________________________

Strata (layers) of the epidermis
1____________________________
2____________________________
3____________________________
4____________________________

c. Smooth muscle attached to hair
1____________________________

d. Other structures of dermis
1____________________________
2____________________________
3____________________________
4____________________________

EXERCISE:

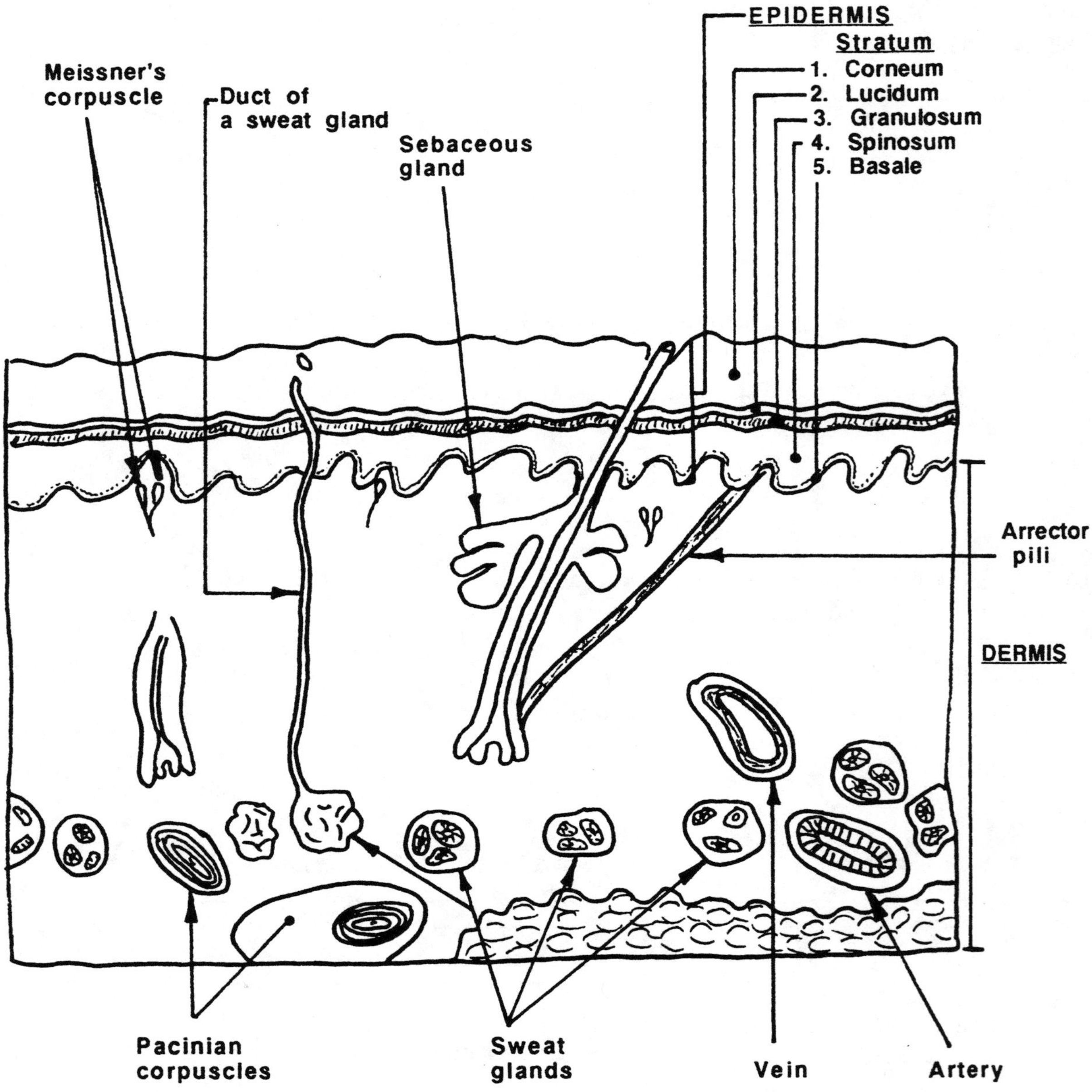
EPIDERMIS
Stratum
1. Corneum
2. Lucidum
3. Granulosum
4. Spinosum
5. Basale
Meissner's corpuscle
Duct of a sweat gland
Sebaceous gland
Arrector pili
DERMIS
Pacinian corpuscles
Sweat glands
Vein
Artery

QUESTIONS TO ANSWER AND THINK ABOUT (Some of these questions can be answered directly from looking at the skin model, but some will require that you consult your lecture notes and reading assignments):

1. What are the two major areas of the skin ?

2. What are the five layers (strata) of the epidermis ?

3. Which stratum of the epidermis is found only in the skin of the palms of the hands and the soles of the feet ?

4. What stratum of the epidermis contains dead cells ?

5. What are the functions of the skin ?

6. What structure of the dermis participates in thermoregulation ?

7. What layer of the epidermis participates in protection ?

8. What area of the skin participates in reception of general sensations ?

9. In what areas of the skin is skin the thickest ?

10. What is the function of the sebaceous gland ?

11. What type of muscle is the arrector pili ?

13. What happens to a hair when its arrector pili contracts ?

14. Describe how the cells of the stratum corneum that are worn away are replaced.

15. Why is the stratum germinitivum so named ?

16. Is the hypodermis part of the skin ?

17. How does sweat made by the sweat glands get out onto the surface of the epidermis ?

18. What is the receptor in the dermis for the sensation of pressure ?

19. What is the receptor in the dermis on the skin model for the sensation of touch ?

20. What is the formula for % chances of survival in a non-burn center ?

21. What is the relationship between age and the chance of survival with skin loss ?

WHAT YOU SHOULD KNOW AND BE ABLE TO DO AFTER THESE LABORATORY EXERCISES

1. Be able to know the structures, areas, etc. on the skin model.

2. **Be able to answer questions concerning lecture and reading assignment information about these structures on the skin model.**

LABORATORY 13 - SPECIAL SENSES

THE EYE

INTRODUCTION: Let's begin with just a few words about **electromagnetic waves (radiation)**, because this is what your eyes are for - the detection of electromagnetic radiation. Electromagnetic waves are everywhere. The waves have a **wavelength**, and as it turns out, the **energy** in electromagnetic radiation is dependent, in a reverse way, on the wavelength. The shorter the wavelength, the greater the energy.

Over the centuries we have learned a lot about electromagnetic radiation. Radiation with very short wavelengths are called **X-rays**, and they have so much energy that they can penetrate through your body, doing great damage to the DNA in your cells. Your skin offers absolutely no protection against this radiation. Only your bones are thick enough to stop an X - ray, and that is why bones show up on X-ray photographs. It seems perfectly odd that you cannot feel these X-rays go through your body. Less powerful, but still highly dangerous, is **ultraviolet (UV) radiation**. UV radiation is not energetic enough to penetrate through to the organs of your body, but nevertheless it still has enough energy to do enormous damage to the epidermis of your skin (a sunburn or tan is evidence that you have been exposed to substantial UV radiation). You cannot "see" X-rays, or UV radiation, but they are there around you nevertheless. You can detect ("see") electromagnetic radiation that is less energetic that UV radiation. The radiation that you see with your eyes is called **"light"**. The wavelengths in light are not damaging to the cells and tissues of your body. You eyes can detect electromagnetic radiation of wavelengths between about 380 nanometers and 800 nanometers. Light of a wavelength of 380 nanometers is **blue**, while that of 800 nanometers is **red**. In between are wavelengths that are **green** and **yellow**. Once above 800 nanometers we get into the **infrared (IR)** range and your eyes cannot detect these wavelengths either, but you skin will notice these wavelengths as heat. Past the IR are the waves for **radio** and **television**. You of course cannot see or even feel these waves as heat, but you know that they are present around you when you turn on your radio or television.

In essence, what this means is that all sorts of electromagnetic radiation are moving through your environment. Your eyes cannot detect all wavelengths of this radiation, only a very narrow range, a narrow range that we refer to as "light". Nonetheless, although you cannot "see" all of the electromagnetic radiation that's out there, some of that radiation can do you great harm.

Your eyes are truly magnificent organs, and if you stop to think about how much information you gather in from the electromagnetic radiation reflected from the objects in your environment, then you begin to understand the importance of vision. Vision is so important that a whole lobe of your brain (the occipital lobe) is devoted to receiving and analyzing visual information sent to it from the retina by way of your optic nerves. No other sensation claims so much brain space.

There are three **tunics (layers)** of the eye, and in going from the outermost to the innermost they are 1) the **sclera**, 2) the **choroid**, and 3) the **retina**. Actually, you can very easily see the sclera of your own eye, when you look in the mirror, or someone else's eye. The sclera is tough, thick, and opaque and it is difficult to imagine that it blends in with the **cornea**, which is thin, fragile, and clear. The choroid is the layer (tunic) of blood vessels, and the retina is the layer (tunic) of the two types of **photoreceptors,** the numerous (130 million / retina) **rods**, and the less numerous (7 million) **cones.**

Although it may be difficult to appreciate on the eye models, the eye is nevertheless divided into two major spaces called **cavities**. Posterior to the **lens** is the large **posterior cavity**. In life, this cavity is filled and inflated with a jelly-like material called **vitreous humor**, and for this reason the posterior cavity is also called the **vitreous chamber**. Anterior to the lens is the much smaller space called the **anterior cavity**. A complication is added to the anterior cavity because the iris is in it, and what the iris does is to divide the anterior cavity into two smaller spaces, the **anterior chamber**, and the **posterior chamber.** The anterior chamber is between the cornea and the iris, and the posterior chamber is between the iris and the lens. These two small chambers communicate through the opening in the opening in the iris called the **pupil**, and are filled with a watery-like fluid termed **aqueous humor**, which inflates and adds fluid pressure to the anterior cavity. You should be certain that you don't get cavities and chambers confused in your thinking. It is easy to do so, so be careful.

In order to focus an image on your retina the shape of the lens must be altered. This alteration is accomplished by the **ciliary muscle** operating through the **suspensory ligament of the lens**.

At the back of your eye are two small areas of the retina that are important. One is the **fovea centralis**, which contains cones only (no rods), and the other is the **optic disk** (**blind spot**), which contains neither rods, nor cones. The optic disk is the place where the retina forms the **optic nerve,** which passes through the optic foramen so that it can attach to your brain. It is of course through your optic nerve that visual impulses from your rods and cones in your retina get to your brain for your conscious awareness.

You move you eyes with great refinement and precision. If you are reading these words then that act alone is a great testimony of just how refined and precise these movements can be. Like all movements in other parts of the body, you eye movements are accomplished by muscles. There are two sets of muscles for the eye, so let's say something about each set. One set of three muscles is inside the eye and for that reason these three muscles are called **intrinsic eye muscles.** You have already met one of these, the **ciliary muscle**, which functions to change the shape of your lens. The other two are located in the iris and function to regulate the size of the opening of your pupil. These two muscles are the **sphincter pupillae**, which constricts the pupil, and the **dilator pupillae**, which dilates the pupil. You should be aware that all three of the intrinsic eye muscles are **smooth (involuntary) muscles**. The other set of seven muscles are outside your eye and for that reason these muscles are called the **extrinsic eye muscles.** Six of the seven extrinsic eye muscles are attached to the sclera of the eye, and one enters the upper eyelid (superior palpabra) and has nothing to do with movements of the eye. The six muscles that move the eye are the 1) **superior rectus** (rotates the eye upward to look at the ceiling or sky), 2) **inferior rectus** (rotates the eye downward to look at the ground or floor), 3) **lateral rectus** (abducts the eye to look laterally), 4) **medial rectus** (rotates the eye medially, toward the nose), 5) **superior oblique** (rotates the eye downward and medially, and abducts the eye), 6) **inferior oblique** (rotates the upward and laterally, and abducts the eye) The **levator palpebra superioris** does not move the eye, and its function is to elevate your upper eyelid ("open you eyes"). All seven of the extrinsic eye muscles are **skeletal (voluntary) muscles**.

OBJECTIVES: The objectives of this laboratory are to:

1. Identify and correctly spell the structures of the eye models
2. Identify and correctly spell the intrinsic and extrinsic eye muscles that are on the eye models

PRELABORATORY PREPARATION: You should complete this worksheet before you come to lab. It will allow you to work more efficiently and to complete your work earlier and with less confusion. By using the diagrams under the heading of **EXERCISES**, fill in the blanks below:

Structures outside the eye
a. Extrinsic eye muscles

1____________________
2____________________
3____________________
4____________________
5____________________
6____________________
7____________________

b. Nerve

____

Cavities of the eye
1____________________
2____________________

Structure that separates anterior / posterior cavities

____

Structures of the anterior cavity
a. Structure that separates anterior / posterior chambers

____

b. Adjustable opening in the iris

____

c. Two intrinsic eye muscles in the iris
1____________________
2____________________

d. Clear, fragile structure that forms the anterior wall of the anterior chamber

____

Structures of the posterior cavity
a. Intrinsic eye muscle that controls the lens' shape

____

b. Ligament that attaches to the lens

____

c. Structures of the retina
1____________________
2____________________

EXERCISES:

1. Non-detachable eye models*

These models are good for seeing external features of the eye and for six of the seven extrinsic eye muscles. The internal structures, such as the cavities, chambers, the two innrmost tunics, etc. cannot be seen on these models.

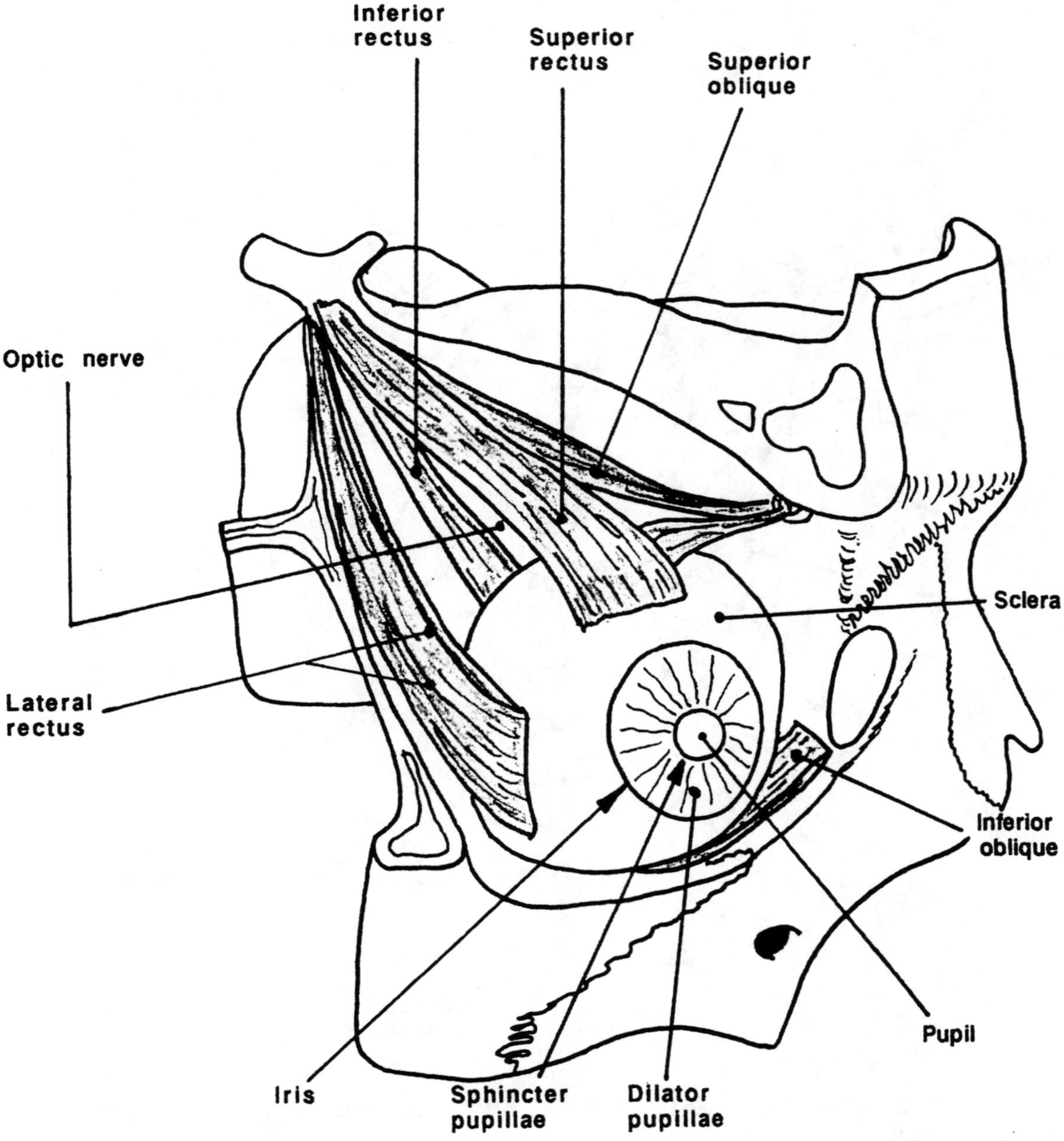

*There are two structures that are on the model but which are not in the diagram. These are the medial rectus muscle and the cornea. Be sure that you find and identify these two structures.

2. Detachable eye models

There are two detachable models and the difference is that on one there is a levator palpebra superioris.

a. With the levator palpebra superioris

1) Outside structures

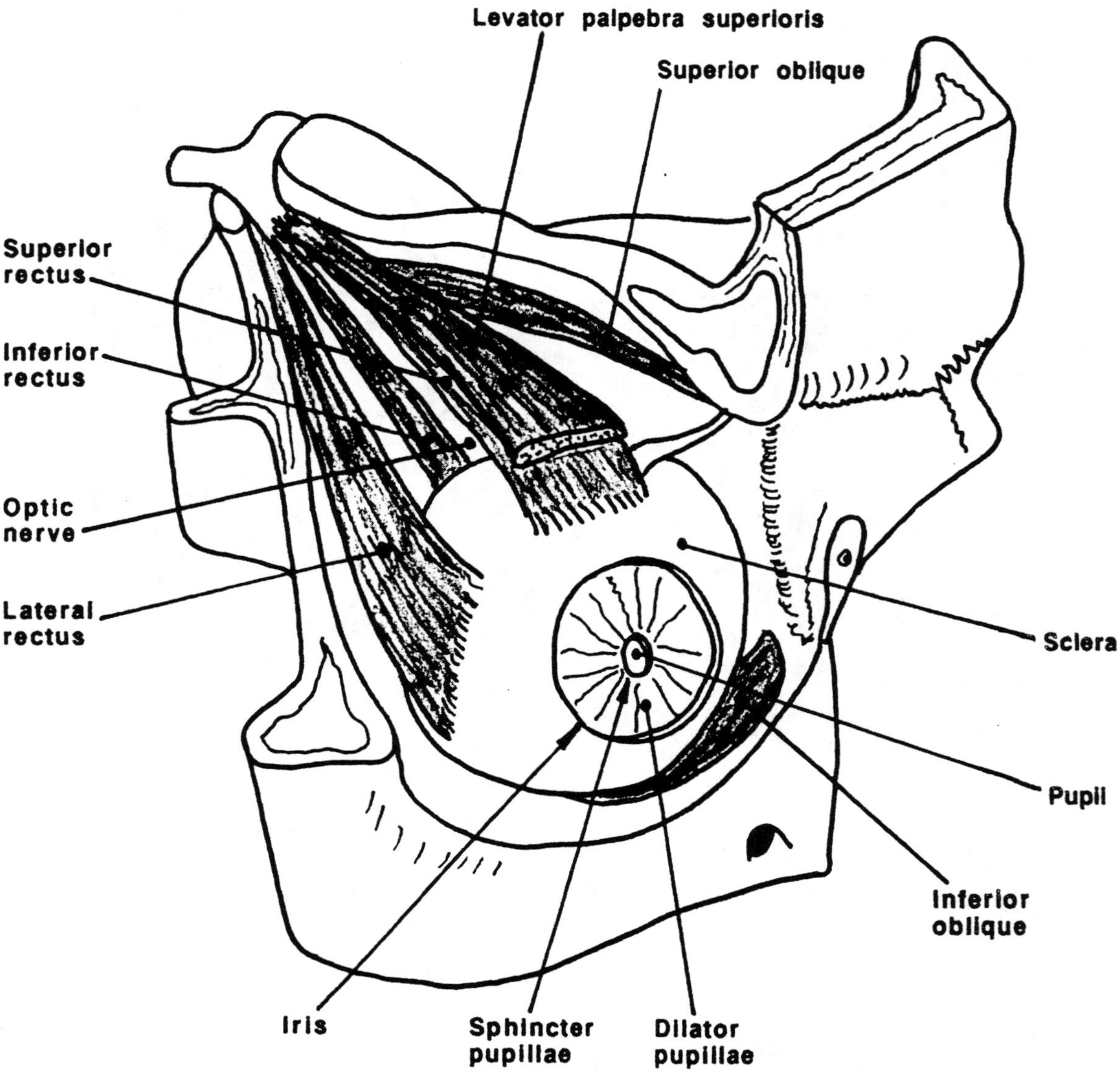

2) Detach the model and find the following structures, which are not shown on the above diagram

a) Choroid
b) Retina
c) Fovea centralis
d) Optic disk
e) Ciliary muscle (body)
f) Suspensory ligament
g) Lens
h) Posterior cavity (Vitreous chamber)
i) Anterior chamber

b. Without the levator palpebra superioris

1) Outside structures

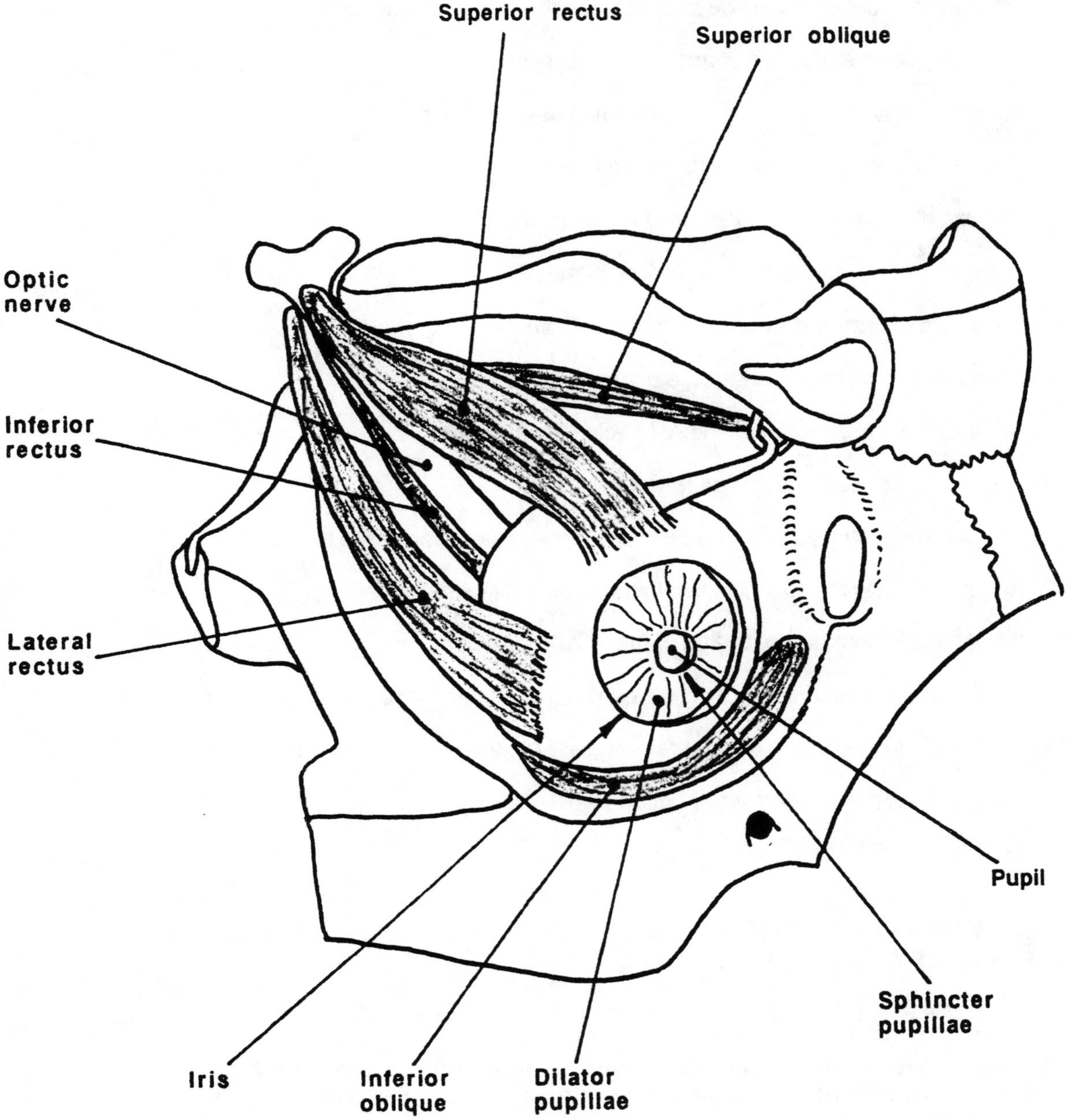

b) Detach the model and find the following structures, which are not shown on the diagram

a) Choroid
b) Retina
c) Fovea centralis
d) Optic disk
e) Ciliary muscle (body)
f) Suspensory ligament
g) Lens
h) Posterior cavity (Vitreous chamber)
i) Anterior chamber

<u>QUESTIONS TO ANSWER AND THINK ABOUT</u> (Some of these questions can be answered directly from looking at the eye models, but some will require that you consult your lecture notes and reading assignments)

1. What tunic of the eye can you see on yourself or looking at someone else ?
2. What four other structures of the eye can you see ?
3. What foramina in the skull do the optic nerves go through ?
4. What is the name of the cavity that each eye is located in ?
5. The extrinsic eye muscles are what type of muscle ?
6. The intrinsic eye muscles are what type of muscle ?
7. What do the terms extrinsic and intrinsic refer to ?
8. What two muscles control the size of the pupil ?
9. Are these two muscles involuntary or voluntary muscles ?
10. In what structure are these two muscles located ?
11. Of the seven extrinsic eye muscles, which one is not attached to the eye ?
12. Are the extrinsic eye muscles involuntary or voluntary muscles ?
13. What muscle controls the shape of the lens ?
14. Is this muscle an intrinsic or extrinsic eye muscle ? A voluntary or involuntary muscle ?
15. The optic disk and the fovea centralis actually have something in common. What is it ?
16. What tunic of the eye forms the optic nerve ?

WHAT YOU SHOULD KNOW AND BE ABLE TO DO AFTER THESE LABORATORY EXERCISES

1. Be able to identify the structures of the eye models.
2. **Be able to answer questions concerning lecture and reading assignment information about these structures on the eye models.**

LABORATORY 14 - SPECIAL SENSES

THE EAR

INTRODUCTION: Like your eyes, your ears allow you to obtain information about events that are going on perhaps far away from your body. Also like your eyes, your ears detect waves, except not light waves, but compressional waves of the molecules in the air. If the compressional waves that enter your ear are of a high frequency, then you "hear" a high pitched sound. Compressional waves of low frequency are heard as low pitched sound. Like your eyes which cannot see all electromagnetic (radiation) waves, your ears cannot detect all compressional waves that are travelling through the air. Just as honeybees can see ultraviolet light and we humans cannot, dogs can hear higher pitched sounds than we can. More than likely, there are other animals that can detect lower pitched sounds than the human ear.

What kind of event produces compressional waves ? Actually, it is **movement**. Any object that moves disturbs the molecules in the air and creates compressional waves for our ears to detect. If there were no movement, then there would be no compressional waves produced and we would "hear" nothing. So, when you "hear sounds" your brain is registering the fact that something out in your environment has moved.

Compressional waves of the atmosphere travel much slower than light waves (1100 feet per second for sound as opposed to 186,000 miles, or about 1 billion feet, per second for light; light is therefore about 1 million times faster than sound). This difference in velocity is the reason that you, for instance, see lightening before you hear the thunder. Moreover, unlike light waves, compressional waves require that there be molecules present in the air. If there are none, as in a vacuum, or outside the upper atmosphere of the earth, then there can be no compressional waves. You could talk all you wanted to in a vacuum, but there would be no waves set up and no one, including you, could hear you.

What is sound, anyway ? Let's define **sound** , for the sake of that age - old question, as a conscious awareness of any animal's brain that a compression of the molecules in the atmosphere has occurred. This definition of sound allows us to give separate definitions to a compressional wave and to sound. So, if a tree falls in the forest and there is no organism to detect the compressional waves created, then does it make a sound ? What do you think ?

The human ear is divided into three major areas, termed simply 1) the **external ear**, 2) the **middle ear,** and 3) the **inner ear**. You can actually see part of the external ear, but some parts of the external ear, and all of the middle and internal ears are in temporal bone and cannot be seen.

The external ear is made up of a cartilaginous flap, called the **pinna**, one on each side of your head. The function of the pinna is to gather in the compressional waves in the air and direct them down the **external auditory canal**. At the end of the short external auditory canal is your thin and flexible **tympanic membrane** (commonly referred to as the "ear drum"). The tympanic membrane is shaped like an open umbrella with the concave surface facing the external auditory canal. Actually, the tympanic membrane is a partition between the external and middle ears and for that reason it is difficult to designate the position of this membrane. However, most texts place the tympanic membrane with the middle ear. Therefore, your external ear has two structures: 1) the **pinna**, and the 2) the **external audutory canal**.

On the medial side of the tympanic membrane is the very small cavity called the **middle ear cavity**. In each of your middle ear cavities are the three smallest bones in your body. These bones are so tiny that they are referred to as the **ear** **ossicles** ("little bones"). The three ear ossicles connect the tympanic membrane to the oval window, which is an opening that connects the middle ear cavity to the vestibule, a space in the inner ear. In going from the tympanic membrane to the oval window, the order of ear ossicles is 1) the **malleus**, 2) the **incus**, and 3) the **stapes**. The malleus is attached to the tympanic membrane and the stapes covers the oval window. The incus forms synovial (diarthrotic) joints with both the malleus and stapes.

Without a doubt you have had the experience many times of changing your altitude (Driving up or down a hill or mountain, during the landing of an airplane that you are in, riding either up or down in an elevator, etc). You may recall that during the change in altitude that you had to move your mandible in some way in order to "clear your ears". What were you really doing, and why did you need to "clear your ears" ? It is necessary that you keep the pressure on either side of your tympanic membrane the same. That is, you need to have the same air pressure on the external auditory canal side as on the middle ear cavity side of your tympanic membrane. Changing altitude increases or decreases the air pressure on the external auditory canal side. So, how do you equalize the new air pressure on the other (middle ear cavity) side ? Fortunately, there is a tube, called the **auditory (eustachian) tube** that leads from the middle ear cavity to the **nasopharynx** . The purpose of the auditory tube is to allow you to get air into and out of the middle ear cavity in order to equalize the air pressure on either side of the tympanic membrane. The auditory tube is normally closed (collapsed) and so that is why you have to quickly move your mandible around in space in a complicated manner, or yawn, because it is the movement that temporarily opens your eustachian tube to let air into or out of your middle ear cavity. That's the good news. The bad news is that the auditory tube also allows bacteria to migrate from the pharynx up into the middle ear cavity thereby producing infection and inflammation of the middle ear (**Otitis media**).

The **inner ear** is complicated, but not impossible to understand. It is located in the petrous portion of the temporal bone, and has two **parts**: 1) the **bony labyrinth**, and 2) the **membranous labyrinth**. It is important to understand right away that the bony labyrinth is nothing more than a group of three interconnected spaces. One space is the **vestibule.** From one end of the vestibule there is a space that spirals around on itself called the **bony cochlea,** and on the other side of the vestibule is a series of three tubes called the **bony semicircular canals**. Therefore, the bony labyrinth is composed of three interconnecting spaces called the vestibule, the bony cochlea, and the bony semicircular canals.

Inside the spaces of the bony labyrinth are some fragile, membranous structures that collectively make up the membranous labyrinth. Inside the space called the vestibule are two small, membranous bags called the **saccule** and the **utricle**. Each of these gives off tubes. The saccule gives off a single tube, called the **membranous cochlea**, that coils around inside the bony cochlea. The utricle gives off three sets of membranous tubes, called the **membranous semicircular canals.** The membranous labyrinth is therefore a set of fragile, membranous structures that are contained inside the spaces that we call the bony labyrinth. In order to prevent the membranous labyrinth structures from collapsing, they are filled with a fluid, the **endolymph**. The spaces of the bony labyrinth are filled with a fluid, the **perilymph**, in which the membranous structures are floating.

The function of the membranous cochlea is for **audition (sound)**, while that of the saccule, utricle, and semicircular canals is **equilibrium (balance)**. These functions are conducted to the brain over the **vestibulocochlear nerve** (Cranial nerve VIII). In order to get from the inner ear to the brain this nerve must go through the internal auditory meatus.

OBJECTIVES: The objectives of this laboratory are to:

1. Be able to identify and correctly spell the structures of the external, middle, and inner ears on the ear model

PRELABORATORY PREPARATION: You should complete this worksheet before you come to lab. It will allow you to work more efficiently and to complete your work earlier and with less confusion. By using the diagram of the ear model given under the **EXERCISE** section, fill in the following blanks:

Areas of the ear
1____________________________
2____________________________
3____________________________

External ear structures
1____________________________
2____________________________
3____________________________

Middle ear structures
a. Bones
1____________________________
2____________________________
3____________________________

b. Tube

____

Inner ear structures
a. Spaces that make up the bony labyrinth *
1____________________________
2____________________________
3____________________________

b. Membranous labyrinth structures
1____________________________
2____________________________
3____________________________

c. Nerve of the inner ear

____

* These spaces may not be readily apparent, but try to find them. If you have trouble, then please do ask your Laboratory Instructor for help

EXERCISE:

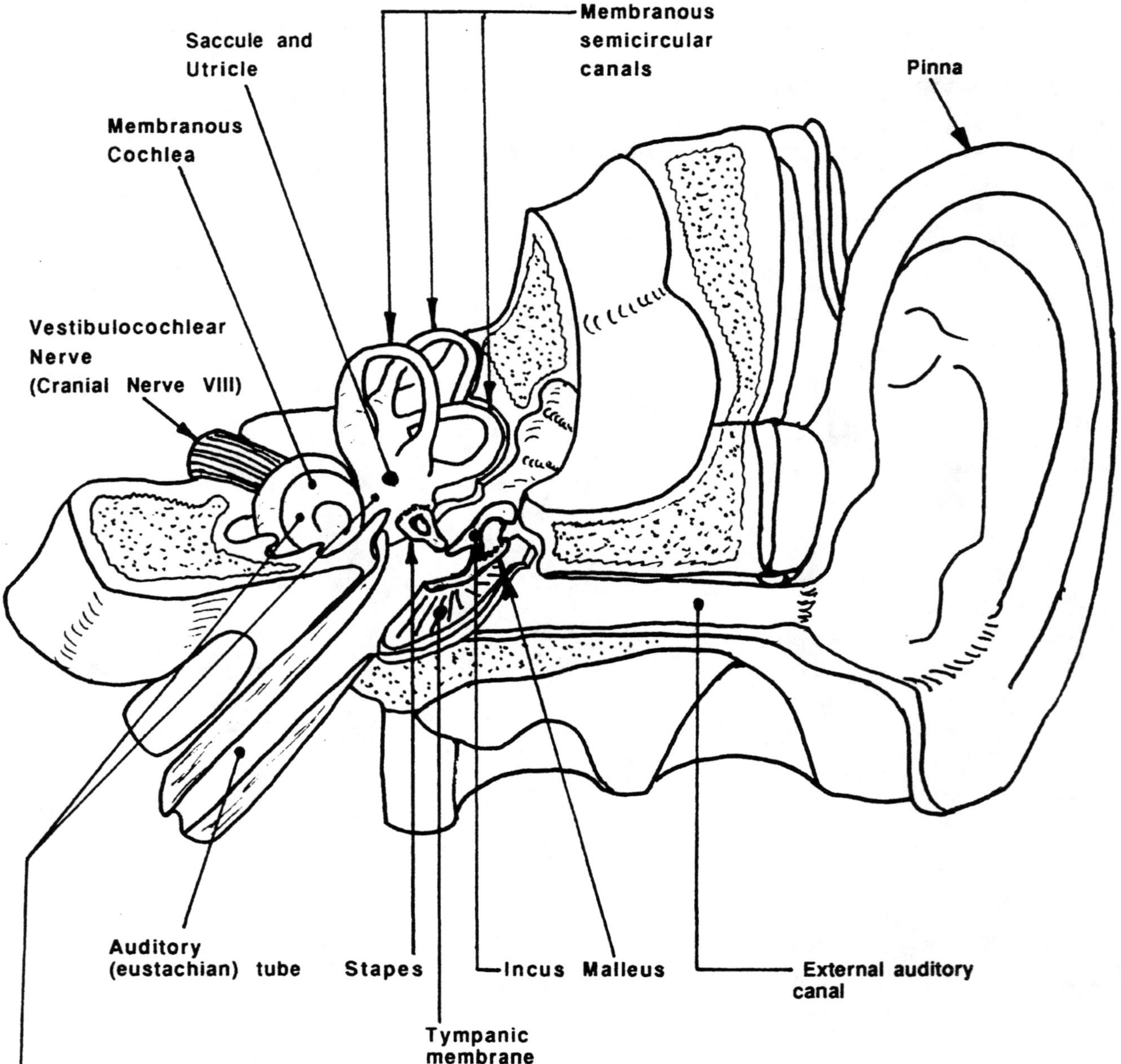

Remove these membranous structures and identify the three spaces of the bony labyrinth

1. Bony cochlea
2. Vestibule
3. Bony semicircular canals

QUESTIONS TO ANSWER AND THINK ABOUT (Some of these questions can be answered from looking at the ear models, but some will require that you consult your lecture notes and reading assignments)`

1. If a tree falls in the forest and there is no living organism present, then is a sound produced ?

2. What are some differences between light waves and sound waves ?

3. What event must occur in order to produce a compressional wave ?

4. Can compressional waves be produced in a vacuum ? Why not ?

5. Why do you see a distant event before you hear it ?

6. Do all organisms hear the same range of compressional waves ?

7. What are the three major areas of the ear ?

8. Which structure of the external ear gathers in the compressional waves and directs them into the external auditory canal ?

9. What thin structure is at the end of the external auditory canal ?

10. Therefore, what are the three structures of the external ear ?

11. Name the three ear ossicles.

12. What are two good and acceptable names for the tube of the middle ear that connects the nasopharynx to the middle ear cavity ?

13. What is the purpose of the auditory (eustachian) tube ?

14. What is one bad consequence of having an auditory tube ?

15. What are the four structures associated with the middle ear ?

16. What does the term ossicle mean ?

17. What are the two major parts of the inner ear ?

18. What is the bony labyrinth ?

19. What are the names of the three interconnecting spaces that make up the bony labyrinth ?

20. List the four structures of the membranous labyrinth.

21. In what space of the bony labyrinth are the saccule and utricle located ?

22. What tube comes out of the saccule ? In what space of the bony labyrinth is this tube located?

23. What tubes come out of the utricle ? In what space of the bony labyrinth are these tubes located ?

24. What is the name of the fluid inside the membranous labyrinth structures ?

25. What is the name of the fluid in the spaces of the bony labyrinth and in which the membranous labyrinth structures are floating ?

26. In what function does the membranous cochlea participate?

27. In what function do the saccule, utricle, and semicircular canals participate ?

28. What are two good and acceptable names for the nerve that conducts auditory and equilibrium information from the inner ear to the brain ?

WHAT YOU SHOULD KNOW AND BE ABLE TO DO AFTER THIS LABORATORY EXERCISE:

1. Be able to identify and correctly spell the indicated structures of the ear.

2. **Be able to answer questions from your lecture notes and reading assignments concerning these structures of the ear.**

LABORATORY 15 - SYSTEMS

THE NERVOUS SYSTEM - SPINAL CORD, SPINAL NERVES, AND PERIPHERAL NERVES

INTRODUCTION: There are all sorts of events occurring in your environment. Some of these events are dangerous and can cause you great harm, and even death. Some events are harmless, some represent food and nourishment, and some represent reproductive opportunity. In order for you to survive the events in your environment and understand them and to take advantage of them, you must first **DETECT** these events (performed by the **receptors** in your skin, retina, membranous cochlea, tongue, and nose), **ANALYZE** the sensations created by these stimuli (performed by the neurons in specific areas of your **cerebral cortex**), and then **RESPOND** to the events (performed by your **effectors - muscles and glands**). Detection, analysis, and response are three things that you must constantly do because events are going on constantly in your environment. When you stop to think about it, the order of what you must do is a real clue to the construction and function of your nervous system.

One question that you will be able to answer while looking at the structures of the spinal cord, spinal nerves, and peripheral nerves is: **How do general sensations get from your receptors to the cerebral cortex of the postcentral gyrus of your brain ?** Another question that can be answered is: **How can your skeletal muscles be willfully contracted by the neurons in the cerebral cortex of the precentral gyrus of your brain ?**

In order to understand how stimulation of your receptors allows your brain to create sensations it is necessary to at least mention some of the characteristics of the nerve impulse and the function of synapses. A nerve impulse is an electrochemical event that travels very rapidly (200 - 280 mph) along the surface of a neuron. When a nerve impulse reaches the end of an **axon** it causes the release of neurotransmitter substances that are stored in the synaptic knobs (Boutons) into the small space, the synaptic cleft, between two neurons . These released neurotransmitter substances then diffuse very rapidly across the synaptic cleft space and come into contact with the next neuron and in doing so create a nerve impulse in that neuron.

As it turns out, **all receptors are attached to dendrites of what are termed first sensory neurons**. Stimulation of a receptor will initiate a nerve impulse in a first sensory neuron.

Let's set up three neurons that we will call, for convenience, **first, second, and third sensory neurons** (sometimes these neurons are referred to as first, second, and third order neurons). It is this set of three sensory neurons that convey, by their nerve impulses, sensory information from the receptors in your skin to the neurons in the cerebral cortex of your postcentral gyrus. How the axons, cell bodies, and dendrites are related to the peripheral nerves, and the structures of the spinal nerves and spinal cord is given in the diagram on the next page.

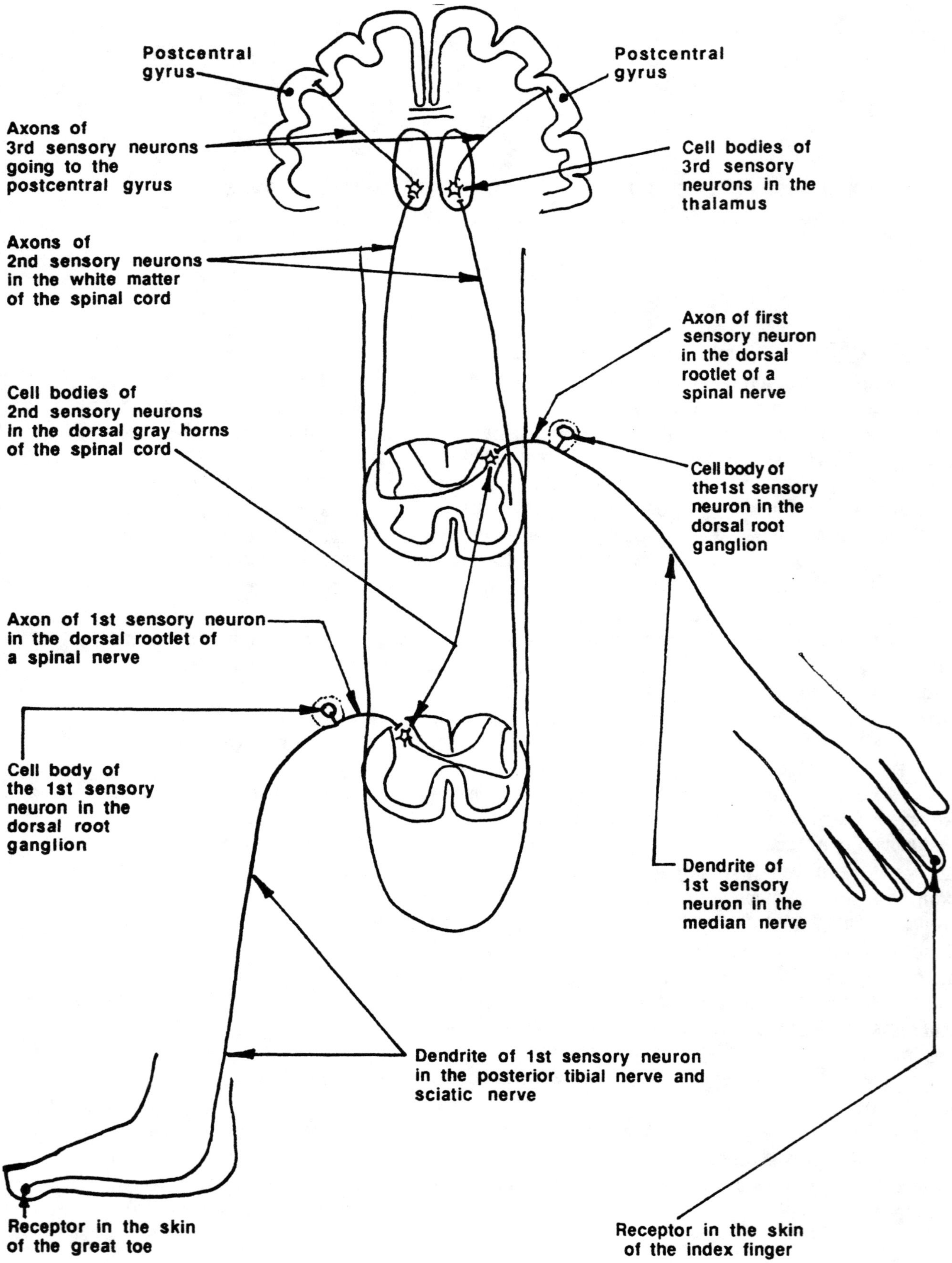
Postcentral gyrus
Postcentral gyrus
Axons of 3rd sensory neurons going to the postcentral gyrus
Cell bodies of 3rd sensory neurons in the thalamus
Axons of 2nd sensory neurons in the white matter of the spinal cord
Axon of first sensory neuron in the dorsal rootlet of a spinal nerve
Cell bodies of 2nd sensory neurons in the dorsal gray horns of the spinal cord
Cell body of the1st sensory neuron in the dorsal root ganglion
Axon of 1st sensory neuron in the dorsal rootlet of a spinal nerve
Cell body of the 1st sensory neuron in the dorsal root ganglion
Dendrite of 1st sensory neuron in the median nerve
Dendrite of 1st sensory neuron in the posterior tibial nerve and sciatic nerve
Receptor in the skin of the great toe
Receptor in the skin of the index finger

The **dendrites of your first sensory neurons** are located in all of your peripheral nerves (the median, posterior tibial and sciatic nerves serve as examples of peripheral nerves in the diagram), the **cell bodies of your first sensory neurons** are located in the dorsal root ganglia of your spinal nerves (there are only two ganglia shown in the diagram, but you have 31 pairs in your body), and the **axons of your first sensory neurons** make up the dorsal (posterior) rootlets of your spinal nerves. These axons penetrate the dorsal (posterior) aspect of your spinal cord and enter the dorsal gray horn where they synapse with the **cell bodies** and dendrites of your **second sensory neurons**. The **axons** of your second sensory neurons cross over to the oposite side of your spinal cord and ascend through the white matter, then up through the lower part of your brain and into the thalamus where they synapse with **cell bodies** and dendrites of your **third sensory neurons**. The axons of your third sensory neurons ascend through the white matter of your brain to synapse with neurons in the cerebral cortex of your postcentral gyrus.

If you take a look at the diagram and follow a nerve impulse from the receptor located in either the great toe, or the index finger, then you will notice that **between receptor and postcentral gyrus there are three sensory neurons**. That is true regardless of where a receptor is located in the body. If you follow a nerve impulse from a receptor, then you will notice that it travels over the first sensory neuron, is transferred to the second sensory neuron at the synapse between the two in the dorsal gray horn, and then travels over the axon of the second sensory neuron to the third sensory neuron, whose cell body is located in the thalamus. The nerve impulse generated in the third sensory neuron by the second sensory neuron travels over the axon of the third sensory neuron to neurons located in the cerebral cortex of the postcentral gyrus. Once the neurons in your postcentral gyrus are stimulated by the nerve impulses of the third sensory neurons you are consciously aware of two important things: **1)** the type of general sensation (pain, temperature, touch, or pressure), and **2)** exactly where on the body that stimulus came from. It is not understood at all how stimulating the neurons in your postcentral gyrus produces conscious awareness.

The nerve impulses that begin at the receptor in your skin reach your postcentral gyrus very rapidly because they move along the surface of neurons on the average of about 200 mph. If the distance between your toe and the top of your brain is about six feet, then how many seconds does it take to get nerve impulses from your foot to you brain for conscious appreciation?

The movement of nerve impulses from your receptors along the three sensory neurons is a **sensory function**. However, not all neurons in your nervous system are devoted to sensory functions. You have hopefully understood and answered the first question of how sensory information gets from your receptors to your postcentral gyrus

In your examination of the structures of the spinal cord, spinal nerves, and peripheral nerves, you can now answer the other question: **How do the neurons in your precentral gyrus cause the willful contraction of your skeletal muscles ?**

Contracting your skeletal muscles , because it involves movement, is called a **motor function**. Let's set up two neurons, called the **upper motor neuron** and the **lower motor neuron**, and see how the cell bodies and axons of these two neurons are anatomically related to the structures and areas that you will see on the spinal cord, spinal nerve, and peripheral nerve models. A diagram of these neurons and structures and areas are given on the following page.

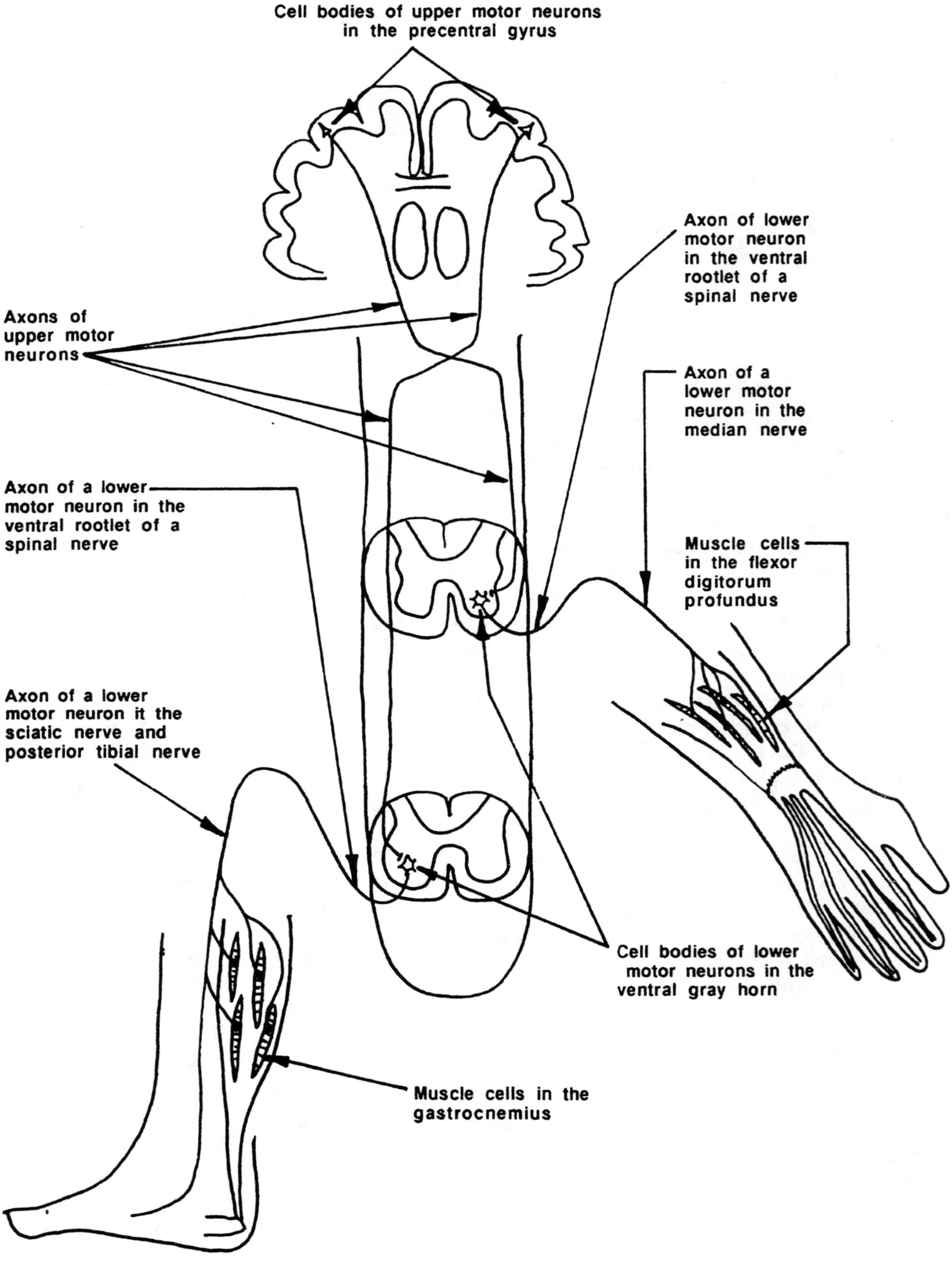

Cell bodies of upper motor neurons in the precentral gyrus
Axon of lower motor neuron in the ventral rootlet of a spinal nerve
Axons of upper motor neurons
Axon of a lower motor neuron in the median nerve
Axon of a lower motor neuron in the ventral rootlet of a spinal nerve
Muscle cells in the flexor digitorum profundus
Axon of a lower motor neuron it the sciatic nerve and posterior tibial nerve
Cell bodies of lower motor neurons in the ventral gray horn
Muscle cells in the gastrocnemius

The axons of your upper motor neurons pass through your brain and down into the spinal cord where they synapse with the cell bodies and dendrites of lower motor neurons located in the ventral gray horn of your spinal cord. The ventral gray horn is the location of the cell bodies of lower motor neurons. The axons of these neurons leave the ventral (anterior) aspect of your spinal cord and make up the ventral rootlet of a spinal nerve. This axon continues into a peripheral nerve (the medial, sciatic, and posterior nerves shown in the diagram are examples of peripheral nerves) and as a peripheral nerve passes by your muscles, such as the gastrocnemius or flexor digitorum profundus, axons of some of the lower motor neurons leave the nerve and penetrate into the muscle to connect to skeletal muscle cells that make up the muscle. The junction between the axon of a lower motor neuron and a skeletal muscle cell is called either a **neuromuscular junction**, or a **myoneural junction**.

When you want to willfully contract a group of muscle cells in one or more of your muscles, you start nerve impulses in your upper motor neurons which travel very rapidly (280 mph) down their axons to the synapses with the lower motor neurons located in the ventral gray horn. The nerve impulse is then initiated in your lower motor neurons which travel very rapidly (280 mph) out to your muscle fibers. When this nerve impulse reaches the muscle cells it causes them to contract.

Perhaps we can now summarize what is in some of the structures that you will see in laboratory. Each of the peripheral nerves that you will see (Axillary nerve, radial nerve, ulnar nerve, median nerve, femoral nerve, sciatic nerve, common peroneal nerve, posterior tibial nerve) are made up of millions of dendrites of first sensory neurons and axons of lower motor neurons. The same is true for the spinal nerve proper. Each dorsal root ganglion of a spinal nerve is composed of thousands of cell bodies of first sensory neurons. The dorsal rootlet of a spinal nerve is composed of thousands of axons of first sensory neurons, and in the dorsal gray horn are the cell bodies of second sensory neurons. The ventral gray horn is the location of the cell bodies of lower motor neurons. The ventral rootlet of each spinal nerve is made up of thousands of axons of lower motor neurons. White matter in the spinal cord is composed of axons of second sensory neurons and axons of upper motor neurons.

So far, we have discussed structures and areas in a functional way. Let's say some anatomical things about these structures. The **spinal cord** is divided into two general areas, gray matter, which is made up of cell bodies and the surrounding white matter, which is made up of sensory and motor axons. The gray matter of the spinal cord is further divided into dorsal, lateral, and ventral gray horns. Likewise, white matter can also be divided into dorsal, lateral, and ventral white matter areas. The spinal cord is in the **central nervous system** and therefore so are the components of its gray and white matters.

The spinal nerves are attached to the spinal cord through dorsal and ventral rootlets. The anatomical components of a spinal nerve include the two rootlets, the dorsal root ganglion and the spinal nerve proper.

Several spinal nerves proper merge to form the plexuses (brachial and lumbosacral) and peripheral nerves. The components of spinal nerves, plexuses, and peripheral nerves belong to the **peripheral nervous system**.

OBJECTIVES: The objectives of this laboratory are:

1. Be able to identify and correctly spell the peripheral nerves on the upper and lower limb models

2. Be able to identify and correctly spell the structures of the spinal nerve and the spinal cord on the spinal cord models

PRELABORATORY PREPARATION: You should complete this worksheet before you come to lab. It will allow you to work more efficiently and to complete your work earlier and with less confusion. By using the diagrams of the upper and lower limbs, and the spinal cord models given under the **EXERCISES** section, fill in the following blanks:

Peripheral nerves of the upper limb
1______________________________
2______________________________
3______________________________
4______________________________

Peripheral nerves of the lower limb
1______________________________
2______________________________
3______________________________
4______________________________

Spinal cord models
a. Gray areas of the spinal cord
1______________________________
2______________________________
3______________________________

b. White areas of the spinal cord
1______________________________
2______________________________
3______________________________

c. Fissures of the spinal cord
1______________________________
2______________________________

Spinal nerve components
1______________________________
2______________________________
3______________________________
4______________________________

EXERCISES:

1. Upper limb model, anterior view*

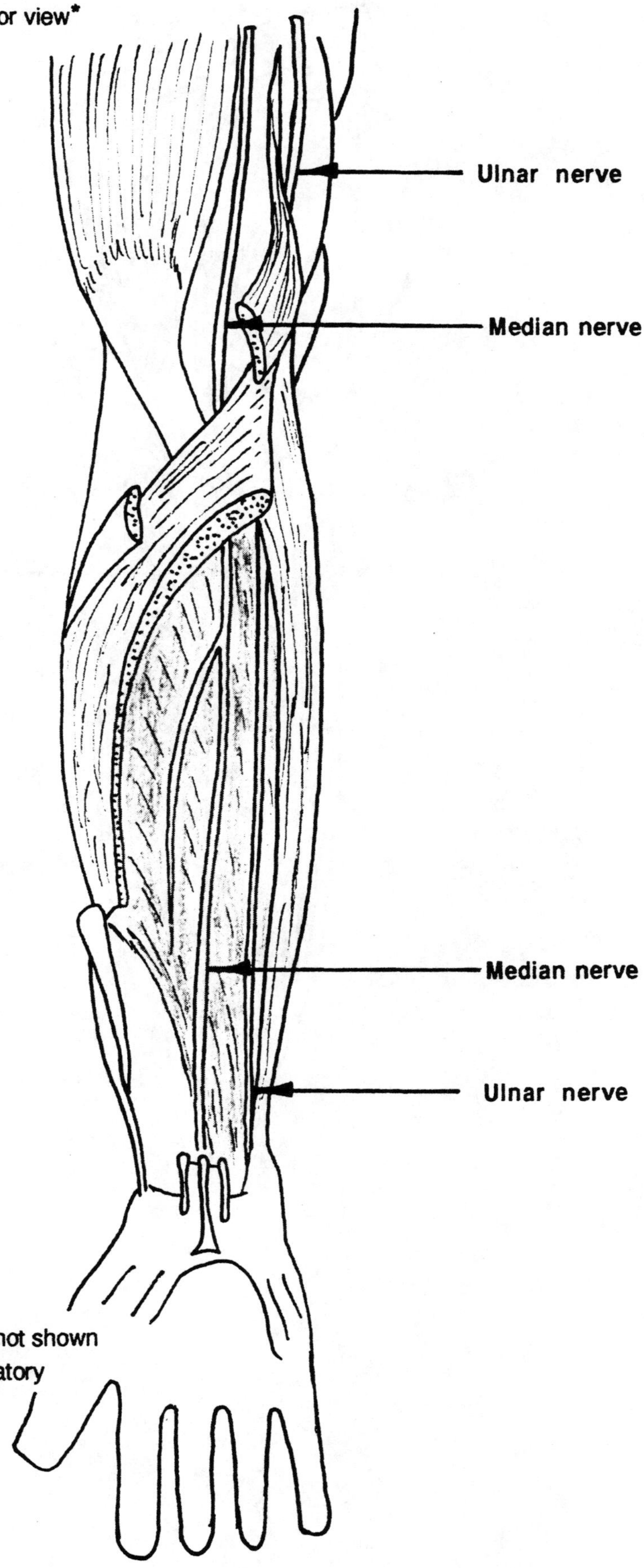

*The axillary and radial nerves are not shown in this diagram. Consult you Laboratory Instructor for their locations

2. Lower limb model, anterior view

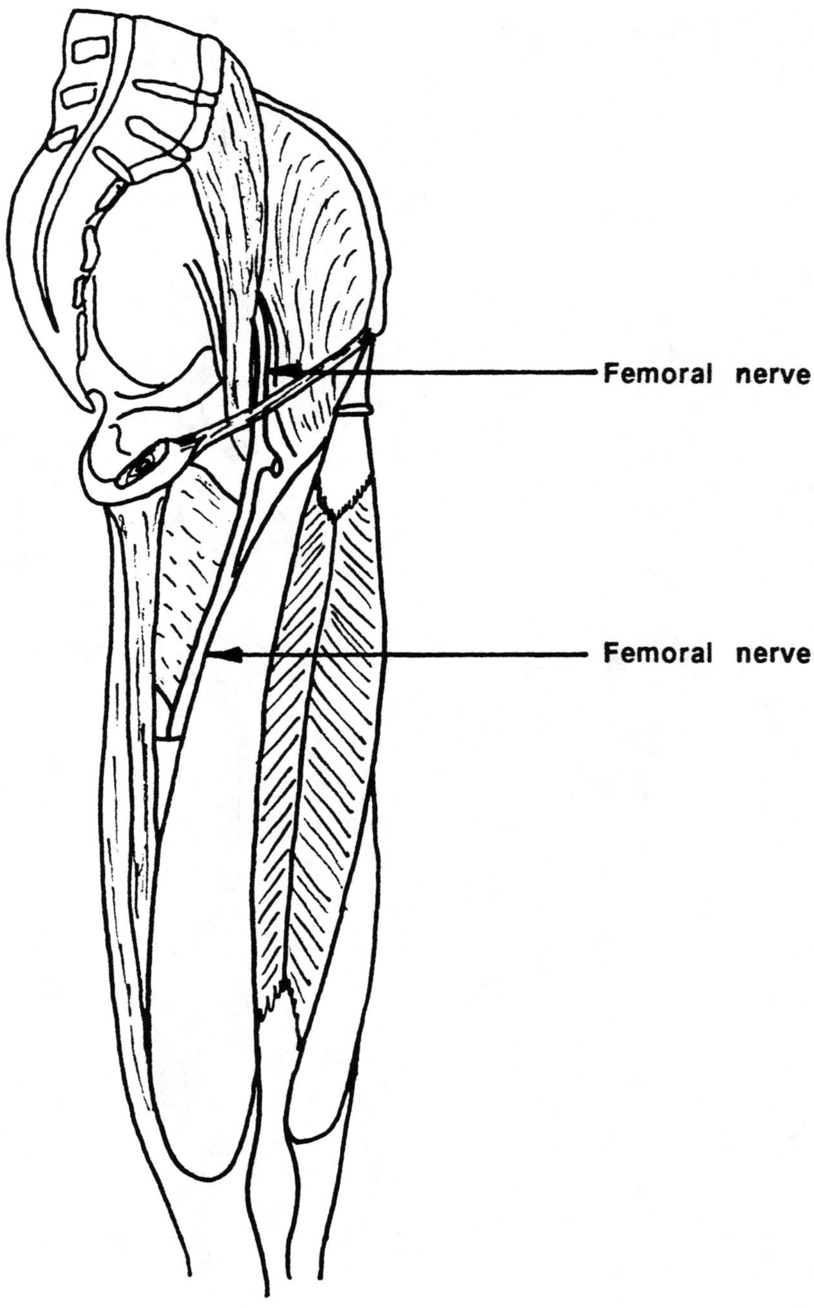

3. The lower limb model, posterior view

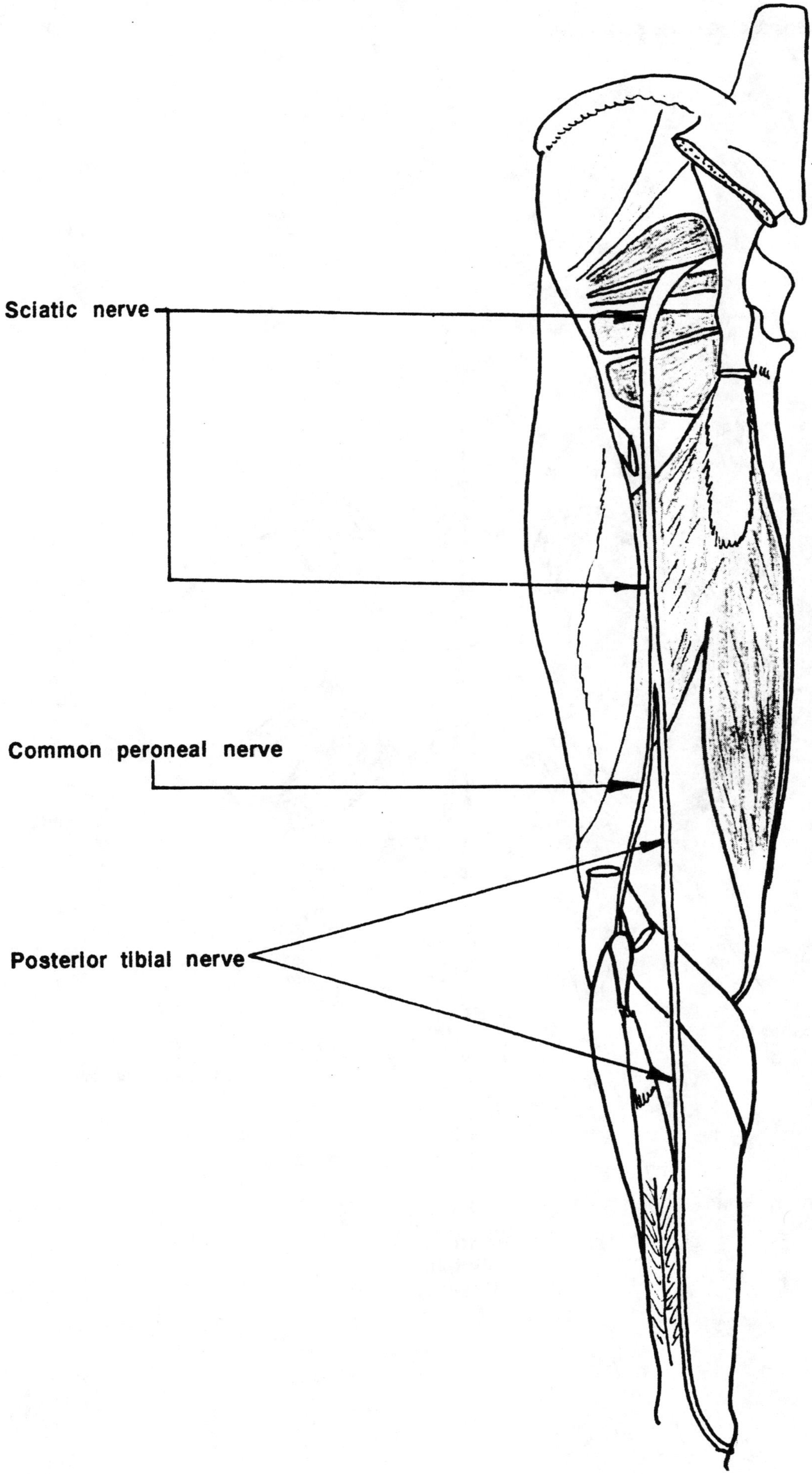

4. Spinal cord cross section model

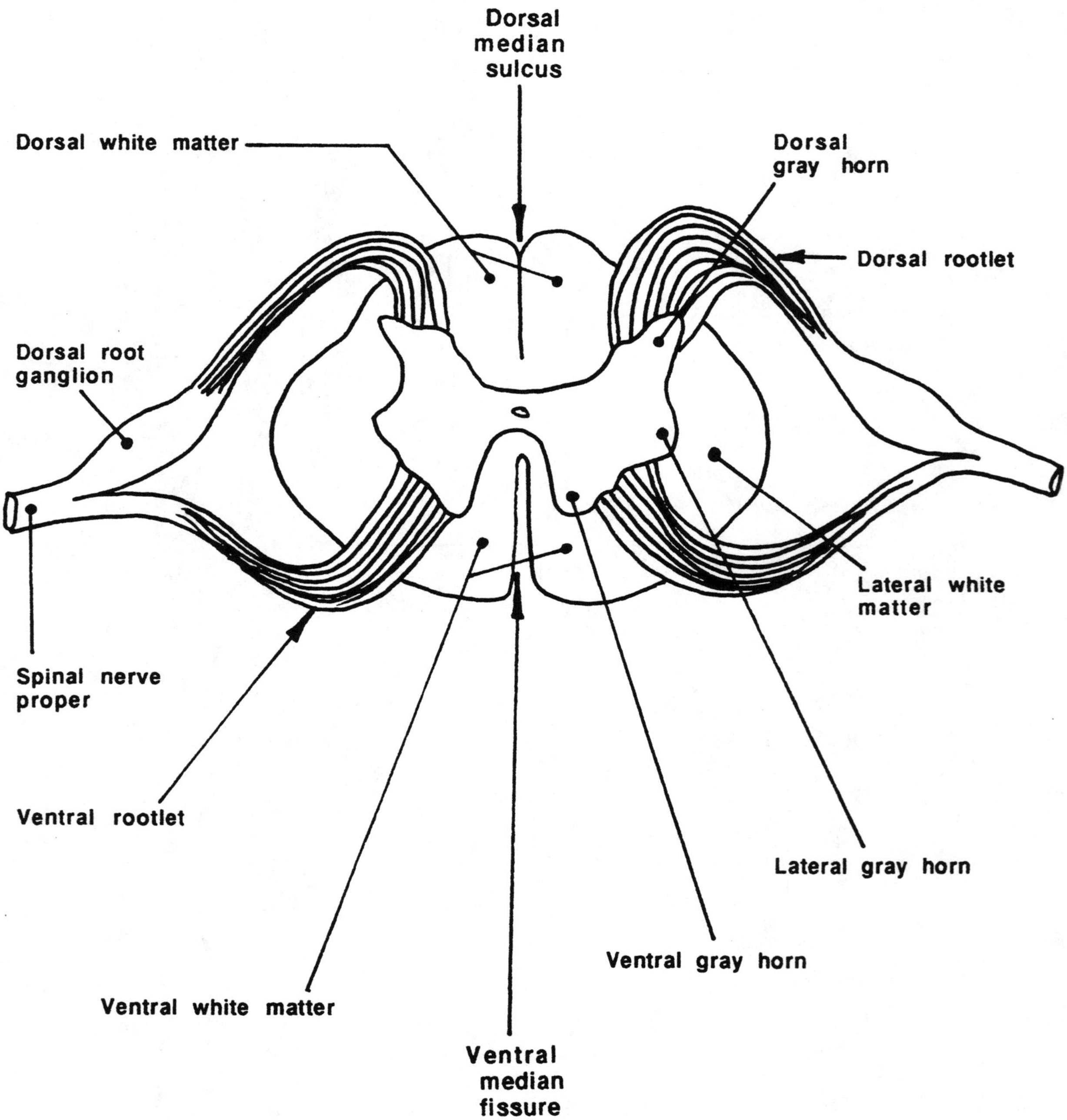

5. Spinal cord cross section model

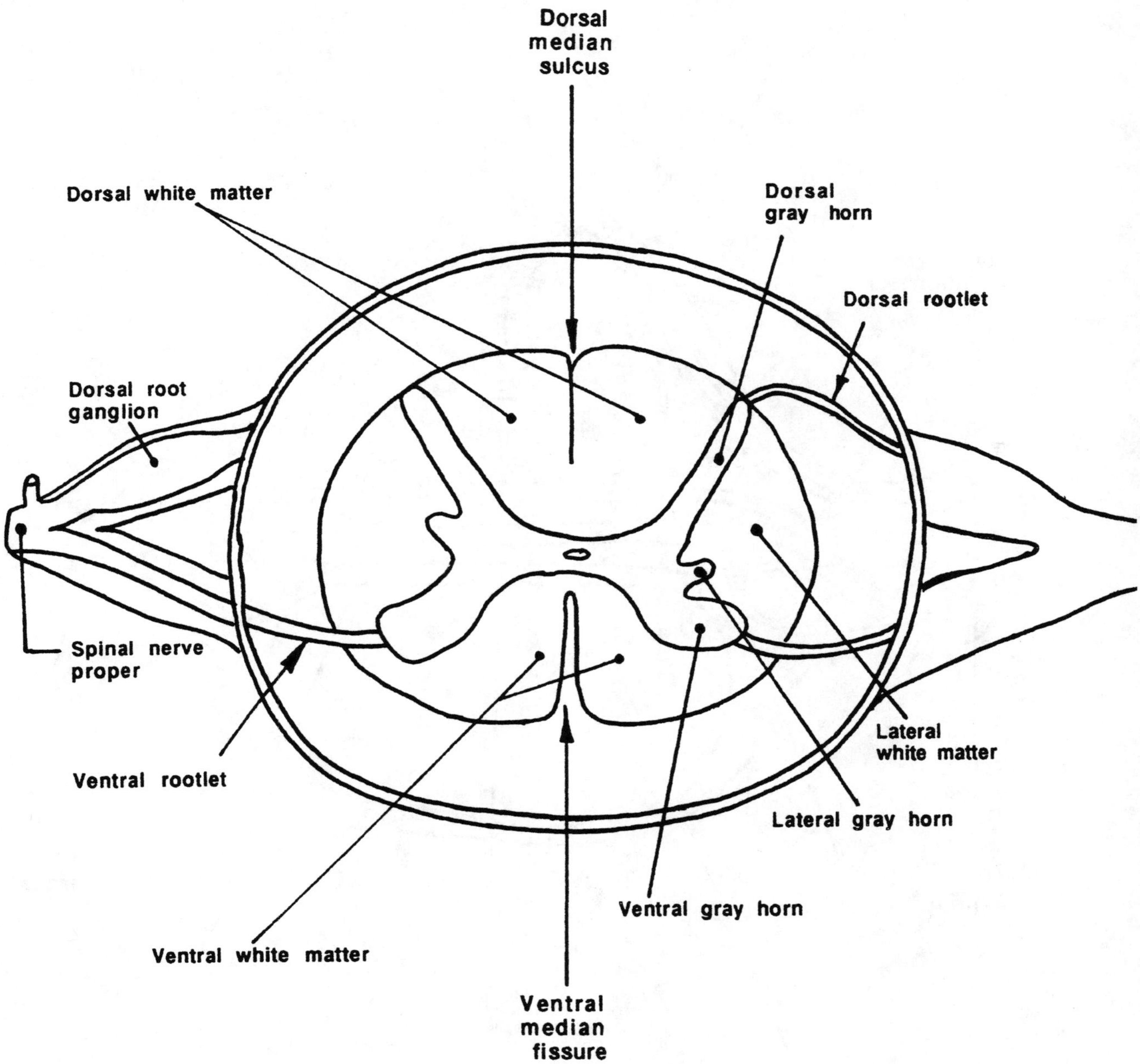

6. Spinal cord cross section model

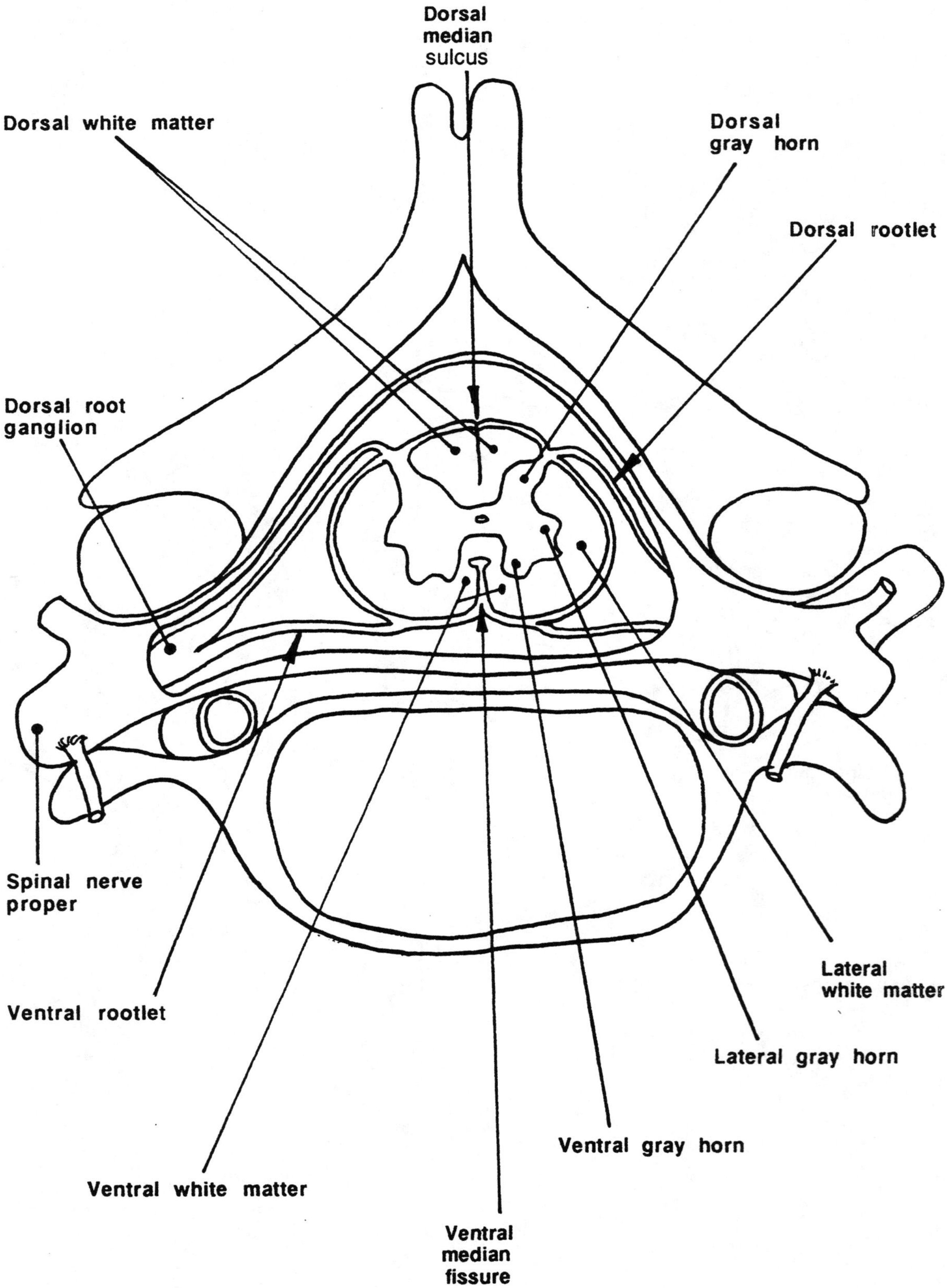

QUESTIONS TO ANSWER AND THINK ABOUT (Some of these questions can be answered from looking at the upper and lower limb models, and the spinal cord models, but some will require that you consult lecture notes and reading assignments)

1. What are the three things that you must do in order to make sense of and survive the events in your environment ?

2. What structures make up your detection system ?

3. What is the difference between a stimulus and a sensation ?

4. What are the eight sensations that you can experience ?

5. Where are your sensations analyzed ?

6. What are the two effectors in your body ?

7. What is a brief description of a nerve impulse ?

8. What will a nerve impulse cause when it reaches the end of an axon ?

9. Receptors are attached to what ?

10. What will stimulation of a receptor cause to happen in a first sensory neuron ?

11. How many neurons are between a receptor and the neurons in the postcentral gyrus ?

12. Name the two structures that are in the peripheral nerves that you learned on the upper and lower limb models ?

13. Where are the cell bodies of first sensory neurons located ?

14. Are the cell bodies of first sensory neurons in the central or peripheral nervous system ?

15. Of what, specifically, is the dorsal rootlet made ?

16. Where are the cell bodies of second sensory neurons located ?

17. Where are the axons of second sensory sensory neurons located ?

18. Where are the cell bodies of third sensory neurons located ?

19. Where do the axons of third sensory neurons go ?

20. What do nerve impulses that travel over the axons of third sensory neuron cause to happen ?

21. What two important things does stimulation of neurons in your postcentral gyrus allow you know ?

22. Where are the cell bodies of upper motor neurons located ?

23. The axons of upper motor neurons synapse with what ?

24. Where are the cell bodies of lower motor neurons located ?

25. The ventral rootlet of a spinal nerve is composed of, specifically, what ?

26. With what do the axons of lower motor neurons synapse ?

27. What are two good and acceptable names for the connection between an axon of a lower neuron and a skeletal muscle cell ?

28. What are the two general areas of the spinal cord ?

29. What are the three areas of the white matter of the spinal cord ?

30. What are the three areas of the gray matter of the spinal cord ?

31. What are the structures that you have seen on the lab models that belong in the peripheral nervous system ?

32. What are the structures / areas that you have seen on the lab models that belong in the central nervous system ?

33. Which peripheral nerve passes behind the medial epicondyle ?

34. Which peripheral nerve is on top of the flexor digitorum profundus ?

35. The femoral nerve emerges from between what two muscles ?

36. What two muscles does the sciatic nerve pass between ?

37. This is a question for the males in this course. What nerve does your wallet press on if you carry it in your back pocket ? You might want to think about this especially if you are experiencing pain in your posterior pelvis and / or posterior thigh region.

38. The sciatic nerve is actually two nerves that just happen to be bound together in the posterior pelvis and the upper part of the posterior thigh. What two nerves are these ?

WHAT YOU MUST KNOW AND BE ABLE TO DO FOLLOWING THIS LABORATORY:

1. Be able to identify and correctly spell the peripheral nerves on the upper and lower limb models.

2. Be able to identify and correctly spell the structures and areas on the spinal cord models.

3. **Be able to answer questions from your lecture notes and reading assignments concerning the structures / areas on the models.**

LABORATORY 16 - SYSTEMS

THE NERVOUS SYSTEM - THE BRAIN

INTRODUCTION: In Laboratory 15 you studied how sensory information from the receptors in the skin of your trunk and limbs goes over the first, second, and thrird sensory neurons to arrive at and stimulate the neurons in the cerebral cortex of the postcentral gyrus. Once these neurons are stimulated, you are consciously aware of the type of sensation and its location. Moreover, you learned about upper and lower motor neurons and how they are anatomically arranged for the contraction of your skeletal muscles. Laboratory 15 showed you the peripheral nerves, and the structures / areas of spinal nerves and the spinal cord that participate in the sensory and motor systems.

With Laboratory 16 we want to take a close look at the brain, especially the surface of the cerebral hemispheres, so that you can basically understand how this amazing organ works. What you will come to appreciate is that the functioning of the brain is anatomically partitioned and that once you learn the anatomical areas, the functions are fairly easy to understand.

Your brain is divided into five **major brain regions**: 1) the **TELENCEPHALON**, 2) the **DIENCEPHALON**, 3) the **MESENCEPHALON**, 4) the **METENCEPHALON,** and 5) the **MYELENCEPHALON**. We can divide your brain another way, into the **CEREBRUM**, which is made up of the telencephalon, and the **BRAINSTEM**, which is composed of the other four major brain regions.

The telencephalon (cerebrum) is by far the most dominant area of the human brain, both anatomically and functionally. Your cerebrum is composed of right and left cerebral hemispheres. When you look at the surface of these two hemispheres you notice right away that their surfaces are not smooth, but are folded. The creases (depressions) are called **sulci** (singular = **sulcus**), and the raised areas between two sulci are called **gyri** (singular = **gyrus**). There are two odd things about these sulci and gyri: **1)** You have the same ones on the surfaces of both hemispheres, and **2)** everyone has the same sulci and gyri. In fact, your brain and Albert Einstein's brain look almost exactly the same on the surface. Also, there is no truth to the old idea that the more you learn the more sulci you have. There does, however, seem to be some correlation between learning and the number of synapses. It appears from studies on numerous types of animals that in an enriched and stimulating environment the brain actually creates more synapses, and a deprived environment results in a diminished number of synapses in the cerebral cortex.

Let's take a look at the anatomy of the cerebral hemispheres of the human brain and try to use this information to assign some functions. We can start big, with the **lobes**, and then become more specific with the sulci and gyri within the lobes.

What is it that the neurons in the cerebral cotex of your cerebral hemispheres do ? Some of them are engaged in receiving nerve impulses from the receptors in your skin, retinas, membranous cochleas, tongue, and nose. It is these neurons that allow you to be consciously aware of the general sensations **(pain, temperature, touch, and pressure)**, and the special sensations **(vision, sound, taste, and smell)**. Being consciously aware of sensations simply is not enough and so the neurons that receive information from your receptors for conscious awareness will send their nerve impulses to neurons in other areas of the cerebral cortes called **analytical (association) areas.** The analytical neurons allow you to add additional information to your sensations (shape, size, weight, texture, shades of color, velocity, understanding of music and language, etc). These neurons allow you to "understand" the meaning of the events in your environment. Other neurons are the upper motor neurons and these neurons send out nerve impulses that go to lower motor neurons and make your skeletal muscles contract.

Each of your cerebral hemispheres is divided into four lobes. Each lobe is named for the flat bone of the skull that covers and protects it. Therefore, there is a **FRONTAL LOBE**, a **PARIETAL LOBE**, a **TEMPORAL LOBE**, and an **OCCIPITAL LOBE**.

Functional assignments can be made for the four lobes. Some of the neurons in the cerebral cortex of the **frontal lobe** are the upper motor neurons that function to initiate contraction of your skeletal muscles. Neurons in the **parietal lobe** are for receiving and analyzing your general sensations (pain, temperature, touch, and pressure). Neurons in the **occipital lobe** are devoted to receiving and analyzing vision. Some of the neurons in the **temporal lobe** function in receiving and analyzing sound. There are two important considerations from this information: **1)** Your brain is involved in receiving and making sense out of the events in your environment, as well as allowing you to respond to the events, and **2)** the functioning of your brain is not haphazard.

If we take a closer look we of course notice that the surface of each lobe is folded and contains its own sulci and gyri. The sulcus that separates the frontal lobe from the parietal lobe is the **CENTRAL SULCUS**, a depression that extends vertically along the lateral surface of the cerebral hemisphere. Anterior and parallel to the central sulcus is the **precentral sulcus**. Posterior and parallel to the central sulcus is the **postcentral sulcus**. These three parallel sulci create two gyri, the **PRECENTRAL GYRUS** and the **POSTCENTRAL GYRUS**. The precentral gyrus and sulcus are both in the frontal lobe, and the postcentral gyrus and sulcus are both in the parietal lobe. Now we can get specific about **upper motor neurons**. **Not all neurons** in the cortex of the frontal lobe are upper motor neurons, only those in the cortex of the precentral gyrus. For that reason the precentral gyrus is very often called the **PRIMARY MOTOR STRIP**. The cell bodies of upper motor neurons are located in the cortex of the primary motor strip. It is their axons that travel down through the brain and into the spinal cord to synapse with **lower motor neurons** located in the ventral gray horns. You no doubt noticed from the **INTRODUCTION** to Laboratory 15 that the axons of upper motor neurons cross over before they descend into the spinal cord to synapse with lower motor neurons. For this reason, the upper motor neurons in the cortex of your right primary motor strip initiate skeletal muscle contractions on the left side of the body, and those in the left primary motor strip initiate muscle contractions on the right side of the body.

The frontal lobe also contains another extremely important motor area called the **MOTOR SPEECH AREA**. This area is located just anterior to the bottom of the primary motor strip, at the posterior end of the inferior frontal gyrus. Oddly, the motor speech area works on only one hemisphere. If you are right handed, then there is about a 90 % chance that your functional motor speech area is located on your left cerebral hemisphere. If you are left handed, then there is about an 80 % chance that your functional area is located on your right cerebral hemisphere. The hemisphere that contains the functional motor speech area is designated as the **dominant hemisphere**.

The postcentral gyrus and sulcus are in the parietal lobe. Coming out of the postcentral sulcus and running posteriorly is the **intraparietal sulcus**. This sulcus divides the parietal lobe posterior to the postcentral gyrus into a **SUPERIOR PARIETAL LOBULE** and an **INFERIOR PARIETAL LOBULE**. The axons of the third sensory neurons for general sensations, whose cell bodies are located in the thalamus, **do not** go up to synapse with all neurons in the parietal lobe, only those in the **postcentral gyrus**. Stimulation of the neurons in the cortex of this gyrus results in conscious awareness on your part that your have experienced general sensations. For this reason, the **postcentral gyrus** is called the **PRIMARY SENSORY AREA FOR GENERAL SENSATIONS**. Once these neurons are stimulated they send nerve impulses to the neurons located in the cortex of the **superior parietal lobule** for analysis, and for this reason the superior parietal lobule is called the **ANALYTICAL AREA FOR GENERAL SENSATIONS**.

As with your general sensations, your other sensations are first received in a specific area and then sent to another cortical area for analysis. The **visual information** that begins with the stimulation of your rods and cones in your retina are sent by a set of first, second, and third sensory neurons to your **thalamus** and then on to the neurons in the cortex in and around the **CALCARINE SULCUS** on the medial surface of your occipital lobe. Once the neurons at the calcarine sulcus are stimulated you are consciously aware that you have seen something, and then these neurons send their nerve impulses to neurons in the rest of the occipital lobe for analysis of vision. The **calcarine sulcus** is known as the **PRIMARY SENSORY AREA FOR VISION**, and the **remainder of the occipital lobe** is the **ANALYTICAL AREA FOR VISION**.

Sensory information for sound that begins in the membranous cochlea of your inner ear travels over five sensory neurons through the **thalamus** to get to the middle area of the **SUPERIOR TEMPORAL GYRUS**. Stimulation of these neurons lets you be consciously aware that you have experienced the sensation of sound and for this reason this area on the **superior temporal gyrus** is called the **PRIMARY SENSORY AREA FOR SOUND**. These primary sensory neurons, once stimulated, send their nerve impulses to the posterior portion of the superior temporal gyrus for analysis of such qualities as loudness, language, music. This area posterior to the primary sensory area for sound is the **ANALYTICAL AREA FOR SOUND**.

Let's leave the telencephalon (cerebrum) and say a few words about the **DIENCEPHALON**. There are four areas to the **diencephalon**, but we will consider only two of them, the **THALAMUS**, and the **HYPOTHALAMUS**. You most likely have noticed that all of the sensations that were discussed above (general sensations, vision, sound) passed through the **thalamus** on their way to their primary sensory areas in the cerebral cortex. One of the major functions of the thalamus is therefore to be a **sensory relay station**.

The **hypothalamus**, although small in size, is nevertheless one of the most important areas on your brain. Among other functions, the neurons here are responsible for **1)** intergrating your emotions, **2)** for regulating the fluid balance in your body, and **3)** for regulating your hunger cycle. Your hypothalamus also is the origin site for a **motor system** called the **AUTONOMIC NERVOUS SYSTEM**, which controls the smooth muscles, cardiac muscle and glands in your body. The neurons in the cortex of the precentral gyrus, you probably recall, are the ones that initiate the contraction of your skeletal muscles. The autonomic nervous system regulates such things as heart rate and cardiac output, pupil dilation and constriction, penile erection in males, lubrication in females, ejaculation and /or orgasms, how blood is shifted around in the body to meet various demands, and body temperature (thermoregulation).

The outstanding features of the **MESENCEPHALON (MIDBRAIN)** are the **cerebral peduncles**. Each of these structures is a collection of millions of axons of upper motor neurons that have come off the cell bodies that are located in the precentral gyrus. These axons continue on down into the spinal cord to synapse with lower motor neurons located in the ventral gray horns.

The **pons** and **cerebellum** are two structures of the **METENCEPHALON**. The **cerebellum** is the structure that is responsible for regulating and coordinating your **skeletal muscle** contractions. With injury to your cerebellum you would still be able to contract your muscles, but your movements would be uncoordinated.

The **medulla oblongata** makes up the **MYELENCEPHALON**. It and the lower area of the pons contain your reflex breathing centers, which allow you to breathe while asleep or unconscious.

Attached to your brain are twelve pairs of **CRANIAL NERVES**. These nerves have the same function as your spinal nerves except that they are conducting sensory information from your head structures (teeth, tongue, nose, eyes, inner ear, and the skin of your face), and sending out motor impulses to your muscles of facial expression, muscles of chewing, muscles of your tongue, extrinsic eye muscles, lacrimal glands, and salivary glands.

The names, numbers, and functions of the cranial nerves are as follows:

OLFACTORY (I) - Sensory for smell. No motor functions

OPTIC (II) - Sensory for vision. No motor functions

OCULOMOTOR (III) - No sensory functions. Motor for five of the seven extrinsic eye muscles, and motor to the sphincter pupillae.

TROCHLEAR (IV) - No sensory functions. Motor to the superior oblique muscle

TRIGEMINAL (V) - Sensory for general sensations from the skin of your face, from your teeth, from the lining of your mouth and from the surface of your tongue. Motor to the muscles of mastication (chewing)

ABDUCENS (VI) - No sensory functions. Motor to the lateral rectus muscle

FACIAL (VII) - Sensory for taste from the anterior 2/3 of the tongue, and motor to the muscles of facial expression, the sublingual and submandibular glands, and to the lacrimal glands

VESTIBULOCOCHLEAR (VIII) - Sensory for the sensations of sound and balance from the inner ear. No motor functions

GLOSSOPHARYNGEAL (IX) - Sensory for taste from the posterior 1/3 of the tongue, and motor to the stylopharyngeus muscle

VAGUS (X) - Sensory for your alimentary tract, and motor to muscles of the pharynx and to the muscles controlling the true vocal cords

SPINAL ACCESSORY (XI) - No sensory functions. Motor to the trapezius and sternocleidomastoid

HYPOGLOSSAL (XII) - No sensory functions. Motor to the muscles of the tongue

OBJECTIVES: The objectives of this laboratory are to:

1. Be able to identify and correctly spell the major brain regions on the brain models

2. Be able to identify and correctly spell the lobes of the telencephalon

3. Be able to identify and correctly spell the major structures in each of the major brain regions

4. Be able to identify and correctly spell the sulci and gyri on the surface of the cerebral hemispheres, and to give their sensory and motor functions

5. Be able to identify and correctly spell the twelve cranial nerves and give the general function of each

PRELABORATORY PREPARATION: You should complete this worksheet before you come to lab. It will allow you to work more efficiently and to complete your work earlier and with less confusion. By using the diagrams of the brain models given under the **EXERCISES** section, fill in the following blanks:

Major brain regions
1____________________
2____________________
3____________________
4____________________
5____________________

Structures of the major brain regions

a. Of the telencephalon
1____________________
2____________________
3____________________
4____________________

b. Of the diencephalon
1____________________
2____________________

c. Of the mesencephalon
1____________________

d. Of the metencephalon
1____________________
2____________________

e. Of the myelencephalon
1____________________

The telencephalon

a. Lobes

1____________________
2____________________
3____________________
4____________________

b. Sulci on the lateral surface

1____________________
2____________________
3____________________
4____________________
5____________________
6____________________
7____________________

c. Sulci on the medial surface

1____________________
2____________________
3____________________

d. Gyri on the lateral surface

1____________________
2____________________
3____________________
4____________________
5____________________
6____________________

e. Lobules on the lateral surface

1____________________
2____________________

f. Gyri of the medial surface

1____________________

g. Functional areas on the lateral surface

1. Primary sensory areas
   a)____________________
   b)____________________

2. Analytical areas
   a)____________________
   b)____________________

3. Motor areas
   a)____________________
   b)____________________

h. Functional areas on the medial surface

1. Primary sensory area
   ____

2. Analytical area
   ____

Cranial nerves

1____________________
2____________________
3____________________
4____________________
5____________________
6____________________
7____________________
8____________________
9____________________
10____________________
11____________________
12____________________

EXERCISES:

1. The five major brain regions and the major structures within each

TELENCEPHALON

a. The cerebral hemispheres
b. Corpus callosum
c. Cerebral cortex (not present on all models)
d. Basal ganglia (not present on all models)

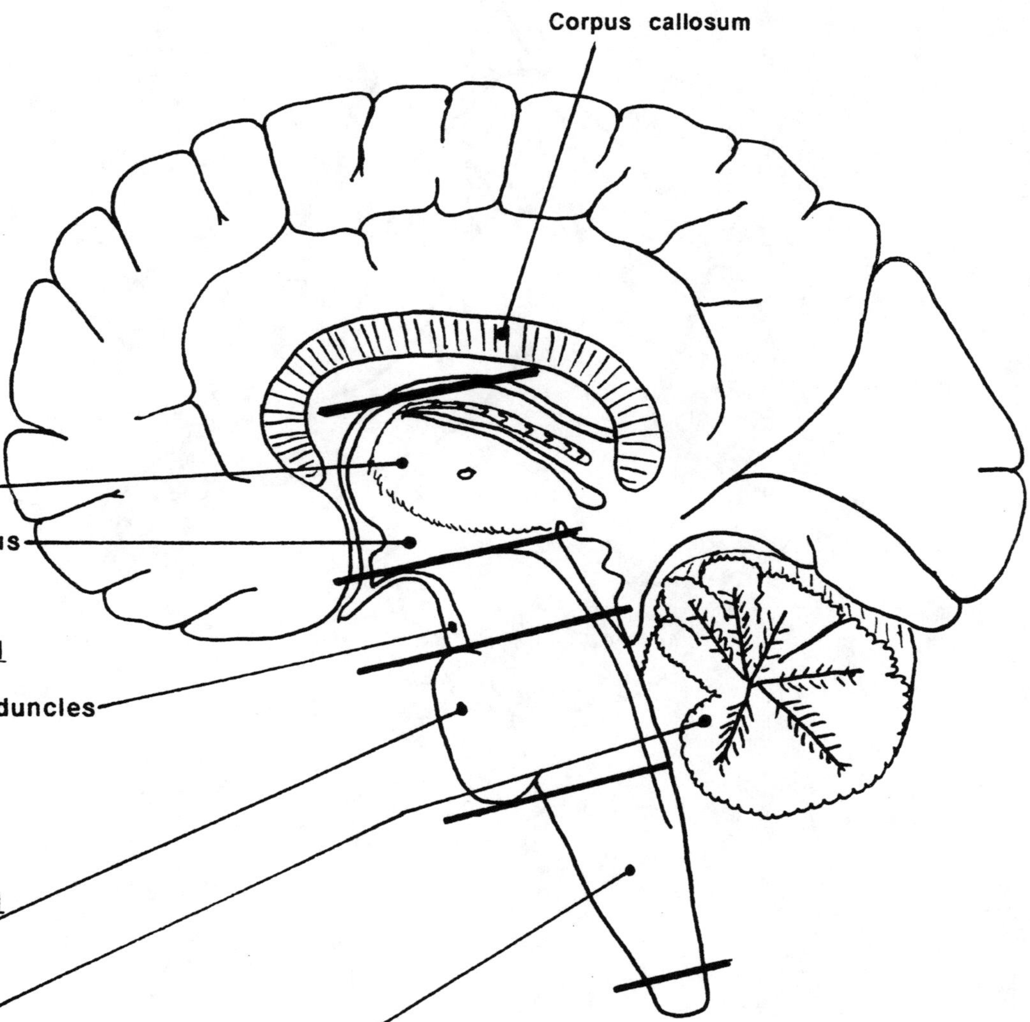

DIENCEPHALON

Thalamus

Hypothalamus

MESENCEPHALON

Cerebral peduncles

METENCEPHALON

Pons

Cerebellum

MYELENCEPHALON

Medulla oblongata

2. Lobes of the telencephalon

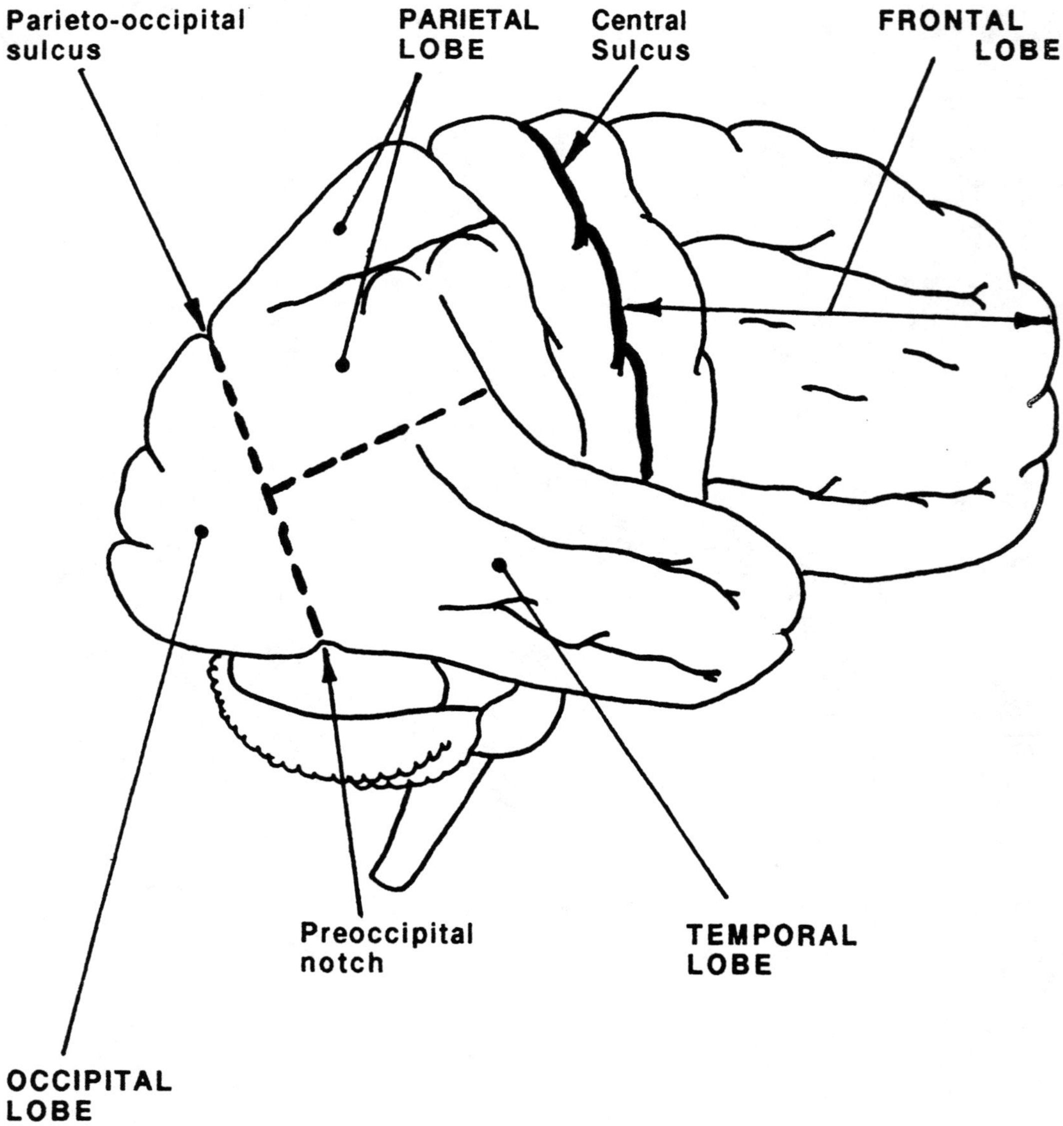

3. Sulci and gyri of the cerebral hemispheres

a. Lateral surface

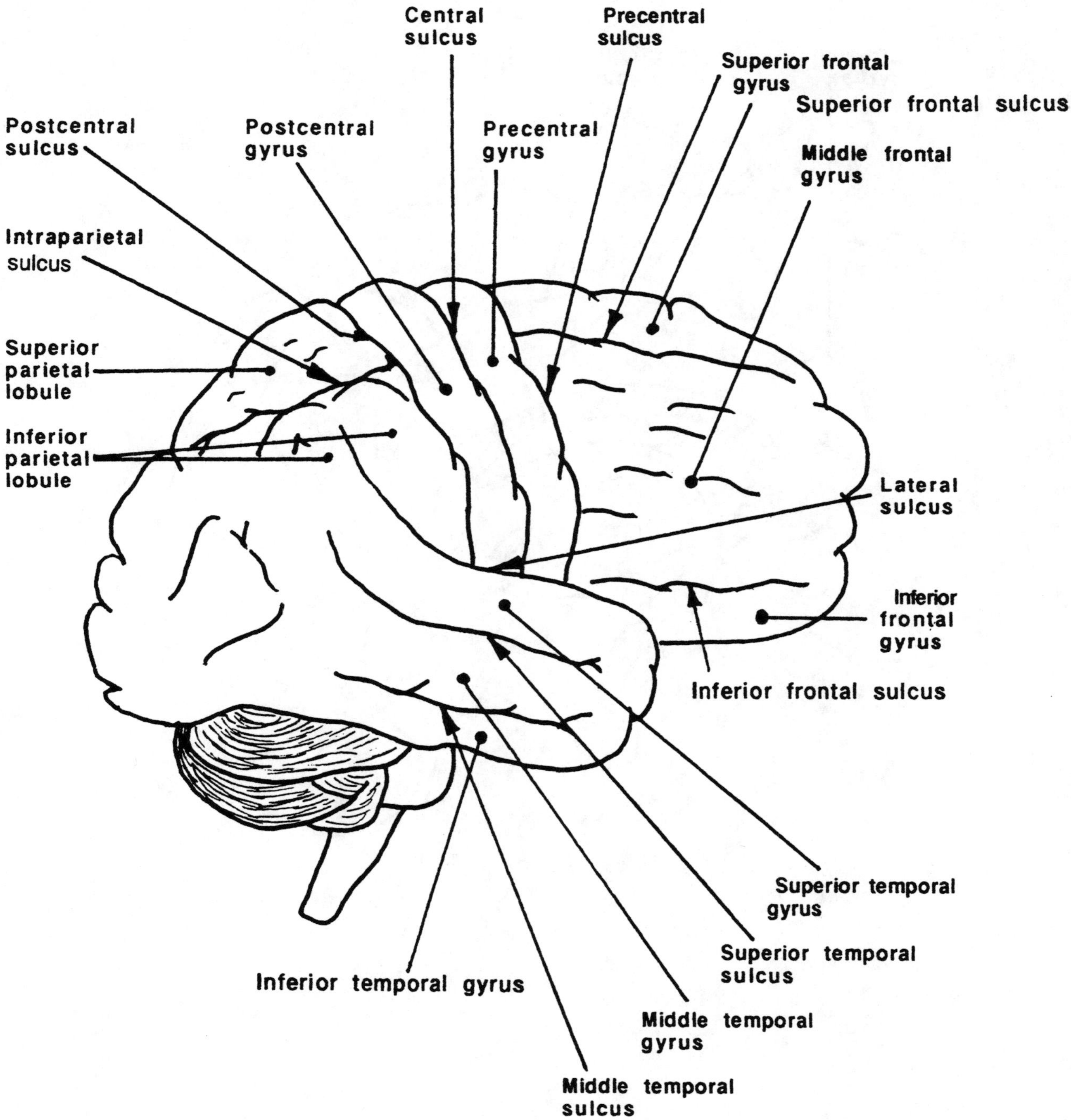

b. Medial surface

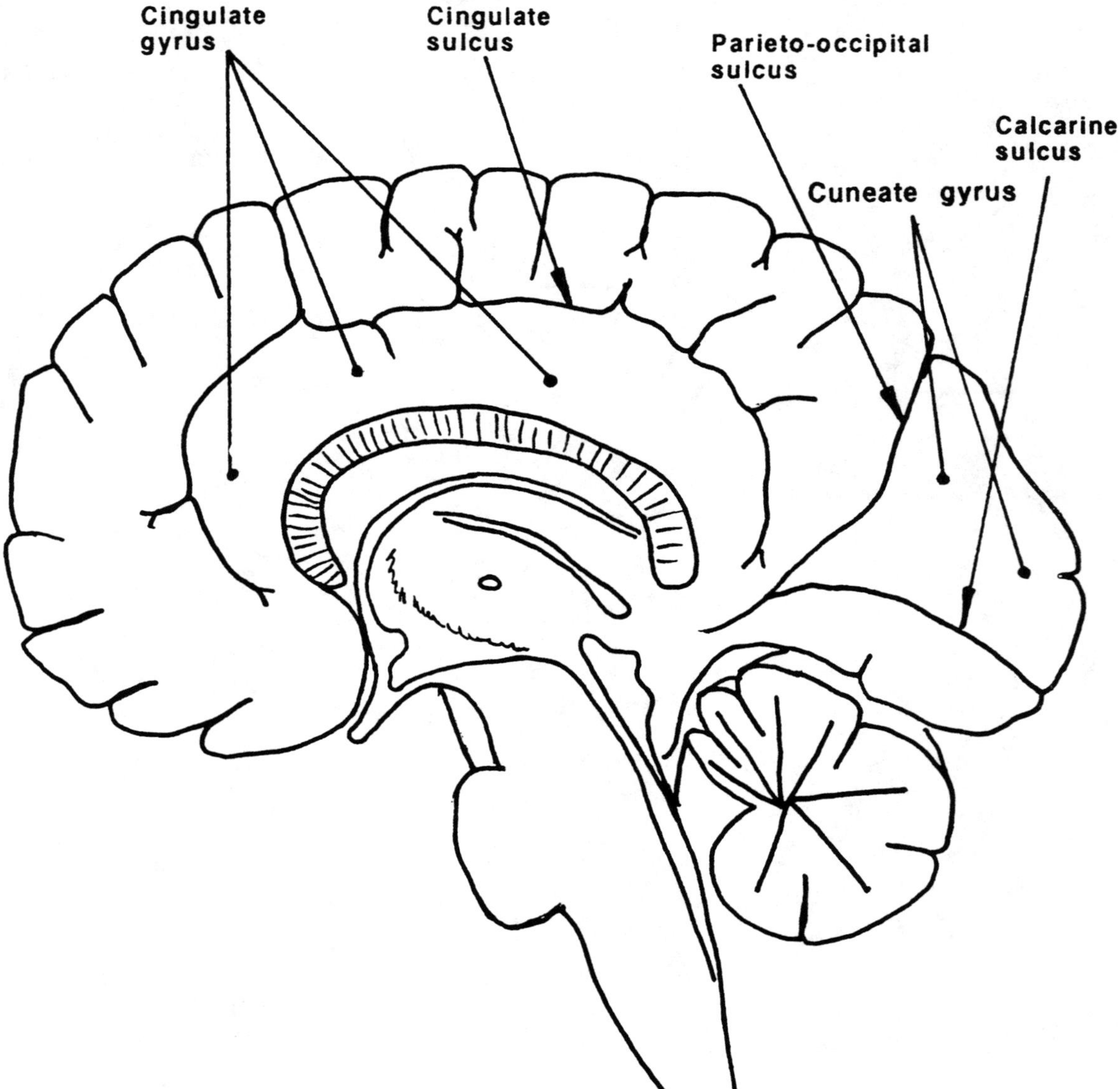

4. Functional areas of the telencephalon

a. Lateral surface

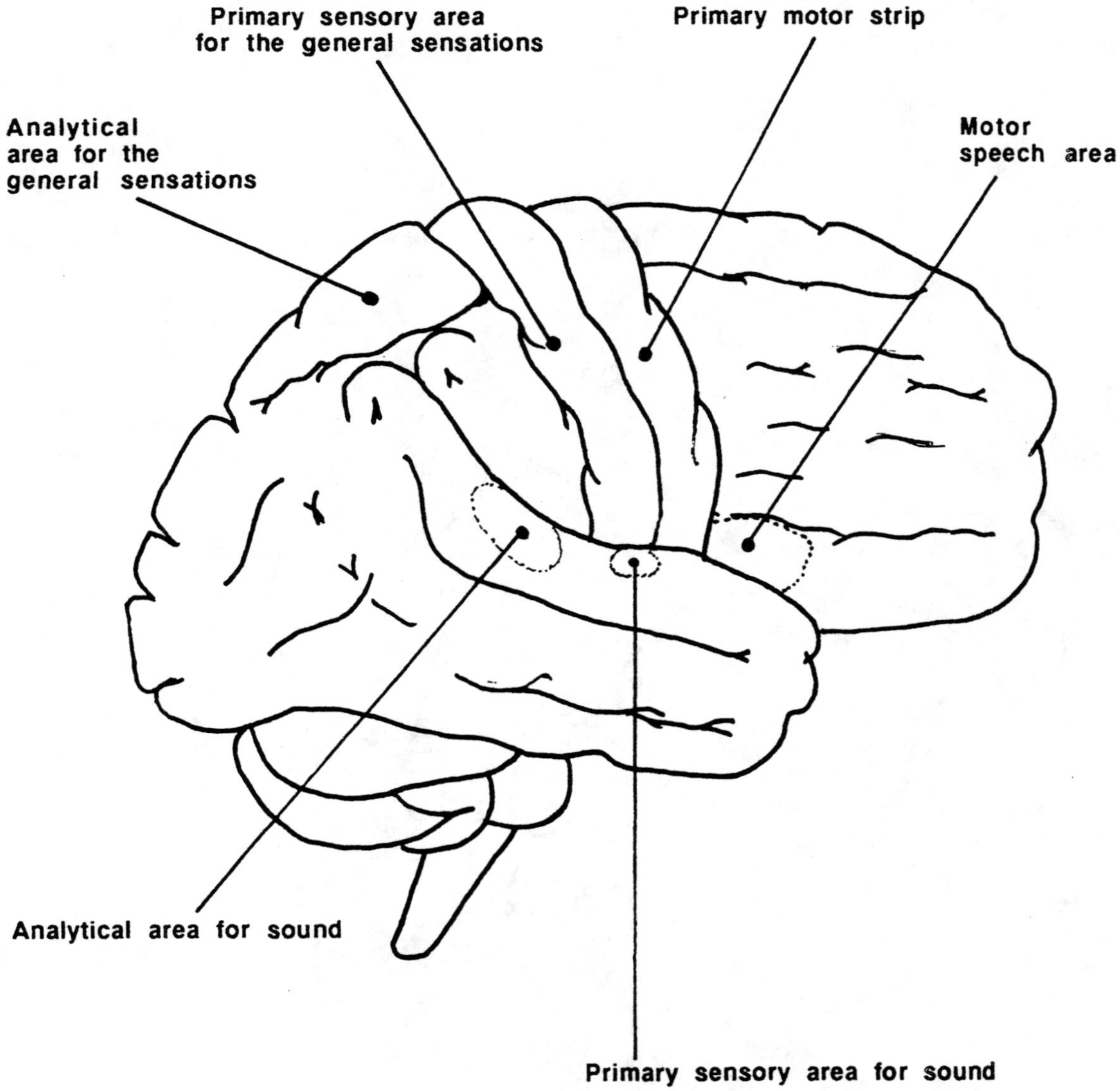

b. Medial surface

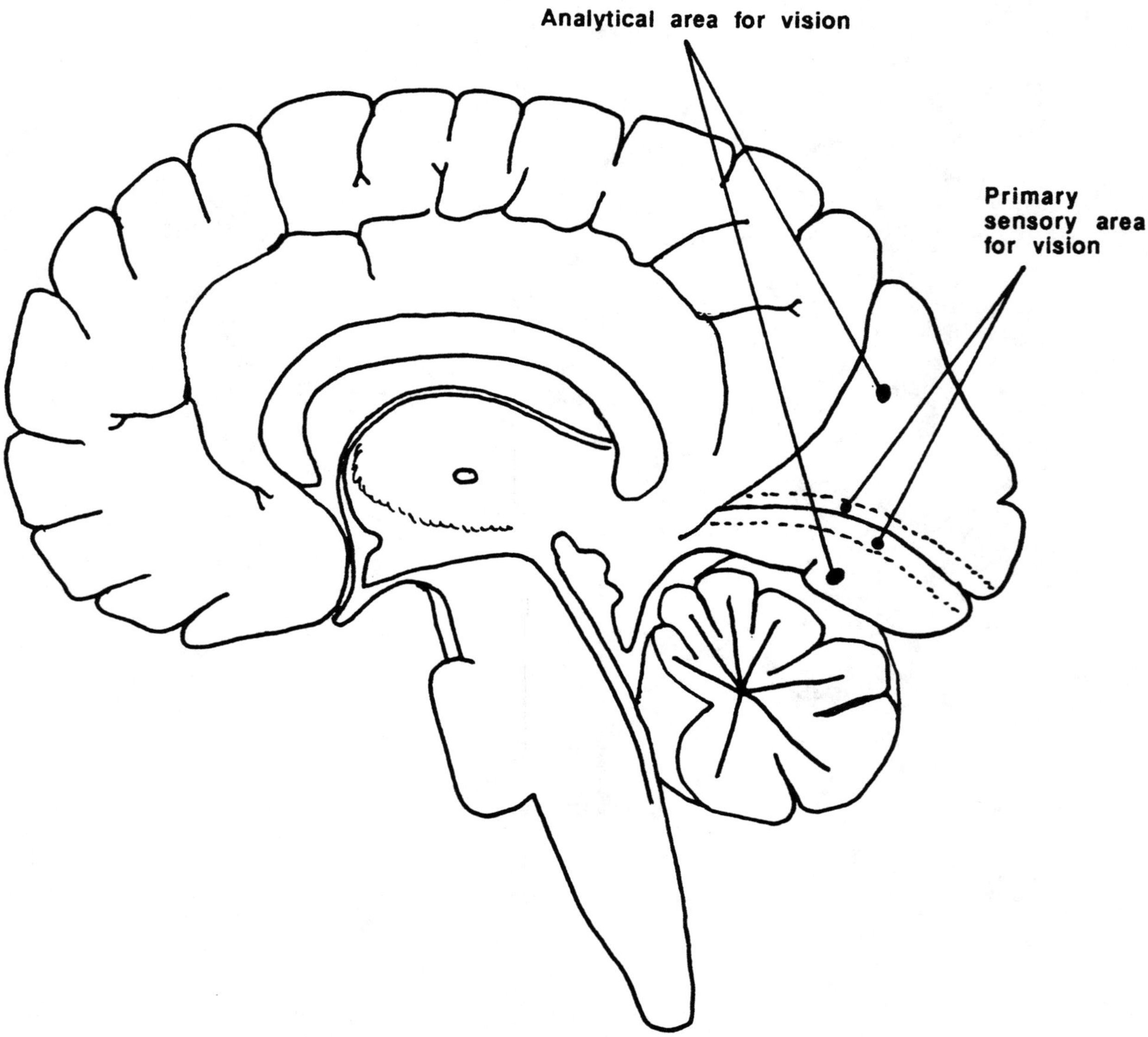

5. Cranial nerves

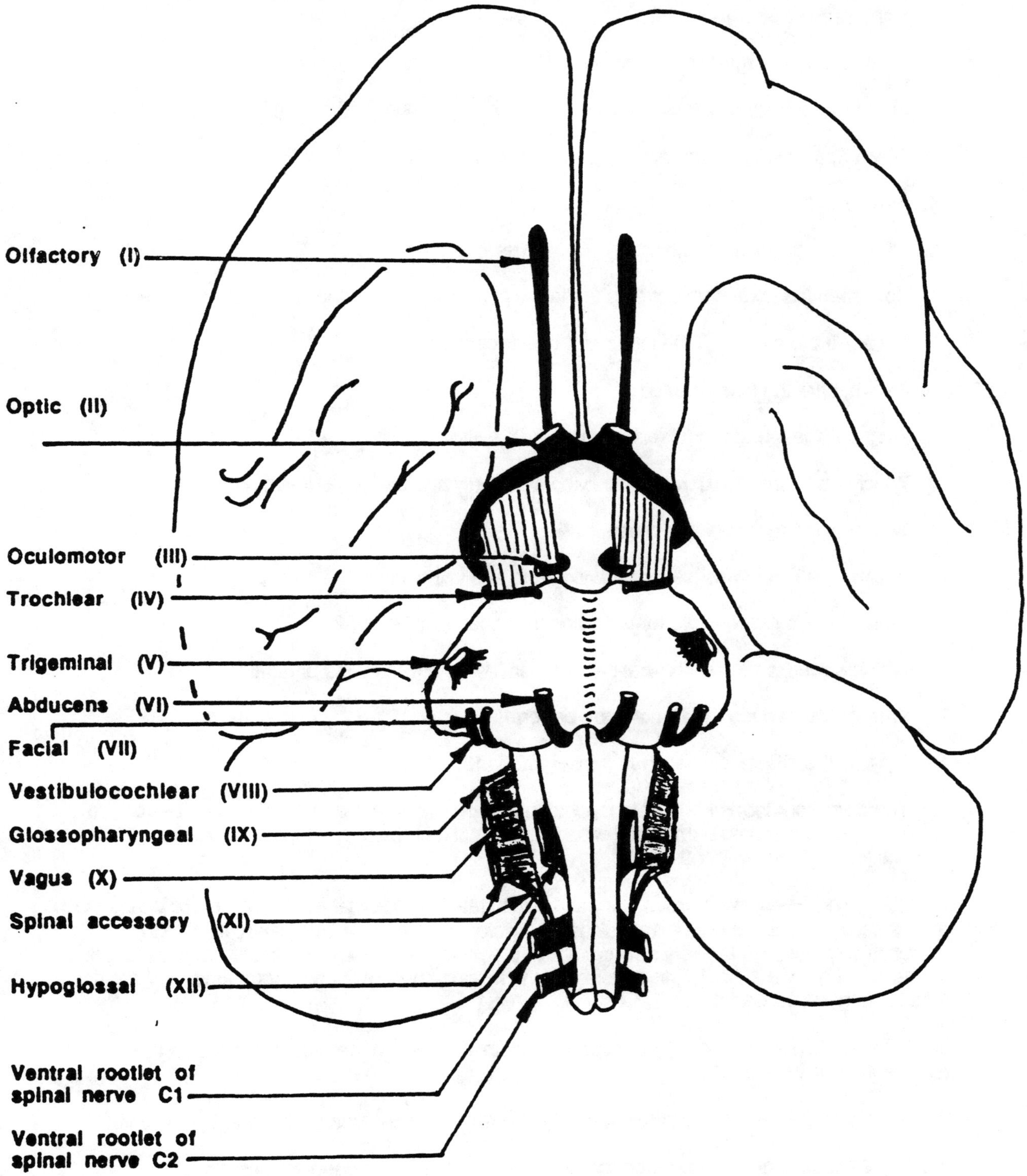

<u>QUESTIONS TO ANSWER AND THINK ABOUT</u> (Some of these questions can be answered from looking at the brain models, but some will require that you consult your lecture notes and reading assignments)

1. What are the five major brain regions ?

2. Which major brain region is the largest in the human brain ?

3. Which major brain region makes up the cerebrum ?

4. Which major brain regions make up the brainstem ?

5. What is the definition of a sulcus ? Of a gyrus ?

6. How many cerebral hemispheres are in the cerebrum ? How many in the telencephalon ?

7. What are the names of the lobes of the telencephalon ?

8. For what are the lobes named ?

9. What are the four general sensations ?

10. Which lobe is responsible for receiving and analyzing the general sensations ?

11. What are the four special sensations ?

12. Which lobe is responsible for receiving and analyzing vision ?

13. Which lobe is responsible for receiving and analyzing sound ?

14. Which sulcus is the anatomical division between the frontal and parietal lobes ?

15. What function does the frontal lobe perform ?

16. What is the difference between sensory and motor ?

17. There are four lobes in the telencephalon. If you had to assign whether each performed sensory motor functions, then which would the frontal lobe perform ? The parietal lobe ? The occipital lobe ? The temporal lobe ?

<u>18. - 29</u>: For the following areas, gyri, and sulci, state the lobe that each is in: a) calcarine sulcus b) postcentral gyrus, c) precentral sulcus, d) motor speech area, e) primary motor strip, f) Postcentral sulcus, g) Primary sensory area for general sensations, h) analytical area for sound, i) Precentral gyrus, j) Analytical area for vision, k) primary sensory area for sound, l) Analytical area for general sensations.

30. Which sulcus separates the parietal and occipital lobes on the medial surface of the telencephalon ?

31. What anatomical area corresponds to the primary sensory area for general sensations ?

32. What anatomical area corresponds to the analytical area for general sensations ?

33. What anatomical area corresponds to the primary sensory area for vision ?

34. What anatomical area corresponds to the analytical area for vision ?

35. What anatomical area corresponds to the primary sensory area for sound ?

36. What anatomical area corresponds to the analytical area for sound ?

37. What anatomical area corresponds to the primary motor strip ?

38. What is located in the cortex of the primary motor strip ?

39. Where do the axons of the upper motor neurons in the cortex of the right primary motor strip go ?

40. How about those from the left primary motor strip ?

41. What is the function of the upper motor neurons ?

42. Neurons, like any other cell in the body, require oxygen in order to live and function. What do you think would be the result if you had a blood clot that suddenly cut off the oxygen supply (a "stroke") to the upper motor neurons in your left primary motor strip ? In your right primary motor strip ?

43. Are you right or left handed ?

44. Considering your answer to question 43, which one of your cerebral hemispheres is your dominant hemisphere ?

45. What do you think you would be unable to do if you had a stroke that involved your motor speech area ?

46. What does stimulation of the neurons in the primary sensory areas accomplish ?

47. What are the two structures of the diencephalon that you studied in the lab ?

48. Give a function of the thalamus ?

49. Give several functions of the hypothalamus ?

50. The hypothalamus is the origin of an important motor system. What is the name of this motor system ?

51. What effectors does the autonomic nervous system control ?

52. What are the structures of the mesencephalon that you studied in lab ?

53. Anatomically, what are the cerebral peduncles composed of ?

54. What are the two structures of the metencephalon that you saw on the brain models in lab.

55. What is the very important function of the cerebellum ?

56 - 58. Which of the cranial nerves have no motor functions (sensory - only cranial nerves) ?

59 - 63: Which of the cranial nerves have no sensory functions (motor - only cranial nerves) ?

64 - 67: Which cranial nerves have both sensory and motor functions ("mixed" cranial nerves) ?

68. What two cranial nerves are involved with the sensation of taste ?

69. From what areas of the tongue do the facial and glossopharyngeal nerves convey the sensation of taste ?

70 - 72: What three cranial nerves control the seven extrinsic eye muscles ?

73. What is the effector of the trochlear nerve ?

74. What is the effector of the abducens nerve ?

75. Therefore, what are the five skeletal muscle effectors of the oculomotor nerve ?

76. What are the receptors associated with the optic nerve

77. Where are the receptors associated with the vestibulocochlear nerve specifically located ?

78. What are the receptors associated with the facial and glossopharyngeal nerves and where are they located ?

79. Which cranial nerve has the tongue muscle as effectors ?

80. Which cranial nerve has the trapezius and the sternocleidomastoid as effectors ?

81. Which cranial nerve has the muscles of mastication as effectors ?

82. Which cranial nerve has the lacrimal glands as effectors ?

83. Which cranial nerve is responsible for general sensations from the face, mouth, teeth, and tongue ?

84. Can you attach numbers to the corresponding twelve cranial nerves ?

85. Which cranial nerve has the sphincter pupillae as an effector ?

86. Therefore, list all six effectors of the oculomotor nerve.

87. Which two cranial nerves have the three salivary glands as effectors ? Which nerve controls the parotid gland, and which controls the sublingual and submandibular glands ?

88. Which cranial nerve controls the muscles of facial expression ?

89. Therefore, what are the effectors for the facial nerve ?

WHAT YOU SHOULD KNOW AND BE ABLE TO DO AFTER YOU COMPLETE THIS LABORATORY:

1. Be able to identify and correctly spell the major brain regions and their components on the brain models.

2. Be able to identify and correctly spell the lobes, lobules, sulci, and gyri on the brain models.

3. Be able to identify and correctly spell the primary sensory areas, analytical areas, and motor areas on the brain models.

4. Be able to identify and correctly spell the cranial nerves and answer questions about their sensory and motor functions.

5. **Be able to answer questions from your lecture notes and reading assignments concerning the structures and areas on the brain models.**

LABORATORY 17 - SYSTEMS

THE CIRCULATORY SYSTEM - THE HEART

INTRODUCTION: Your heart is a truly amazing muscular organ. Its function is simple and that is to pump blood through the arteries, capillaries and veins in your body. The purpose of moving blood around in your body is **to deliver nutrients** to your cells from your your digestive tract and liver, **to deliver O_2** to your cells from the alveoli of your lungs, **to deliver CO_2** from the cells to the alveoli of your lungs, **to deliver hormones** from the endocrine glands where they are made to the cells that they affect, and **to deliver urea** from your liver to the glomeruli of your kidneys. Clearly, one of the main functions of your circulatory system is to serve as a **transport system**.

The pumping function of your heart is, like the functions of other organs, derived from its structure. The heart is composed of four chambers. The walls of these chambers are made of muscle that is categorized as cardiac muscle. This muscle is also, like smooth muscle, involuntary muscle because you cannot contract it at will. The cardiac muscle in your heart is ofter referred to as the **myocardium**. During your cardiac cycle, the chambers of your heart fill up with blood, and then the cardiac muscle in the walls of these chambers undergo contraction and pump the blood out.

The models of the adult heart that you will see and study in the laboratory are static and rigid. Nevertheless, what we will do in this introduction is to briefly review the cardiac cycle and the three circulatory cycles (systemic, pulmonary, and cardiac) so that you can better understand how the heart structures and chambers participate in the cardiac cycle and the three adult circulatory routes.

Your 100 trillion or so **cells** in the **SYSTEMS** of your body and those cardiac muscle cells in the walls of your **HEART** are literally bags of **enzymes** that are very rapidly catalyzing thousands of **chemical reactions**. Collectively, these reactions in a cell are called **metabolism**. Many, but not all, of the chemical reactions of your cellular metabolism require a source of energy in order to occur, and if there is no source of energy, then metabolism stops and the cell dies. The chemical compound that the cell makes and uses for its source of energy for these energy - requiring reactions is **Adenosine Triphosphate (ATP)**, which is made in the **mitochondria** of your cells. All cells get their energy to make ATP from the breakdown of the nutrients **glucose or fatty acids** to **CO_2** and water. Importantly, **in order to make ATP, O_2 is absolutely required** because cells can live only a matter of minutes without O_2.

What all this means is that each of your cells, in order to stay alive, must have a constant supply of nutrients and O_2, and that each of your cells is constantly making CO_2. The O_2 that is constantly used up by the mitochondria of your cells to make ATP (which is constantly used up to drive the energy - requiring chemical reactions) must be constantly replenished, and the CO_2 constantly being generated by the break down of glucose and fatty acids is toxic (posionous) to the cell and must be eliminated from the cell. So, the problems of getting O_2 to your cells on a constant basis and removing CO_2 from your cells, and actually out of your body on a constant basis are two critical problems, because if you fail for very long in doing either one of these, then .... Well, let's not think about it.

What is true for the O_2 supply / CO_2 removal in the cells in the systems of your body is also true for the cardiac muscle cells in the walls of your heart. They , too, require their O_2 supply and need to eliminate their CO_2.

These O_2 / CO_2 requirements by the cells in the systems and by the muscle cells in the heart lead us to three circulations in your body : 1) **SYSTEMIC CIRCULATION**, 2)

CORONARY CIRCULATION, and 3) PULMONARY CIRCULATION. The purpose of **SYSTEMIC CIRCULATION** is to deliver oxygenated blood from the heart to the cells in the organs of the systems of the body, to pick up the CO_2 from the cells of the body, and to return the unoxygenated blood containing CO_2 back to the heart. The purpose of **CORONARY CIRCULATION** is to deliver oxygenated blood to the muscle cells of the cardiac muscle (myocardium) in the wall of the heart, to pick up the CO_2 from these muscle cells and return the unoxygenated blood containing CO_2 back to the heart. Obviously, with these two circulations your blood has made a trip from your heart out to some organ in your body and then back to your heart, and in doing so **your blood loses its oxygen and picks up CO_2**. Bummer.

Losing O_2 and gaining CO_2 creates two problems for your blood: **1)** How do you get O_2 back into (reoxygenate) your blood, and **2)** how do you eliminate the CO_2 from your blood. Having CO_2 in your blood is actually a critical problem because CO_2 causes the acidity of your blood to increase, leading to a dangerous condition known as **acidosis**. In order to solve these two problems you have a **PULMONARY CIRCULATION**. It is the purpose of **PULMONARY CIRCULATION** to deliver unoxygenated blood containing CO_2 to the capillaries around the alveoli so that CO_2 can leave the blood and O_2 can enter it for reoxygenation, and then be returned to the heart to re-enter either systemic or cardiac circulations.

There are four chambers in the heart that include a right and left **ATRIUM** and a right and left **VENTRICLE**. In order to understand the relationship of these four chambers with the three circulations, it is necessary to divide the heart into a **RIGHT HEART (right atrium and right ventricle)** and a **LEFT HEART (left atrium and left ventricle)**.

Both systemic and coronary circulations begin with oxygenated blood in the left ventricle. Blood from the left ventricle that enters the aorta and then flows into the arteries of such organs as the kidneys, brain, skeletal muscles, liver, stomach, intestinal tract, uterus, etc. is in **systemic circulation**. Blood from the left ventricle that enters the aorta and then flows into the coronary arteries that lead into the myocardium is in **coronary circulation**. The blood that returns from the organs and from the myocardium is unoxygenated and contains CO_2 and enters the right atrium. So, in systemic and coronary circulations, blood leaves the left ventricle and returns to the right atrium. The blood starts out oxygenated and returns unoxygenated. From the right atrium the unoxygenated blood is pumped into the right ventricle. Therefore, **in the adult, the chambers of the right heart contain unoxygenated blood**.

The right ventricle begins pulmonary circulation because blood in this chamber is pumped from this chamber to the lungs where , at the alveoli level, O_2 is taken into the blood and CO_2 enters into the alveoli. Oxygenated blood from the lungs returns to the heart through the pulmonary veins and enters the left atrium, which represents the end of pulmonary circulation. From this chamber, oxygenated blood is pumped to the left ventricle from which it can re-enter systemic and coronary circulations. Therefore, **in the adult, the chambers of the left heart contain oxygenated blood.**

Another way to look at the heart's chambers is as upper chambers (the right and left atria) and lower chambers (right and left ventricles). This division into upper and lower chambers is especially convenient for understanding the **cardiac cycle**. When you take your pulse what you are measuring is the number of cardiac cycles per minute. The average number of cardiac cycles / minute in women is about 84, and in men it is 72. This means, of course, that your heart undergoes one cardiac cycle in about 0.75 seconds.

What goes on during one cardiac cycle ? Basically, **1)** the musculature of the **two atria contract at the same time**. The **unoxygenated blood** in the right atrium is pumped into the right ventricle at the same time that **oxygenated blood** in the left atrium is pumped into the left ventricle. Then **2)** the **atria relax and the musculature of the two ventricles**

contract. The **unoxygenated blood** in the right ventricle is pumped into the pulmonary trunk and to the lungs, and the **oxygenated blood** in the left ventricle is pumped into the aorta and then on into systemic and coronary circulations. During the time that the ventricles are contracting and emptying their blood into arteries, the relaxed right atrium is refilling with unoxygenated blood from the organs and from the myocardium, and the relaxed left atrium is refilling with oxygenated blood from the lungs. Then **3) the ventricles relax** and it is then the atria's turn to contract and begin the cardiac cycle all over again. This cycle, if you live your four score and ten years, will repeat itself some three billion times **!!!!!!!!!!!!!!!!!!!!!!!!!!!!!!!!!!!!**, a remarkable testimony to your cardiac muscle and your coronary circulation.

When your ventricles contract, why does the blood go into the pulmonary trunk and aorta and not go back into the atria where it it has just come from ? The reason is that in your heart there are **two atrioventricular valves** which are located between the atria and the ventricles (the **mitral valve** in the left heart and **THE tricuspid valve** in the right heart). These two valves are are kept **"closed"** by the action of the papillary muscles and chorda tendineae when the ventricles contract . The blood in the ventricles then has no choice but to enter the arteries. Once the ventricles relax, what prevents a substantial amount of blood from re-entering them from the two arteries (pulmonary trunk and aorta) ? The answer is the **semilunar valves (aortic and pulmonary valves)**, which are forced **"closed"** when the ventricles relax and catch the returning blood thereby preventing blood from re-entering the ventricles. **The important function of the atrioventricular and semilunar valves is to keep blood flowing in one direction in your body**.

OBJECTIVE: The objective of this laboratory is to be able to identify and correctly spell the structures / chambers of the heart model

PRELABORATORY PREPARATION: You should complete this worksheet before you come to lab. It will allow you to work more efficiently and to complete your work earlier and with less confusion. By using the diagrams of the brain models given under the **EXERCISES** section, fill in the following blanks:

Chambers
a. Upper chambers
1____________________
2____________________

b. Lower chambers
1____________________
2____________________

c. Chambers of the right heart
1____________________
2____________________

d. Chambers of the left heart
1____________________
2____________________

Veins attached to the right atrium
1____________________
2____________________
3____________________

The set of veins attached to the left atrium

____

Arteries attached to the ventricles
a. The artery attached to the right ventricle

____

b. The artery attached to the left ventricle

____

Valves
a. Atrioventricular (A-V) valves
1____________________
2____________________

b. Semilunar valves
1____________________
2____________________

Structures associated with the A - V valves
1____________________
2____________________

Structure that was formerly the ductus arteriosus

____

Coronary circulation arteries
1____________________
2____________________
3____________________
4____________________
5____________________
6____________________

Arteries and veins of pulmonary circulation
a. Arteries
1____________________
2____________________
3____________________

b. Veins

____

Areas of the aorta
1____________________
2____________________
3____________________

Respiratory system structures
1____________________
2____________________
3____________________

Digestive system structure

____

EXERCISES:

There are several heart models in the lab, and they are termed: 1) large, nondetachable heart model, 2) large, detachable heart model, and 3) small heart model. You will not be able to see all of the structures that you listed in the **PRELABORATORY PREPARATION** section.

1. The large, nondetachable heart model

   a. Anterior view, chambers closed

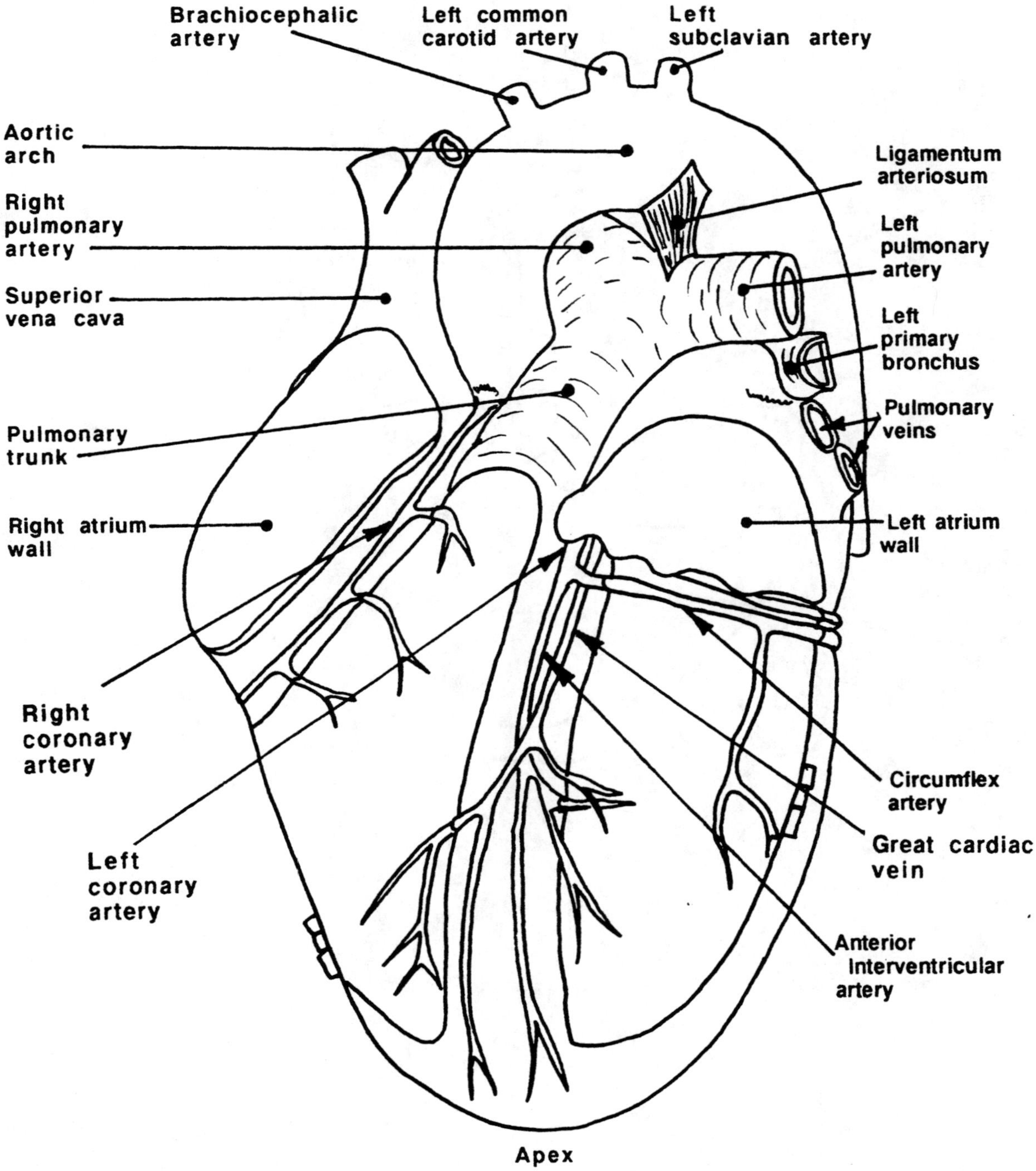

b. Posterior view of the large, nondetachable heart

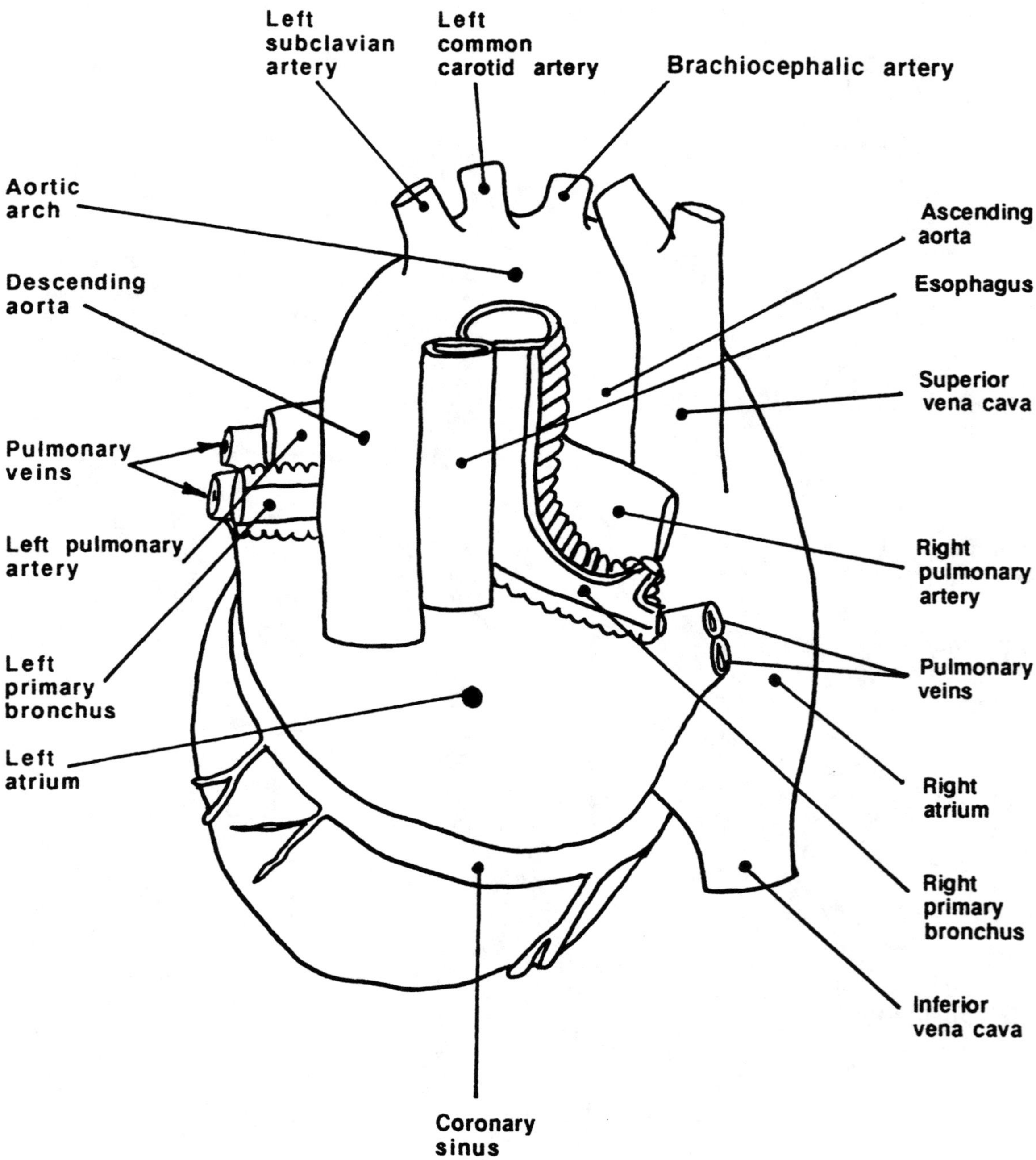

2. The large, detachable heart model

a. Anterior view

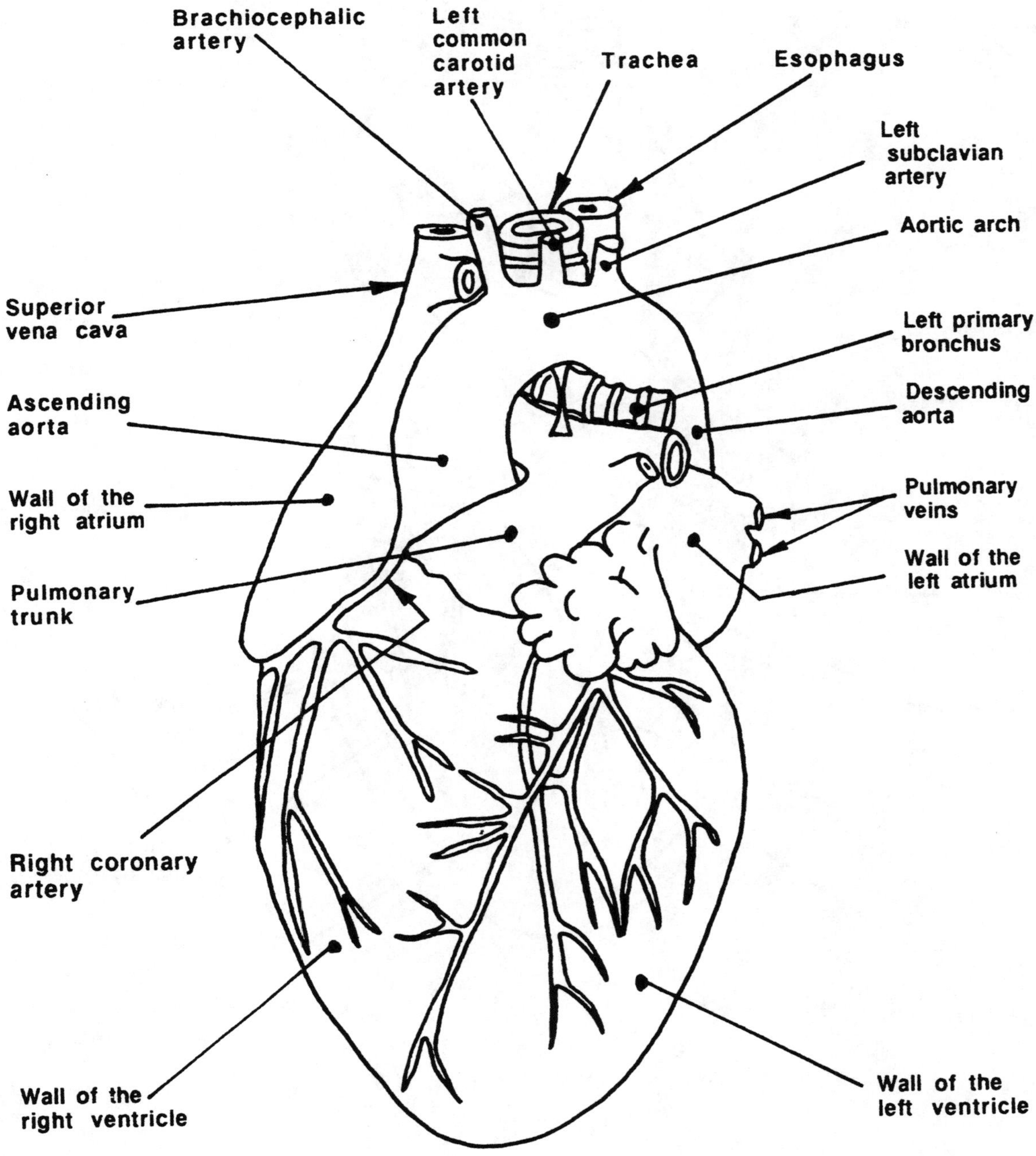

b. Anterior view of the interior of the large, detachable heart model

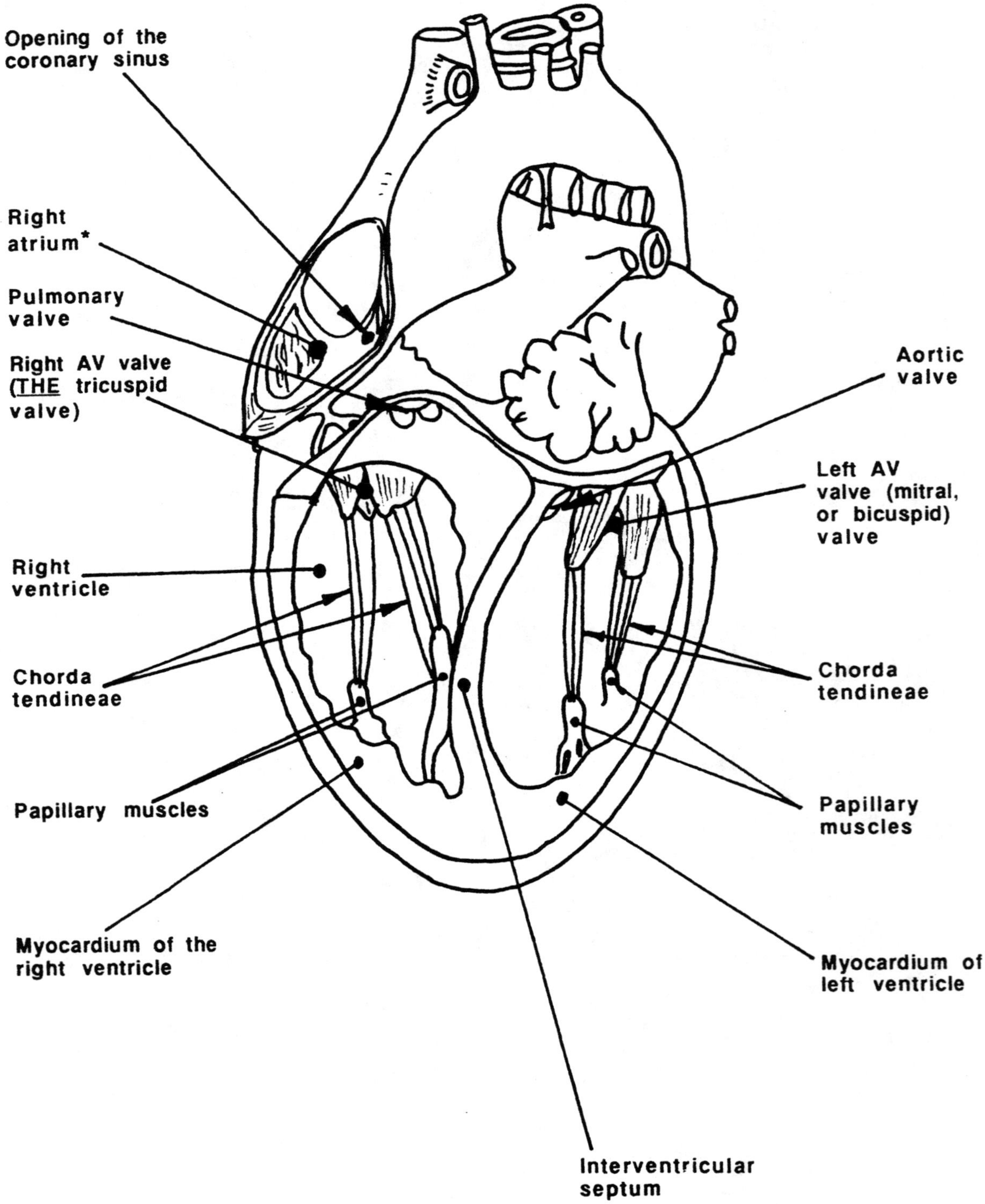

* Be certain that you find and identify the left atrium

3. The anterior and posterior views of the small heart model

a. Anterior view

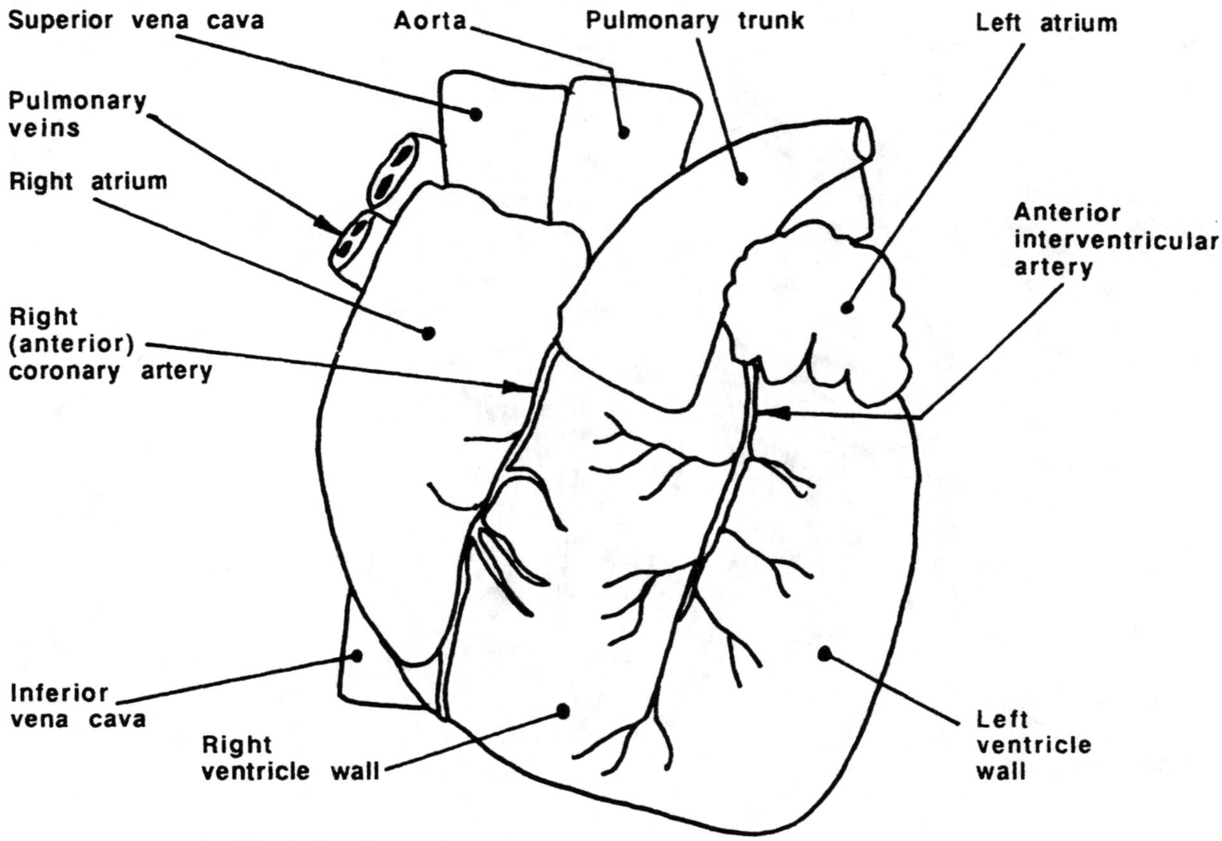

b. Posterior view

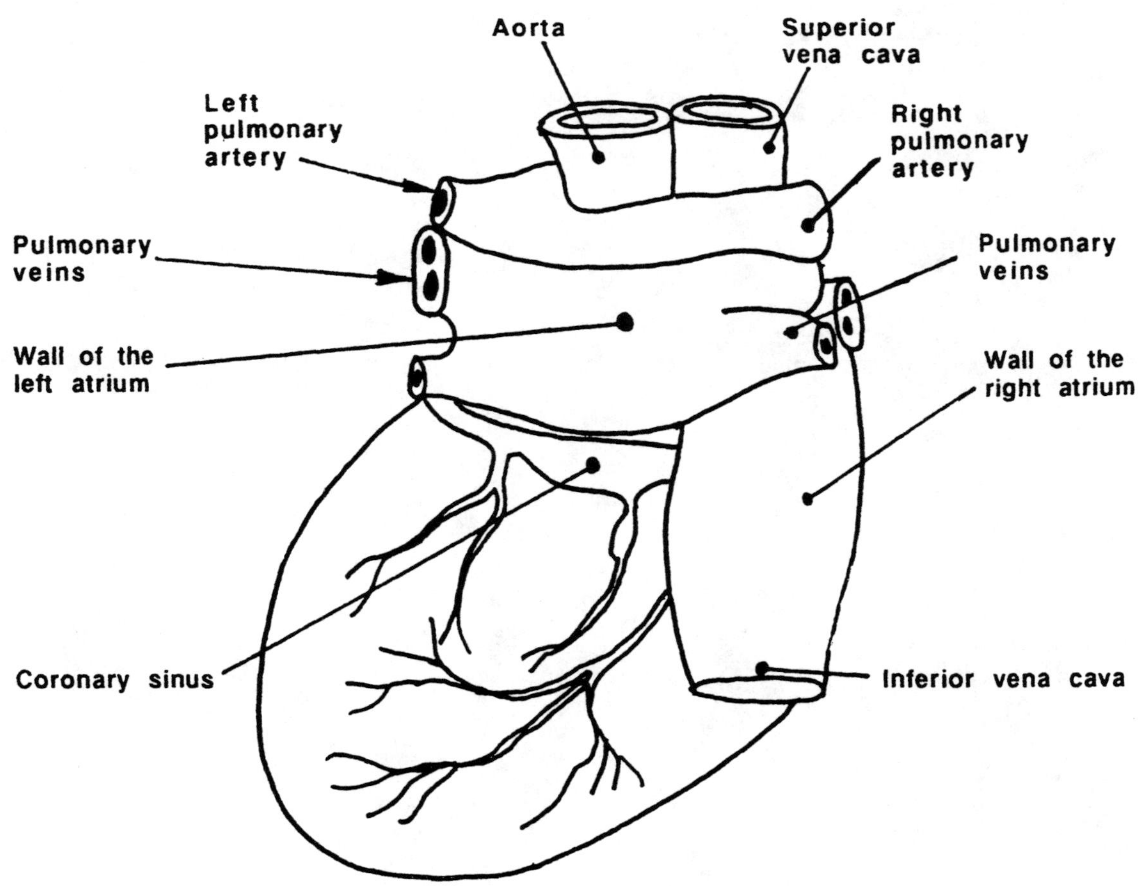

c. Anterior view of the small heart model showing the interior

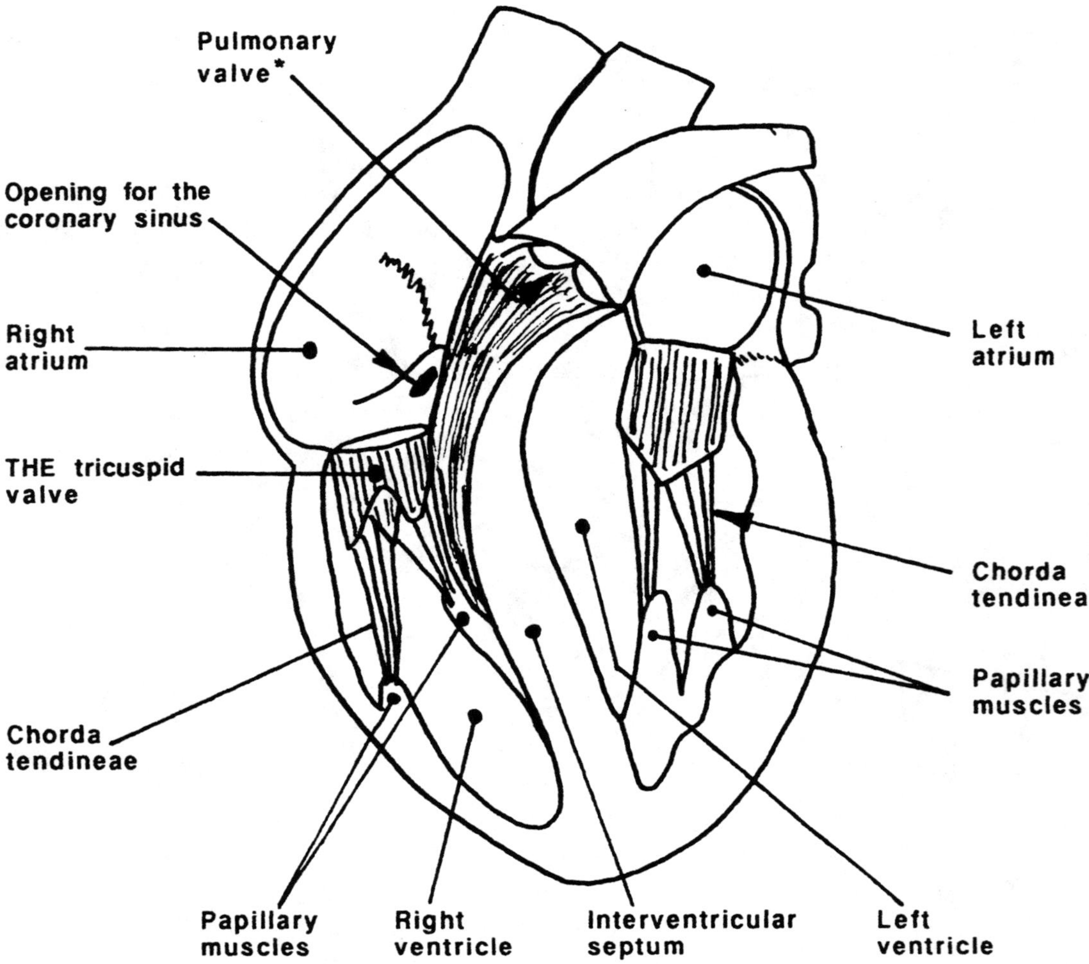

* Be certain that you also find and identify the aortic valve

QUESTIONS TO ANSWER AND THINK ABOUT (Some of these questions can be answered from looking at the three heart models, but some will require that you consult your lecture notes and reading assignments)

1. What is the function of the heart ?

2. One of the main functions of your circulatory system is for transport. What are the items that your circulatory system transports and from where to where are these items transported ?

3. What type of muscle is heart muscle ?

4. Give a very brief definition of metabolism.

5. Why do your cells need ATP ?

6. What two things must a cell take into itself in order to make ATP ?

7. What is the normal, but nevertheless toxic, byproduct of the breakdown of glucose and fatty acids in a cell ?

8. What is the problem of having too much CO_2 in your blood ?

9. What is systemic circulation ?

10. Where does systemic circulation begin and where does it end ?

11. What is true about the O_2 content of the blood that leaves the left ventricle and enters systemic circulation and that which returns to the right atrium ?

12. What is true about the CO_2 content of the blood that leaves the left ventricle and enters systemic circulation and than which returns to the right atrium ?

13. Where does coronary circulation begin and where does it end ?

14. What is true about the O_2 content of the blood that leaves the left ventricle and enters coronary circulation and that which returns to the right atrium ?

15. What is true about the CO_2 content of the blood that leaves the left ventricle and enters coronary circulation and that which returns to the right atrium of the heart ?

16. What are the similarities between systemic and coronary circulations ?

17. What are some differences between systemic and coronary circulations ?

18. Where does pulmonary circulation begin and end ?

19. What does pulmonary circulation accomplish as far as O_2 and CO_2 are concerned ?

20. What are the four chambers of the heart ?

22. Considering the three circulations, what is a convenient way to divide the heart ?

23. What are the two chambers of the right heart ?

24. What are the two chambers of the left heart ?

25. Which side of the heart contains oxygenated blood in the adult ? Why ?

26. Which side of the heart contains unoxygenated blood in the adult ? Why ?

27. Considering the cardiac cycle, what is a good way to divide the chambers of the heart ?

28. What are the two upper chambers of the heart ?

29. What are the two lower chambers of the heart ?

30. What are you really measuring when you take some's pulse rate ?

31. Does it take, on average, more, or less than one second to complete a single cardiac cycle ?

32. Describe the cardiac cycle.

33. Where does blood in the right atrium go ?

34. Where does blood in the left atrium go ?

35. Why can blood in the right ventricle not be pumped back into the right atrium ?

36. Why can blood in the left ventricle not be pumped back into the left atrium ?

37. How are the two atrioventricular valves kept closed when the ventricles contract ?

38. What is the function of the papillary muscles and chorda tendineae ?

39. What is happening in the atria when the ventricles are contracting ?

40. Are the semilunar valves open or closed when the ventricles contract ?

41. Are the semilunar valves open or closed when the ventricles relax ?

42. What does having the semilunar valves closed when the ventricles contract accomplish ?

43. How is blood prevented from re-entering the ventricles from the aorta and pulmonary trunk when the ventricles relax ?

44. When the right ventricle contracts where does the blood go ?

45. When the left ventricle contracts where does the blood go ?

46. What is the overall function of the atrioventricular and semilunar valves ?

47. What are the three areas of the aorta that you can see on the heart models ?

48. What are the two very first branches of the aorta ?

49. What are the three branches that come off of the aortic arch ?

50. What are the two branches of the pulmonary trunk ?

51. There are three veins that empty their blood into the right atrium. What are these three veins?

52. Is the blood in these veins oxygenated or unoxygenated ?

53. What set of veins empties its blood into the left atrium ?

54. Is the blood in these pulmonary veins oxygenated or unoxygenated ?

55. Which artery receives blood from the right ventricle ?

56. Is the blood in the pulmonary trunk oxygenated or unoxygenated ?

57. Which artery receives blood from the left ventricle ?

58. Is the blood in the aorta oxygenated or unoxygenated ?

59. What is the general relationship between the attachment of arteries and veins to the heart and the chambers to which they are attached ?

60. What did the ligamentum arteriosum used to be ?

WHAT YOU MUST KNOW AND BE ABLE TO DO AFTER THIS LABORATORY:

1. Be able to identify and correctly spell the structures and chambers of the three heart models.

2. Be able to answer questions from your lecture notes and reading assignments about these structures and chambers on the heart models.

LABORATORY 18 - SYSTEMS

THE CIRCULATORY SYSTEM - ARTERIES AND VEINS

INTRODUCTION: **The heart**, which you studied in **Laboratory 17**, is the marvelous muscular pump that moves blood around in your blood vessels.

There are three types of blood vessels - **ARTERIES**, **CAPILLARIES**, and **VEINS**. **Arteries** contain blood that is flowing away from the ventricles of the heart and toward the capillaries. **Veins** contain blood that is flowing away from the capillaries and toward the atria of the heart.

The arteries and veins that you will study in this laboratory exercise are the major ones that are large enough to be seen and therefore given names to. But not all arteries and veins can be readily seen because some are inside organs, and indeed some of them are microscopic. The smallest vessel that can be considered an artery is called an **arteriole**, and the smallest vessel that can be considered a vein is called a **venule**. These are microscopic vessels.

As an artery enters an organ it undergoes branching into smaller arteries, which in turn penetrate further into the organ and branch into smaller and more numerous arteries. And so on. This branching into more numerous and smaller arteries continues inside the organ until there are millions of microscopic arterioles permeating the substance of an organ such as the brain, or the myocardium of the heart, or a given muscle such as the biceps brachii. Ultimately, **each arteriole will branch into smaller vessels called capillaries.** Capillaries are not considered to be arteries and so the arterioles represent the end of the arterial system. Capillaries are microscopic vessels that surround the cells of an organ and it is at the capillary level that chemical exchanges (O2 / CO2 / nutrients) with your circulatory system and your cells occurs.

The capillaries created by the branching of arterioles are very short and on the other side of the cells they merge. **Several capillaries will merge to form a single microscopic vein, the venule.** If we stop to consider what has happened with blood flowing around the cells in the capillaries, then we see that arterioles contain little CO2, and lots of O2 and nutrients, but that the venules contain little O2 and nutrients and lots of CO2.

The venules can be considered to be the beginning of your venous system. Venules merge to form larger veins, which in turn merge to form even larger veins. This merging continues until veins large enough to give names to have been formed. The blood in the venules and veins is flowing away from the capillaries around the cells **in the systems** of your body and toward the right atrium of your heart.

The foregoing describes what happens in systemic and coronary circulations. But in your lungs something different occurs. The pulmonary arteries enter the lungs and, like the major arteries entering all other organs, begin to branch. The smaller and more numerous branching arteries permeate out into the substance of the lungs creating an enormous number of capillaries. **These capillaries surround the alveoli (air sacs).** There are approximately 300 - 350 million alveoli in each of your lungs, each being surrounded by capillaries. The branching of the single pulmonary artery that enters a lung is therefore truly enormous.

At the alveoli / capillary level the CO_2 in the blood of the capillaries leaves and enters the alveoli, and the O_2 that is in the alveoli enters the blood of the capillary and binds to the hemoglobin molecules inside your erythrocytes (red blood cells: RBCs). On the other side of the alveoli the capillaries merge into venules, which merge into larger veins, and so on until finally the pulmonary veins are formed, which attach to the left atrium of the heart.

As far as your blood is concerned, what happens at the alveoli / capillary interface in your pulmonary circulation is just the opposite of what occurs at the cell / capillary interface in systemic and coronary circulations. Your body wants only the O_2 from the air that you breathe into your alveoli. Nevertheless, you should be aware that almost everything that you breathe into your alveoli will also enter your circulation and be distributed to the cells of your body. During pregnancy, these pollutants are also distrubuted, through the placenta, to the cells of the fetus, sometimes with detrimental consequences to the developmental processes going on in the fetus.

There is a **flat circulatory system model** and **several torso models** that will be necessary for you to study. However, in the **EXERCISES** section the flat circulatory system model is presented several times in order to prevent confusion, and on these the arteries and veins have been grouped according to:

1. **SYSTEMIC CIRCULATION VESSELS**

    a) Branches of the aorta

    b) Arteries and veins of the upper limb

    c) Arteries and veins of the lower limbs

    d) Arteries and veins of the head, face and brain

    e) Arteries and veins of internal organs

    f) Arteries and veins of pulmonary circulation

2. **PULMONARY CIRCULATION VESSELS**

You have already seen some of the arteries of coronary circulation on the heart models in Laboratory 18. **Please review these vessels**.

OBJECTIVES: The objectives of this laboratory are to learn the arteries and veins on the flat circulatory models and the torso models

PRELABORATORY PREPARATION: You should complete this worksheet before you come to lab. It will allow you to work more efficiently and to complete your work earlier and with less confusion. By using the diagrams of the flat circulatory system models given under the **EXERCISES** section, fill in the following blanks:

FLAT CIRCULATORY MODEL ARTERIES AND VEINS

Branches of the aorta
1____________________
2____________________
3____________________
4____________________
5____________________
6____________________
7____________________
8____________________

Arteries and veins of the upper limb
a. Arteries
1____________________
2____________________
3____________________
4____________________
5____________________

b. Veins
1____________________
2____________________
3____________________
4____________________

Arteries and veins of the lower limbs
a. Arteries
1____________________
2____________________
3____________________
4____________________
5____________________

b. Veins
1____________________
2____________________
3____________________

Arteries and veins of the head / face / brain
a. Arteries
1____________________
2____________________
3____________________
4____________________

b. Veins
1____________________
2____________________
3____________________
4____________________

Arteries and veins of internal organs
a. Arteries
1____________________
2____________________
3____________________
4____________________
5____________________
6____________________

b. Vein
1____________________

Arteries and veins of pulmonary
a. Arteries
1____________________
2____________________
3____________________

b. Veins
____

Additional arteries and veins
a. Arteries
1____________________
2____________________
3____________________

b. Veins
1____________________
2____________________
3____________________

TORSO MODELS ARTERIES AND VEINS

a. Arteries

1__________________________

2__________________________

3__________________________

4__________________________

5__________________________

6__________________________

7__________________________

8__________________________

9__________________________

10__________________________

11__________________________

12__________________________

13__________________________

14__________________________

b. Veins

1__________________________

2__________________________

3__________________________

4__________________________

5__________________________

6__________________________

7__________________________

8__________________________

EXERCISES:

1. Branches of the aorta

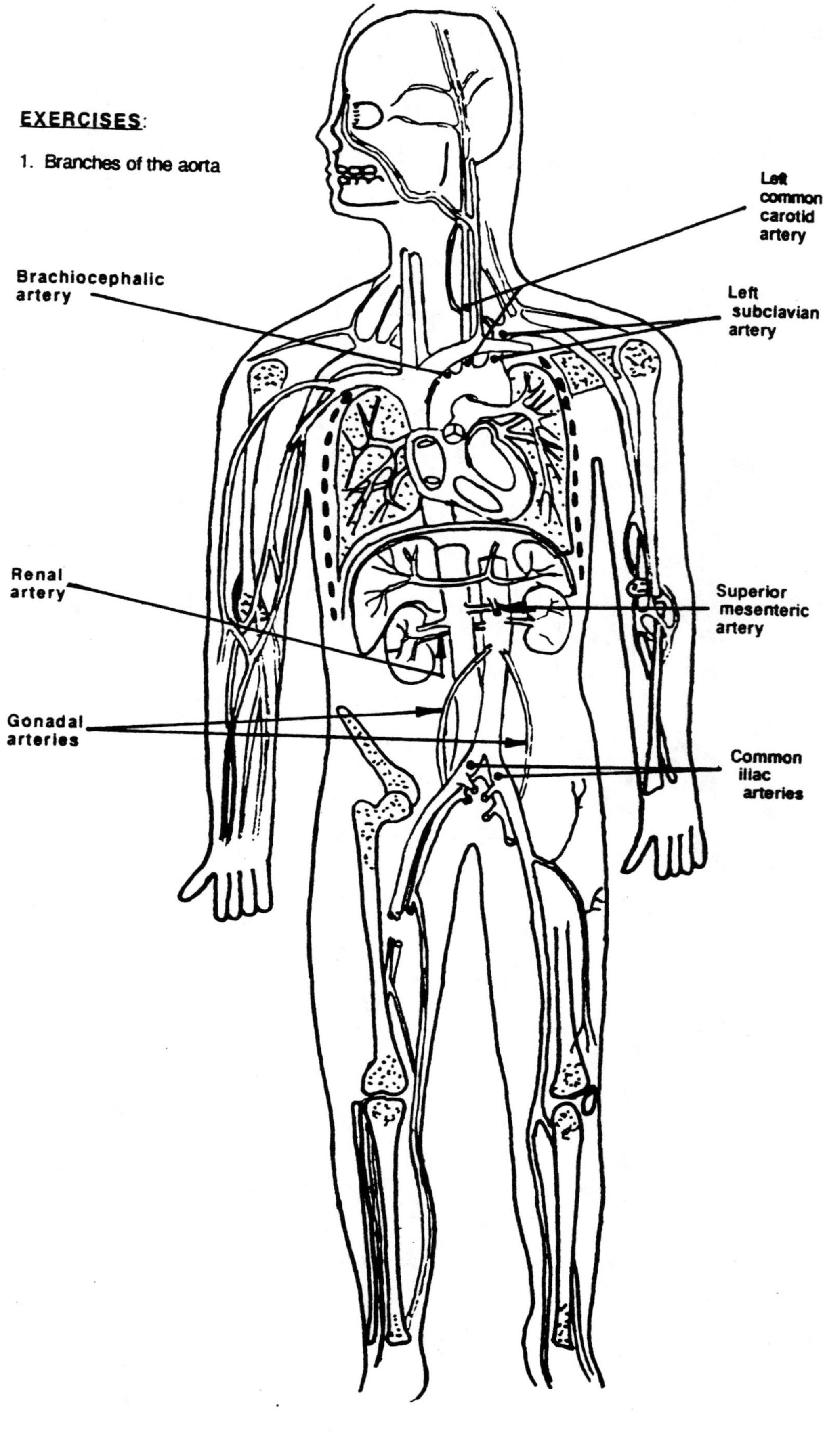

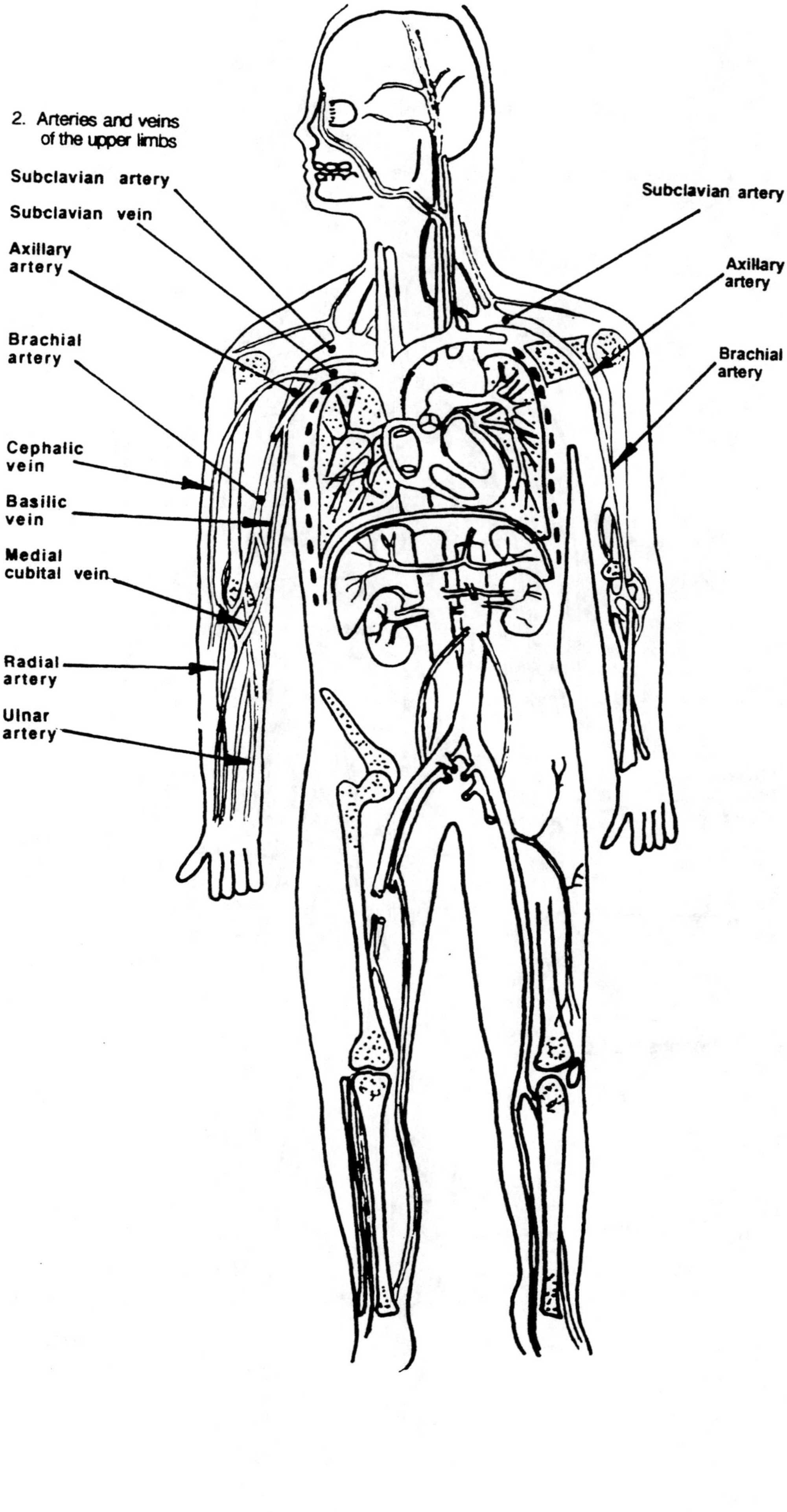
2. Arteries and veins
of the upper limbs
Subclavian artery
Subclavian vein
Axillary
artery
Brachial
artery
Cephalic
vein
Basilic
vein
Medial
cubital vein
Radial
artery
Ulnar
artery
Subclavian artery
Axillary
artery
Brachial
artery

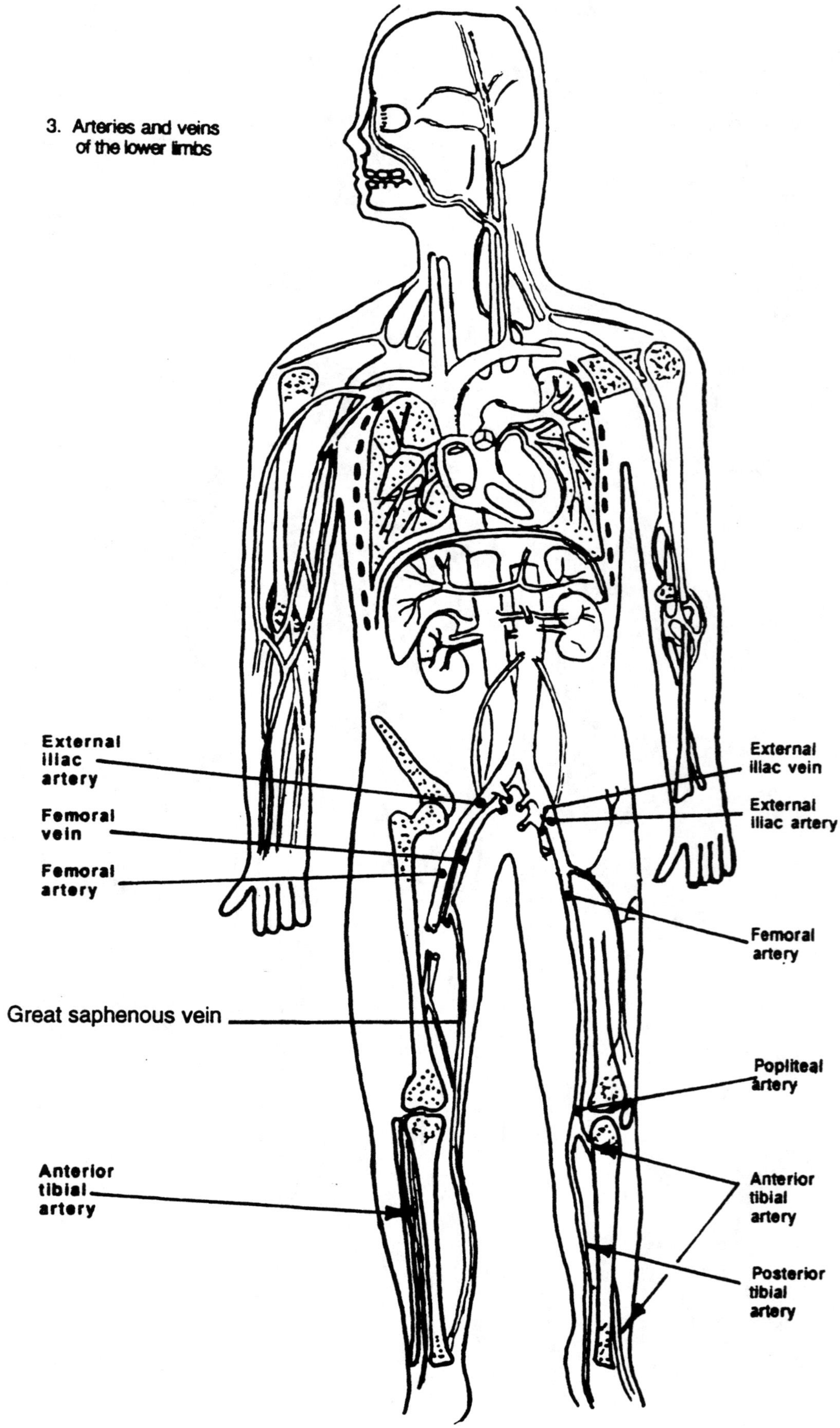
3. Arteries and veins
of the lower limbs
External
iliac
artery
Femoral
vein
Femoral
artery
Great saphenous vein
Anterior
tibial
artery
External
iliac vein
External
iliac artery
Femoral
artery
Popliteal
artery
Anterior
tibial
artery
Posterior
tibial
artery

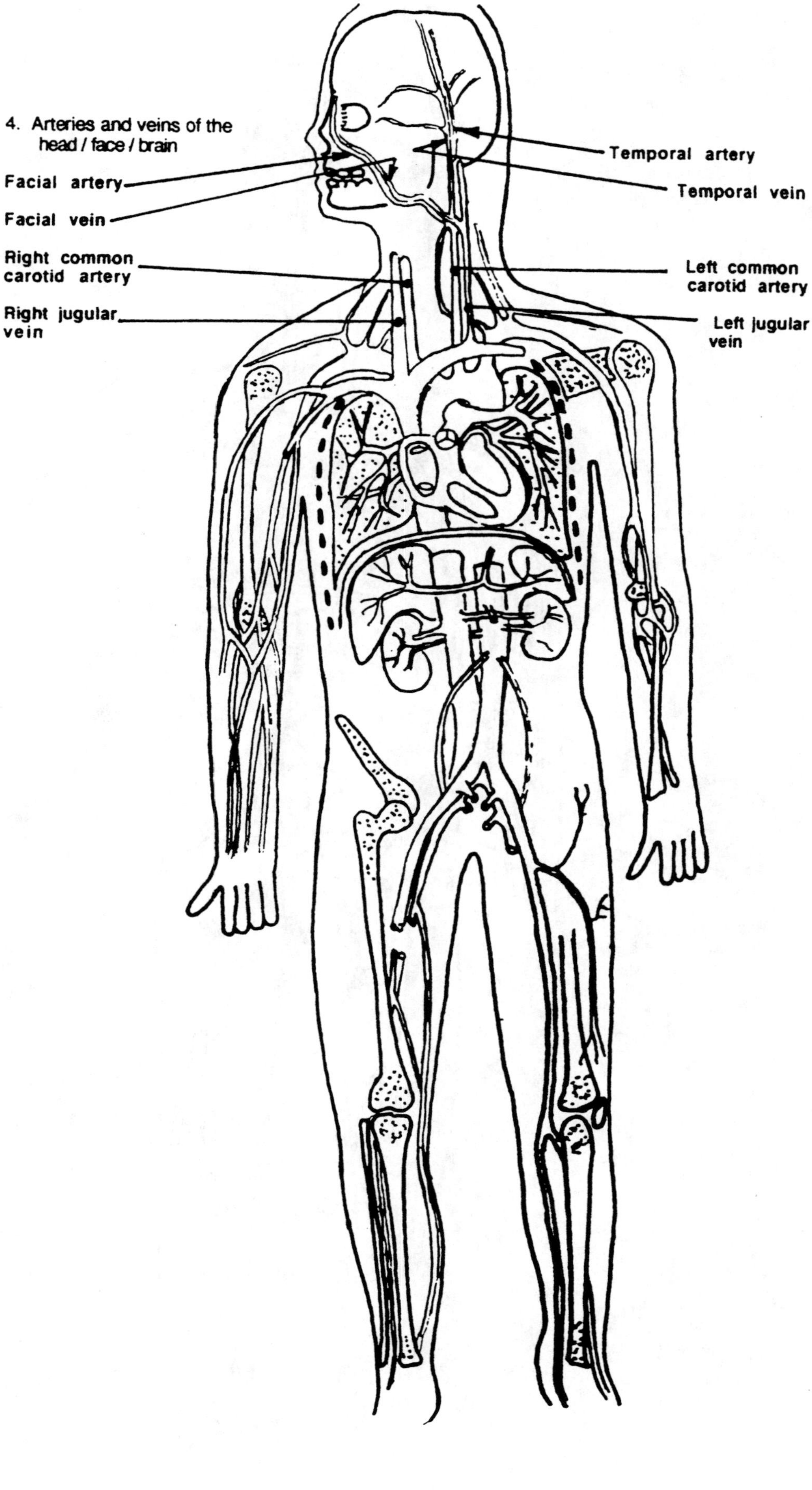
4. Arteries and veins of the head / face / brain
Facial artery
Facial vein
Right common carotid artery
Right jugular vein
Temporal artery
Temporal vein
Left common carotid artery
Left jugular vein

5. Arteries and veins of the internal organs

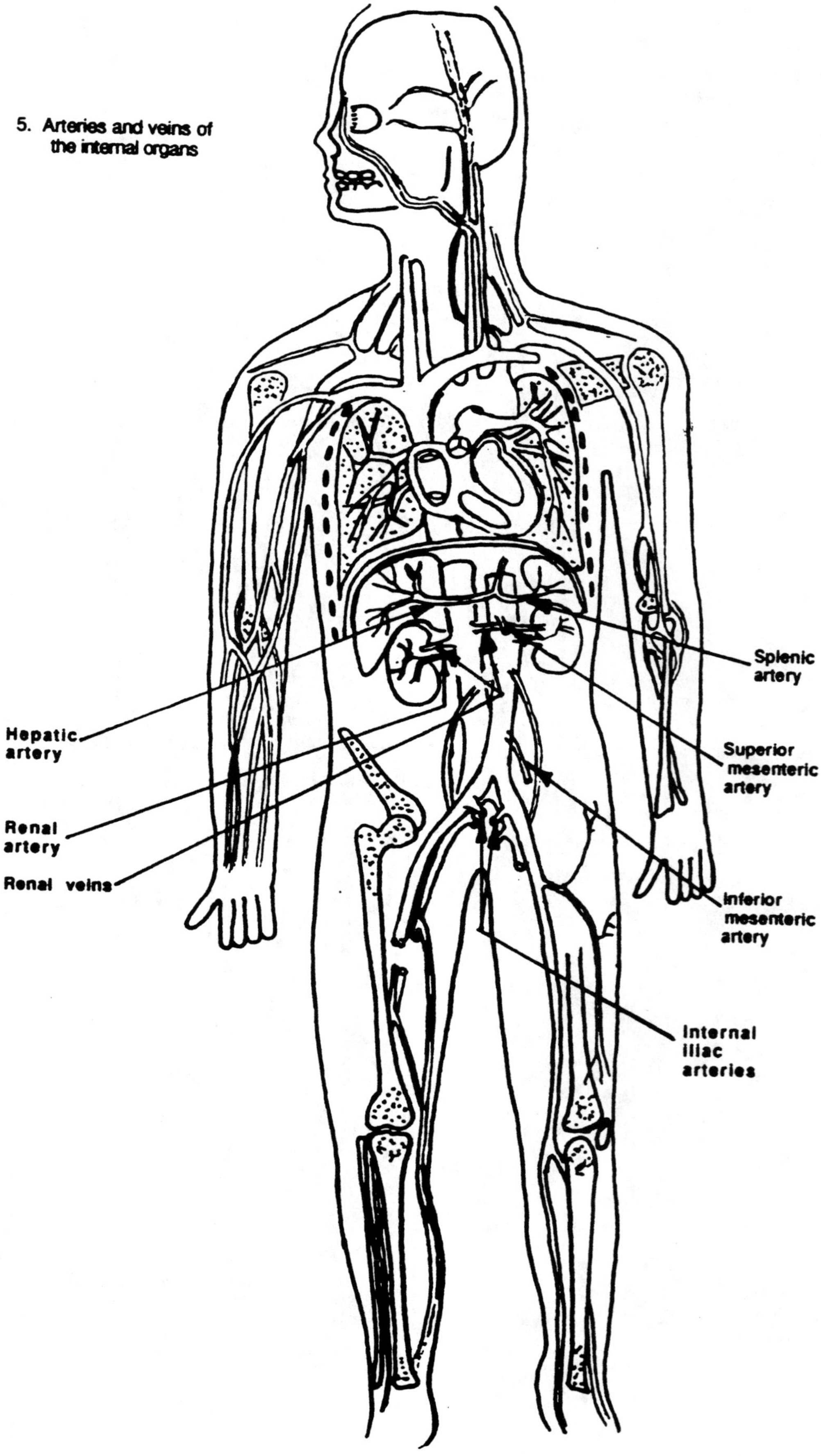

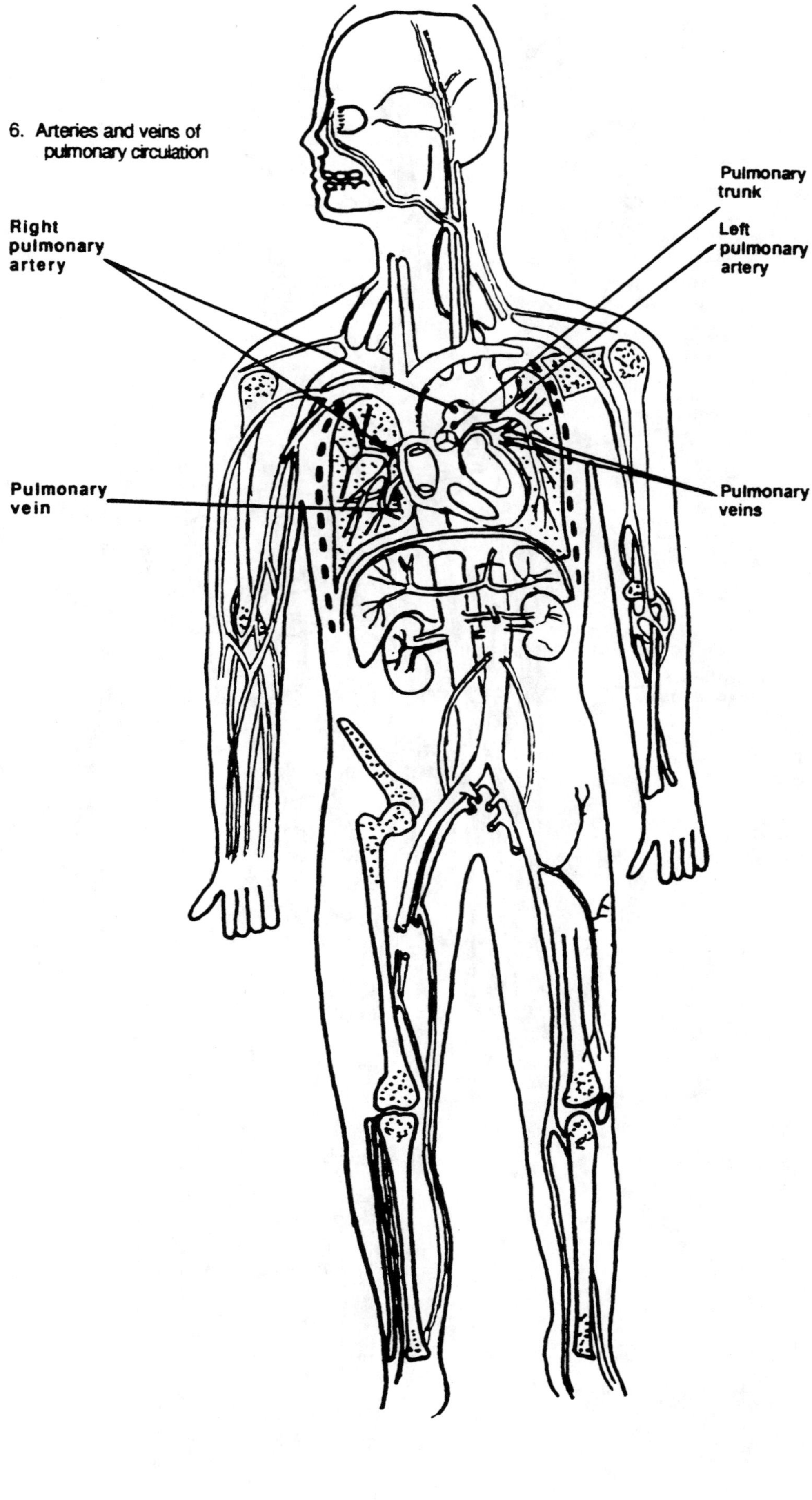
6. Arteries and veins of pulmonary circulation
Right pulmonary artery
Pulmonary trunk
Left pulmonary artery
Pulmonary vein
Pulmonary veins

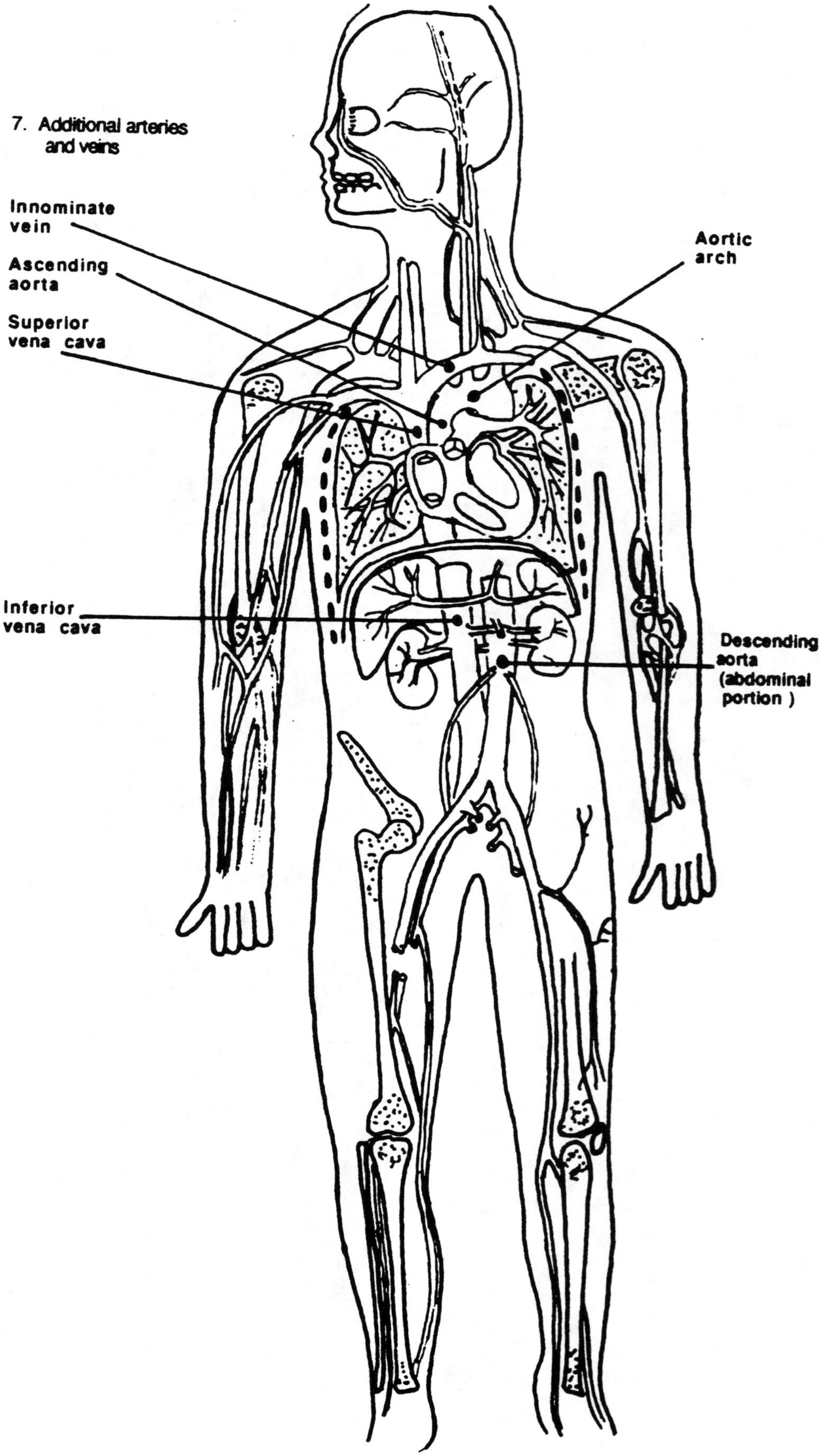
7. Additional arteries and veins
Innominate vein
Ascending aorta
Superior vena cava
Aortic arch
Inferior vena cava
Descending aorta (abdominal portion)

8. Arteries and veins on the torso models

The torso models are not drawn in this manual. Please consult your textbook and your Laboratory Instructor if you need help in identifying the following arteries and veins:

a. Arteries*

1) Brachiocephalic
2) Right common carotid
3) Left common carotid
4) Left subclavian
5) Descending aorta (thoracic portion)
6) Descending aorta (abdominal portion)
7) Right and left common iliac
8) Right and left external iliac
9) Right and left internal iliac
10) Renal
11) Superior mesenteric
12) Splenic
13) Inferior mesenteric
14) Gonadal

b. Veins*

1) Jugulars
2) Right and left innominate
3) Superior vena cava
4) Inferior vena cava
5) Right and left common iliac
6) Right and left external iliac
8) Right and left internal iliac
9) Renal

* Not all of the arteries and veins can be found on a single torso model, so it will be necessary for you to look at all of them

QUESTIONS TO ANSWER AND THINK ABOUT (Some of these questions can be answered from looking at the circulatory and torso models, but some will require that you consult your lecture notes and reading assignments)

1. What are the three types of blood vessels in your body ?
2. What is the definition of an artery ?
3. What is the definition of a vein ?
4. What is the name of the smallest vessel that can still be an artery ?
5. What is the name of the smallest vessel that can still be a vein ?
6. What is the definition of a capillary ?
7. What happens at the capillary level in systemic and coronary circulations ?
8. How are venules formed ?
9. What occurs at the capillary level in the pulmonary circulation ?
10. The arterial system of systemic and coronary circulations begin from what chamber of the heart ?
11. The venous system of systemic and coronary circulations end in what chamber of the heart ?
12. What is the difference in O_2 / CO_2 content between arterioles and venules in systemic and coronary circulations ?
13. What is the difference between O_2 / CO_2 content between arterioles and venules in pulmonary circulation ?
14. What are the three arterial branches off of the aortic arch ?
15. What is the name of the main artery into the kidney ?
16. What is the name of the artery into the liver ?
17. What is the name of the artery into the spleen ?
18. What is the name of the artery that passes between the tibia and fibula ?
19. What is the name of the artery that is behind the knee ?
20. What are the last two branches of the aorta ?
21. What are the first two branches of the aorta ?
22. The external iliac arteries continue out of the anterior pelvis and then their name is changed to what artery ?
23. Is the blood in hepatic, splenic and renal arteries flowing toward or away from the liver, spleen, and kidneys ?
24. Which vessel is a blood sample usually taken from ?

25. There are many places where arteries come close enough to the surface so that a pulse can be taken, but nevertheless, which artery is used most for taking someone's pulse ?

26. Is the femoral artery located in the anterior or posterior thigh ?

27. Is the blood in the pulmonary veins oxygenated, or unoxygenated ?

28. How about the blood in the pulmonary arteries ?

29. You can take a pulse at several places, not just at he wrist with the radial artery. If you place your fingers just to the side of the bridge of your nose you can detect a fairly strong pulse. What artery is giving you that pulse ?

30. Likewise, if you place your fingers just in front of you external ear you can pick up a pulse. What artery are you getting the pulse from ?

31. When your blood pressure is taken, a stethoscope is held over the front of your elbow (cubital region) in order to listen to your pulse. Which of your arteries is being listened to ?

32. On the models are the pulmonary veins red or blue ? Why ?

33. How about the color of the pulmonary arteries ? Why ?

34. What two arteries does the popliteal artery branch into ?

35. From where do you think the radial and ulnar arteries get their names ?

36. Each common iliac artery branches into what two arteries ?

37. The popliteal artery is actually a continuation of what artery ?

38. If you take your pulse just behind your medial malleolus, then what artery is giving you that pulse ?

WHAT YOU SHOULD KNOW AND BE ABLE TO DO AFTER THIS LABORATORY:

1. Be able to identify and correctly spell the arteries and veins on the flat circulatory system model.

2. Be able to identify and correctly spell the arteries and veins on the torso models.

3. **Be able to answer questions form lecture notes and reading assignments about these arteries and veins.**

LABORATORY 19 - SYSTEMS

THE CIRCULATORY SYSTEM - FETAL CIRCULATION

INTRODUCTION: Since the moment you were born you have obtained the O_2 for the cells in your systemic and coronary circulations from the atmosphere. The cells in your systemic and coronary circulatory routes take O_2 from your blood, and it is your pulmonary circulation that replenishes the O_2 to your blood.

If we again take a general view of these circulations then we note that oxygenated blood leaves the left heart and enters systemic and coronary circulations and unoxygenated blood is returned through these circulations to the right heart. Now we arrive at an important observation and that is that **for blood in the right heart to get back to the left heart in order to re-enter the systemic and coronary circulations the blood MUST FIRST GO THROUGH THE LUNGS**. Going from the right heart to the left heart in the adult circulation represents pulmonary circulation. **Reread the two previous sentences and really think about them.**

Before you were born you were surrounded by a fluid called **amniotic fluid**. Your lungs were, under these circumstances, absolutely useless insofar as O_2 / CO_2 exchange with the blood is concerned. You could get no O_2 directly from the atmosphere. Yet, your cells were growing and dividing and therefore in dire and constant need of both nutrients and O_2. The O_2 and nutrients that you could not directly obtain from the atmosphere were obtained from your placenta, a joint structural venture between you and your Mom, where your fetal blood (fetal circulation) and Mom's blood (maternal circulation) exchanged O_2 / CO_2, nutrients, and urea.

As an adult the total volume of your blood is circulated through your lungs many, many times each day in order to eliminate your CO_2 and to replenish your O_2. But as a fetus, before your birth, your lungs did not function as they do now, and so there was no reason to send large quantities of blood to your fetal lungs.

There are therefore two anatomical devices that allow the diversion of large quantities of blood away from the fetal lungs. One of these devices is inside the fetal heart and is an opening between the right heart (right atrium) and the left heart (left atrium). This opening is the **FORAMEN OVALE.** Therefore, in the fetal heart, which is going through the same cardiac cycle as the adult heart, the **BLOOD, UNLIKE IN THE ADULT HEART, CAN GO FROM THE RIGHT HEART TO THE LEFT HEART WITHOUT GOING TO THE LUNGS FIRST**. During the cardiac cycle, a significant amount of blood in the right atrium of the fetal heart passes through the foramen ovale into the left atrium and immediately past the mitral valve and goes into the left ventricle. A significant amount of blood in the right atrium does not pass through the foramen ovale, but passes the tricuspid valve and goes into the right ventricle.

The other device that diverts blood away from the lungs is outside the heart and is a short arterial shunt between the two largest arteries in the body, the pulmonary trunk and the aorta (aortic arch). This is the **DUCTUS ARTERIOSUS**. As blood comes up into the pulmonary trunk due to contraction of the right ventricle musculature, most of it passes through the ductus arteriosus and enters the aorta where it joins the blood that is entering the aorta due to the contraction of the left ventricle (don't forget that during the cardiac cycle both the ventricles contract at the same time).

At birth, both of these devices disappear. The foramen ovale closes and becomes a depression in the wall of the right atrium known as the **fossa ovalis**. The ductus arteriosus closes and becomes a ligament between two arteries called the **ligamentum arteriosum**.

The structure that connects the placenta to the body of the fetus is the **umbilical cord.** Within the umbilical cord are **two umbilical arteries** and **one umbilical vein.** In the umbilical vein is oxygenated blood that is flowing away from the capillaries in the placenta toward the fetal heart. If we follow this vein we see that it enters the fetus at the umbilicus, passes under the fetal liver and connects to the inferior vena cava through a short vessel called the **ductus venosus.**

The umbilical arteries are continuations of the two hypogastric arteries that branch from the internal iliac arteries and enter the umbilical cord at the umbilicus. These arteries transport unoxygenated blood from the fetus to the placenta so that it can be reoxygenated.

OBJECTIVES: The objective of this laboratory is to be able to identify and correctly spell the structures and vessels associated with fetal circulation

PRELABORATORY PREPARATION: You should complete this worksheet before you come to lab. It will allow you to work more efficiently and to complete your work earlier and with less confusion. By using the diagram of the fetus given under the **EXERCISE** section, fill in the following blanks:

Arteries that become the umbilical arteries

____

Vessels in the umbilical cord

____

____

Vessel between umbilical vein and the inferior vena cava

____

Opening between the right and left atria

____

Vessel between the pulmonary trunk and the aortic arch

____

Structure for O_2 / CO_2 exchange

____

Structure containing two umbilical arteries and umbilical vein

____

EXERCISE:

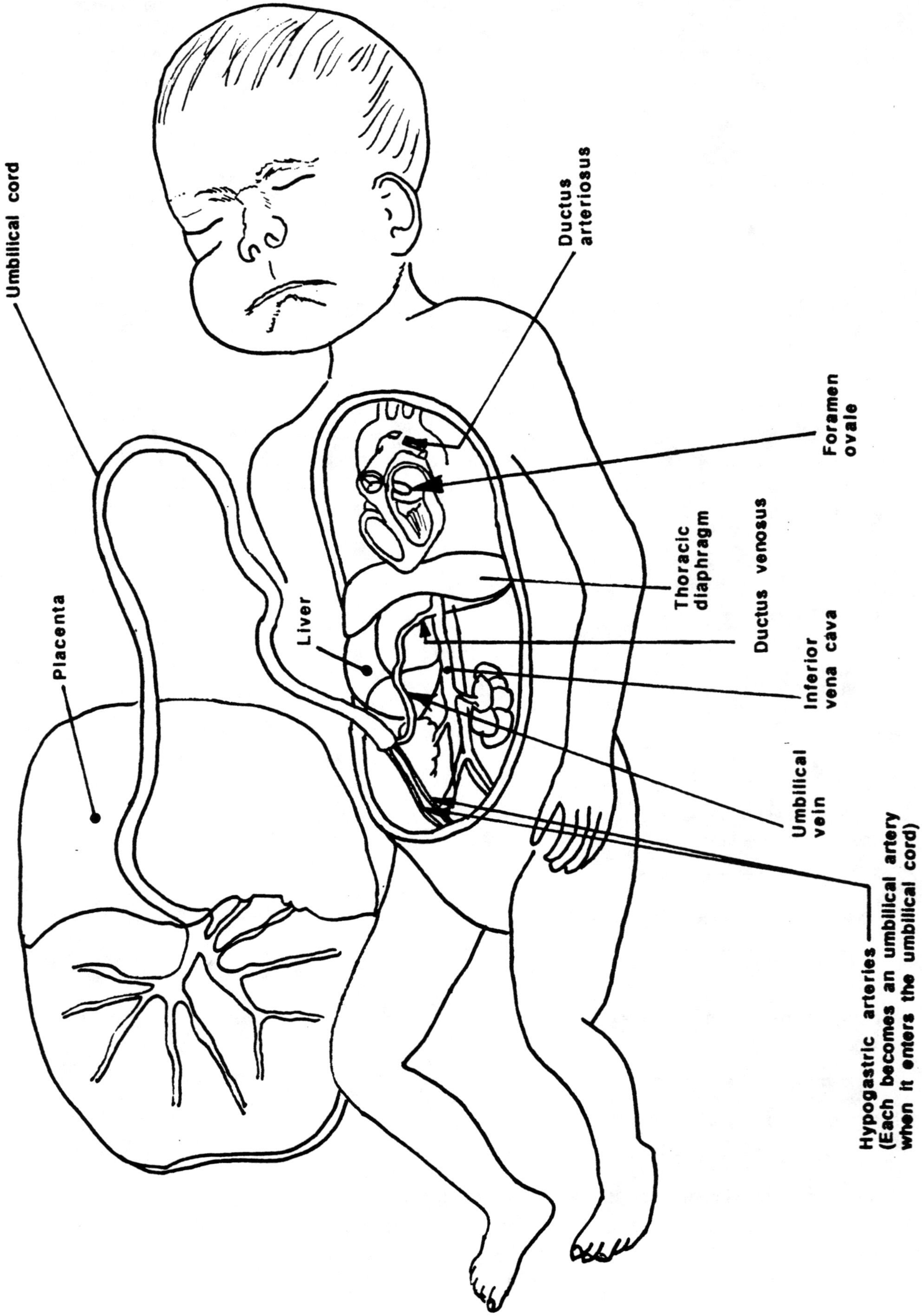
Umbilical cord
Placenta
Liver
Ductus arteriosus
Foramen ovale
Thoracic diaphragm
Ductus venosus
Inferior vena cava
Umbilical vein
Hypogastric arteries
(Each becomes an umbilical artery
when it enters the umbilical cord)

QUESTIONS TO ANSWER AND THINK ABOUT (Some of these questions can be answered from looking at the fetal circulation model, but some will require that you consult your lecture notes and reading assignments)

1. In the adult, where must the blood in the right heart go before going to the left heart ?

2. What circulation is involved in going from the right heart to the lungs and then to the left heart?

3. What are the two anatomical devices that divert large quantities of blood away from the fetal lungs ?

4. Trace a red blood cell from the right atrium of the fetus to the aorta.

5. Why does the fetus need a foramen ovale and a ductus arteriosus ?

6. In the fetus the right heart, unlike that of the adult heart, contains oxygenated blood. How can this be ?

7. The ductus venosus is a connection between what two veins ?

8. Which two arteries become the umbilical arteries of the umbilical cord ?

9. Is the blood flowing through the umbilical vein oxygenated or unoxygenated ?

10. What two arteries does the ductus arteriosus connect ?

11. What does the ductus arteriosus become shortly following birth ?

12. What does the foramen ovale become shortly following birth ?

13. What is the name of the fluid that the fetus is floating in ?

14. Is the blood that is flowing in the umbilical arteries oxygenated or unoxygenated ?

15. From what arteries do the hypogastric arteries branch ?

WHAT YOU MUST KNOW AND BE ABLE TO DO AFTER THIS LABORATORY

1. Be able to identify and correctly spell the structures give in the diagram in the **EXERCISE** section.

2. Be able to answer questions from your lecture notes and reading assignments concerning the structures in the laboratory exercise.